Tales from an
Uncertain World

TALES FROM AN

UNCERTAIN

*What Other
Assorted Disasters
Can Teach Us About
Climate Change*

L. S. Gardiner

WORLD

University of Iowa Press | Iowa City

University of Iowa Press, Iowa City 52242
Copyright © 2018 by L. S. Gardiner
www.uipress.uiowa.edu

Printed in the United States of America
Design by April Leidig

The University of Iowa Press is a member of Green Press Initiative
and is committed to preserving natural resources.

Printed on acid-free paper

Cataloging-in-Publication data for *Tales from an Uncertain World:
What Other Assorted Disasters Can Teach Us About Climate Change*,
by L. S. Gardiner, is on file at the Library of Congress.

ISBN 978-1-60938-553-8 (pbk)
ISBN 978-1-60938-554-5 (ebk)

Contents

Acknowledgments

Writing a book that knits together many disciplines requires getting to know research in many different fields. Many people let me pepper them with questions as I researched the stories in this book. Many others have helped me learn through their peer-reviewed papers and books — including climate scientists, volcanologists, social scientists, historians, educators, communicators, psychologists, marine scientists, ecologists, geologists, and other researchers. Without their brilliant work, there would have been little to knit together. If there are any errors in this book, they are in no way related to these researchers.

My first ideas about this project started to take shape during the 2010 Wildbranch Writers Workshop, organized by the Orion Society. I am so glad I had the opportunity to learn from inspiring instructors and writers during that rainy week in Vermont.

The creative and inspiring community at the Goucher College MFA program provided much-needed guidance, support, insight, and cheerleading during this project. I'm grateful especially for the advice and support of my mentors Madeleine Blais, Richard Todd, Suzannah Lessard, Tom French, and Joy Truela, as well as Patsy Sims, who created an incredible program, and Leslie Rubinkowski, who now leads it. The community of students and alumni continue to inspire me each day.

Thanks to Elisabeth Chretien at the University of Iowa Press, who was incredibly patient with me while I figured out what it was that I was trying to say through these stories and who gently helped me see the direction I was heading through several drafts. And, of course, thanks to Catherine Cocks, Susan Hill Newton, Faith Marcovecchio, and the team at the University of Iowa Press for their interest and support of this project.

Acknowledgments

Several people gave me incredibly helpful feedback on the manuscript along the way, including Adam Holloway, Jen Henderson, Julie Malmberg, and two anonymous reviewers organized by the press. I am so grateful for the ways they helped guide me, from the sentence and word level to the major organization and themes.

Many others have been enormously helpful in the encouragement department, including Sara Gardiner, Clarisse Hart, Becca Hatheway, Moira Kennedy, Lee Patton, Jen Christiansen, Joel Tolman, Beth Landau, Al Curran, Sally E. Walker, George Ware, Abby Evans, and Mila, the world's best fluffy orange dog.

And the world's largest thank you to Adam Holloway for his unwavering support as I obsessed over disasters, organized trips to catastrophe locales, and sat hunched over my computer for hours on end. I'm thankful that I get to explore the world with him. Sara Gardiner and Justin Canada also let me explore with them and were remarkably understanding when I pulled out my computer to write during a trip to the beach.

Several years ago, Tracy Seeley (RIP) helped me start thinking about what I was writing as a book instead of an odd collection of words and sentences. I will forever be grateful. I wish she were still around to see what those words and sentences became.

Introduction

W
ile E. Coyote looks into the camera, so to speak, his yellow eyes open wide, and time stands still in the cartoon desert as he recognizes the impending catastrophe. Maybe he's falling over a cliff while wearing a pair of skis outfitted with wheels. Maybe an anvil is about to fall on his head. It is this little moment — the one that separates the time before catastrophe from the time when catastrophe has arrived — when I pause the cartoon, the roadrunner a blur, the coyote stock-still. The coyote's stunned expression is the same look I see in many people's faces, including my own, although our twenty-first-century version lacks the whimsy of a midcentury cartoon. At this time in history, many of us are sharing this same moment of realization: that climate change is no longer hypothetical. It's no longer a possibility in the distant future. It is the catastrophe of our time.

At the end of this century, when today's kindergarteners are white-haired old folks, the world may be as warm as it was thirty-five to forty million years ago in the Eocene, when warm, wet subtropical forests thrived in the Arctic. That's a warmer world than humans have ever known. Global average temperatures would be 5 to 9°F (2.6 to 5°C) warmer. This is the worst-case scenario described in the 2014 report by the Intergovernmental Panel on Climate Change (IPCC) — an international group of thousands of experts who assess our current state of knowledge about climate. According to the IPCC's best-case scenario, if we stop emitting greenhouse gases in the next few years, those kindergarteners will live in a world that has warmed less than 3.5°F (2°C), a world that we can keep livable.

Like the coyote, we are falling over a cliff, and we need to use physics, cunning, and ingenuity to get ourselves out of this pickle. Like the coyote, we've created this situation. Mr. Coyote can't blame a speedy bird for his predicament. He was the one who bought the anvil from Acme. In the case of climate change, we can't blame the sun, the earth, volcanoes, or faulty thermometers, as some do. We are the ones who filled the atmosphere with a surplus of greenhouse gases by burning fossil fuels. We got ourselves into this mess. It's time for some quick thinking to get ourselves out of it. Unlike the cartoon world, time is not standing still while we figure out what's going on. Instead, it feels like time is speeding up.

Seven out of ten Americans agree that climate change is happening, according to a 2017 poll by researchers at Yale and George Mason University, and over half (57 percent) said they were worried. In August 2015, President Obama announced his strongest plan to transform the US energy infrastructure to curb greenhouse gas emissions, saying, "We only get one home. We only get one planet. There is no plan B." The climate talks in December 2015 convened by the United Nations in Paris led to a monumental agreement among world leaders about action to reduce climate change. Understanding of the catastrophe is building.

However, in 2016 the United States elected a president who not only thinks that climate change is not a problem, he has indicated that he thinks it does not exist. He and his appointees threaten to undo the progress that the United States has made on climate change, extracting the US from the agreement reached in Paris and promising to bring coal and greenhouse gas emissions back into vogue. As I write this, I am still trying to adjust to the new reality, trying to gauge which proclamations will become reality.

President Trump is not the only one who is misinformed about climate. Only 58 percent of the Americans who participated in the 2017 poll knew that humans are the cause of current climate change, and only one in eight people knew that virtually all scientists agree that climate change is happening and that humans are the cause. According to an examination of peer-reviewed scientific articles on climate change by geologist James Lawrence Powell, of the 69,406 scientists who published articles about climate change in 2013 and 2014, only four rejected the idea that climate change is human-caused. Four out of nearly seventy thousand scientists (0.006 percent) disagree.

For many, the climate crisis does not seem like an emergency. Climate change is often communicated through statistical probabilities. Just as a certain coyote would jump to action when there was a falling anvil but dismiss reports about a future increase in the probability of falling anvils, it's easier to realize what climate change can do when Hurricane Sandy floods New York City than it is when the impacts seem distant and indirect.

Our brains are adapted to react to situations that are personal and local. This has been a helpful survival instinct for more typical perils. It keeps us safe. But climate change is usually viewed as global, the opposite of personal and local. A couple of years ago, when asked to drum up a single graphic about climate change for another book cover, my first thought was to find an image of the earth from space. Any other picture or illustration — a polar bear clinging to a shrinking iceberg, a flooding coastal city, sweltering masses coping with a heat wave — would only capture a portion of climate change. Climate change is global. Not local. Then again, everything local is part of global. If you look close enough at a picture of the earth from space, you will see the polar bear, the flood, the heat.

Trained as a scientist, I used to study the clams, snails, and other creatures that made their homes in coral reefs about a hundred and twenty thousand years ago. Now their shells are fossilized in rocky outcrops in the Bahamas. They lived during a time of change; sea level was fluctuating, their reef eroded, and then their descendants — their great-great-great-grandmollusks, if you will — came back once conditions became livable again. This led me to a PhD in geology and the realization that evidence from the past can offer a new perspective on present and future crises.

Since 2001, I've been communicating the science of weather, climate, and the earth system to students, teachers, and the general public. I got into the business of earth science education because I wanted to help people feel closer to nature, but I wound up helping people learn about some of the world's scariest things. Writing for educational websites, I covered climate change, sea level rise, Hurricanes Katrina and Rita, the tsunami in Southeast Asia, various volcanic eruptions and earthquakes, melting sea ice, dying polar bears, and other assorted catastrophes. I wrote a book for kids about natural disasters in Colorado. Working with a team to develop a museum exhibit about climate change at the National Center for Atmospheric Research, I created colorful graphics and edited text about rising temperatures and shrinking ice, about the environmental impact of the food we eat, about the amount of carbon dioxide that spews from tailpipes. On my good days, I tell myself that knowledge is power and that the scary things in this world are less scary once one knows how they work. One might even be in awe of the planet, and humbled by it, when one knows how it works. On the days when I feel ineffective, I fear that I'm coming across like that guy on the corner with a giant sandwich board that reads "The end is near." That's not really the message I hope to convey.

■ ■ ■

In a new and scary situation — whether it is a war that seems unjustified, a riot in the streets of a city, or an economic downturn — I find comfort in the fact that while we think we are in unfamiliar territory, in actuality we have some experience in these departments. I find myself turning to historic or present-day analogs and asking, how have we dealt with this before or elsewhere? There is always an analogy to something we have done before. I suppose this is not comforting in the case of urban riots — it is history repeating itself — but at least we are on familiar ground. It is this coping strategy that leads me to explore disasters. Because in the face of global climate change, a seemingly insurmountable situation, I find it comforting to know that we have made it through other situations that seemed difficult to surmount. As Theodore Roosevelt once put it, "Full knowledge of the past helps us in dealing with the future."

I've found that my interest in disasters is not shared by all. Generally speaking, mentioning that millions of people are vulnerable to various hazards causes silence around the table at a dinner party and depression or panic among the guests. These vulnerabilities do scare me, but they also awe me, and, in a glass-half-full way, they are proof that humanity has the ability to endure.

I am able to find disaster everywhere. I spiced up a California vacation with plate tectonics tourism, heading well off the beaten path to stand on the San Andreas Fault. Wandering San Francisco, I was more interested in snooping around for signs of the 1906 earthquake and fires than I was in cable cars and fine dining. California's Division of Tourism does not usually highlight these bleak and existential attractions, but they carried the most meaning to me. When visiting a tropical island, I was more interested in the invading killer fish than sipping fruity cocktails and basking in the sunshine. On a Cape Cod beach, I was focused on massive coastal erosion rather than building sand castles. And when a historic rainstorm and floods washed through my city, I was as thrilled by the power of water as I was traumatized by the destruction it caused. I see disasters all over the place. And I'm looking at them to learn how we react.

It occurred to me that this fascination with natural disasters and how people cope started around the same time that I began working as a climate change educator. The more I explained the science of climate change to the public, students, and teachers, the more I wanted to understand how we humans recognize that things are not normal. When do we decide to do something about it? In some ways, climate change is a disaster like no other, but in other ways, it shares similarities. Perhaps we can learn from how we react to other disasters, putting our reactions to climate change into perspective.

This book explores that moment when we face the camera, when we realize that there is a catastrophe afoot, or when we fail to realize it. Learning the stories of what humans have gone through can help us confront contemporary problems. We are, in some ways, facing the unknown as we start to deal with unprecedented climate change along with a host of other environmental problems like pollution, but we are bringing with us our greatest tool kit — our wealth of experience with disaster and our capacity to endure change on earth.

To recognize a catastrophe, we must be able to perceive risk. And how we perceive risk depends on cultural and psychological factors, our values, and our way of living. This is true for all types of risk, including climate change. The more we understand how we perceive risk, the more we will be able to identify why so many people have blind spots when it comes to assessing the risks of climate change.

There are two main ways that people perceive risk and decide whether to take action: they can use the analytic part of the brain, or they can use the emotional and instinctual part of the brain.

The analytic part of the brain deliberates about a possible risk using data and facts to guide the process. This part of the brain is slow and methodical. It is not quick to take action. It stays cool and sensible, logical.

Using feeling and instincts to assess risk is an evolutionarily older mechanism. This part of the brain is intuitive and fast, prone to snap decisions, which was beneficial when our hunter-gatherer ancestors were faced with a saber-toothed tiger or other prehistoric threat that required quick thinking. This part of the brain depends on our beliefs and values to make decisions. According to psychologist Paul Slovic, using feelings and instincts is the dominant way that people assess risks. This is because the emotional part of our brain helps us navigate uncertainty.

■ ■ ■

When doctors in scary exam rooms shake their heads and say, "We just don't know," or when biologists who study the fate of polar bears shake their heads and say, "We just don't know," or when plumbers looking at my sewer line shake their heads and say, "We just don't know," I shed my anxiety in the face of their uncertainty and buck up for the medical treatment, altered ecosystems, or large plumbing repair and say, *I'm sure it will be fine.*

I started using this mantra in college when my limited cash supply made the cheap Shurfine brand at the supermarket the most practical option. Shurfine products were not always reliable, but they were affordable. For example, choosing Shurfine marshmallows for a camping trip and uncertain whether they would roast properly, I looked deep into the Shurfine logo and said, "I'm sure it will be fine." As I recall, the Shurfine marsh-mallows refused to roast and instead melted into a sort of non-Newtonian fluid. Proclaiming that something will be fine does not make it so.

I rely on this mantra to remain calm when facing change. It helps me transform from wide-eyed and frozen, like Wile E. Coyote, into action as I recognize a problem. However, I'm realizing that the same expression can be used as a way to avoid taking action in the face of uncertainty and change, in the hope that everything is normal and nothing needs to be done. This can be hazardous to our health. With climate change, peo-ple often use uncertainty to justify inaction; however, there will never be complete certainty about how the climate disaster will play out until we are looking back at it from the future.

There are still people, lots of people, who deny that climate change is happening, or who agree that it's happening but rationalize why it is not a problem, or who agree that something should be done and then hop into a gas-guzzling SUV. It's the *I'm sure it will be fine* mentality applied on a global scale. In the 2017 survey of the American public by Yale and George Mason University researchers, only 7 percent of people surveyed were op-timistic that humans can and will reduce the amount of climate change. People are uncertain whether we can fix the problem.

■ ■ ■

Most of the ways that we feel the effects of climate change now are through smaller-scale catastrophes such as stronger hurricanes, floods, droughts, wildfires, and heat waves. The timing of these extreme weather events is also affected by a natural unpredictability of events, known as chaos. Chaos in the atmosphere is natural, but it now has a sidekick, or an evil twin — human-caused climate change. Both of them are at work in the world today. Sometimes the combined effect of natural chaos and human-caused climate change compounds, making these smaller-scale ca-tastrophes larger, and making the effects of the catastrophe much worse.

Knowing how we respond to catastrophic weather is practical, since we are now facing more of it. But we can also look beyond extreme weather events to other types of disasters to understand how we respond to all types of change on earth. Our actions in the face of earthquakes, erupting volcanoes, eroding coasts, invading species, as well as weather disasters are allegorical of how we are dealing with, or not dealing with, climate change.

Because this is a time in history when it is clear that humankind is having a large impact on the planet, discussions about our control over nature or nature's control over us keep coming up. Can we avert disaster? Can we filter carbon dioxide out of the atmosphere to slow or stop climate change? Can we divert a river to stop flooding a town? Can we grow enough crops to feed ten billion people? Or does nature control our lives, leaving us vulnerable? In this time of environmental change, it would be helpful to know who is in control.

We have a tremendous impact on nature even when we lack control over it. We are warming the atmosphere with emissions from fossil fuels. We change the land surface with agriculture and housing developments. We are capable of making a hole in the ozone layer even though we rarely travel that high in the atmosphere. We are reminded that we are vulnerable to nature when a blizzard dumps three feet of snow, lightning strikes, or an earthquake shakes the ground.

The field trips into our planet's chaos that comprise most of the chapters of this book explore how people react to, and make sense of, uncertainty — from small and mundane examples, like tomorrow's weather forecast, to large and unlikely examples, like earthquakes and flash floods. They look at how we control nature and how nature controls us. Most are not about climate. They are all about change — whether to the land, water, atmosphere, or life, because what we learn from other disasters can apply to how we handle climate change. It will only be when we get over the shock and shake off our Wile E. Coyote expressions that we will be able to start solving the climate change emergency in large ways. And when we do that, it will be the way we view nature and uncertainty that will determine the amount of risk we face and guide the decisions we make and the path we take in this century and beyond.

Chapter One

In Uncertain Terms

ook – Listen – Respond. That's the triptych of instructions provided on most In Case of Emergency signs. The instructions are clear and terse, but the signs assume we know what constitutes an emergency. In the case of climate change and other hazards that hit us obliquely, it can be particularly difficult to discern whether this is, in fact, an emergency. Of course, most In Case of Emergency signs are made for situations like fires, subway accidents, or plane crashes, situations in which declaring an emergency might be more straightforward. Declaring an emergency because of potentially dangerous changes in the environment is often not as clear.

Unlike disasters that slap you in the face — earthquakes and tsunamis, for example — climate change is communicated through the language of statistical probabilities and scientific uncertainty. Uncertainty can make it seem abstract. Climate change is a large, abstract problem, but what we do about it will need to be concrete.

Let's say you are in a house and a fire starts in the kitchen. You would do something, right? Perhaps you are the hero who would put the fire out single-handedly, or perhaps you are the hero who would ensure that everyone gets out of the house unscathed. Perhaps you are not the hero and would knock over Grandma in your attempt to get yourself out of the house. In all these cases, you would do something. You wouldn't just stand there.

But let's say there is not a fire; instead, radon is slowly leaking into your house. Radon is a naturally occurring radioactive gas that comes from decay of elements in rocks and soil. It is invisible and odorless. According to the US Environmental Protection Agency, "Exposure to radon in the home is responsible for an estimated 20,000 lung cancer deaths each year." You are standing in the kitchen with Grandma. All is quiet. Would you know what was happening? Would you know it was a problem? You probably wouldn't knock over Grandma to get out of a house filled with radon. You might look at the number on the radon test result and wonder whether that number was particularly bad, especially if the number was at or near the safe threshold. You might wonder, is this an emergency?

The EPA scientists cannot say for sure that if you live in a house with radon you *will* get lung cancer. They can only say you *might* get lung cancer. (Similarly, they cannot say that if there is no radon in your house, you will not get lung cancer.) The difference between knowing what *will* happen and what *might* happen is uncertainty. We are not sure what the future will look like. But we do have a good idea. The statistical uncertainty in the radon example crops up because of chaos — unpredictability because of nature's complexity. There is another type of uncertainty that is important when assessing risk posed by disasters — the uncertainty that is an inherent part of the process of science. Scientists communicate uncertainty to indicate how much we know about a situation and what we still need to learn. The latter we can decrease with more research, but it will always exist in some small amount because we will never know everything. The former — uncertainty due to chaos — we have to acknowledge and live with.

■ ■ ■

Once, I was at a local television station meeting with a TV meteorologist and several others about weather education projects. Our meeting was held at a conference table within the news studio. I was familiar with the look of the news desk and other sets from watching television, but I had never considered that on the other side of the room, off camera, there might be something as mundane as a conference table. At one point during our meeting, the meteorologist excused himself to film a series of short teasers to be broadcast during the commercial breaks of Oprah's show. I was tickled. I am not usually at meetings where someone excuses himself to be filmed. This particular meteorologist was one of those people who radiates so much charisma that when he left the table, the light seemed to dim. He strode over to the blank wall — the green screen — and a camera was trained on him. With natural confidence he spoke in quick haikuesque phrases about the evening's clear skies and full moon, bite-sized morsels of the future. He returned to the table muttering that in terms of predicting the future, at least the full moon was a sure thing.

Weather is the most common way that we experience environmental change. Weather prediction is much more accurate than it used to be, but there are still days when the future doesn't come out the way we thought it might. And when that happens, it is common to blame the meteorologist, to shake a fist at the cloud-covered sky and curse his or her name. But is that justified? The job of predicting future change is not an easy one. According to a meteorologist interviewed in *The Weather Book* by Jack Williams — an excellent handbook on weather and forecasting for curious novices — if there is not a cloud in the sky, you can bet there will be no rain in the next half hour. By limiting the distance into the future you forecast, you limit the number of chance events that could affect it. Basically, a prediction of no rain for half an hour under blue skies is only slightly more uncertain than a prediction of a full moon.

For a weather forecast that extends a day or a week into the future, you'll want to know what the weather is like over the whole country or even the world, because weather moves around. It helps that we can look at our planet from the vantage point of space. Weather radar and satellites are like the security systems in malls, and meteorologists are like the security guards, except that while the security guards in malls watch mallgoers and shoplifters, meteorologists watch water vapor, temperatures, and wind around the world. But even with excellent surveillance of present weather, it's a challenge to predict how weather will change.

Weather models use math equations to describe all sorts of processes that happen on earth every day — how sunshine heats land and water, how clouds form, and how differences in air pressure cause winds to blow. The more we know about how the planet works, the better our math, the better our models, and the better our predictions. If you feed a model the current weather, it can extrapolate into the future based on what it is programmed to know about how our planet works. You can extend beyond tomorrow to the next day and the day after. However, there are limits to how far out in time the weather model can predict. That's why you see a seven- or ten-day forecast instead of a one-year forecast. The farther into the future you are looking at weather, the more uncertainty there is because of the possibility of chance events — of chaos.

If there's an antihero in the world of weather prediction, it's chaos. It limits our ability to see the future. Chaos theory is the idea that when added up over time, small things affect large things. It makes predictions uncertain.

The father of chaos theory, Dr. Edward Lorenz of the Massachusetts Institute of Technology, explained chaos in the atmosphere with a metaphor: a butterfly flapping its wings may cause a tornado thousands of miles away. He wasn't blaming the butterfly. There might be another butterfly with flapping wings that would stop a tornado from forming. And neither butterfly would be trying to change the weather. They would just exist, as butterflies do. Extrapolated into larger and more human terms, it's a good example of how we all make a difference in the world even if we don't feel we're accomplishing anything.

With chaos, small changes in the starting conditions can lead to a wide variety of outcomes. Much like a Choose Your Own Adventure story, in which there are many possible stories in one book, chaos means that there can be a whole variety of outcomes when a hurricane hits the coastline, a tornado forms, or an earthquake rattles.

Put another way, chaos theory is like the difference between a day's to-do list and what one actually does that day. Each morning over breakfast, I scribble on a notepad a list of things that I intend to do, but I know there is a chance that my plans will change. My morning list is a prediction of the future. When things come up during the day that weren't in my prediction, that's chaos. It might be something good, like free ice cream. It might be something bad, like a car accident. Either way, it's a little slice of chaos. Sometimes the things that come up are so large that they make the rest of my list irrelevant. Other times they cause only slight changes to my list. Anything can happen. Often, it does.

If you don't like the weather, just wait a few minutes. Mark Twain once said this about New England, and Will Rogers once said it about Oklahoma. Countless other people have said it about countless other places. I've noticed that it's often said with a boastful tone about a specific place, as if the weather in that place is far more cunning than weather anywhere else. But weather changes everywhere. That's what weather does. And chaos is often at the root of the changes. And while chaos affects everything, weather is often the part of our planet where it is most obvious. It affects us every day, and most of us try to prepare for it, wearing sweaters, coats, and galoshes and then finding ourselves inappropriately dressed when a butterfly flaps its wings and the predicted cold front moves in a different direction. Weather is especially affected by chaos compared with other parts of the planet, because it happens in the atmosphere, a giant sloshing fluid that is prone to perturbations.

■ ■ ■

In 2009, the National Oceanic and Atmospheric Administration (NOAA) started to describe uncertainty due to nature's chaos in weather forecasts, including the probability that specific weather would occur. For example, instead of a forecast stating that the low temperature in central Florida will be 36°F tonight, the new forecast includes that there is a 30 percent probability that the temperature will dip below freezing, 32°F. With a forecast that includes the uncertainty in the form of the probability of a freeze, Florida orange growers can decide whether they need to take precautions to keep their groves from freezing. NOAA issued a statement to explain how they would transition from providing forecasts as a single value of the most likely scenario to reporting both the most likely value and the probabilities of other values. "NOAA's constituents are now requesting uncertainty information for weather, water, and climate scenarios for better risk-based decision making," read the statement.

In 2011, the United Kingdom Meteorological Office (the Met Office) did the same, describing the probability of precipitation in weather forecasts. This was done in an effort to provide transparency, to let people who hear the forecast come to their own conclusion about whether to bring an umbrella. However, this change was met with outrage from a UK organization called the Plain English Campaign, a group that lobbies against what they call "gobbledygook" in communications of all sorts. In 2011 the Plain English Campaign gave the Met Office their Golden Bull award for the year's best example of gobbledygook. According to the Plain English Campaign, the award was given for "empowering people to make their own decisions by using the technical systems for the probabilities of precipitation." I'm fond of alliteration, so I'm a fan of the phrase "percent probability of precipitation," but I can understand that this seems confusing to some. "Often people want to make a decision, such as whether to put out their washing to dry, and would like us to give a simple yes or no. However, this is often a simplification of the complexities of the forecast and may not be accurate," the Met Office explained in a rebuttal. "By giving [probabilities of precipitation], we give a more honest opinion of the risk and allow you to make a decision depending on how much it matters to you."

Forty percent probability of precipitation means that if today were repeated ten times, rain, snow, graupel, or hail would fall from the sky during four of those times. The weather forecasters run their models many times to figure out these odds. Because we all have different tolerances for risk, some people might carry an umbrella if there was 30 percent probability of precipitation. Others might not carry an umbrella unless the probability was over 50 percent, or only if they were wearing a shirt with a Dry Clean Only tag. Knowing about the tendency for rainy weather in the UK, and my own aversion to risk, I'd carry an umbrella all the time.

Researchers Rebecca Morss, Jeff Lazo, and Julie Demuth have found that some people interpret weather uncertainty incorrectly. Some interpret a forecast of 40 percent probability of precipitation as rain falling during 40 percent of the day or rain falling over 40 percent of the land area. In contrast, most people interpret uncertainty in temperature forecasts correctly as being a range of degrees instead of one average number.

Providing only an average hides the amount that something can differ from the average. For example, if you were to describe what dogs look like to someone who had never seen a dog — a visiting alien, perhaps — you'd need to communicate that there is a wide range in size, shape, color, fur texture, and friendliness. If you only described the average dog, the alien might imagine a four-footed animal that's about thirty pounds. When encountering a two-pound Chihuahua tucked into someone's purse or a two hundred–pound Saint Bernard pulling a sled, the alien would not think that either was a dog. In the case of dog breeds, the variability is not natural; it's because we have chosen to breed dogs that have certain characteristics. In the case of a weather forecast, variability in the prediction is largely due to nature's chaos, and this uncertainty is communicated as a probability. While it is unlikely that visiting aliens will ever ask you to describe dogs, it is likely that you will encounter statistical uncertainty in other realms.

In the case of weather, models are run over and over again to identify the amount of variability in outcome. Climate models are looking at a broad scale instead of details, so they are not affected by chaos, but there are some chaotic elements — the timing of El Niño events, for example. Climate models are also run multiple times to help us understand the range of possible changes in temperature that we are facing in the future. However, most of the range in temperatures is not because of chaos and uncertainty. It's because of us. We don't know how much greenhouse gas we'll be adding to the atmosphere. We don't know whether we'll change our ways quickly enough to stop catastrophic levels of climate change. The outcome is contingent upon our actions now and in the future. Just as the looks and temperaments of dog breeds are due largely to human decisions about which dogs to breed over many generations, our planet's warming climate is due largely to the decisions we make about whether to burn fossil fuels.

■ ■ ■

There is another source of uncertainty that affects every branch of science. This uncertainty is due to the fact that we don't know everything.

Scientists work at the outskirts of our understanding, on the border between known and unknown. On one side of this boundary is all the science that we know. On the other side of the boundary is an endless plain of knowledge that we do not yet know. And the science that researchers are exploring now is right on the boundary.

Over a century ago, the idea that carbon dioxide traps heat was on the outskirts of our understanding, but today it is far inland, because many scientists have tested it with different methods. About seventy years ago, the idea that humans were changing the amount of carbon dioxide in the atmosphere was at the boundary of our understanding, but today that's understood, too. For decades, scientists around the world have been measuring the amount of carbon dioxide in air collected each day. They have also been measuring carbon dioxide from ancient air bubbles extracted from glaciers and the amount of carbon dioxide we emit into the atmosphere. Over time, understanding of our ability to change carbon dioxide levels and warm the planet has grown. That is no longer on the border of what's known.

Today, scientists are working at the outer reaches of our understanding, learning details that will let us make better predictions of future climate, such as the rate that methane gas is released as frozen soils melt in the Arctic and the impact of different types of clouds on climate. These topics and many others are on the boundary of what's known. The answers will help us know more specifically how much climate change we should expect and how climate change will affect other parts of the planet.

Scientists need to communicate where on the boundary they are located so that others understand what we know and what we don't know. Talking about what we don't know lets us understand what we do know. It defines the location of the border.

For example, many scientists develop the math equations that make up a climate model. They need to communicate with each other about the equations they are adding to the model. And by describing the uncertainty in each addition they propose, they are identifying what still needs work in order to make the model more accurate. It's as if the model is a pot of soup and hundreds of scientists are the chefs. They are all adding spices to the soup and need to make it clear what's been added and what still needs to be added; otherwise, the soup is going to taste horrendous.

. . .

There are aspects of our planet that are way inland from the border of our understanding. Dropped objects fall because of gravity, for example. These concepts are not likely to change with more knowledge, so they fill science textbooks, which need to last for many years. And because they fill textbooks, it can appear to students that everything in science is known. If that were the case, then scientists would stand around in lab coats with nothing to do. They might dissect the occasional fetal pig or identify the same rock types over and over like students do, but they would always find the same organs within the pig, and the rock types would never change.

This is not the case.

"Science is a constant process of narrowing uncertainty and gaining improved understanding. It is not static," explained climate scientist and psychologist Jeffrey Kiehl in his book, *Facing Climate Change: An Integrated Path to the Future.*

Not everything is known. There will always be more unknowns in the universe than knowns. Scientists have an eye for unknowns and an internal reservoir of curiosity that drives them to figure things out. Science involves wondering why things happen. It also involves coming up with possible answers to the "why" questions and testing those possibilities. It involves looking at all the evidence. This can narrow down the reasons why something happens, but often does not totally eradicate the uncertainty.

"Science is uncertain," wrote biochemist and writer Isaac Asimov. "Theories are subject to revision; observations are open to a variety of interpretations, and scientists quarrel amongst themselves." Asimov was referring to the reason that many people prefer the "rigid certainty of the Bible" to the science of biological evolution, but the same could be said for any variety of science.

Understanding the way scientists approach their work can be helpful for understanding how they navigate uncertainty and how we all can do the same. To describe how science works, I will share an example of a fictional situation in which characters did not think like scientists and consider how the situation would have been different if the characters were thinking like scientists.

In *The Little Prince*, by Antoine de Saint-Exupéry, the narrator, a pilot stranded in a desert, recounts a story from his childhood. When young, he made a drawing titled *Drawing Number One*. He showed his drawing to grown-ups, and they thought it was a hat. That was not what he had intended. He created *Drawing Number Two*, a cross section view, and showed that to the grown-ups. What appeared to be the space that a head would occupy in a hat was instead an elephant, which was within a snake. (Note: For copyright reasons, I am unable to share these two drawings with you, fair reader. But due to the Internet's complete disregard for copyright, you can find the images easily by putting the words *prince*, *hat*, and *elephant* into a Google search. Go ahead. Try it. And then keep reading.)

If the grown-ups were thinking like scientists, they would have looked at *Drawing Number One* and made observations. When they exclaimed that it was a hat, they would have been making a hypothesis, a testable statement. *Drawing Number Two*, the cross section that shows what is below the surface, would be new information that would allow them to refute the hat hypothesis.

However, the grown-ups were not scientists. When presented with *Drawing Number Two*, which clearly presents more evidence in support of an alternate hypothesis — this was not a hat, but instead an elephant that was consumed by a snake — the grown-ups dismissed the new data. Having come to a conclusion early on, they were not receptive to more information provided by *Drawing Number Two*. This is happening in the world today among people who refuse to accept climate change. Many people, including a few people who call themselves scientists, have been shown information about what is happening to the planet's climate and refuse to accept it. If we all went around making hypotheses instead of conclusions about the first facts that we learn, we would be more open to learning new things.

The narrator in *The Little Prince* was not a scientist either. He judged the grown-ups for their hat hypothesis. He stopped showing grown-ups *Drawing Number Two* and instead made the assumption that these grown-ups were not worth his time, that they were narrow-minded and had limited imagination. A hat wasn't a bad guess, and thus it was a good hypothesis (except for the fact that the hypothetical hat had an eye).

Had the grown-ups rejected the hat hypothesis when they were shown *Drawing Number Two*, they might have formed a new hypothesis that this was an elephant that had been eaten by a snake. According to the scientific method, they would not know that this hypothesis was correct. There would still be uncertainty. The grown-ups would need to be skeptical. They would want to gather more data. For example, a grown-up might ask to see *Drawing Number Three*, in which a penny is drawn next to the animals to give scale to these creatures. Another grown-up might look for eyewitness accounts of the snake eating the elephant. They would be working on the outskirts of our understanding. And as scientists, they would share their new data and hypotheses with each other to get critical feedback from other scientists.

When a hypothesis is tested in different ways, we narrow down the reason for a particular phenomenon, but there will always be some uncertainty. Perhaps we will find that it is a toy elephant inside a sock puppet. In the case of climate science and other types of environmental change, there will also always be some uncertainty because we will never know everything.

Environmental changes may be slow or fast, large or small. One change may cause another. With environmental change, uncertainty crops up repeatedly, and whether the uncertainty is due to nature's chaos or because we are on the edge of what's known, we will never be totally sure.

Uncertainty about environmental changes can make it particularly diffi-cult to recognize an emergency when we see one. It is challenging to make decisions when we are faced with uncertainty, but it's everywhere, even if we ignore it. If we understand where the uncertainty is coming from and why it is there, we can make better decisions in the face of it to keep ourselves safe and our world habitable. There may not be an In Case of Emergency sign when we find ourselves in harm's way, but we'll still need to take action, despite the uncertainty.

Chapter Two

When Sands Shift

My nephew Justin was three years old when my sister, Sara, and I took him on his first Cape Cod road trip. We were at the beach to be beachgoers, to splash in the waves, stare out at the ocean, make castles of sand, and soak in the sunshine, but instead, we craned our necks and looked up to where a small, weathered house was perched at the edge of a sand cliff.

The sand cliffs are a common sight along the easternmost part of the Cape. Sand grains refuse to stack themselves to form a truly vertical cliff, so this wall of sand was actually more of a steep sloping hill. But I like thinking of it as a wall or a cliff. That's how it seemed to me. Dramatic. Not a molehill. A mountain.

Home sizes in the area had inflated enough for this older saltbox to look like a tiny outbuilding of one of the Cape's large, modern homes. Over decades, its grassy, sandy yard cleaved off into the sea.

This is an example of gradual change. We are not shaken alert by a disaster until the house falls. Instead, minor adjustments over many years leave the house increasingly vulnerable, and we may not even notice because our brains evolved to handle change that is quick, not slow. With slow change it's easy to become "change blind," unable to notice what we don't want to see. But the changes add up. Or, in the case of the cliff, the change is subtracted.

As we looked up at the eaves of the precarious little house, sand flowed down in rivulets — sand rivers flowing through sand valleys on the side of the sand mountain. While Sara and I focused on the house teetering on the top and considered its options, Justin saw the mountain of sand.

"I want to climb that," he proclaimed, a toddler planning his Everest expedition.

I wanted to explain to my nephew the process of erosion. I wanted to explain that the reason there is a cliff of sand here is because the land is moving. That year after year, storm after storm, the cliff face is etched away and the sand flows elsewhere. I wanted to mention that climate change is causing sea level rise and stronger storms, which has an impact on erosion here, too. I wanted to know when the little house would be rubble at the bottom of the slope. Would Justin be an adult by then? Would he be telling his nephew about a time when he was young and the pile of broken timbers and weathered shingles was a tiny house at the top of the cliff?

But I didn't say any of that. I looked down at the little person tugging my hand with his two sandy ones and told him that if he climbed up, the sand would come down.

He nodded and slackened his grip. He seemed to accept my logic, but I was not sure I did. I looked back up the sandy cliff, squinting to avoid errant grains in my eyes. The sand is going to come down at some point. It was coming down as we stood there. The cliff was eroding. Change was happening. We can be certain that the house will fall eventually if it is left in place. We cannot be certain when. The rate of erosion — the amount of sand lost in each storm and during each year — will determine how long the house will remain on its perch. But that rate is not consistent. Average rates of erosion give a general idea of when the house will fall, but we can't say for sure. If there are more storms and more erosion, then the little house will be toppling into the ocean very soon. If there aren't as many storms, it might have many good years left. Like a patient with a terminal disease, the little house asks, "How long have I got?"

▪ ▪ ▪

We were at the seashore because I had begged to go there. Sara and Justin love the beach too, but I was the instigator of that trip. I live a mile above sea level at the base of the Rocky Mountains. When I need perspective, I head uphill to mountain tundra, where I can see for miles. I breathe thin air under clear blue skies. But when I'm anywhere near a coast, the ocean is where I go for perspective. Said Henry David Thoreau, "The sea-shore is a sort of neutral ground, a most advantageous point from which to contemplate this world."

On maps, this far eastern part of Cape Cod, which I like to think of as the forearm, looks almost narrow enough to shout across, but in actuality the surface of the land, made entirely of sand and mud, flows up and down in ancient hillocks and little ponds formed as glacial ice melted about fifteen thousand years ago. The pattern of sediment released from the melting ice created a land filled with nooks and crannies, places where today people get lost for a weekend, a summer, or forever.

We traversed the nooks and crannies as we biked north, following the Cape Cod Rail Trail from Orleans almost ten miles to get to the beach, the wall of sand, and the house in peril. A former train track, the rail trail is immaculately paved. It follows the Cape's forearm like a vein, passing marshes and ponds, small houses tucked into the scrubby trees, and a few small towns. Chipmunks darted across the path, causing Sara to slam her squeaky bicycle brakes. A bright red cardinal flew across too. We ducked. Justin laughed at us from behind, where he rode the path in a rented trailer, a little tent on wheels he called his spaceship.

We were on the Cape in early season, early enough for people in the service sector to still smile, early enough that we were wrapped in sweaters and coats during the first stormy day we were there. A little weather wasn't going to keep us from the beach, and we resembled classic New Englanders in wool and rain gear, looking pensively out to sea as sand grains, driven by wind, hopped along at our feet and dark clouds loomed overhead. We stared at gulls that appeared to hover in place as they rode the storm's wind.

Because we came to Cape Cod before the bulk of the crowds, we felt like pioneers for that season's beachward migration, but we were pioneers in no other way. People have been coming to Cape Cod for hundreds of years. Indians were living here when the Pilgrims arrived on the *Mayflower* in

1620. More than two hundred years later, in 1849, Thoreau first wandered the outer part of Cape Cod, which he described in a book called, simply, *Cape Cod*.

I don't think people usually say this of transcendentalist writings, but Thoreau's *Cape Cod* is hilarious. He poked fun at the characters he met on the Cape. He seemed to find humor almost everywhere. But, on a more serious note, he noticed then, as I noticed now, that the sands of the Cape seem temporary.

"Perhaps what the ocean takes from one part of the Cape it gives to another,— robs Peter to pay Paul," Thoreau speculated. "On the eastern side the sea appears to be everywhere encroaching on the land."

While Thoreau and I may have noticed the same geological instability on the Cape, that is where our shared experience ends. Thoreau walked up the shore from Nauset to Provincetown. Sara and I biked on a paved path. We took turns on the bike that was towing Justin in his rented spaceship. Thoreau and his hiking companion were not traveling with a preschooler, so I'm guessing they did not sing songs about sleepy bears, birdies, and Mr. Sun. They probably toted fewer snacks.

Thoreau and his friend started out walking along an unpaved, sandy road in which "a horse would sink up to the fetlocks." It was raining and windy as they walked in October. In June, more than a century and a half later, the day after the storm was sunny and hot as we rode our bikes. I'm guessing we stopped for ice cream more often than they did.

Thoreau also did a bit of complaining, describing the Cape landscape as "bare swells of bleak and barren-looking land." He mentioned that there were no trees in this part of the Cape. "The trees were, if possible, rarer than the houses," he wrote after noting how rare it was to see a house. Today the Cape is shaded with scrubby forests and peppered with summer homes.

On their way to Nauset Light on the eastern shore, Thoreau described the terrain as "an apparently boundless plain, without a tree or a fence or, with one or two exceptions, a house in sight." His hiking partner described the land as looking like the rolling prairie of Illinois. Today, there is very little Illinois left on Cape Cod. Yet one of the most prairie-like parts of the Cape is the area where we saw the small house teetering at the edge of the cliff. Alongside the house, the land rolls with light grasses, shrubs, and other weathered cottages.

According to the National Park Service, which manages Cape Cod National Seashore, big storms during the winter before we arrived undermined the sand cliffs, making the sand more likely than ever to topple down. For a few months, police tape was strung up like a crime scene to keep people out of the zone of falling sand. But in June, when we wandered the beach, the crime scene was gone. The spring's gentler ocean was slowly supplying sand to the beach, which, park officials hoped, would buttress the cliffs somewhat.

The cliff has lost about three feet per year on average over the past century according to the US Geological Survey, but nature is rarely uniform. In some locations, the cliff is losing much more land than in others. In 2015, the cliff lost eighteen feet to erosion at Nauset, a hot spot for erosion. The park service is working with coastal geologists to understand the reasons for rapid erosion at Nauset. In general, along the coast more sand is lost during some decades than others. The rate of erosion is related to the shape of the adjacent seafloor, which is ever-changing as sandbars form and move over time and storms vary in number and size. In general, the cliffs lose more sand during storms than at quiet times. Erosion from these areas of Cape Cod is not something that can be fixed. It has been happening for thousands of years. In fact, geologist Arthur Strahler estimated that the eastern shore of the Cape has eroded back as much as two miles in the past thirty-four hundred years. In this case, change is normal and, for the most part, natural.

■ ■ ■

To get a sense of how the sands had shifted around Cape Cod, Thoreau consulted a historian. He told Thoreau that in the village of Eastham, the sand in some places "has been raised into hills fifty feet high, where twenty five years ago no hills existed."

The historian's words reminded me that the sands of Cape Cod are relocating, not disappearing. The sand that falls from the cliffs in Nauset and Wellfleet travels either north, joining the dunes near Provincetown, or south, to the elbow of the Cape — near the town of Chatham — and a dune-covered spit of sand called Monomoy Island, which juts out eight miles south of the Cape's elbow. The moving sand forms bars in shallow water, making the sea look tropical blue. Around the same time that Thoreau was wandering the Cape, scientists were watching Monomoy Island grow more than 150 feet longer each year with all the added sand.

Sand scoured away from one place is deposited someplace else. The little grains of sediment come to rest, falling out of air or water that has slowed and can no longer carry the load. When I took a college archaeology class, one of the things that confounded me most was that all the civilizations we were studying had been covered up by sand and mud. I asked the professor where all the sand and mud came from and why brilliant civilizations always wound up covered in the stuff. As I recall, he didn't have a great answer. He was much more interested in the artifacts, while I was focused on the sand, which seemed to be magnetically attracted to ancient towns in the Middle East or the American Southwest. It was a bit like that picture in which some people see two faces and other people see a vase. I was seeing the stuff between artifacts while we were supposed to be seeing the artifacts. A couple of years later, I learned in a sedimentology course that there are countless ancient towns we know nothing about because, rather than getting covered, the tablecloth was pulled out from under them. The sands and rocks below those towns eroded over time. Buildings crashed into streambeds. Pottery sherds weathered into sand grains. The remnants of many ancient villages have been destroyed by erosion instead of preserved by deposition.

Look at rocks made of sand from the side and you will see lines between the layers. Each is a bit like the ribbon of frosting that separates two layers of cake. Like frosting, the line is actually a plane — a two-dimensional surface spread between the layers. Unlike frosting, the plane between sediments has no thickness. It is just the boundary between two times when sand was deposited. This surface without substance could represent a day or it could denote thousands or millions of years — time for which we have no record in the stack of rock layers — when sediments were not deposited or were laid down but then eroded, erased from the record of what happened in that place. Someday, Cape Cod, with its eroding cliffs, might be one of these lines between layers of rock, with nothing to mark its existence other than a surface between layers. In the future, no one might know that this cliff was here and that a small house once counted its final days on the edge.

Following Thoreau's lead, I, too, consulted a historian: Bill Burke at Cape Cod National Seashore. When I talked to him in 2016, he'd worked at the park for twenty-eight years, first as a member of the interpretation staff, which leads educational programs, and later as the historian. Burke told me that while land is added and lost as the sand moves, overall the Cape loses about five acres of land each year.

"It's a little strange, working in a park that gets smaller every year," said Burke. He told me that no one takes much notice of erosion when it happens to areas where there are no buildings that could be lost. "It's out of sight, out of mind," he said. But when erosion starts to damage buildings, people get frustrated and concerned.

When Burke first got to the park in 1988, he developed a coastal erosion education program for visitors. Erosion was happening then. And it's happening now, worsened by the impacts of climate change and sea level rise.

■ ■ ■

Thoreau wrote about a lighthouse called Highland Light, which today is often called Cape Cod Light. It is located in Truro, a few miles north of where Sara and I spotted the little house on the edge of a cliff. Thoreau expressed the same wistfulness for the lighthouse that Sara and I had for the little house, reflecting that its days were numbered. "It stands about twenty rods from the edge of the bank, which is here formed of clay," he wrote.

Highland Light,
Cape Cod. (Courtesy
of Adam D. Holloway.)

Rod. What's a rod? After consulting a massive, dusty dictionary, one that was probably older than Thoreau, I calculated that Thoreau's 20 rods add up to 330 feet. The lighthouse was 500 feet from the edge when built in 1797 and 330 feet from the edge when Thoreau was measuring it in rods around 1850.

Thoreau cobbled together surveying tools borrowed from a carpenter and got to work figuring out where exactly the lighthouse stood and how precarious this situation was. The lighthouse keeper told him that the sand in that area was taking off in both directions — some headed north and some headed south. "Ere long, the lighthouse must be moved," said the lighthouse keeper.

Modern estimates put the erosion at a bit less than Thoreau's estimate of six feet per year, although it's not the same everywhere and it changes year to year. In total, the cliff at Truro eroded back about four hundred feet in the two hundred years since the first incarnation of the lighthouse was built, in 1797.

An old Cape Codder (as the people who live on the Cape are called) told Thoreau that when the lighthouse was built, it was calculated that it would stand only forty-five years on that spot before the cliff eroded enough that it would fall into the sea. The implication was that the effort to build a lighthouse lasting just forty-five years was worth it to keep people safe at sea. The lighthouse Thoreau stayed in during the 1850s was not the original, but it was located in the same spot as the original lighthouse, and that place had not yet eroded into the sea. The wood structure of the original 1797 lighthouse had to be replaced by brick for structural stability in the early 1800s to defeat the incessantly huffing, puffing wind.

Both Thoreau and the old Cape Codder were thinking along the same lines: that the edge of the cliff was creeping closer to the lighthouse each year, meaning that the lighthouse wasn't going to last for long unless it could run away.

That's what it did. Well, it didn't run. But it did move. Over eighteen long days in 1996, movers slowly slid the 430-ton lighthouse 450 feet to the west along steel beams that had been slicked up with Ivory soap (99.44 percent pure!), putting a total of 560 feet between the lighthouse and the ocean. Nearly twenty years after the move, the lighthouse and keeper's house are 470 feet away from the cliff's edge in a northeast direction and 512 feet from the cliff's edge in a due east direction. The location of the original lighthouse is marked with a boulder, which, I estimate, is only a couple of rods from the edge.

What caused people to recognize the risk and take action? I wondered, thinking that might help us understand how we can reach a similar point of recognition with climate change. I asked Bill Burke how the decision was made to move the lighthouse. He told me that it was a grassroots effort. Local residents formed a group called Save the Light in 1990. The Save the Light members valued the lighthouse and keeper's cottage as more than just an aid to navigation. It was a relic, an important part of our history. The group recognized that for the lighthouse to be saved, it would have to be moved long before it was teetering over the edge of the cliff. Moving the lighthouse could only happen when it was on land solid enough for hefty earthmoving equipment to get under it. Their goal was to move the lighthouse before it was in grave danger. It was moved when it was about 110 feet from the cliff's edge.

With climate change there have been numerous goals set to help us avoid grave danger. Yet they have not provided the momentum needed for humanity to stop adding extra greenhouse gases to the atmosphere. The organization 350.org, for example, advocates for stabilizing carbon dioxide in the atmosphere to 350 parts per million (ppm) in order to preserve a livable planet, but the carbon dioxide concentration rose above 400 ppm in 2015 and could be as high as 1,000 ppm by the end of the century if we don't stop burning fossil fuels. Numbers don't always work. Or perhaps the difference is that a lighthouse threatened by a cliff is visual. Climate change is largely invisible. Its impacts can be visual — such as coastal flooding or polar bears on dwindling ice — but, like the Dementors in Harry Potter books, it's easy to have some doubt about what caused the impacts when the foe can't be seen. Humans are poorly adapted to handle invisible foes and slow change. People are more likely to fear extreme threats that are very unlikely than understated threats that creep up on us. Social scientists Susanne Moser and Lisa Dilling call climate change a "creeping problem," explaining that because it is a slow threat, there is not a sense of urgency. It will happen someday, but future risks do not seem as bad as if they were happening right now, even if the consequences are the same. Humans have a tendency to undervalue the future.

■ ■ ■

To understand why this sand is here at all, you have to look further back in time. Cape Cod is a product of glacial ice that was once ten thousand feet thick. As this ice melted, it left piles of sand and gravel and even some huge boulders behind. The sediment that the glacier bulldozed was left behind, forming the western part of Cape Cod. The sediments that make up the eastern part of the Cape were likely once caught up within the ice itself. As the ice melted, the sediments flowed down rivers of meltwater until they dropped out into plains.

The bedrock of the rest of New England is much, much older than Cape Cod — hundreds of millions to 1.4 billion years older. Just as older people often think younger people are flighty and make snap decisions without planning, the old rocks that make up the rest of New England must look at the newcomers on Cape Cod and the islands of Martha's Vineyard and Nantucket as unpredictable and ever changing.

After Sara, Justin, and I turned our backs to the wall of sand and the teetering house for a couple of hours of beach time, Justin revisited his plan to climb to the top of the wall. This time he had spotted a staircase made of rickety wood that scaled the entire height, from the beach to the grass-covered top, a bit of a walk down the beach from the house at the edge. The wall seemed particularly high, or perhaps that was an illusion prompted by the thought of walking up all those steps. The staircase had a railing made of rope, and the spacing between the stairs seemed enormous for three-year-old legs. We decided to give it a try anyway. Slowly we made our way up the stairs. As we climbed, rivers of sand flowed down beneath and beside the staircase.

At the top, we emerged into a neighborhood incongruous with the coastal landscape. There were yards with trimmed hedges and flowers. A guy washed his car. Kids biked in the street. A big, clumsy dog ran up to greet us. We had been scaling the wall of sand for quite some time and were sweaty, out of breath, and gritty. We had conquered a mountain. Yet we found ourselves in an entirely different context. It looked so, well, normal. It was what one might expect of any suburban neighborhood, except that where there should be a cul-de-sac there was a cliff. What's actually normal in this environment is change — shifting sands and evolution of the landscape. What's normal is instability. But we long to control that,

Stairs up the cliff of
sand on Cape Cod.
(Courtesy of the author.)

keeping the coast permanently as we imagine it. We think of normal as
things staying the same. If everything is normal, then there are not cottages dangling at the edge of cliffs.

The people around us didn't seem too concerned with the wall of sand,
even though it is impossible to ignore. The idea that this neighborhood
might not be permanent didn't seem to be of great concern on a sunny afternoon. I suppose the same is true of a neighborhood moments before an
earthquake. But unlike rapid environmental change, slow change — and
a slight increase in risk over time — is often ignored by those who, instead,
focus on the present. In 2006, insurance companies reassessed risk for
Cape Cod homes in light of sea level rise and more severe storms caused by

climate change, making premiums much more expensive. In 2014, FEMA flood zone maps were redrawn and more Cape Codders found their homes officially in harm's way. Local news stories periodically profile houses that have lost their footing and fallen off the edge of Cape Cod and homeowners going to great lengths to move their homes away from the edge, rebuild, or put in deep footings to moor their house in the sand.

I looked down from the top of the staircase to where tiny people romped on the beach below. I clung to the rope railing, a little dizzy. It's not brazenly apocalyptic to forecast that the forearm of the Cape, including that neighborhood, will continue to narrow as a bit more of the wall of sand falls into the sea each year. That falling sand will allow the Cape to reshape itself over time as long as waves come crashing to its shore and winds blow over its surface. Of this, we can be certain. It is the nature of slow environmental change.

Nearly six months after Sara, Justin, and I saw the little house on the cliff's edge in June 2010, the *Cape Cod Times* reported that the front of the house slid down the cliff as the sand gave way during a winter storm. The house was torn down in early 2011. According to the assessor's records, the property was worth $1,180,000 in 2010 and 90 percent of that was the value of the land, not the two-bedroom cottage. In 2015, four years after the cottage was demolished, the assessor valued the vacant and unbuildable lot at $43,000. Over a million dollars had eroded away. When Sara and I visited in 2016, a rocking chair sat in the sand and grass at the cliff's edge. Rocking the chair might prove calamitous, but it would be an incredible place to look out to sea, at least until the cliff crumbles another foot.

Chapter Three

When Ground Shakes

Once upon a time, dapper Victorians gathered around a dining table adorned in white linen, topped with fancy china. They sat upright in spindly chairs, politely prodding food with silver. On an adjacent sideboard covered with books rested one of the men's hats. Someone snapped a photo, a copy of which I found in a San Francisco bookshop.

I tacked my copy of this black-and-white photo to a kitchen cabinet several years ago so that I'd see it every day. The scene portrays typical domestic life with an incongruous backdrop. The table, the linens, the china, and the Victorians are carrying on with life soon after a massive earthquake and fires shattered their city, San Francisco, in the spring of 1906. Beyond the table dressing, a charred landscape of rubble stands like an elephant in the room — a room that isn't really a room at all. It's like a play performed on the wrong stage, a graphic reminder that human society and nature coexist, even if they don't usually run into each other like this.

I have more photographs from the time stuck in strategic locations around my home, places where I see them as often as I see my own face. Some people put up pictures of friends or relatives or vacation destinations. I have pictures of people unknown to me, clinging to the trappings of society as their world crumbles into nature. I'm not against other types of pictures. I have some of those too. Yet I find these eerie images of the 1906 disaster intriguing, even comforting. They show nature as dramatic and powerful, but any old waterfall or volcano photo could show that.

Dinnertime on Franklin Street near Fulton shortly after an earthquake and fires ravaged San Francisco in 1906. (Bear Photo, courtesy of California Historical Society.)

What they also show are people existing with that power. Like a geologist's photo of a rock outcrop or fossil that includes a pen or penny for scale, humans give a sense of scale to a disaster. In fact, earthquakes and other natural disasters aren't even considered disasters if humans are not in their way. They are only called natural disasters if they have disastrous consequences for people.

What started with a few photos tucked around my home has evolved into a full-fledged quest to understand the San Francisco earthquake and fires and what happened when this formal and orderly Victorian society fell vulnerable to nature. I wanted to know how this catastrophe happened, but I also wanted to know how people reacted, why they might dine amid the rubble, and what they thought about nature when their city came crumbling down. How do we rationalize the decision to build on a fault line or an eroding coast? And how do we react when the ground starts to shake?

The events of the past can be thought of as completed experiments. We know what happened. Maybe the results can tell us something about human nature when faced with chaos and uncertainty. Looking at photographs of Victorian splendor amid the San Francisco rubble, I wanted to know how those people's lives were affected, how they reacted as their world changed, how they made decisions about whether to stay or flee. Did they know that their city was built in raw wilderness, as all urban places are?

. . .

The shaking started in San Francisco at about 5:12 on the morning of April 18, 1906. Rocks more than six miles below the earth's surface released their pent-up energy in waves that spread through the ground at speeds over ten thousand miles per hour. The earthquake is estimated to have had the explosive force of five hundred megatons of TNT — a magnitude of 7.8 on the Richter scale.

A block from Golden Gate Park, Emma Burke, her husband, and their son clutched the frame of a door within their apartment while listening to the deafening noise of destruction as the ground shook — crashing dishes, rattling roofs, and a piano hurling across the parlor. The collective cacophony "made such a roar that no one noise could be distinguished," Burke wrote in an article a few months after the disaster. She continued:

> We never knew when the chimney came tearing through; we never knew when a great marine picture weighing one hundred and twenty-five pounds crashed down, not eight feet away from us; we were frequently shaken loose from our hold on the door, and only kept our feet by mutual help and our utmost efforts, the floor moved like short, choppy waves of the sea, crisscrossed by a tide as mighty as themselves. The ceiling responded to all the angles of the floor. I never expected to come out alive. I looked across the reception-room at the white face of

our son, and thought to see the floors give way with him momentarily. How a building could stand such motion and keep its frame intact is still a mystery to me.

Stand in front of your clock and count off forty-eight seconds, and imagine this scene to have continued for that length of time, and you can get some idea of what one could suffer during that period.

In less than a minute, most of the shaking had ended. This is rapid change. It can't be ignored like the slow erosion along a coast. Many of the buildings covering San Francisco's hills had crumbled into bricks and dust that morning. Chasms had opened in the middle of some streets. Most of the people who were not in bed were selling and buying produce in the market district when the earthquake hit. These early birds would wind up trapped under mounds of bricks, fruit, and vegetables.

During an earthquake, the ground doesn't move in exactly the same direction and at exactly the same speed. This causes cracks to form in brittle ground. In San Francisco, the cracks were one to five feet wide and up to twenty-five feet deep. DeWitt Baldwin was eight years old at the time the earth shook. Freed from his mandatory early morning piano practice (his family's piano was in no condition to be played), he and other boys roamed the city once the shaking stopped and before the fires started, eager to see the damage. "Sometimes when I dared to peer down the fissure I would see fallen things inside," he reflected. "At times I couldn't see anything because the crack was frightfully deep and dark."

It was just before the sun climbed above the horizon when the earth shook. Most people were in their pajamas. The general reaction was to run into the street and away from unsteady buildings. Many of these buildings would catch fire later that day. Thus, many people would be left with only the pajamas on their back, and they remained dressed that way days longer than they intended. In his memoir of the experience, photographer Arnold Genthe described the post-earthquake fashions that he witnessed in the streets outside his studio:

The streets presented a weird appearance, mother and children in their nightgowns, men in pajamas and dinner coat, women scantily dressed with evening wraps hastily thrown over them. Many ludicrous sights met the eye: an old lady carrying a large bird cage with four kittens inside, while the original occupant the parrot, perched on her hand; a man tenderly holding a pot of calla lilies, muttering to himself; a scrub woman, in one hand a new broom and in the other a large black hat with ostrich plumes; a man in an old-fashioned nightshirt and swallow tails, being startled when a friendly policeman spoke to him, "Say, Mister, I guess you better put on some pants."

There may not be anything more vulnerable than urban Victorians on unsteady ground, surrounded by unsteady buildings, without pants. Most of Genthe's description emphasizes the mixture of vulnerability with trophies of society — a hat with ostrich feathers, a dinner coat, calla lilies, and the man in the nightshirt wearing a tailcoat, or swallowtails, as they were known — these material objects give us a sense of our culture, a sense of ourselves, and a connection to our normal reality. Did they still matter as the earth shook?

Across the city, Emma Burke hurried to dress after the shaking subsided, and she realized the clothing she had valued the day before was no longer of such value. She chose a coarse wool skirt and a long coat lined with white silk. She reflected, "I had no thought for the dress I had cherished the day before, I was merely considering what was warmest and most substantial." For Burke, fashion was less of a consideration in the face of natural disaster than it was for Genthe, who roamed the streets in his khaki riding outfit, which he had deemed to be his most suitable "earthquake attire" as he was dressing and preparing to flee his building.

■ ■ ■

It wasn't the earthquake that caused the most destruction in San Francisco. It was the fires that engulfed the city after the ground calmed —

fires that proved very difficult to stop. It's easier to stop a fire when you have water to douse the flames. But the earthquake had caused the city's water supply to fail, an irony for a city on a peninsula, almost completely surrounded by water.

Some disasters, like an earthquake, happen all at once. They come out of the blue, making whatever to-do list you had made that morning entirely moot. Some are incremental, like the gradual warming of the atmosphere from additional greenhouse gases or the erosion Sara, Justin, and I witnessed on the eastern edge of Cape Cod. They are so slow that it can be difficult to see the effects until the situation is dire. Other disasters are in between: rapid compared to recent climate change, but gradual compared to an earthquake. A fire that eventually engulfs a city is in between. A single spark may start a fire so small that it could be quelled with the sole of a shoe. The risk increases as the fire spreads.

Given exactly the same information about a situation, some people will assess the risk as high and others will assess the risk as low. Some might be certain the flames will be controlled, while others are certain the flames will grow unchecked. Because we evaluate risk more by intuitive feelings than by logic and reason, we all perceive risk differently, and uncertainty makes our perception of risk even more varied. The fire could be stopped. There is also a chance it could grow. Some of the uncertainty is due to nature's chaos. Some is due to the limits of our understanding of the science of fire. We all live with uncertainty, even if we don't live on a fault.

The 1906 earthquake was quick — less than a minute — and there was little time for such assessments of risk. The fires that followed the earthquake grew over days. And when a disaster grows in size, it's difficult to know whether to react, difficult to know when it becomes an emergency. It's difficult to know whether we can quell the fire or whether nature will win out.

That morning after the earthquake, the sky was clear and blue. The wind was calm. The city was quiet. The people gathered in the streets and parks watched with calm wonder as a column of smoke rose a thousand feet in the air. Within thirty-six hours, the calm wonder had been replaced by panic. The fires had spread. They were everywhere.

Genthe wrote:

> I have often wondered, thinking back, what it is in the mind of the individual that so often makes him feel himself immune to the disaster that may be going on all around him. So many whom I met during the day seemed completely unconscious that the fire which was spreading through the city was bound to overtake their own homes and possessions. I know that this was so with me. All morning and through the early afternoon I wandered from one end of the city to the other, taking pictures without a thought that my studio was in danger.

Perhaps they didn't know about the fires, but it's also possible that they were experiencing optimism bias, which causes some people to see themselves as immune to a disaster's effects. It's the "it won't happen to me" philosophy. It's a relative of my "I'm sure it will be fine" mantra. Also, Genthe and the others had survived one disaster already that day, and research has found that people who survive one disaster feel that they are unlikely to see another.

Perhaps that calm is what allowed Genthe to take a break in his day and join friends for a glass of wine. As they raised their glasses, an aftershock rumbled. The ground shook again. Everyone ran outside except for Genthe and the hostess. The two raised their glasses, and Genthe gave a toast: "And even if the whole world should collapse, he will stand fearless among the falling ruins" — a line from the Roman poet Horace.

■ ■ ■

In San Francisco today, as in many other places in the country, little red boxes shaped like small savings and loan company buildings with Roman architectural details are mounted atop poles on street corners. Below each box's roof is a handle that when pulled, will summon the fire department. In the era before cell phones and the Internet, these things had purpose, but today they look merely sculptural — totems in tribute to the paradox that the security of organized society exists within the uncertainty of the wild. The little red boxes, beacons of security, can be found every few blocks in the older neighborhoods of San Francisco.

Most tourists wandering a city spend their time in museums or ogling historic buildings. Maps or cameras obscure their faces. They rely on smartphones as tour guides. In San Francisco, they walk to Alamo Square for that classic postcard view of ornate Victorian houses with modern skyscrapers beyond; they seek out paintings in art museums, food and drink, or sea lions on piers. As a rule, they do not stand on street corners gazing at metal boxes on poles. But that is where I find myself in San Francisco. Walking through the city's hilly streets past myriad houses built in 1906, just after their predecessors were reduced to rubble, I am fascinated that these boxes still reside in the city. Perched on poles, they look more like birdhouses than protectors.

Peering at the box, just above the little handle, I read five simple words with no punctuation, the instruction:

In case of fire
Pull

They were originally a telegraph system, which, to me and my lack of technological know-how, sounds little more advanced than the system of Dixie cups and string my sister and I used as the communications infrastructure in our childhood backyard. Modern San Franciscans might prefer a cell phone, but Victorians must have seen this as an almost magical means of summoning help.

These alarm boxes were installed on San Francisco street corners in the 1860s, about four decades before the earthquake and fires that would send citizens clambering to pull their little handles all over town. Unfortunately, the shaking earth disturbed the telegraph wires, and they were unable to sound the alarm as fires spread.

Being able to cry for help does not stop whatever torment is going on. But it's comforting to be heard. The emergency boxes on the San Francisco street corners look civilized, like wise old ancestors to which you can run when times are bad. I wonder if they still work.

As I stood on the corner, peering at the box, considering how we try to civilize the wild in cities (and taking notes, I must admit), a group of tourists congregated around me. They removed their noses from their maps and guidebooks long enough to deduce that there might be something interesting in front of me. Acting like kids on an Easter egg hunt, they swarmed the little box. I stepped back to let them have a look and to prevent them from seeing my scribbled notes. To my surprise, they looked beyond the box to the landscape of the city seen so clearly from on top of hills on a blue-sky day. They pointed to Coit Tower and remarked on the view. Then one member of their group who was carrying a guidebook suggested a destination downhill, and they were off to swarm other, more exciting tourist attractions. I remained with the little emergency box, still wondering: When a disaster happens slowly, like climate change, and some of us see the risk that others don't, when do we pull the handle? In the case of the Cape Cod lighthouse, it took community recognition that something of value was in jeopardy. The longer I stood before the box on the street corner, the more I itched to pull the handle.

. . .

After writer Jack London roamed the aftermath in San Francisco, he wrote, "Not in history has a modern imperial city been so completely destroyed." In 2005, New Orleans could probably have claimed this title after Hurricane Katrina broke levees and flooded the place (although I'm not sure New Orleans would consider itself a "modern imperial city"), and in 2012, New York City could have claimed the title after Superstorm Sandy, but in 1906 not many Americans had seen an urban house of cards toppled by a single blow from nature. It was the first time that we were able to manufacture whatever we needed. The Industrial Revolution was in full force, and we were starting to live beyond nature, or so we thought. "All the shrewd contrivances and safeguards of man had been thrown out of gear by thirty seconds' twitching of the earth-crust," wrote London amid the smoldering rubble. It is estimated that at least three thousand people lost their lives in the disaster, and two hundred twenty-five thousand people, more than half the city's population, were left homeless.

When catastrophe hit, people acted in all sorts of ways. There were heroes who ran into burning buildings to save people. On the other end of the spectrum, there were looters. I'd always thought of looting as a relatively recent phenomenon, an act that required props not found until the latter half of the twentieth century, such as plate-glass windows to smash and electronics to cart away. But it turns out that there are stories of looting after the 1906 earthquake too. I suppose looting goes with the territory in a city where so many people had lost everything.

Yet most of the stories I've read about the reactions of San Franciscans to the disaster in 1906 were about how calm and orderly everyone was. London wrote that "never, in all of San Francisco's history, were her people so kind and courteous as on this night of terror." A week after the earthquake, University of California, Berkeley professor Samuel Fortier, sounding a bit like an elementary school teacher praising his class, wrote, "There is no lack of confidence. The courage of the people is simply remarkable. The thousands who have lost about all they possessed are wonderfully cheerful, and one seldom hears any whining."

■ ■ ■

There is a way that the slow Cape Cod disaster and the rapid San Francisco disaster are similar. In both cases, while really big things are happening, while the sky is falling, the small stuff is still going on. The photograph of the Victorians eating dinner amid the rubble reassures me. I glance at it as I eat cereal in the morning, and it reminds me that the mundane and the extreme are always happening simultaneously. Arctic sea ice is dwindling while I wash the dishes. Pollution pours into the ocean while I feed the dog. Species become extinct while I eat ice cream with Justin and Sara on Cape Cod. While I am living my small life, and while a lot of other people are living their small lives, really big things are going on.

Why dine amid the rubble? Well, you've got to eat. But that's not why you trot out the fancy tablecloth. We long for our routines when they have been disrupted by a traumatic event. Humans tend to fear change, which is why traditions can help us feel stable. Traditions may have helped these people regain their tether to the Victorian version of normal. At a time when the world is a mess, we long for normal. In this case, the earthquake was normal for the earth, but dining in style was normal for Victorian San Franciscans.

The firsthand accounts of the disaster often mix late-Victorian routine with Wild West pull-oneself-up-by-the-bootstraps survivalism. Peering into cavernous fractures in the earth occurred in lieu of piano practice. Avoiding falling bricks and losing one's home were combined with contemplating the proper earthquake attire.

I suppose that mix of the proper and the wild is what drew me to the photo of Victorians dining amid the rubble. The contrast is so striking between the reserve of what is often known as civilized society and the uninhibited earth acting without reserve — with shaking ground, with fire. The convergence of the human-built and natural worlds allowed some urban Victorians to see their world differently. Reflecting on his childhood experience in 1906 San Francisco, DeWitt Baldwin wrote:

> As the fires continued, we were ordered to evacuate our homes and find shelter on the hills. At this point the gravity of the situation began to dawn on me. "This is getting serious," I said to myself. The fire had threatened the very place we live. The whole situation set me to thinking of the frailty or the incompleteness of the power of man relative to the power of nature.

▪ ▪ ▪

The San Francisco History Center is a quiet, light room in the city's main library. The walls are lined with books, but most objects in the collections have to be requested by a little paper form and then brought from somewhere deep within the building by a librarian pushing a special cart. In the online catalog, I notice a number of scrapbooks about the 1906 earthquake and fires. These early scrapbook-makers pasted the articles and pictures they had about the destruction into large hardcover books. They aren't quite as festive as scrapbooks created by today's crafters, but the topic isn't festive either, so it works.

The requested scrapbooks are brought to me. I'm amazed by their size, each as large as my coffee table. Each is wrapped like a present in archival brown paper. The librarian who brought them out tells me in a whispered voice that I may carefully unwrap them one at a time. I nod solemnly and then eagerly yet gingerly unwrap my first present once his back is turned.

One of the giant scrapbooks includes an article from *Harper's Weekly* advocating for steel structures as the solution to earthquakes and fires. Perhaps we can't stop the quakes, the author argues, but we can engineer ourselves out of the destruction. I skim halfheartedly; I'm used to hearing about engineering solutions to counteract our battles with the planet — that we can redirect the Mississippi to go where we wish, that we can pull greenhouse gases out of the atmosphere and store them at the bottom of the ocean. Some see a world in which we can engineer anything to make ourselves invulnerable to nature. It's a *Three Little Pigs* philosophy: with the right engineering, the big bad wolves of the world will never affect us. Sometimes this philosophy works, but I'm skeptical of the quick fix.

In the article is a picture of a steel building standing tall amid crumbled stone and brick on Market Street in San Francisco — evidence that, indeed, steel is superior to stones. The steel building is a striking contrast to the ruins, and in an age before Photoshop, I am obliged to believe it is true. But my eyes widen as I see the photo's caption: "A Monument of Steel in a Wilderness of Masonry Debris."

A wilderness? The ruins of the city described as wilderness? I have such a different definition of that word. Wilderness is the absence of human influence. Some argue that no area of the world is beyond human influence at this point, and thus there may not be true wilderness, or that

From stereograph copyright, 1906, by H. C. White Co.
A Monument of Steel in a Wilderness of Masonry Debris
The " Call " building on Market Street, a modern steel-frame sky-scraper

Image from a *Harper's Weekly* article published May 26, 1906, with a caption that refers to the ruins of the city as wilderness.

humans are a part of wilderness, but I've never heard the remains of a destroyed city described as wilderness and the undestroyed part described as a monument.

This is not the only reference I find to the ruins of the city as wilderness. Charles Kendrick, a San Francisco businessman, described what he observed in San Francisco just after the disaster by saying, "The wilderness of ruins was beyond words — thousands of blocks of complete desolation."

Again. There's that word, wilderness.

Was anything inhospitable to humans or untamed dubbed wilderness? It makes me wonder what we mean when we use the term *wild*. Earthquakes are certainly wild.

Today, people sign up for wilderness treks and wilderness experiences, which generally means they will be wearing hiking boots and heading to areas where their cell phones don't work, where social media is a campfire. Some of us crave these experiences with nature, with wilderness. We wish to commune with wilderness. We hope that no one sets up a tent near ours so that we can better connect with nature as our only neighbor.

The Wilderness Act of 1964 defines wilderness in more formal terms. Areas that are designated as wilderness are more protected than any other federal lands. There are different rules and policies in wilderness than in the often adjacent national forests. The main difference that I've noticed is that a large sign in the woods indicates to a hiker that he or she is entering wilderness. It's a classic wooden US Forest Service sign with an amorphous amoeba shape engraved with the name of the wilderness area in all caps. I imagine the wood engraving was made by a gigantic avocado-green machine in a dusty shop run by someone in a khaki ranger uniform. You might have hiked five miles along a trail and passed no one along the way, aside from the occasional deer or bird, but when you pass the sign, you enter wilderness. It may not look any different. The deer and birds don't know the difference. But there is a difference in the way the land is protected and managed. The wonder I feel as I pass these signposts in the middle of the wild is because someone had to carry those things all the way there. No roads. No mechanized vehicles. Just a heavy wooden sign on a heavy wooden post.

Today the word *wilderness* is taken to mean "nature" by most people: the wild, without human influences. But my *American Heritage Dictionary* reminds me that there is another definition of wilderness, one that I don't hear often. It also means a broad, barren wasteland. This definition of wilderness was the most common usage of the word until the nineteenth century. Somewhere along the timeline of the last century, the definition of wilderness shifted. And the value we placed on nature changed. When the earthquake struck in 1906, wilderness was moving its image from desolate and worthless to revered and protected. The first national parks and the first wildlife refuge had formed. Zoos and aquariums were evolving from private menageries to public spaces that brought people and animals together. Land was starting to be set aside as national forests. The environmental movement was in its infancy, but growing. President Theodore Roosevelt kept nature high on his agenda. Just a few years after the earthquake in San Francisco, Girl and Boy Scouts would start trooping into wilderness for fun.

Wilderness was being redefined as something worthwhile at the same time that the people of San Francisco experienced the extent to which it could leave urban dwellers vulnerable.

"Earthquakes exemplify nature's terrifying randomness — and also people's hubris in pretending that rare, irregular events can safely be ignored simply because they can not be predicted," notes historian and environmentalist Dr. William Cronin in his introduction to the book *Uncommon Ground*.

Catastrophic earth events are a reminder that we humans are not, in fact, running the planet. Chaos in nature means that we can't be entirely certain of the size of the next earthquake, whether a fire will spread, or when a house will fall off a cliff. It's a reminder that we are vulnerable in ways we don't often consider.

You can't stop an earthquake, yet we've invented ways to create strong buildings that can hopefully stand up to the shaking. In the book *Peace of Mind in Earthquake Country*, author Peter I. Yanev advises homeowners who have found themselves near a fault not to panic. The risk of a house falling down, the homeowner squashed under the rubble of his or her investment, can be minimized with knowledge of geology and proper construction techniques. "Earthquakes are a fact of life," the author states. "But you do not need to live with fear and uncertainty about your safety." The same holds true for many of our interactions with the planet. In the case of disasters that cannot be stopped, like earthquakes along the San

Andreas Fault, efforts to prevent destruction of buildings and the lives within are called mitigation. Because climate change is a disaster that can be stopped, the term *mitigation* is used to mean efforts that prevent the disaster by avoiding warming the atmosphere more than we have.

Using engineering to avoid vulnerability isn't ignoring the chaos. It's finding a way to live with it. Engineering helps people decrease their vulnerability to earthquakes in California. In the case of climate change, engineering is helping people cope with impacts such as virulent heat waves, higher sea level, and stronger storms, because we'll have to adapt to some amount of warming even if we do stop it.

. . .

When San Francisco's streets were planned in the mid-nineteenth century, some advocated for curved streets that skirted the sides of steep hills, matching the shape of the land, allowing for gradual grade of streets. Instead, a grid of streets at right angles was developed, which allowed the land to be subdivided into squares and rectangles. The city was growing, development was happening, people were moving into town, into squares and rectangles.

It wasn't just San Francisco planners who partitioned the land into squares and rectangles. Most city planners of the time were laying a grid atop the flat paper map of the land, placing streets at right angles. That works in flatter places like Denver, so why not San Francisco?

The worst effects of an earthquake are often in places where features of the local landscape were ignored by planners and builders. That was the case for the damage during the 1994 Northridge earthquake in Southern California, according to Cronin. San Francisco ignored the landscape in its urban planning, building a grid of streets upon topography that fails to conform.

I've been to San Francisco more than a dozen times. And I walk endlessly around the place when I have the time. Yet I can never remember when looking at a map where the hills are located. I'll plan my route to this park or that pier or that café and then turn a corner to find that the road soars up or down. Occasionally a road appears to end at a fence that overlooks a cliff. Eventually I learned that what looks like a fenced balcony is often a block with a staircase instead of a street. The houses and apartments are aligned as if it is a regular block of squares and rectangles, tucked close together with a space between that would, in another block, be covered with asphalt.

From the perspective of a pedestrian, the staircases are so much better than asphalt streets. They feel like tunnels into another world. Since the staircases are generally narrower than streets, there is space alongside for trees, ferns, and mossy rocks. It is a piece of what I consider wilderness peeking out from within the urban grid.

Perhaps San Francisco's philosophy is not so much an effort to control nature but rather a recognition that nature can do its thing while people do their own thing. This seems like a very San Franciscan idea, and an American ideal. But sometimes, like in 1906, what nature does controls what people are able to do. Sometimes the earth shakes and fires start that cannot be extinguished. Sometimes our actions only make the situation worse. Sometimes we can't avoid being vulnerable; we just forget about it when the world is calm. The reverse is true as well in the modern world; we are causing drastic changes to the environment, and yet, because much of that change is invisible, all seems calm. It's easy to forget the catastrophe. Sometimes people must face threats in nothing but a nightshirt and tailcoat. Yet eventually, they find pants and, hopefully, put them on.

Chapter Four

When Fish Invade

We might be trying to do one seemingly harmless little thing, and because everything on the planet is connected, that one thing causes several other changes to the environment that we never imagined. We might think we have everything under control until we see a snowballing of unintended consequences leading toward disaster. At times like that, we are the butterflies flapping our wings. We often don't know what the impacts of our actions will be.

Aside from climate change, the environmental changes we've examined so far in this book have been natural — or at least they used to be. Today we are changing natural processes. The character of extreme weather is changing as we change climate. The erosion on Cape Cod is changing as more storms pound the coast and as warming causes sea level rise. Earthquakes, which I thought could not possibly be affected by humans when I started writing this book, now appear to be changing by our actions in places where wastewater from oil and gas drilling and fracking is injected deep underground, causing rocks to shift. This process led to three times as many earthquakes in Oklahoma as in California during 2014, according to the US Geological Survey, even though the former sits on tectonically stable ground.

Our influence on the earth can be roundabout and difficult to discern. Teasing apart the whodunit of unusual storms or long heat waves is challenging because our influence on the climate, while clear, is indirect. But other times, our impact on the environment is obvious. Such is the case with invasive species. Although it is the invaders that appear to be the problem, we are usually the ones who have aided and abetted, transporting these animals and plants from one place to another.

Invasive species, sometimes called alien species, multiply rapidly when they get to a new territory. They thrive and grow like weeds. (In fact, in certain locations, many of the weeds that people pull from their gardens each summer are invasive species.) Invasive species compete with native species for food and living space. They can outcompete the natives, reducing the number of species that can live in an area. They can cause extinctions of native species.

In their native ranges, the same animals or plants don't multiply out of control. They don't take over. Something keeps their numbers in check. It might be a predator that eats them or a limited amount of food.

Not all species invade when released into a new place. Tulips, for example, have been introduced to backyards all over the world, yet they have not taken over. How do we know if a species is going to invade? We can't be certain. But we can predict.

Species are moving into new areas as we change the climate, and often, recent arrivals thrive while natives languish. Yet most invasions are caused when people transport life from place to place.

With the increase in shipping and world travel since the start of the Industrial Revolution about 150 years ago, species are moving around the world faster than ever. They travel via FedEx and UPS. They hitch rides in container ships. Today, they show up in places that a century ago would have been impossible to reach. More traveling species have led to more invasions.

The ballast water of ships — taken in at one port and released somewhere else in the world — can move aquatic species from one place to another. Snakes have been known to cling to the wheel wells of airplanes traveling from one tropical Pacific island to another, spreading snakes from island to island. In Hawaii, airport employees check the wheels of planes for snakes, attempting to prevent the invaders. The soles of shoes trotting through the airports of the world carry seeds and microbes. Then there are the species that are intentionally brought into a new place to serve a purpose such as (ironically) keeping other species in check, controlling erosion on hills, or just looking pretty.

Invading aliens can cause major problems for ecosystems and financial and logistical headaches for people. For example, the Formosan termite, native to China, was accidently brought to the United States on ships. As goods labeled "Made in China" were unloaded, the termites debarked on their own and settled in the southeastern United States, where they reproduced quickly and munched wood frame homes. Compared with other termites, Formosans consume buildings as if they are winners in a pie-eating contest. According to the Louisiana Department of Agriculture and Forestry, the state spends over $500 million a year battling Formosan termites.

One of my favorite invasive species, a fast-growing vine called kudzu, was intentionally brought from Asia to the United States to help control erosion on slopes. It was thought to be a possible food for farm animals too. But kudzu grows so quickly in the Southeast that it takes space from other, native species with incredible efficiency. It will cover your house if you don't attack it regularly with sharp garden tools. When I lived in Georgia, there were stories (possibly tall tales) of people who left on vacation with no kudzu management plan in place and returned to find their house so completely covered it looked like a topiary. The vine may be an invader, but it's also beautiful, with broad leaves and clusters of small pink flowers in summer. The demure flowers are replaced by grotesque, hairy seedpods, which dangle from the vines in the fall. By winter, the leaves have fallen and the vines are not as visible. It's easy to forget about the alien invasion. Then, in spring, the leaves regrow and the South is, once again, coated in green. Where the vines climb telephone poles and reach out laterally along the wires, the kudzu looks like human bodies with arms outstretched. I call them kudzu creatures and like to imagine that they are coming alive like zombies, a little horror movie thrill.

These alien invasions are not like the Hollywood version. The attack is quiet, very quiet. It looks like everything is all right, but it's not. Often, by the time people realize that the invasion is taking place, it's too late to stop it. Invasive species rarely make a scene until they are so populous that their effects cannot be ignored. That's what happened when lionfish got into the Caribbean. We lost control of the fishes.

Lionfish, native to tropical waters of the western Pacific, Indian Ocean, and the Red Sea, are an invasive species in the Caribbean and western Atlantic. The total number of lionfish has exploded in the new environment, to the detriment of native species. "Lionfish are eating their way through the reefs like a plague of locusts," said Dr. Mark Hixon, a coral reef ecologist at Oregon State University, in a 2008 interview with the *Times* (London). "This may well become the most devastating marine invasion in history."

Lionfish are found in many public aquariums, and the signs next to their tanks have had to be replaced over the past decade to both describe a venomous fish from the Indian and Pacific Oceans and to say that they are invading the Atlantic and Caribbean too. What started as a few fish out of place two decades ago has escalated to a massive threat to the health of reefs, which are already stressed by environmental change. It's the most documented marine invasion in history, capturing not only the attention of marine scientists but also recreational divers, citizen scientists, and fishers. The story of the lionfish is a story of rapid ecological change, yet it might also be a story of recovery and the lengths we will go to correct a mistake. Many people are hopeful that we can overcome the lionfish threat, the thought being that if we had the power to cause the invasion, we might also have the power to control it. Some are uncertain whether it's possible. We may have hesitated for too long and let the problem grow beyond our control.

■ ■ ■

No one knows exactly when the first lionfish made its way into the Atlantic from the western Pacific. But we do have a good idea of how it made the journey. It is most likely that lionfish were released from people's fish tanks into the ocean — either accidentally or intentionally.

One of those releases started with a storm.

Adult lionfish (*Pterois volitans*) from an 1881 engraving.
(iStock.com/Graffissimo)

It was 1992. Hurricane Andrew was bearing down on southern Florida. It had been a sleepy hurricane season that summer until Andrew hit Florida in late August. A Category 5 storm, this was the strongest of hurricanes. Floridians braced themselves and their homes as reports came in about how Andrew had ravaged the Bahamas. The wind from the storm was so powerful that it destroyed weather stations in South Florida. Before the wind-recording anemometers broke, they measured gusts of 150 to 200 miles per hour. The ocean swelled up to seventeen feet higher than normal along the coast because of the storm's intense low pressure, temporarily drowning seaside homes.

At the time of the storm, six tropical Pacific lionfish were residing in a large fish tank perched on the porch of a Miami home overlooking Biscayne Bay. The porch was atop a seawall intended to stop the ocean should the need arise. The lionfish lived within sight of an ocean, yet they were trapped in a glass box. They made their escape as the ocean level rose during the storm. This could be a happy *Finding Nemo*–like story if the fish had escaped into the same ocean they had come from. But it's hard to swim from Florida to the Indo-Pacific. Instead, the lionfish set up residence in Biscayne Bay and started eating vast numbers of other fish.

These particular lionfish were not the only ones released by aquarists, nor the first. In 1985, years before Hurricane Andrew, a Florida fisherman caught a lionfish. Lionfish were spotted in the waters off Boca Raton, Palm Beach, and Miami in the late 1980s and early 1990s. When many people are responsible for a problem, there is a diffusion of responsibility. No one person feels responsible. If many people set their pet lionfish free in the ocean, then each is only a little bit at fault. No one can point a finger at any particular amateur aquarist, but we can point a finger at a place. Whatever the exact date lionfish got into the Atlantic, it is clear that they came from Florida.

Genetic studies on lionfish invaders have found that the gene pool is very small, meaning that all the lionfish currently roaming the Caribbean and Atlantic are descended from just a handful of individuals. In fact, one study found that as few as eight females might have been the roots of the population. Of course, there would also need to be some male lionfish for the population to grow, but the genetic analyses focused on female DNA. We know that the genetic makeup of the lionfish in the Atlantic and Caribbean is most like lionfish that live near Indonesia, where almost all lionfish in US fish tanks are from — further evidence that these invaders came from land.

Over the last decade of the twentieth century, they spread north to the Georgia coast, then South Carolina and North Carolina — about eight hundred miles north of Miami. Juvenile lionfish have been found as far north as Cape Cod, although no adults have ever been spotted there. The young fish that make it so far north probably die in winter because of the cold waters north of Cape Hatteras.

Lionfish showed up in Bermuda in 2000 and were widespread in the Bahamas by 2004. South of the Bahamas, they appeared in the Turks and

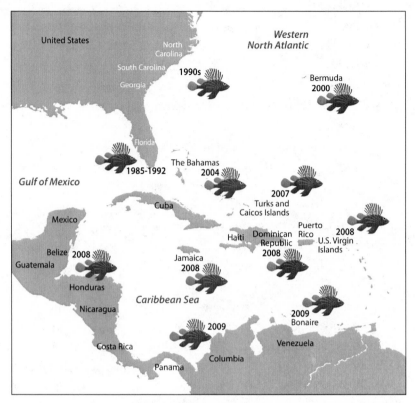

Map showing when lionfish spread through the Caribbean and Atlantic. (Courtesy of the author.)

Caicos Islands by 2007 and within a year were also found in the Cayman Islands, Jamaica, Belize, the US Virgin Islands, and the Dominican Republic. In 2009, the first lionfish was spotted in Bonaire and off the coast of Colombia and Venezuela, over twelve hundred miles south of Miami. It's easier to see the disaster that was building in hindsight than it was at the time.

While adult lionfish tend not to stray far from wherever they call home, the population is able to spread because lionfish eggs can travel large distances. The eggs swirl through the Caribbean and the Gulf of Mexico in currents. They travel up the East Coast in the Gulf Stream current. A female lionfish can release thousands of eggs several times each month, year-round. The eggs are buoyant and float upward in a single mass. En route to the surface, a male lionfish fertilizes the eggs. Eventually they become embryos, then larvae, and then miniature versions of their parents, all while floating at the surface with the currents.

. . .

Other people go to a Caribbean island to relax on a beach. I go to visit old friends — ones that live underwater. I used to study coral reefs in college and graduate school, scuba diving as a researcher with sample bags and clipboard. Now I dive as a tourist. Visiting the reef was the reason I went to Bonaire in the southern Caribbean. This was eighteen months after the first lionfish was spotted there, on October 26, 2009. I would discover firsthand that the invaders were everywhere. Soon after I dropped from the ocean surface to a reef on my first dive, I saw one. I gawked at it through my mask. I had seen this fish, but never in the Caribbean.

The Bonaire reefs were filled with brightly colored fish, yet they didn't stand out in quite the same way as the lionfish. Perhaps that's because the other reef fish blended in with each other, or perhaps it is simply because I knew the lionfish was out of place. Bright swarms of native fish swam past me while the lionfish stayed still, placid and slow, watching. Hovering in the warmth of the Bonaire reef amid algae-covered rocks, the lionfish was accessorized with striking red and white stripes and showy, feathery spines. In contrast to the pizzazz of his color and spines, his downturned mouth and plodding pace gave him an Eeyore-ish demeanor.

Back above water and on the dock, I fumbled with my gear to get a new tank of air properly assembled so I could again jump into the turquoise water. In the reef off this particular dock on the west side of Bonaire, more fish species have been found than in any other place in the Caribbean. There are more species counted here than are listed in some fish identification guidebooks (the lesser guidebooks, anyway). Coral reefs are like cities, supporting tremendous biodiversity. That's one reason why there are so many types of fish here. But the lionfish are a threat to biodiversity. Lionfish have a voracious appetite and an ability to consume multitudes of reef fish. They are one of many factors that make the future of Caribbean coral reefs uncertain.

From the dock, you'd never know what is going on below. Reflected sunlight sparkles on the ripples at the water's surface. The drama of the ocean is mostly cloaked from view. The reef starts about forty feet below the surface and drops off to deeper than it is safe to dive. It continues until it is so deep that sunlight can't penetrate. Lionfish have been found that deep too.

A lionfish is protected by a Mohawk of spines along its back, plus more on its belly and radiating from its sides. Most of these spines, while beautiful and showy, also contain a reservoir of venom. Get too close to a lionfish, and a spine will prick you, allowing the poison to diffuse into the puncture. The venom contains neurotoxins that affect the nervous system and muscles. In humans, they cause anything from mere pain and swelling at the site of the attack to advanced paralysis. Most swimmers and divers who have encountered lionfish venom wind up in an emergency room, no matter the severity of their reaction. Animals that are much smaller than a human have a much lower chance of survival when hit with the venom from a lionfish. In small animals, neurotoxins can stop a beating heart. Interestingly, and as evidence that nothing is pure evil, the venom also appears to contain antitumor agents. Some scientists speculate that an aspect of the venom might be used in a future cancer-fighting drug.

Until recently, it was thought that no other fish would eat lionfish, meaning that predation by other species couldn't keep lionfish populations in check in the Caribbean and Atlantic. The poisonous spines make the lionfish unappetizing to potential predators, although ecologist Peter Mumby has found evidence that groupers sometimes eat lionfish.

■ ■ ■

In 2002, researcher Paula Whitfield and her colleagues studying lionfish off the coast of North Carolina noted,

At this time, negative impacts on the ecosystem are unlikely as the number of lionfish observed is relatively few, but future effects on tropical and subtropical reef communities in the western central Atlantic may occur if reproduction and dispersal of this species result in population growth.

Notice that they used terms such as "unlikely" and "may occur." This is the language of uncertainty. In this case, it is both the uncertainty that is an inherent part of science and the uncertainty that is due to the chaos of our planet. Some species don't invade when they are introduced to a new area. Others wreak havoc. By communicating the uncertainty, Whitfield and her colleagues were letting people know the location of the boundary between known and unknown, and that this was right over the border on the unknown side. For example, these researchers mentioned that an invasion may occur if the rates of reproduction and dispersal were high, meaning that we need to learn more about the rates of reproduction and dispersal. Because we don't know everything, it's helpful to communicate areas where we need more information. In 2002, we couldn't know for sure that this invasion was going to happen. However, that is no reason for inaction, because we didn't know that it wouldn't happen either.

The lionfish population started out small, just like the fires in San Francisco after the 1906 earthquake did. In both cases, the threats grew out of control and became catastrophes, and in both cases, swift action early on could have prevented some amount of the catastrophe. Similarly, politicians have been hesitant to take action to stop the greenhouse gas emissions that are changing climate, which has allowed that problem to grow as well.

The number of lionfish swelled in some invaded areas off the North Carolina coast between 2004 and 2008, according to estimates by the National Oceanic and Atmospheric Administration (NOAA). For every lionfish found in 2004, researchers were finding about fifteen lionfish in 2008. Similar numbers were reported in the Bahamas in 2009. By that year, lionfish were eight times more common in the Bahamas than in the Pacific, where they are supposed to live. The population was, and is, growing. Studies have found that the fish are eating more in the Caribbean than they do in the Pacific. In the Atlantic and Caribbean, lionfish eat more than forty types of other fish. They also attack crustaceans such as shrimp and crabs. Lionfish stomachs can expand thirtyfold in volume when the fish are eating a large meal.

Divers and snorkelers in the Bahamas have noticed that as the number of lionfish is increasing, the number of other fish is decreasing. Especially at risk are fish species that have a small range. A study in Belize found that nearly half the fish in the stomachs of lionfish were critically endangered. Lionfish are now the most common predators on reefs in some areas of the Caribbean. Like public aquariums, tropical tourist destinations rely on reefs filled with a plethora of colorful fish, and they are struggling to find a solution. Unlike at aquariums, humans do not control the reef biodiversity, but they do affect it.

Tourists like to see healthy corals as well as fish. Scientists suspect that lionfish might be having an impact on the health of corals too. As these predators eat fish that graze on algae, algae can proliferate, growing over areas covered in corals, sponges, and other living things. Researchers have calculated that lionfish are present on many reefs in the Caribbean in such large numbers that they are capable of eating all the small foraging fish in the area. This could cause huge changes in reef ecology, but we just don't know yet.

Lionfish also eat cleaner fish in Caribbean reefs — small fish that clean parasites off larger fish. These small fish provide a helpful service to larger fish, since it is good to be clean, and they get to eat all the parasites they find. It's not uncommon to see cleaner fish zooming in and out of the mouth of a large grouper that is sitting still on the reef. A lionfish opens its mouth and the cleaners come to work, then the mouth closes. The lionfish swallows the cleaners.

If lionfish were the only cause for the decline of reefs, perhaps we could face it more easily. But lionfish are not the only problem. There are at least half a dozen additional environmental changes that have plagued reefs for decades. Warming waters due to climate change are causing corals to bleach, turning them ghostly white and weak. Pollution and sediment washed into reefs from shoreline development are clogging the mouths of the tiny coral animals as they attempt to sift nutrition from water. Overfishing, coral diseases, too many nutrients, acidic water, oil spills, even mining the reef for construction material — the number of threats to corals reefs in the Caribbean is vast. Lionfish are able to topple the delicate balance because it was already wavering out of equilibrium. But unlike many of the other hazards, lionfish are a clear enemy, and it's easier to attack what you can see.

■ ■ ■

Unless we do something, the number of invading lionfish in the Caribbean and western Atlantic will continue to grow. Today we can say with more certainty that lionfish will expand into new areas wherever the water is warm. That certain threat has spurred many people to take action to attempt to eliminate lionfish from the Atlantic and Caribbean.

"Eliminate" is a discreet way of saying "kill." Often after the kill, lionfish are eaten, but the purpose of hunting them is not primarily for food. In an effort to save marine ecosystems from this alien invasion, conservationists — people who spend their lives trying to protect living things — are hunting and killing as many lionfish as possible.

As I watch a YouTube video that demonstrates how lionfish can be successfully hunted with spears, I feel a familiar twinge. It is the guilt of the nature lover who is so fond of marine creatures that she has trouble eating them. How can I root for killing this beautiful fish? The lionfish star of this video appears to have no idea that it is about to be stabbed to death. It is quietly hovering between the rocks of a reef, fluttering its fins and feathered spines. It looks so docile and wide-eyed. I can't help but feel sorry for it as the spear lunges. The rational side of my brain knows the very real threat this beautiful fish poses to the health of Caribbean reefs. I know that it is a fish out of place. But I am one of those people who traps a bee that has gotten into the house in a drinking glass topped with a piece of junk mail and then releases it out the back door rather than slapping it with a shoe. I once found myself facilitating a debate among Home Depot employees on the ethics of mouse entrapment. I am just not a killing sort of person. What makes this instructional video even more horrifying is that while the lionfish is being hunted to death, a happy song from the soundtrack of the *Curious George* movie, one of Justin's favorites, plays in the background. Where I would expect the ominous soundtrack from *Jaws*, there are instead upbeat lyrics by Jack Johnson and the mental image of a carefree cartoon monkey. It seems disrespectful to the lionfish that lost his life during the filming of this video.

The video is called *Spearing Lionfish Underwater*, so I knew what I was in for when I started watching (aside from the incongruous soundtrack). The host is from a website called lionfishhunter.com in case the premise was not clear from the video title. Yet I could not help but feel a twinge of concern for the fate of the lionfish. Couldn't we just put them into glasses topped with junk mail and take them to the tropical Pacific?

Just as islands in the Caribbean have their own cultures, they also have their own strategy about what to do with the lionfish. Lionfish derbies in Florida and the Bahamas tap into people's competitive streaks, awarding prizes to those who catch the most lionfish during the event. Often a food made from lionfish is featured during the festivities. There is a lionfish cookbook so you can make a tasty dinner out of the invasive species you kill to save the reefs. In Bonaire, I heard they make an excellent ceviche.

■ ■ ■

Bonaire National Marine Park includes the entire ring of ocean surrounding the island. What's in the waters within the park is as important to the Bonaireans as what is on land. It's illegal to hunt, fish, or harm any kind of marine life in the waters surrounding Bonaire. However, after the invasion began, one exception was made: the lionfish are not protected. The people of Bonaire, fond as they are of marine life, are taking spearguns to the reefs to kill the lionfish.

In November 2010, for the first time in forty years, the use of spearguns was allowed on Bonaire's reefs and one hundred ELF spears were imported to the island. ELF stands for "eradicate lionfish" and is the weapon of choice for lionfish hunting. One hundred Bonaireans were trained to use the spears, a scuba diving army. All other marine species remained safe. It was just lionfish they were after.

The spear-wielding hundred became known as "the Eliminators" on Bonaire. Eliminators are regularly sent lists of lionfish locations in the coastal reefs based on sightings contributed by scuba divers. Then, they head out with their spears. They dive around the island, sometimes in groups and sometimes in pairs, and kill as many lionfish as possible.

In Bonaire, divers who do not wish to wield a spear are instructed to tag the location of lionfish underwater. Bonaire National Marine Park provides tags: brightly colored flagging tape, like the type used to mark trees that will be cut down, with a wine cork tied to one end. The cork floats, buoying the flagging tape vertically in the water when the other end of the tape is tied to rocks near the lionfish. The Eliminators remove the markers from the reef when they remove the lionfish.

Marking the location of lionfish seemed more my style than shooting them with a speargun, so I filled the pocket of my scuba equipment with corks and flagging tape before I hopped into a reef-bound boat. And just like that, I was transformed from common tourist into a special agent supporting Bonaire's army of lionfish Eliminators. As we sped toward a dive site offshore, it occurred to me that the number of flagged lionfish must be tied to the wine consumption on Bonaire — one cork, thus one bottle, per fish. Cheers.

The boat deposited six of us in the ocean and we descended to the sea-floor, more than sixty feet deep. I saw dozens of tags. By each was one or two lionfish. It became a game. I'd see a cork hovering atop a little piece of colored flagging tape and swim over to check on the lionfish that were invariably sitting at the base of the flag. They don't move around much.

When I saw a pair of lionfish loitering unmarked, I pulled a tag from the pocket of my vest and unfurled the flagging tape. The cork was from a bottle of Italian wine. I tied the end of the tag to a nearby algae-covered rock and looked to the lionfish to make sure the marker was close enough to them. I fumbled with the tape, scared to get too close to the venomous spines as I tied the flag.

These lionfish didn't know the significance of the marker. They were marked fish, I thought. Maybe I felt a little sorry for them. Placidly fluttering their glitzy ruffle of fins and hovering next to the flag, they looked vulnerable, almost.

Back at the dock, I reluctantly added my lionfish sighting to a log maintained by the marine park using a blunt pencil, trying not to get the paper wet with the salt water dripping from my hair and trying to not feel like a tattletale. This was one of many handwritten logbooks at dive shops on the island. The entries are transcribed and sent electronically to the Eliminators so they know where to find marked fish.

As I flipped pages and looked at the hundreds of entries in the log, I lost a little sympathy. The log illustrated the problem of an invasion. It reminded me that while an individual fluttery fish with bulging eyes is vulnerable to a spear, the population seems mighty and resilient. It's a matter of scale. Individual fish are not much of a problem. The massive and growing population is a problem. The same is true for all types of disasters. For example, one person burning fossil fuels will not have much impact on climate, but billions of people burning fossil fuels are capable of changing global climate.

One reason that people don't take action to address a problem is because they feel their efforts will make little difference. As I flipped through the log, I realized that the pair of lionfish I tagged would not make much of a difference to the invasion. Similarly, when I have the thermostat set so low in the winter that I need a wool hat and coat indoors, I know that my effort to avoid burning natural gas in the furnace will not stop climate change. It can seem futile.

This is a collective action problem, meaning that it can be difficult to take action when your action alone cannot solve the problem. Compounding this is the predicament that when many people cause a problem, no one person feels responsible for it. If we are all responsible for a mess, then you, personally, are only a little bit at fault. The finger cannot be pointed directly at you. In relation to climate change, psychologist Robert Gifford described this as one of twenty-nine psychological barriers to behavior change that he called "dragons of inaction." Wrote Gifford, "Because climate change is a global problem, many individuals believe that they can do nothing about it as individuals."

Can lionfish be stopped? It's possible. There are hundreds or thousands of pairs of human eyes looking for lionfish. Yet we are probably only finding a small portion of the fish. Think about all the ones we haven't seen, all the others that are out there. And the colorful swarms of fish that they will eat.

Eliminating all the invaders from the Caribbean and Atlantic might not be possible, but there is evidence that the population can be controlled enough to prevent ecosystem destruction — at least if people agree to eat them for dinner. In Bermuda, for example, government policies encourage fishermen to catch and sell lionfish. Bermuda's Eat 'Em to Beat 'Em campaign, started in 2010, helped develop a market for lionfish as seafood. Eat 'Em to Beat 'Em events have gotten spear hunting volunteers to kill the invaders. Collective action has reduced the number of lionfish in Bermuda's reefs.

Humans are good at depleting a species of fish from the ocean. Atlantic cod, orange roughy, toothfish — overfishing has caused massive declines in the populations of many fish worldwide. This is where we can play to our strengths. The lionfish may be a threat in the Caribbean and western Atlantic, but we humans can be threats too. In this case, eating the problem can help. And if we are able to defeat the lionfish, then I'm hopeful that we will be able to fix the other environmental catastrophes we've caused. In the case of the lionfish, we are seeing that coordinated efforts are able to make a dent in the lionfish population. One person is not going to get them out of the Caribbean, but many spear-wielding people can.

Although there will be little opportunity to wield a spear as we attempt to solve climate change, the same thinking may apply. One person

Poster advertising a 2016 Eat 'Em to Beat 'Em lionfish eliminating event in Bermuda. (Courtesy of Groundswell; created by artist Stratton Hatfield.)

reducing greenhouse gas emissions is not going to solve the problem, but many people will. We will always be somewhat uncertain about whether we will be able to make enough of a difference, but we need to act before climate change grows beyond our control. We need learn how to make a tasty ceviche out of greenhouse gases.

Chapter Five

On the Dry Side of the Glass

Below the glittery surface of the ocean there is a catastrophe in progress that we have caused. Releasing invasive lionfish into non-native environments is just a small part of the problem. Around the world, the ocean is in trouble. We are causing the water to warm and become more acidic as the carbon dioxide we add to the atmosphere seeps into the ocean. We have depleted fish, added plastic, and spilled oil. The ocean is vulnerable.

In terms of the lionfish invasion, it has been mostly scuba divers, people who feel at home below the ocean surface, who have led the effort to save reefs from the invasive species. But for most people, the ocean is the blue part on a map. It is somewhere to gaze at sunset. It is a place where most of us stay at or near the edge. Many people don't think much about life below the surface. This concerns me because if people only save what they can see, understand, and value, then the ocean is not much of a priority. However, there are places where anyone can glimpse the marine world: aquariums.

Public aquariums reveal our impressions of the ocean, and they are the only opportunity many people have to get to know the creatures that are affected by our actions. They are places created by humans yet devoted to nature. They can indicate how we land-dwellers coexist with our marine neighbors, how we react to change in the seas, and whether we will be able to take care of the ocean. It's possible that public aquariums motivated some of the passionate people who are slaying lionfish in the Caribbean.

In an aquarium, the interesting living parts of the ocean are concentrated and humanity is invited to come on in. Aquariums make for a less messy experience than going out to find marine life yourself, and you can often find animals that live in various habitats or even widely separated parts of the world right in the same building. It's like the United Nations of marine life. And for many people, this is the closest they will get to the ocean.

Animals in some aquarium exhibits hardly seem to be on display. They are ambassadors of their kind. Yet it is not always like that. In some aquariums, I've had an uneasy feeling and sudden urge to set all the animals free. They appear to be freaks on display or hostages from the natural world. Sometimes that's not the fault of the displays but of the visitors' reactions, like the two southern fishermen I once saw at the Tennessee Aquarium salivating over a freshwater exhibit, boasting about their latest catches, and exchanging recipes for a good fish fry.

Why is it that some aquariums help connect people with the ocean and empathize with marine life but others either whet people's appetite for seafood or make the ocean seem like a foreign and scary place? The same thick, impenetrable glass is used in all types of aquarium fish tanks, creating a division between people and nature. Yet in some, the glass seems like a window into the undersea world. In others, it seems like a barrier — a line between us and them.

■ ■ ■

I once dove into a tank of sharks, eels, and other marine life at a small aquarium in South Carolina's Riverbanks Zoo. I'd pleaded with the right people to get permission to join the volunteers and staff who actually care for the animals. But I wasn't there to help with their work. I wanted to take pictures with a fancy underwater camera that I'd borrowed from one of my professors. I wanted to teach myself how to photograph marine life in an aquarium tank before I tried it in the open ocean with full scuba gear.

Inside the tank, a small shark soared past the lens of the bulky camera so close that it was just a fuzzy, gray blur. A large green eel was threatening to

Visitors on the dry side of the glass look at marine life in Denver's Downtown Aquarium. (Courtesy of the author.)

emerge from his rocky cave, his mouth open and teeth exposed. Small blue fish darted around my head like flies. Beyond the glass, there was an entire school field trip of awed kids with open mouths and hands suction-cupped to the glass. Many of them appeared to be talking in outdoor voices, but I could hardly hear them through the glass and water. I turned off the crane-shaped flash so it would not reflect. I took pictures of people on the dry side of the glass, their noses smooshed. I documented their awe from inside the tank. When I got back to my college campus and developed the film in the art department darkroom (this was at a time in history when cameras were filled with film), I realized how many of my blurry pictures were of aquarium visitors and how few were of marine life. That's what it looks like to be a fish peering into the human world, I thought.

Some researchers watch the behavior of animals. Other researchers watch the behavior of people. And other researchers watch the behavior of people who are watching animals. The last is called visitor behavior research, and it is helpful for understanding what types of exhibits are most successful. I am not a specialist in this type of research, yet at aquariums I find it to be a barometer for how humans connect, or don't connect, with the ocean and whether we see a role for ourselves in this different, neighboring world.

Visiting animals at a zoo or aquarium might be a contrived encounter with nature, but it still has an impact, according to a recent study by the Association of Zoos and Aquariums, the Institute for Learning Innovation, and the Monterey Bay Aquarium. They surveyed fifty-five hundred visitors to zoos and aquariums over a three-year period and found that these places helped people consider the role they play in environmental problems. People walked away from the experience with a way to see themselves as part of the solution too. However, the people who have decided to fork over the admission fee probably already have a greater understanding of nature than people who choose to spend their free time and money in other ways. Zoos and aquariums support and reinforce the value that visitors already place on nature. In a way, a zoo or aquarium is preaching to the choir.

According to surveys by the Ocean Project, people's understanding of the ocean is low. About a third of the American public surveyed couldn't name an environmental issue that threatens the ocean. They are also unable to explain why oceans are important.

"The American public possesses significantly greater literacy about topics such as college football, the Academy Awards, luxury automobiles, casino gambling, and video games than it does the ocean," the Ocean Project authors state in their report. While this might come as no surprise to many Americans, to me and to the authors of the report, it seems like a great injustice.

■ ■ ■

In California's Monterey Bay Aquarium, we softly ooh and aah as an octopus makes her way to the other end of her tank, slowly and gracefully propelling herself. Eight orange arms glide along, one after another, flowing like ribbons in a breeze. Light pink suction cups adorn the underside of each arm. These little cups are so sensitive that they allow an octopus to tell one person from another by touch. I will not get to introduce myself to this octopus so closely. None of us will. There is a thick piece of glass between the humans and the octopus. A man puts his hand on the glass opposite one of her arms. His stroller-bound toddler babbles to the octopus with the familiarity of an imaginary friend.

"I've gotta go," mutters a man into his cell phone, punctuating our calm as he enters the dimly lit area near her tank. He slips the phone into a suit pocket and joins our silence.

The octopus has nestled cozily into a nook in her faux-rock exhibit. Her licorice jellybean eye sits above a pendulous orange avocado-shaped body that houses her brain and organs. It must be huge, her brain, and I wonder if she is smarter than those of us on the far side of the glass. She has curled the ends of her arms like fiddleheads, making them more compact. This octopus seems gentle and shy, a giant, gangly marine introvert. However, octopi used to be called "devilfish." They were once thought to be monsters. She rises and falls slowly with each breath, sending water past her gills. She turns from orange to deep red as her black eye narrows and she drifts to sleep.

Huge, powerful, and *mighty* are some of the adjectives that Ocean Project survey respondents offered when asked to describe the ocean. Adjectives fit for a superhero, but not for the octopus with the jellybean eye. The ocean is huge and it can be powerful, but it is not immune to harmful environmental threats. Pollution, runoff, climate change, overfishing, invasive species, and acidic waters are all caused by our actions. Fish become infused with mercury from industrial pollution. Seawater turns murky with algae fed by fertilizer from farms on land. There are dead zones in the ocean where the food chain has collapsed and life can no longer thrive. Oil spills, floating plastic, *E. coli* bacteria outbreaks — the ocean's glimmering surface gives no sign of its vulnerability. If anything, we see ourselves as vulnerable to its power, its storm surge flooding and undertow, its devilfish.

How we view nature affects whether we take action on environmental issues. People who perceive nature as mighty are more likely to think nature is able to bounce back, and thus they are less likely to see a need to do anything. People who view nature as vulnerable are more likely to jump to action when nature is in jeopardy. But no one is likely to take action if we don't see ourselves as having an impact on the ocean.

No one owns the ocean, and yet everyone is responsible for it. The ocean and the atmosphere are shared resources, unaware of political boundaries. The same water that laps up on the coast of Maine also laps up on the coasts of South Africa, New Zealand, and China. Whenever a resource is shared, it is in danger of becoming a tragedy of the commons. This term, popularized in the scientific community by UC Santa Barbara biology professor Garrett Hardin, describes how a shared resource is depleted or destroyed more easily than a resource that is owned. That's an elaborate way of saying that we take care of our own stuff but don't take care of stuff that belongs to everyone. Hardin described the tragedy of the commons in detail in a *Science* magazine article in 1968, with an example of cattle herders who share the same land, concluding that:

> Each man is locked into a system that compels him to increase his herd without limit — in a world that is limited. Ruin is the destination toward which all men rush, each pursuing his own best interest in a society that believes in the freedom of the commons. Freedom in a commons brings ruin to all.

However, one does not need to herd cattle to relate to this concept. Anyone who has ever noticed a putrid smell coming from their office refrigerator (shared by everyone, cleaned by no one) has firsthand experience with the tragedy of the commons.

In the case of the ocean, we take out too many fish and put in too much pollution. This happens when people and corporations act in their own best interests instead of the best interests of the planet as a whole, an instinct rooted in natural selection. As Hardin eloquently put it, we are "fouling our own nest." Since he wrote about the tragedy of the commons in 1968, we have found that we are fouling our own nest in new and different ways. For example, some of the carbon dioxide we put into the atmosphere diffuses into the ocean, making seawater slightly acidic, which is an inhospitable environment for marine creatures that have shells and other types of exoskeletons. While Hardin emphasized that acting in our individual self-interests was natural, he also pointed out that "education can counteract the natural tendency to do the wrong thing."

According to the Ocean Project report,

> Perhaps the most significant implication of the survey findings for aquariums, zoos, and museums is that the public expects and trusts our institutions to educate regarding environmental and conservation issues and to provide guidance about how to address them personally and societally. Collectively, however, we are not meeting these expectations, and the vacuum is being filled by corporate messaging that does not always promote the best interests of conservation.

■　■　■

Across the aquarium from the quiet octopus corner, I stare at a jellyfish whose pink bell pulses rhythmically. I have to remind myself that the pulsing is not breathing; it's just a way of moving around. I find myself breathing in time with the pulsing nonetheless. Dainty red tentacles trail after the bell. They look so innocent, but their sting can be horrible. This is an animal with only one end to its digestive track, thus contents that go in the mouth are also evacuated from the same orifice. I used to tell college students taking their required science class this little factoid, and their faces would scrunch up into the "that's gross" expression. Yet the way I see it, the jellyfish is evidence that we are all weird in our own special way, a way that works just fine for each of us.

Catty-corner to where the jellyfish dance their ballets in the Monterey Bay Aquarium, an immense exhibit introduces the fish that live in the outer reaches of the bay. It is such a large fish tank that if I look directly at it from the center, I can see nothing else in my peripheral vision. It is like being in the water with these fish, except much less wet. This illusion is periodically interrupted as the silhouettes of other aquarium visitors wander into my field of view. While large silver tuna, smaller mahimahi, and a few others glide swiftly through the water, two sunfish move slowly, spending a considerable amount of their time lolling in parts of the tank that are not visible to the people peering through the glass. All the other fish are shaped as you might expect. But the sunfish are complete oddballs. They look a bit like pieces of flying toast propelled by two fins, one on each margin. They don't have a caudal fin, the triangular fin that gives a fish that archetypal fishy shape. The face of a sunfish looks as if it were poorly carved into Mount Rushmore.

"It looks like a moon rising," a teenage boy mumbles as the smaller of the two pieces of flying toast emerges from the depth. He's right, maybe it's not a piece of toast but instead like a lumpy moon, and there it rises slowly through the gray-blue water. Why is it called a sunfish? I wonder.

When the larger sunfish, a five-hundred-pound giant, makes its slow way up from the bottom of the tank and into view, boisterous school groups, weary parents, jabbering toddlers, pensive retirees, pierced and tattooed teenagers, notebook-toting college students, docents in khaki pants, aloof professionals, bearded scientists, European and Indian and Japanese tourists stop and stare. Mouths fall open. Animals, even the odd ones, do not offer explanations for who they are. They make no apologies for their differences and do not defend their lifestyle or appearance. The sunfish makes its slow, quiet progress toward the center of the tank as a great collective gasp releases from those of us on the dry side of the glass. For that little moment, despite our differences of culture, race, and politics, we, the humans at this exhibit, have something in common. We are united. We are of one mind, one single thought. We have set aside our opposing views on war, religion, education, gun control, and the environment in order to share this thought: that this stony-faced fish, this odd piece of toast, this giant monolith rising slowly from the depths is awesomely and astonishingly weird.

I've heard critics of the Monterey Bay Aquarium question whether it is good ocean public relations to have the creatures from the surrounding bay in the spotlight instead of the most charismatic or beautiful of the world's marine life. But this weird sunfish has captivated dozens of us who are standing in front of the tank. It has rendered most of us speechless, even without charisma or beauty. An aquarium that displays the creatures that inhabit a nearby ocean, as if the animals just hopped out of the water and landed in glass-sided tanks in order to visit with us, is more likely to connect us to an adjacent ecosystem that would otherwise be inaccessible.

Being in a habitat that suits one's adaptations is very important, whether the creature in question is an octopus that prefers quiet caves, a sea otter playing in a canopy of kelp, a cell phone addict wearing a suit to an aquarium, or a troop of wide-eyed kindergarteners with name tags shaped like fish. By portraying the nearby habitats in Monterey Bay, the aquarium allows us to look over the watery horizon outside the aquarium and imagine what is living below its surface. We can see how the animals are adapted for that place — and for just a moment, we are connected to that watery wilderness. We empathize. We may even be stewards, too. But we are dry, on the other side of the glass, mere spectators of nature, or at least a good facsimile of nature.

■ ■ ■

About a thousand miles east of Monterey, and a mile above the surface of the ocean, sits a very different aquarium.

The Downtown Aquarium, near the South Platte River in Denver, Colorado, was built as a nonprofit aquarium called Ocean Journey, the largest between Chicago and California. In August 1998, the fish started arriving. One thousand tropical fish found themselves at high altitude and almost as far from the ocean as it is possible to be within the North American continent. The aquarium opened in 1999 but was unfortunately not pulling in enough money and only lasted a few years. In 2003, it was sold to the for-profit Landry's seafood restaurant chain. Today it's an aquarium and seafood restaurant.

The mission of the Downtown Aquarium is not entirely clear from its website. Is it a place that helps people see fish or a place that feeds people seafood? I look at the aquarium's Facebook page and find the following:

> We have a world of ocean and sea waiting for you. Visit us and dive into over one million gallons of fun. From gliding stingrays to majestic tigers, you don't want to miss a thing.

Why are tigers in an aquarium? I wonder. Yet overall, the focus of this marketing effort seems to be on fostering a connection with nature.

After the customary Facebook request to "like" this page (to promote the aquarium to all your Facebook friends), the message continues:

> Traveling the seven seas can make a person hungry. Join us for a meal.

Humph. There it is — the seafood side of the aquarium. At least it was divided from the message about connecting with the ocean. And at least the ocean message got top billing in the advertisement. But still, I have my doubts. Is this a place that connects humans to nature? Or is this a place that helps people eat nature?

When I heard the aquarium in Denver was in financial trouble several years ago and then learned that it was bought by a company that runs seafood restaurants, I could not help but anthropomorphize, wondering whether the fish and other marine life might see the writing on the wall. What if they knew that the same people who had been coming to peer into their tanks, marine voyeurs, were soon to be dining on their kind? Trapped in tanks in the center of the country, they would have no *Finding Nemo*–style rescue. There would be no escape.

When the aquarium opened as Ocean Journey, the theme of its exhibits was the connection between mountaintops and the ocean — the journey of rivers.

A 1999 article in *Sunset* magazine describes Ocean Journey thus:

One exhibit traces the Colorado River, which begins in nearby Rocky Mountain National Park; another follows distant Indonesia's Kampar River. Linked by their common destination, the Pacific Ocean, the two rivers depict very different ecosystems: the arid countryside of the Colorado and the lush rain forest around the Kampar.

Sounds educational, doesn't it? The idea that the natural worlds on land and in the ocean are connected is a big shift in understanding for many. And it explains the presence of the tigers. But from the start, there were concerns about the mix of real and false within the exhibits. There were real trees in the Colorado River section of the exhibit and fake trees in the Indonesian section of the exhibit, blurring the lines between real and fake and prompting visitors to question the authenticity of nature. I found that some approximations of nature could be convincing, but others fell flat.

■ ■ ■

I wander into the rain forest exhibit in the current-day Downtown Aquarium and discover that since its Ocean Journey days, animals from South American jungles have been added to the Indonesian rain forest. From Southeast Asia there are Sumatran tigers and a bird with a giant yellow beak called a wrinkled hornbill. From South America, piranha that are found in the Amazon River sit placidly in a glass tank set between pillars of mock trees that grow from the concrete floor.

"Gary!" A woman squeals. "Do you see the orangutan?"

Gary and I turn around from the piranha, following her voice and her outstretched arm. There, in the corner of the room, sits a quiet wallflower of an Asian orangutan holding a baby orangutan. They are not in an enclosure and shock people who see them close and at eye level. Another woman gasps. But the orangutans are not real. A docent tells us that they are motorized and do occasionally move. We stare for some time. The mock orangutan stays still. Its baby stays still. We stay still too. Was the docent joking? Or did someone unplug the primates? Eventually, each of us gives up and moves on to animals that do move—because they are alive, not because they are properly motorized.

■ ■ ■

"When Colorado's Ocean Journey co-founders Bill Fleming and Judy Petersen-Fleming moved to town in 1992 with an idea for an aquarium in the Platte Valley, they appeared to be a couple of pipe-dreaming flakes," wrote journalist Michael Paglia in a 1999 article in the local alternative paper *Westword* shortly after the aquarium's grand opening attracted nearly half a million visitors in just a few months. He continued, "It turns out that the husband-and-wife team were, in fact, visionaries."

Less than four years later, Michael Paglia wrote in *Westword*, "In what could only be described as a far-fetched treatment for an upcoming *Simpsons* episode, Landry's, the Houston-based seafood-restaurant chain, purchased Ocean Journey."

Landry's bought Ocean Journey in 2003. Ripley Entertainment was also bidding for the aquarium. Imagine you are a fish. You are left high and dry, a mile above the sea. There is a bidding war going on for you and your fellow fish between a seafood restaurant and an entertainment company whose motto is Believe It or Not! The outcome of this bidding war will cement your reputation as either food or freak. Which bidder would you prefer?

■ ■ ■

A woman and man race up to the glass, pointing at sharks in the Downtown Aquarium, and shout to a girl and a boy a few steps behind, "Look! Sharks!" The kids follow.

"Whoa, did you see that one?"

"Awesome!"

The sharks are behind eleven inches of glass and don't seem to mind the shouting, or the *Jaws*-like soundtrack that is playing in their display.

Their mother sits down cross-legged on the floor, which does seem inviting. Other kids are lying on their stomachs nearby, peering into a porthole in the floor that is over a deeper part of the same tank. Occasionally a shark trundles past the glass just above the sandy bottom. Most of the time, the kids see only sand as they peer through the porthole, entranced by the possibility of seeing a shark.

The family sitting up against the glass of the tank has a third child, whom I hadn't noticed until I felt eyes on my notebook as I scribbled about the humans and the sharks. The girl is school-aged, but just barely, maybe first or second grade. Her round glasses have been borrowed from either Harry Potter or Harriet the Spy and are surrounded by a bob of blond hair. She is more curious about what I'm doing than about the sharks. She could have been me long ago. She takes a seat next to me on a bench that looks like stone. It is really a bench, but it's not really stone.

Her mother twists around and calls to her from the glass. "Do you see this?" she asks and then turns to watch, exclaiming about the size of the sharks and the size of their teeth with the rest of her family.

The girl turns to see her mother and cringes at the sharks beyond. She turns away from the glass.

"Do you like sharks?" I ask, knowing the answer.

She shakes her head back and forth. No.

"Most sharks don't hurt people," I offer.

"Maybe," she replies slowly, looking at the sharks in her peripheral vision while grasping the false stone of the bench. It's not even cold like real stone.

She might have truly reconsidered her feelings about sharks, even had a little empathy, if the scary *Jaws* music had not grown louder at that moment.

Da da. Da da. Da da.

The music fades into the background again and a booming prerecorded voice begins, "Ladies and gentlemen! Boys and girls!"

The voice tells us about the sharks and other fearsome fish we see in the exhibit. The ominous music continues in the background. The voice is that of a game show host.

"These are the largest sharks," says the voice. The crowd gasps and peers deeper into the tank. "...in Colorado," the voice continues. There are a few chuckles and a little eye rolling amid the crowd. The disembodied voice has lost some credibility.

"Sawfish are found worldwide *and* on the endangered species list!" The voice continues as if it is a prize or an honor to be endangered as well as to be found worldwide. Everything the voice says ends in an exclamation point. And after each exclamation point, I anticipate hearing the words of a circus ("Feast your eyes on the center ring!").

■ ■ ■

The aquarium cost $93 million to build. It sold to Landry's for $13.6 million in bankruptcy court. When the city of Denver valued it at $30 million, Landry's didn't want to pay their property tax. They threatened to tear down the aquarium and sell the land if the city didn't back off on the tax bill.

"How do you address comments/criticisms about selling seafood at an aquarium?" reporter Janet Forgrieve asked Tilman Fertitta, Landry's CEO, in a 2005 interview for the *Rocky Mountain News.*

"I really don't hear it that often," Fertitta responded. "I think whenever somebody says anything like that, they're trying to think of themselves as a comedian."

"We like to stress the education side (of the aquarium)," said Fertitta in another interview with the *Rocky Mountain News.* "But we also add the fun — it's what people want."

Yes, there is boring education. (We've all been stuck in that classroom at some point.) But I contend that boring education is not really education, because not much learning happens without fun. The type of education that gets one actually learning about things has to be engaging, or learning doesn't happen. But I'm biased. I earn a living as a science educator and I, myself, am a great fan of education, returning to school every few years for a course or degree. I find education entertaining. And yet, promoters of edutainment and places like the Downtown Aquarium make it sound like education is tolerated as long as there is a dose of fun, just as vegetables are often begrudgingly consumed when there is a promise of chocolate cake.

. . .

"Hi, mermaid!"

"Joey, look at the mermaids!"

Two young women have appeared in the large tank of turtles, pancake-shaped rays, and skateboard-sized fish. They are clad in bikini tops and, from the waist down, wear sparkly tail fins, their legs bound together. Both have long hair that streams out in the water as they hover near the bottom of the tank, smiling and waving to small children beyond the glass.

Music, vaguely calypso in origin, plays as the mermaids swim. Both women have extraordinary swimming skills. They must flap their tail fins back and forth to reach the bottom. They must hold their breath for long periods of time as they entertain the audience and swerve to avoid the fish and turtles. Every so often, one of them goes to the surface of the water and rests on a rock, presumably to catch her breath. I try to imagine what a Help Wanted ad might look like for a mermaid.

This is how humans have a role in the aquarium exhibit. Will the small children glued to the glass grow up to believe that mermaids really do swim in the ocean? Will they value the ocean's mermaids more than other, less fictional marine life?

A sea turtle comes up to the glass with one of the mermaids. This endangered species appears to be hamming it up with the crowd. It's a coincidence, not a trained turtle. Everyone laughs.

About a half hour after the mermaid show, I see a utility cart passing through the exhibits, pushed by a bored employee in khaki pants. The two mermaids are on the cart. While these women are extra sparkly and appear to have permanent smiles and tans, and somehow they have perfect hair even though they have been swimming, they are unable to walk in their costumes. Graceful underwater, they would just flop about on land, and so must be moved around on the utility cart. People gasp, turning away from the real fish to see the mermaids. Small girls point. The mermaids beam as the cart's wheels rattle over the concrete floor.

All visits to an aquarium are at least somewhat contrived experiences with nature. There are always fake rocks somewhere, and red lights that illuminate the jellyfish pulsing through the water of their tank. Some of these adjustments seem subtle, introducing people to the ocean in a more inviting way. Others seem like breeding grounds for misconceptions about nature.

For clarity, wild sharks do not have a *Jaws* soundtrack and mermaids are not a part of any marine ecosystem.

■ ■ ■

Both the Monterey Bay Aquarium and the Downtown Aquarium are accredited by the Association of Zoos and Aquariums, a nonprofit organization founded in 1924 that is dedicated to the advancement of conservation, education, science, and recreation at zoos and aquariums. According to the organization's long-term vision statement, the AZA "envisions a world where, as a result of the work of accredited zoos and aquariums, all people respect, value and conserve wildlife and wild places." There are over three-quarters of a million animals in AZA accredited US zoos and aquariums. There are forty-seven accredited aquariums in the United States, and seven additional aquariums that are combined with zoos.

Although I was in some ways horrified by the Downtown Aquarium, I was also heartened by the way people were connecting with wildlife. No one strives to protect what he or she does not know or understand. And the people I observed at the Downtown Aquarium were getting to know sea creatures. Plus, the animals seemed to be well cared for, which is reassuring. Despite the seafood restaurant foundations of owner Landry's Inc., the spotlight was on living animals, not their sautéed, grilled, and deep-fried counterparts. Perhaps people visit the Downtown Aquarium for many of the same reasons that they visit the Monterey Bay Aquarium. But at the former, the ocean's glitzy pizzazz is on display, and the latter portrays a more everyday ocean.

"Is it better than no aquarium at all?" asks my friend Jen after I explained my mixed feelings about the Downtown Aquarium.

That is the same sort of question posed about all sorts of edutainment. These quasi-educational experiences attempt to balance fun and learning. I am not against fun, but when it comes to breeding misconceptions about nature, when it comes to valuing the drama of an experience more than its truth, when it comes to inaccuracies, I have a hard line. If people are walking out of the aquarium with more misconceptions about the ocean than they had when they walked in, then no aquarium would be better.

In some senses, an aquarium resembles an art museum more than a nature experience. One does not get wet or sandy in an aquarium. The buildings are stroller friendly. The glass tanks exhibit beautiful multimedia works. Aquariums allow us to be a part of the ocean — to see that we are connected to this underwater world that is coping with extreme environmental impacts. The most effective aquarium exhibits activate the emotional side of our brains, which is the part most likely to motivate us to take action. Then again, we are actually appreciating at a distance the way we do when a red rope or silent alarm separates us from paintings and sculpture.

I asked Richard Louv, founder of the Children & Nature Network, whether he thought aquariums connected people with nature. Louv's book *Last Child in the Woods: Saving Our Children from Nature-Deficit Disorder* galvanized a movement often called "no child left inside" that aims to connect children with nature with the premise that exploring and playing in nature is healthy for child development. Yet aquariums are typically indoor places. If no child is to be left inside, do aquariums have a place in terms of fostering connection with nature?

"I do think they can play a role," Louv responded. "One way is simply to inspire interest in the ocean or other aquatic environments — if they stimulate the imagination enough to encourage a child and parents to actually engage in nature, even if there's no ocean nearby."

I agree. One does not start exploring the oceans by scuba diving or taking a trip via submarine. An aquarium can be a way to start. Aquariums are one reason that I got interested in the creatures that live in the ocean. I grew up going to a small aquarium in Woods Hole, Massachusetts. It claims to be the oldest in the country. Run by the National Marine Fisheries Service, it was free and low key. There were no bells and whistles, no shark music. There were just animals, the smell of salt water and algae, the bubbling of tanks. That was enough magic to fuel my lifelong ocean addiction.

Maybe no aquarium would be worse than the Downtown Aquarium, despite its inaccuracies. Maybe helping people feel connected with the ocean when they are a thousand miles away is good, even if they walk away thinking that mermaids swim the oceans with turtles. Maybe the connection with real marine life will be what they take home and what they might respond to when they learn that the oceans are in trouble.

It can be difficult to perceive the impacts of environmental change when the changes are distant. This is true for change in the ocean. It's also true for the impacts of climate change. During International Polar Year, which was actually two years long, between 2007 and 2008, there was a great deal of press about the impacts of climate change on the Arctic and Antarctic. But polar bears floating on sea ice are about as relevant to life in, say, New York City as polar bears on the moon. Communications experts, such as those with the National Research Council, recognized this and advocated for closer-to-home examples of the impacts of climate change. Recently, examples have been handed to us by the earth itself in the form of unusual hurricanes, coastal flooding at high tide, drought and deluge in Texas and California, and other extremes.

■ ■ ■

An aquarium can be a gateway, fostering a fondness for the ocean, which could, hopefully, lead to taking care of the ocean and the rest of our planet. Is it concerning that children might walk away from the mermaid show with the understanding that mermaids roam the seas? Yes. But perhaps we long to see some flippered version of ourselves join the turtles and fish, swimming freely through the water and belonging to the underwater world. It makes the ocean surface look less like a reflective barrier separating us from them. It connects us all. We are all in this together.

"Wildness reminds us what it means to be human, what we are connected to rather than what we are separate from," wrote Terry Tempest Williams. I'd agree; however, I'd amend her statement. Seeing the incredible diversity of animals in an aquarium, even an aquarium with a mermaid show, reminds me that we are all animals — that perhaps we humans are not so disparate from the marine life. I am not pro-mermaid or anti-mermaid, but I am starting to understand what glues people to the glass when the mermaids swim around. They are certainly more glamorous than I was when I dove in the shark tank with an underwater camera, but maybe they're not so different. Maybe a human (or a mermaid) in the ocean world is more compelling than the ocean world on its own. We are more likely to see the risks to the ocean if we have a personal connection to the place, if we take some ownership. The mermaids give people a frame of reference, a sense of scale, a way to imagine themselves as stakeholders, and a reason for taking care of the ocean. We want to see the ocean as beautiful and powerful, not vulnerable to our actions and in the midst of catastrophic change. Perhaps we want to be in this together as long as the story is a happy one. We are connected to the ocean in large ways — through the impacts of pollution, climate change, and other environmental changes, but those things are harder to visualize than a mermaid next to a turtle, and the story is much less cheerful. It's a story of disaster and how we neglect the ocean, not one in which we all swim together. It's not the story we hope for.

Chapter Six

Ashes to Ashes

A mid pastoral fields of corn and bales of hay in northeast Nebraska, an unassuming barn contains evidence that the Cornhusker State was once a subtropical savannah disrupted by a disastrous volcanic eruption. The barn is full of rhinoceroses and other animals. They have been dead for years, nearly twelve million years, to be specific.

Walk inside the 17,500-square-foot rhino barn at Ashfall Fossil Beds State Historical Park and you follow a pathway that hugs the wall around an outcrop of volcanic ash. When I visited in 2015 with my partner Adam, we watched an intern named Erin diligently scrape ash off the surface of the outcrop in the hope of finding bones. In the areas where the ash had been scraped away, the skeletons of over two dozen rhinos were visible. They'd look like they were napping on their sides if they had flesh on their bones.

Today, rhinos are native to Africa and southern Asia, but twelve million years ago they were in North America. Ancient rhinos were miniature by today's standard, only about three and a half feet tall and about two thousand pounds. That's large, but much smaller than the rhino that you might find at your local zoo, which is as much as seven thousand pounds and about six feet tall at the shoulder. There are other animals preserved in the ash too, including ancient and now extinct species of horses, camels, birds, saber-toothed deer, turtles, and snakes. There is evidence of wild dogs and other predators too. These animals once lived by a watering hole in the Nebraskan savannah.

That was before the volcano erupted.

The volcanic ash that the animal fossils are preserved within traveled nearly a thousand miles, from the Bruneau-Jarbidge volcanic field in southwest Idaho. This was a massive eruption spreading ash across nearly the entire continental United States. In Nebraska, the ash was about two feet deep. The wind piled it into drifts. Eruptions of this size are hard to fathom.

Signs on the walls of the rhino barn profile each species that has been found at the site. And for each species, there is an indication of whether its bones were found in the watering hole layer, the disaster layer, or the recovery layer. The watering hole was the time before, when all was calm. The disaster layer was when animals faced the rain of volcanic ash. After the disaster layer, a layer formed without much in it, signifying that the time right after disaster was desolate. And then, life returned. The eruption did not cause the extinction of any species, even if it did cause the death of many individuals. (There were local extinctions.) Eventually, plants rooted in the compacted ash, and soil formed. Rhinos and other grazing animals wandered in from the Dakotas or perhaps Canada. Their bones are found in the recovery layer.

One could imagine a similar stratigraphy for the other disasters we have explored in this book. The disaster layer could contain San Franciscans fleeing in nightshirts, lionfish decimating reefs, cottages falling off a crumbing cliff of sand. The disaster layer is the time that we are living in now as climate is changing and we are deciding what to do. But often, disaster is a time of erosion — of burning buildings and rubble. At Ashfall, it was a time of accumulation. Ash preserved an impression of what life was like during the disaster.

In the rhino barn, disaster is summed up in a stratigraphic snapshot. There is before, during, and after. And that's something to remember as we face climate change and the host of smaller-scale disasters that it is projected to intensify. Whatever happens, there will always be before, during, and after. This sounds comforting, until you realize that most of the animals in Ashfall died during the disaster. For them, there was no after. In

Rhino fossils inside the barn at Ashfall Fossil Beds State Historical Park in Royal, Nebraska. (Courtesy of Adam D. Holloway.)

regional disasters, individuals might not make it, but if the species exists elsewhere, it survives. This is not necessarily the case with a global disaster like climate change. There will be nowhere that is safe for some species. Life on earth will survive, but many living things will not.

The turtles and the birds died right after the volcanic explosion. A rapid change in temperature may have killed the turtles first. Many of the birds that were wading in the watering hole look like they fell face-first into the muck with wings outstretched. They likely died as their feathers filled with ash in the hours after the eruption began. Then, within a few weeks, the camels and horses died. A few weeks later, the rhinos died. Imagine being a rhino and seeing the others die around you. You see birds flop unceremoniously in the muck, the bodies of horses and camels crumpled in the mud. You look out toward the massacre caused by the eruption and are either unable to get out of harm's way, unable to see the writing on the wall, or both.

"They were not instinctively prepared for what happened," Rick Otto, Ashfall superintendent, told me when, finding it challenging not to anthropomorphize, I blurted a question about why these animals didn't do anything about their situation. The rhinos and other grazing animals had their muzzles to the ground to eat grass. They wound up breathing in a lot of ash as they ate. "They were power-snorting it into their lungs," said Otto.

He brought out a small jar of ash and told me and Adam to get some on our fingers. By the door to the barn, where the light streamed in, the ash sparkled on our skin. The ash was tiny shards of glass. It's not healthy to power-snort glass.

The small, jagged glass shards that make up ash are the reason so many animals died at this former watering hole. Breathing in the ash compromised the animals' lung function, which led to Marie's disease (hypertrophic osteopathy), in which bones develop extra growths and limbs swell, becoming stiff. The camels, horses, and rhinos at Ashfall all show evidence of Marie's disease on their bones. They would have been feverish and thirsty. Wading in the watering hole may have provided some relief.

Erin uncovered a small articulated leg in the ash. It may be from a wild dog, she speculated. There are lots of traces of wild dogs at the fossil site, but few bones. There are burrows where they used to live, fossil poop that they left behind, bones that they gnawed, and even a footprint, but few skeletons. I asked her why their skeletons are not preserved in this massive bone bed. She told me that there is evidence that some dogs lived in burrows where they could breathe cleaner air and come up only occasionally to eat. With animals dying at the watering hole, there would have been lots for them to eat. Maybe their lifestyle helped them avoid danger. Perhaps the dogs walked away from the disaster. Perhaps they survived. We can't say for sure.

When field trips of students tour Ashfall, Otto notices the looks on their faces when they realize what happened there. "They are terrified if not traumatized," he said, "because the same thing could happen again." The volcanic hot spot that caused the Idaho eruption is now located under Yellowstone National Park. Another eruption could occur, and if it does, we humans are now in harm's way.

. . .

But humans are different. It's somewhat unfair to compare rhinos with humans. We have large brains that are accustomed to problem solving and assessing risk. We would have the ability to plan our escape. And even if our escape attempts were thwarted, at least we do not snuffle the ground to find food, inhaling ash along the way. We could change how we were living or what we were eating. When humans face change on earth, we don't just stand there. We do something.

All of these actions — problem solving, risk assessment, planning, changing our ways — require an ability that may be unique to humans: the ability to take our minds into the past and into the future. We do this each time we make a grocery list and consider what we will need for the week ahead, each time we smell freshly mown grass and are transported to a childhood summer day when we jumped through lawn sprinklers. The ability to virtually take yourself into the past or the future is formally called chronesthesia. It's known informally as mental time travel, and I prefer that name. Chronesthesia sounds like a medical procedure. Mental time travel sounds like an adventure.

While we have not developed actual time machines to ship us into the future and the past, our brains can do this for us. I'd always thought of my inability to live in the present as a fault. If I were able to live in the present, I might be able to stop fretting about the future. I try to learn from our dog Mila, as she seems to always live in the present. But looking into the scientific literature about mental time travel, I am realizing that the human ability to focus attention on the future or the past can be extremely useful. And it might help us survive the climate catastrophe.

I sit in climate science presentations where numerous graphs on projected PowerPoint slides are really timelines into the future — opportunities for mental time travel. The graphs show changes in global temperatures, sea level rise, the amount of carbon dioxide in the atmosphere, or population growth. In these cases, a line trending upward is not good news. It means flooding coasts or sweltering heat waves, rapid climate change, and an increasingly crowded planet. Sometimes Mila comes to work and attends these presentations. She is usually asleep at my feet, blissfully unaware that people in the room have been transported into a future of climate change and the myriad impacts it will cause.

The idea that humans are the only species on earth with mental time travel abilities is somewhat controversial. It's tough to test, since we cannot ask other animals what's going on in their heads. Researchers observing animal behavior have found evidence that certain birds, such as crows and ravens, and a handful of primates have some mental time travel abilities too. Crows are able to hide food away for later, for example, implying that they are planning for a time in the future when they will be hungry.

Although some animals may have mental time travel capabilities, they, and most humans, are not discussing graphs that run forward in time or developing supercomputers to compare multiple scenarios of what might happen in the decades to come. Those are topics that might fill the minds of climate scientists, but there are more everyday applications of mental time travel. One might buy concert tickets months in advance. One might save for retirement, or at least worry about saving for retirement. All humans, aside from the very young, travel into the future and the past in their minds. This enables us to cope with disaster in unique ways.

■ ■ ■

During a somewhat less ancient volcanic eruption, there were humans on earth and they did have to face disaster. Mount Vesuvius, located on the west coast of Italy, erupted in AD 79, and it's a well-worn story that the city of Pompeii was decimated during that eruption. Like the Nebraska watering hole, the city of Pompeii was covered in ash. A snapshot of the

Roman Empire was preserved. However, unlike the Nebraska watering hole, there is evidence that most people skipped town when the volcano started to erupt.

The bodies of those who died at Pompeii are well preserved, and extensive excavations have given us a snapshot of the city at that time. Archaeologists estimate that only about three thousand of the city's approximately fifteen thousand residents died there when Mount Vesuvius erupted. What happened to the other twelve thousand people? It's possible that they made it to safety. It's possible that they died on their way out of town. But in either case, it appears that 80 percent of the people in Pompeii made a choice to leave. The rhinos at the Nebraska watering hole did not have this ability. According to Rick Otto, the entire herd died at the watering hole. When they saw other animals start to die, they didn't develop a plan to get themselves to safety. It appears that the majority of humans in the city of Pompeii did have that ability.

After surviving seven or eight feet of ash and rocks falling on the city, the remaining three thousand people in Pompeii were killed in a matter of seconds or minutes during the eruption as an avalanche of heat, ash, and toxic gases called a pyroclastic flow raced down the flanks of the volcano. It was the heat that killed most people, not the ash that covered their houses, according to a recent study by Italian volcanologist Giuseppe Mastrolorenzo and his colleagues.

What would you do if you saw a dark cloud growing above Mount Vesuvius? Would you run away? Would you stay put, assuming everything would be fine? Would you head toward it to get a closer look? These are questions that we cannot ask of the former residents of Pompeii, but we do have an account of what a few people did when they saw the volcano coming to life.

■ ■ ■

Pliny the Elder and his nephew Pliny the Younger saw the explosive cloud form above Mount Vesuvius on the west coast of Italy nearly two thousand years ago. They were at Pliny the Elder's home on the coast in Miseno, which in Roman times was called Misenum. Across the Gulf of Naples they could see the volcano, about twenty miles to the east. Years afterward, Pliny the Younger wrote about what he and his uncle did at the time in letters to a Roman historian. In a way, Pliny's letters about his family's experience during the eruption of Mount Vesuvius are like a memoir. The truths that he took away from the disaster say a lot about human nature amid the uncertainty of earth's chaos and the types of decisions that people make when faced with catastrophic change.

It was around lunchtime on August 24 of the year AD 79 when they saw a cloud with a strange shape form over the volcano. Today, we'd call it a mushroom cloud, but in the time before nuclear explosions, Pliny the Younger described it as shaped like a pine tree. A long trunk of ash extended upward from the volcano and then expanded outward in the stratosphere until it reached an altitude where the ash and gas had the same density as the surrounding air.

This column of ash is characteristic of explosive volcanic eruptions in which there is less lava and more falling rocks, ash, and hot pyroclastic flows. When the heat causes glaciers and snow to melt rapidly, there are also mudslides called lahars. As the hot lava moves underground, earthquakes rumble like geologic indigestion.

Surrounding Mount Vesuvius was the powerful and growing Roman Empire. The Romans were gifted engineers who figured out how to harness water and wind. Romans had learned how to exploit nature, and the landscape of Europe was transformed into urban centers and farms with roads, bridges, and aqueducts connecting different locations.

When Pliny the Elder saw this explosive cloud, he wanted to get a closer look and ordered a boat to be readied that would take him across the bay to where the volcano met the sea, not far from Pompeii. He was the author of an encyclopedia of natural history, which, to me, explains this decision entirely. In the same way that today's meteorologists hop into cars and trucks to chase tornadoes, Pliny the Elder sought to chase the erupting volcano, to get a close look, and to learn. In cases like this, learning can be hazardous to your health. As he was leaving, a desperate message reached him from a woman named Rectina who was trapped at her villa at the base of the volcano and needed help. Because of this letter, Pliny the Elder's expedition became part geology field trip and part rescue mission. More ships were prepared for the voyage, anticipating that Rectina was probably one of many people who would gratefully scramble aboard when the ships arrived at the base of the volcano.

"He hurried to a place from which others were fleeing and held his course directly into danger," wrote his nephew.

Pliny the Elder and his crew departed in quadriremes, ships with many pairs of oars and large square sails. According to reports from the crew, ash and large blocks of pumice rock fell on the ships as they crossed the bay. They landed in Stabiae, about ten miles south of the volcano, twice as far from the volcano as Pompeii, farther than they had intended because their route was blocked by floating pumice (volcanic rock with a density so low that it actually floats). There, Pliny the Elder found that the people of Stabiae had loaded up ships and were desperate to leave but could not sail away because the winds were coming into the port. Flames were visible on the volcano. Ash and stone continued to fall. It was dark during the middle of the day, and the land shook with earthquakes. Naturally, the time was right to take a bath. That's what Pliny the Elder did.

"He bathed and dined, carefree or at least appearing so (which is equally impressive)," recounts Pliny the Younger. After dinner, he went to sleep, snoring loudly as ash and rocks fell from the sky and accumulated outside his room. His nephew suggests that he showed a lack of concern to lessen the fears of others. That's possible. It's also possible that he was unconcerned because he was excited to get up close to the volcano. He may have just not realized the risk. Or perhaps he needed to rationalize that Stabiae was safe, not that they were trapped. The ability to rationalize is another quality that distinguishes humans from rhinos.

In the morning, although it was still as dark as night with ash obstructing the sunlight, Pliny the Elder and his crew tied pillows to their heads to protect against falling rocks and ran outside. They were concerned that the earthquakes had made the buildings unstable, so they fled to their ships to assess whether the winds would allow them to leave. It did not look promising.

Pliny the Elder was resting in the shade and drinking water when the air began to smell like sulfur. The others fled, but Pliny was weak. He was a large man, according to his nephew, and two small slaves were tasked with supporting him on either side, like crutches. Pliny the Elder collapsed and died. His crew survived and told this story. His nephew wrote it down.

"As I understand it," wrote Pliny the Younger of his uncle's death, "his breathing was obstructed by the dust-laden air, and his innards, which were never strong and often blocked or upset, simply shut down."

The sulfur smell the crew reported may have been hydrogen sulfide (H_2S), a colorless gas that smells like a swamp. Released by volcanic eruptions, it affects a person's upper respiratory tract and bronchi, just like ozone and other air pollutants do. Some people are more susceptible than others to its effects. Perhaps Pliny the Elder was quicker to succumb to the poor air quality than his crew, just as the horses and camels at the watering hole in Nebraska were quicker to succumb than the rhinos. People with asthma, for example, are more vulnerable to hydrogen sulfide. Could Pliny the Elder have had asthma? I suppose we cannot know.

■ ■ ■

Meanwhile, back in Misenum, the younger Pliny was making a different suite of decisions when faced with rapid environmental change.

Once Pliny the Elder left with his ships, Pliny the Younger, seventeen years old, was left with his mom at the house in Misenum. The house was on the coast, within sight of the erupting volcano. His uncle had invited Pliny the Younger to come along on his swashbuckling geological adventure, but the young bookworm had decided to study instead. He read, bathed, dined, and then went to sleep, a common pattern for Romans. But he awoke in early morning to violent earthquakes. His mother ran to Pliny's room, and the two went to the courtyard between the house and the shore to consider their options.

"I don't know if I should call this courage or folly on my part," wrote

Pliny the Younger. "I called for a volume of Livy and went on reading as if I had nothing else to do." Livy was the author of a multivolume work about the history of Rome. Pliny the Younger was reading about Roman history when one of the biggest historical events to rock the Roman Empire was in progress. Our normal routines are comforting in disastrous times, but they can also cause us harm when we use them to avoid environmental change and the hazards it presents.

A friend of his uncle's from Spain showed up at the house and was flummoxed that Pliny and his mother were just sitting there reading when a natural disaster was in progress. He perceived the risks as much greater than Pliny did. He scolded them — Pliny for reading and his mother for allowing the teenager to do so. "Nevertheless, I remained absorbed in my book," Pliny remarked.

Pliny the Younger, his mother, and the Spaniard were a bit farther from the volcano than the residents of Pompeii or Pliny the Elder and his crew. And since they were a bit farther from the volcano, they sat in the gray area where it's more difficult to decipher the level of danger. Was this an emergency for them? Was this a disaster? Pliny thought it was not. The Spaniard thought it was. The mother seemed to be persuadable either way. This gray area exists in a wide range of disasters. Much as I wondered how one knows when it's time to pull the lever of a San Francisco alarm box, these Romans were unsure of the appropriate level of response. We all perceive risk differently.

The earthquakes continued. The buildings became unstable. Pliny the Younger and his mother decided it was time to leave town. They were not alone. "The panic stricken crowds followed us, in response to that instinct of fear, which causes people to follow where others lead," wrote Pliny. This might be an example of the chameleon effect: when people are not sure what to do, they take their cues from others. Perhaps the townspeople were taking their cues from Pliny. Perhaps he was taking his cues from them.

The Spaniard ran into Pliny and his mother again and told them that they couldn't stop. They must escape.

It's to be expected that Pliny's firsthand account of what he and his mother endured would be much more detailed than the story of his uncle's journey. He saw the water of the Mediterranean recede quickly, stranding sea creatures, a phenomenon that can precede a tsunami (although Pliny the Younger does not report a tsunami). Dense black clouds loomed low. It was "the darkness of a sealed room without lights," he wrote. He described what people did in the darkness as the land shook and fires from the volcano blazed in the distance. He said that there was fear, hope, and prayers to the gods. He heard cries in the darkness. There were heavy showers of ash. Pliny and his mom would lie down to sleep outdoors, awake under a layer of ash, shake off the ash, and lie down to sleep again. The ash formed something resembling snowdrifts, according to Pliny. It did the same in Nebraska about twelve million years before.

They returned to Misenum and "spent an anxious night alternating between hope and fear. Fear predominated." The earthquakes continued. The two decided to stay at the house until hearing from Pliny the Elder. But they would never hear from him again.

■ ■ ■

In retrospect, perhaps the two Plinys did not make the best decisions. But they also did not have access to modern science, eruption predictions, and disaster management planners. Today we know more about the science of volcanoes and the medley of hazards that occur during eruptions — molten rock in red hot lava flows, the ballistic projectiles and rain of rock fragments, pyroclastic flows, earthquakes, mudslides, and lightning generated by the eruption, to name a few. We also know that there are warning signs that an eruption is likely, and our methods for detecting those warning signs have improved over time. We may not know the scale of the eruption or when exactly it will happen, but we can tell when a volcano becomes active, when the land swells with the heat and pressure beneath, and when lava moves underground.

Today, Mount Vesuvius is monitored closely in the hope of catching those early warning signs and preventing disastrous loss of life, mitigating the danger. However, the population in the area has grown. In satellite pictures, the volcano looks like an island of wilderness surrounded by the roofs and roads of the human-built world. If Vesuvius erupted today, as many as three million people would be in harm's way. That's the worst-case scenario, scientists say, including people up to nine miles away from the volcano. That is the scale of eruption that happened in AD 79.

The National Emergency Plan developed and maintained by Italy's Department of Civil Protection is not designed to protect against a worst-case scenario. It is designed to evacuate half a million people who are closest to Vesuvius (within five miles) in seventy-two hours, hopefully before the eruption starts. Those are the people at risk from pyroclastic flows of heat, ash, and toxic gases. People farther away, who might get rained on by ash and rocks, would be evacuated after the eruption starts, once the direction the ash travels in the wind is determined and it's known which communities are most vulnerable.

There are debates about whether the region around Mount Vesuvius should be planning for a worst-case scenario in which all three million people are evacuated instead of the phased evacuations. There is also concern that the current plan is not realistic. During a trial evacuation drill in 2006 in which only a small portion of residents were told to evacuate, massive traffic jams locked cars on roads. But despite these potential problems, one thing is clear. Unlike the rhinos that suffocated twelve million years ago, we humans make plans when confronted with the possibility of natural disaster. Our mental time travel abilities make such planning possible. Our abilities to prepare for the future are unparalleled in the animal world. We try to figure out when environmental change is happening and whether it's hazardous to our health. We seek to reduce risk. We prepare. Or at least we try—usually.

Scientists monitor and report when the earth changes. If Vesuvius rumbles awake, volcanologists tell Italy's Department of Civil Protection, which has the power to put the emergency plan into action. Similarly, as our climate warms, scientists who monitor earth's climate tell the governments of the world (via the Intergovernmental Panel on Climate Change). Those governments have the power to take action, although many governments, including the United States, have been reticent to make the necessary changes until recently, and those actions are fickle, subject to change with the election of new government leaders. The reports from both volcanologists and climate scientists are about changes in the environment, but they are really about the risk we face. The same is true of many other types of scientific reports — from weather forecasts to medical test results. In all cases, it is not the report itself but rather how people perceive risk and interpret uncertainty that determine whether they jump into action, remain absorbed in a book, or take a bath.

Chapter Seven

We Are Not Waterproof

On September 12, 2013, it looked like a slurry of chocolate milk was flowing under the bridge near my home on the north end of Boulder, Colorado. A couple of days earlier, there was only a dry creek bed, a bike path, a few shady characters, and a strong smell of marijuana under this bridge. I regularly passed by en route to a park at the foot of the Rocky Mountains. But on September 12, there was no way to get through. There was only muddy water, at least six feet deep, and the rain continued to fall. Flood tourists like me gathered at the water's edge as it raced toward Kansas and Nebraska. I stood mesmerized by an inner tube and a couch cushion that were swirling in an eddy. A single ski boot passed by, its point of origin unknown. Another waterlogged pair sat in a pile of debris on the sidewalk above the raging torrent, a mix of tree branches, trash, and possessions no longer possessed.

Fast-flowing water has the power to pick up large, heavy sediments — like rocks the size of cantaloupes or a ski boot — and carry those items along with the mud. It can carry the sediment until it slows down, which, in this case, we saw happening where the water could spread out onto roads, into basements, or over the playground of a nearby elementary school.

When half a year's rainfall comes in a week, as it did in September of 2013, the water can sweep homes and lives away, incapacitating towns. It reminds us that something as simple and ubiquitous as water has the ability to trump our attempts to transform the land, sending us into the wilderness.

Here I was, writing about how we react as the planet changes when, lo and behold, I was in the middle of it. This is what I'd been training for, and yet I wasn't entirely sure I could recognize disaster when I saw it. During the days of rain, I would oscillate between being certain that this was a disaster and being certain that it was not. *I'm sure it will be fine.* There might not be anything more humdrum than a rainy day, right? A little rain can't hurt anything. In fact, we could use the moisture; we were in a drought. I'm not quite sure at what point the number of raindrops hit a tipping point, transforming a simple storm into a natural disaster — but it did happen. After that point, I wasn't at all sure it would be fine.

It was not immediate like an earthquake or erupting volcano. It was not as slow as erosion, which takes decades or centuries to topple a home over a cliff. The massive rainstorm and flooding were somewhere in the middle, and my reaction time was somewhere in the middle as well. The planet was doing something outrageously cool — something we had never seen before — which, naturally, led to giddy excitement. On the other hand, homes were swamped and people were in danger — which led to fear and anxiety. This seemed like a humans-versus-nature situation, which confused me because I usually root for nature, and yet I am human. I didn't want my house to flood, but I was excited by the power of flooding. Part of me wanted to applaud the water for being so extraordinary, and for that, I felt guilty.

Daily life merged with epic concerns. The city proclaimed we should stay off roads. I did laundry. Large military vehicles passed my house. I tapped laptop keys. Drywall contractors arrived as scheduled to fix a wall, soaking wet and giddy with stories of flood-related detours they had to take to get into Boulder. I had called the lead drywaller the previous evening to suggest he might want to reschedule. I had become convinced that this was shaping up to be a disaster, but he seemed as unconcerned as Pliny the Younger absorbed in his history book. After they left in the afternoon, I learned that all roads into and out of Boulder had closed. I admired the wall they patched while wondering if they would make it home. I might as

The upper photo shows Boulder Creek flooding in September 2013, while the lower photo shows how the area usually looks. (Upper courtesy of the author; lower courtesy of Adam D. Holloway.)

well have been setting the dining table amid the rubble of San Francisco's 1906 earthquake.

As rain fell between September 11th and 15th, we smashed records for the wettest day, the wettest month, the highest rate of rainfall we'd seen. Rainfall is measured in inches (or millimeters if you are a fan of the metric system). All in all, over seventeen inches of water fell during the storm. If the land were flat and impervious, knee-deep water would evenly cover the land. But land at the edge of the Rocky Mountains is not flat. Where rain fell on narrow valleys, the water flowed to the lowest point — the creeks on the valley floors — and then flowed downslope as flash floods, turning minor creeks into raging rivers. All of this is complicated by the porosity and permeability of the land — the amount of water the land can hold. Many of these valleys are made of solid igneous rock that can't hold on to water like spongy layers of soil and sediment can.

Each day during the rainstorm, a collection of North Boulder neighbors could be found where Broadway, a north-south thoroughfare, intersects with Fourmile Canyon Creek. In our sopping wet jackets and shoes we'd stare in awe or agony at the rush of light brown water and wonder, could this be normal? Somewhere below the water, the bike path where I usually walk Mila was breaking into chunks of concrete, each about the size of a queen mattress. I wouldn't learn that until the floodwater subsided.

A trailer park and self-storage units sit in the creek's floodplain on the west side of Broadway. On a typical day, kids from the trailer park would play on the boulders in the creek bed, a trickle of water beside them. Landscapers and engineers placed those rocks, but the flood brought other rocks, not prescribed by humans, down from the mountains and they filled the creek bed.

During the flood, both the trailer park and storage units were inundated. People living in the trailers had plenty of time and warning to get to higher ground, and no one was hurt as far as I can surmise, but belongings were strewn everywhere. The waters coming from the storage units carried a particularly diverse assortment, including a boxy old computer, multicolored clothing, a file cabinet, artwork, and myriad papers — detritus from anonymous lives whisked downstream.

■ ■ ■

There are over a dozen scientific research institutes in Boulder, and most focus on some aspect of earth science. As a consequence, the small city is home to thousands of scientists, including many who research weather and climate.

As the rain started to fall in September, many of these scientists were heading outdoors in the spirit of Pliny the Elder, eager to document with photos or reports the status of their backyard rain gauges on Facebook. One of my coworkers was making plots of cumulative daily precipitation. Another was battling the raccoons that toppled her rain gauge. When a storm is occurring somewhere else, these scientists are often called upon to provide calm guidance and interpretation from afar. They are the voice of reason. But it was happening on our home turf. Once taking Mila around the block became harrowing, I found it impossible to have the academic detachment that comes with geographic distance. The panic about personal safety combined with excitement about the earth's power can be an intoxicating cocktail.

Boulder is also home to the Natural Hazards Center at the University of Colorado. And it's thanks to the center's founder, Gilbert F. White, that the rainfall and flooding in Boulder were not more disastrous for the community. White, a geography professor, is known as "the father of floodplain management."

There is a memorial to Gilbert F. White on the banks of Boulder Creek at the center of town. Shaped like a spire, the tribute indicates the water level of different types of floods, from more common smaller floods near the base of the spire to less common larger floods farther up. On the day I visited the sculpture during the September flooding, the creek had flooded the base, but it was nowhere near the top. The sun was shining even though there was more rain on its way. People were trying to get back to normal, biking through floodwaters that were spilling over onto the paths, snapping photos, eating ice cream. White died in 2006, seven years before this big storm. What would he have thought of the event and its aftermath?

White studied what people do when faced with risks from the natural world. We all live with risk every day. Some risky behavior we are taught to avoid. Some risky behavior we avoid instinctively. There are risks that we don't realize. And there are the risks that we know we live with, but they have a low probability of occurrence. Chances are we will never face them. That's the case for an eruption of Mount Vesuvius in Italy. It's the case for a flash flood in the Colorado Front Range, too.

Quantified, risk is calculated as the probability of a chance event multiplied by the consequences if that event happened. For example, the probability of an asteroid impact in Bangkok, Thailand, is very low, but the consequences if that were to happen would be dire because fifty million people live in the city. If the asteroid hit an uninhabited area, the risk would be low.

Hollywood thrillers are often based on the risk equation — a low-probability event occurs (a tornado filled with sharks, for example) that has huge consequences (the shark-filled tornado hits a large city such as Los Angeles). These movies would not keep people on the edge of their seats if the event that occurred were not unusual and the consequences were not dire. But the risk equation does not take into account that the way we perceive risk is more complex than probability and consequences. The emotional and instinctual factor cannot be summed up in the equation, and it is those emotional and instinctual factors that help people jump to action in a risky situation.

"Boulder is the No. 1 flood-risk community in Colorado," Cristina Martinez, a city civil engineer, told the *Boulder Daily Camera* five years before the great floods of 2013. She was not a prophet — just stating facts. All was calm on the mid-June day in 2008 when Martinez made this statement. The temperature had climbed to 86°F. There was not a drop of rain, nothing that might indicate the drama of a flash flood, nothing that might trigger the emotional side of risk perception. But Martinez made this statement as Boulder residents were seeing horrific news reports about flood impacts in Iowa so extreme that some called the event "Iowa's Katrina." The City of Boulder mounted an information campaign at that time and updated maps so that people would understand the area's complex floodplain and what to do when emergency sirens start wailing.

There is high uncertainty about the timing of flash floods, but we do know that they can happen although the probability of occurrence is low. It is easy to become complacent about a low-probability event. Odds are it's not going to happen this year, or next year. Perhaps it will not even happen in your lifetime. If Hollywood made moviegoers watch footage of all the humdrum daily life that happens before a tornado filled with sharks ravages a city, people would get bored. They'd demand refunds.

The flash flood in Boulder in AD 2013 and the volcanic eruption in Italy in AD 79 were both low-probability events. One huge difference between these two events is that the recent one happened at a time in history when we have warning systems. We have tasked certain people with watching the weather radar and the stream flow rates in Boulder. In other areas of the country, people are tasked with watching other parts of the planet — how magma is moving underground, the path of a hurricane, the amount of sea ice floating in the Arctic, the global climate. The modern world is filled with monitoring systems. We may not be paying attention to them, but we assume that somebody is. And we assume that we'll hear the alarms if something dire were to happen.

Three years before the Boulder floods, researcher Rebecca Morss and others from the National Center for Atmospheric Research interviewed twenty people in the Boulder area who are tasked with sounding the alarms (literal and metaphorical) when we are in danger of a flash flood. These were weather forecasters looking at rainfall data, public officials who have roles in emergency management, and television and radio news directors and meteorologists who communicate the information about the risk to the public. All of these people knew that there was potential for a catastrophic flash flood in Boulder. As one public official put it:

> We're all sort of wondering when the next big one is going to come, and knowing that it's inevitable but not knowing particularly when, because it could be next year or it could be in 50 or 100 years.

Forecasters, public officials, and broadcasters have to assess the threat of a flash flood and make decisions about warnings quickly, while there is still a lot of uncertainty about whether the flood will happen. They reduce the uncertainty as much as they can (by checking data from multiple sources, for example), but they can't eliminate it. "You've got to act before you have all the information you need. Or all the information you wish you had," said one weather forecaster. If they wait too long to issue a warning, to cut into a television program with a jarring alarm, or to evacuate an area of town, people might die. If they wait until the flash flooding is certain, it is happening, and it's too late for a warning. If they are too hasty and issue too many warnings for floods that never materialize, they risk losing credibility. People will tune out the alarms. Said one public official, "There is a hesitancy to pull the trigger . . . especially when it comes to evacuating a busy residential and retail area. You don't want to do that prematurely . . . so uncertainty is going to pull you apart in hesitancy." But uncertainty is unavoidable. The people who sound flash flood alarms must make decisions despite it.

■ ■ ■

Forty-six years before I was gawking at the churning water in 2013 and wondering how I should react, Gilbert F. White was taking steps to protect Boulder from such an event. White didn't know when floods would hit, but he did know that there was a 100 percent chance that they would hit eventually. That was enough for him to take action to decrease flood danger. In 1967, he told the City of Boulder that the hazard near Boulder Creek at the center of town was growing. The risk to property and life would continue to increase with development in the floodplain, said White. Unlike many engineers at the time, White advocated for adapting to the flood hazard rather than building dams and levees. He knew the floodwaters would always win. Without his efforts to help plan the town, the flooding in Boulder could have been much worse. He knew floods were going to happen eventually, and he got the city to react before residents were in harm's way.

White didn't get hung up with chaos and uncertainty. He took action even though no one can say when this type of disaster might strike. That type of planning is not the sort of hero's journey that we expect from our narratives of American innovation. We tell stories of last-minute saves — those who protect a town or a president or the world at the eleventh hour, when our hero confronts the enemy directly. We have a tendency to undervalue future risks. Biological instincts tell us to take action when the threat is clear and present. That's the type of story that keeps an audience engaged. It's emotional risk assessment. Careful planning that takes into account the potential risks according to the analytical risk equation isn't going to keep anyone on the edge of their seat. Perhaps that's why we've been slow to make plans to deal with our warming world — to reduce the risks and stop the warming. The threat isn't right in front of us. The risk equation doesn't tug at our emotions.

We can't entirely stop most natural hazards. As one of the broadcasters said to Morss, "There's only so much any city can do . . . and nature will win ultimately."

But we can make ourselves less vulnerable. When it comes to potentially catastrophic natural events such as earthquakes, landslides, tornadoes, hurricanes, avalanches, and flash floods, the half of the risk equation that we control is the consequences. Some of these events, like hurricanes and heat waves, are changing their character due to climate change, a compounding factor we do control. Decreasing the consequences means building outside the floodplain in the case of flood risk areas. It means engineering for earthquakes in areas along the San Andreas Fault. It means building houses that use less energy from fossil fuels to quell climate change. It means doing our best to not make things worse.

■ ■ ■

When the rain ceased and the clouds lifted, the huge scars of landslides and slumps appeared on the foothills, the paths of mudflows tracked down their sides. For days, huge black military helicopters thumped through the sky shuttling stranded residents of mountain towns down to the plains. There were no passable roads into or out of the mountains on the western side of Boulder County in the weeks during and after the storm. Once the rain stopped, earthmoving equipment headed into the mountains to reconstruct roads.

I walked through North Boulder and saw people shoveling grit out of basements and garages. A few unfortunate people were shoveling it from their living rooms. A crew of energetic volunteers wielding shovels roamed town, helping anyone in need. FEMA trucks were on the roads that were passable. Fleets of trucks from disaster recovery businesses were roaming too, their license plates from all over the country. Piles of muddy carpet lined the streets in flooded areas alongside soaked cardboard boxes of discarded treasures pulled from basements. I wandered these streets with a touch of survivor's guilt. None of my immediate neighbors were affected, and neither was I. Before the floods I did not realize that we were on a high point. I did not notice the efforts that had gone into creating drainage ditches when our area, a former drive-in movie theater, was turned into a subdivision. I had thought a low area in a nearby park was a sunken garden, not a space designed for floodwater. When it filled with water during the storm, children with small inflatable boats took the opportunity to call it a pond.

I had lugged a few vulnerable boxes to higher ground from my garage on the first day of the rain and then checked the garage every day with anxious concern. Some water came in, and I'd restacked plastic bins and moved bicycles around it, but the amount of water was nothing compared to what many homes suffered. I opened one particularly heavy box that was hard to move, wondering what was in there, and found it full of copies of a children's book I had written several years earlier called *Catastrophic Colorado*. The book highlights the science of some of Colorado's most exciting natural hazards. There are sections about tornadoes, avalanches, lightning, wildfire, and of course, flash floods. Amused by the irony, I left the box where it was, assuming that if the books were flooded, it would be a suitable end to them. Plus, it was too heavy to move.

I wrote that children's book because Colorado is chock full of natural hazards. This state has a piece of the Great Plains, a piece of the Rocky

Mountains, and some southwestern desert too. The diversity of environments leads to a diversity of disasters. For example, tornadoes, common to the plains, are not at all common in the mountains. Avalanches, common to the mountains, are unheard of on the plains.

"Everything burning to the ground must not seem like the worst-case scenario anymore," my friend Kevilyn — referencing Colorado's recent wildfires — speculated from the safety of her home in upstate New York, "just one slice of a worst-case scenario pie."

Worst-case scenario pie. I was down to popcorn and oatmeal, and a river flowed between home and the grocery store, so pie sounded tasty. A worst-case scenario pie seemed to sum up my experience in this disaster-prone state. Some areas of the country specialize in a particular type of disaster — Tornado Alley, for example. But Colorado has a diversified portfolio of disasters, which virtually guarantees that Colorado will consistently outperform less diversified states. I know nothing about investing, but it seems that Colorado would be a relatively safe bet for those who seek a consistent rate of disaster. On a more practical note, Colorado's high disaster rate resulted in some of the highest home insurance rates in the country as of March 2015.

It is possible that the worst-case scenario pie allows us to be more calm about any single disaster. With so many things to fear, there is only so much fear to go around. Andrew Revkin, journalist and blogger, has called this a "Finite Basket of Worry." Once the worry basket is full, says Revkin, people don't have room to worry about additional things. They prioritize worries, and only certain things make it into the worry basket. He was referring to climate change and the fact that people are not necessarily able to worry about the plight of the polar bear if they are worried about their heart conditions, their children, getting laid off, and other such concerns. But perhaps in an environment where there are so many natural hazards, there is only so much worry that can be associated with each. As the floodwaters rise, we exclaim that at least there isn't a risk of wildfire. Shifting one's perspective and contemplating how things could be worse is a path toward happiness, according to the Dalai Lama's book *The Art of Happiness*. The "it could be worse" perspective makes things seem better. Our worst-case scenario pie tastes sweet.

Psychologists refer to a similar prioritization as Maslow's hierarchy of needs, which can be found as a diagram in any introductory psychology textbook. It bears a striking resemblance to the old food pyramid from the USDA, although the categories are different. Our basic needs — shelter, food, and water — sit at the base of the pyramid. This is what we need to take care of first. Farther up are friendship, romance, and cognitive needs. On top of the hierarchy of needs pyramid, in the region reserved for dessert in the old food pyramid, is self-actualization or personal growth. Just like the food pyramid, Maslow's hierarchy of needs has been renovated over the years since Abraham Maslow made it in the mid-twentieth century, and the concept has its proponents and critics. Some say it is too rigid, that there is actually more mixing of priorities, that our needs change depending on what is motivating them. But whatever the nuances, the concept that some needs are of a higher priority than others puts worries about natural disasters and climate change into context. Most of these disasters disrupt the base of the pyramid, affecting our food, water, and shelter. Yet if they are not happening, are not perceived to be happening, or are perceived to be unlikely, then people move freely farther up the pyramid to fall in love or be creative or grow as a person. Problems brewing in the foundation of their pyramid are not a priority until the pyramid crumbles.

■ ■ ■

Growing up in Massachusetts, I thought of hiking as walking through calm and gentle woods. I would go out walking without a backpack or water, with the knowledge that eventually I'd run into a 7-Eleven or a friend's house. The places left to nature are surrounded by civilization. Even when Henry David Thoreau lived alone in the wild near Walden Pond more than 150 years ago, a train would rumble his peace. He could wander into town to dine with his mother should he choose. When I moved to Colorado, I learned that a day hike could be deadly without preparation and planning. I learned that one must pack the Ten Essentials (map and compass, sunscreen, warm clothes, flashlight, first aid kit, matches, food, water, emergency shelter, and a Swiss Army knife). One must leave information about the location of the trailhead and expected time of return with a friend who would call Search and Rescue should you fail to return. There were countless places to get lost in the West, to disappear, to die. You could die from drinking contaminated water, from not drinking water at all, from extreme heat or extreme cold, altitude sickness, bear attacks, mountain lion attacks, drowning, suffocating in an avalanche, or being zapped by lightning.

The potential for harm in Colorado made me look at nature differently. This is a place where rushing water may not allow safe passage, where ice may not be thick enough to cross, where animals may not live and let live. And this is a place where talk about weather is not neutral chitchat; it is talk about risk and the dangers we face.

Far from mountaintops, in Front Range cities, we live on islands within the wilderness. Occasionally that wilderness creeps into the urban islands we've built — a bear is spotted on a Denver street corner or a blizzard incapacitates the city. But in Boulder, these intersections between the human-built environment and the wild seem to happen more often. Some of this is because Boulder is a much smaller city than others in the Front Range, such as Denver and Colorado Springs, but it's also because there is an abrupt line between the city and the wild. This line, on the city's western edge, is defined by altitude. It was called the "blue line" in 1959 when Boulder voted to limit land development in locations above 5,750 feet in elevation. Today it looks like manicured landscaping abutting jagged foothill rocks, fences that separate backyards from wilderness, and irrigated green soccer fields adjacent to dusty prairie dog towns. It's not uncommon for deer, bears, and coyotes to cross the blue line and wander into town. It's very common for people to cross the line to wander into the wild. Streets end. Trails begin.

Travel brochures boast about Boulder's proximity to wilderness — the pristine nature-y definition of the word. But when the floodwaters were tearing through town, the antiquated nineteenth-century definition of wilderness seemed more apropos — as Charles Kendrick described San Francisco in 1906, "the wilderness of ruins was beyond words."

Then again, perhaps there is a third definition of *wilderness* — one that speaks to the wildness and power of nature, the force that can make our attempts to control the land seem futile, reminding us that we exist at the whim of wilderness. This is a scary version of wilderness, one that is highlighted by TV shows about the world's scariest predators, and one that can turn a trip to a national park into an anxious and fearful ordeal. We place more value on wilderness than we did a century ago, but perhaps we also fear it more when we are not in control. Like an earthquake, a flash flood reminds us that we are vulnerable in ways we don't often consider.

■ ■ ■

Six months after the flood, there were a few low single-story homes in my neighborhood that were uninhabited, debris from the flood stuck in the fences indicating the level where water once flowed. Other houses were in the midst of repairs. The creek babbled innocently beside raucous construction machinery. A nearby elementary school was open once again, the layer of sediment and possible contaminants stripped from the playground; carpet, grit, and crayfish removed from the classrooms; and all the sopping books taken out of the library. The bike paths that follow creeks, which were originally designed as flood control, were rebuilt.

Rebounding has not been quick in other areas of the Front Range affected by flooding, such as the town of Lyons, twelve miles to the north of Boulder, which suffered extreme damage. Looking at a Google map of Lyons in 2013, one could see two creeks coming down from mountain valleys into the west side of town. One could see waterways leaving on the east side of town. But in the town itself, Google Maps did not indicate any waterways. The creeks were rerouted as the town grew, with homes built in the floodplain. During the September 2013 rainstorm, the creeks swelled, they filled all available space and were no longer contained in their prescribed paths. Six months after the flood, part of Lyons was still a ghost town. The water carried sediment that abraded home siding, knocked trailer homes off their bases, ripped off windows and doors, and toppled trees. Construction permits were posted on facades alongside the X diagrams left by rescuers — each quadrant indicating an aspect of the home and its occupants, the safety of the structure, the number of people rescued. Gawking at all the building permits, I wondered whether I would choose to rebuild in the floodplain when there is the possibility that this can happen again. One home under construction was covered in plastic house wrap from the home improvement warehouse Lowe's; the store's slogan, NEVER STOP IMPROVING, covered the home like a command for the neighborhood.

Many people had no choice but to repair their homes. A home is the most expensive thing most of us ever buy. Walking away from it is not economically possible for many people. In the end, a buyout program allowed a handful of those with severely damaged homes to walk away with a check, but it took more than two years, and during that time many people were both homeless and paying the mortgage on homes that had drowned.

■ ■ ■

"We want to know — what did we just survive here?" meteorologist Kelly Mahoney said at a workshop for Boulder science teachers about rainstorms and flooding held in response to the disaster. Mahoney's area of research is extreme precipitation events that cause flooding. She works at the National Oceanic and Atmospheric Administration (NOAA) in Boulder. Until September 2013, she thought she'd spend her career studying records of past rainstorms and floods. Now she can focus on the deluge of data collected during this storm. Mahoney is one of many researchers assessing the probability that this type of storm could happen again.

Was it a one hundred–year flood? There are no simple answers, Mahoney said. Some creeks overflowed their banks more than others. Because we were in a drought when the storm hit, there was lots of space to fill in creeks and reservoirs. The flood could have been much worse if waterways had already been full.

We do know that the storm was a one thousand–year rainfall event, Mahoney explained. This is determined by comparing the amount of rainfall with annual exceedance probability estimates — the probability of a given amount of rain occurring in a given period of time. A one thousand–year rainfall event means that there was a tenth of a percent (0.1 percent) chance that a rainstorm like that could happen next year or the year after. It's not impossible, just very unlikely. It's about as likely as winning eight hands of blackjack in a row, which I'm told is extremely unlikely, but it could happen, and it does.

Extreme rain events have happened in Boulder before, said Mahoney. In May 1894, more than eight inches of rain fell during one storm. In September 1938, a storm dropped ten inches. And in May of 1969, there were more than thirteen inches of rain during a storm. As she rattled off the dates of these storms and of local floods, our seventeen inches of rain in September didn't seem like such an aberration.

What was unusual was that the storm occurred over a large geographic area. Often, flash floods happen in this region when a thunderstorm stalls over one valley. The storm drops all its water there, causing one creek to flood. This storm was unusually large and wet, flooding every valley in the area. It was not the typical isolated thunderstorm. Warm, moist air flowed into Colorado from the tropical Pacific and the Gulf of Mexico. When it hit the Rocky Mountains and was forced upward, the air let go of the moisture.

Bob Glancy of the NOAA National Weather Service described how forecasters saw the rainstorm and floods. He reminded us that weather models had predicted that the first week of September 2013 would be hot and dry. It was. In the days before the raindrops started to fall, Adam and I took a couple of end of summer hikes in the hills above Boulder. We set out early in the morning to beat the heat of day, and it was still too hot and dry. The trails were hard and dusty. Mila panted nonstop at my side in her thick, fluffy coat. We were in a drought. A week later, we were swamped.

The models showed that we were in for a change — the weather would turn wetter and cooler, they predicted. Models indicated that there would be a rainstorm over the region that would stick around for days. That is what happened, but the part that the weather models did not get correct was the rainfall total. The models indicated two to four inches of rain, which is a large amount for September, one of this region's drier months, but nowhere close to what happened.

"Humans and computer models are reluctant to forecast something that hasn't happened before," said Glancy. Weather models tend to fail in rare events. Even though there is a history of extreme rainfall here, it is so rare, so unusual, that models don't see it coming. We know that this type of event has happened before, but computer models aren't equipped with the evidence to say that this could happen again.

The rainstorm was normal in the sense that we know this type of low-probability event occurs in our region. But that doesn't necessarily make it natural. We know that climate change is increasing the chances of heavy downpours globally. It's also increasing the frequency and intensity of drought in some areas. I hadn't really thought of those two occurring in the same place, but we were in a drought before the heavy downpour. All weather events are now happening in a world with warmer average global temperatures. Human-induced climate change contributed to this weather event just as it contributes to all weather events now, but could it have been the cause?

A new branch of climate science known as "attribution" is trying to figure out whether specific extreme weather events are impacted by climate change or if they are due primarily to normal chaos in the atmosphere. A couple of decades ago, this seemed impossible, but today the science of attribution is attempting to figure out the whodunit of extreme weather. The boundary of our understanding has shifted.

"Now it is widely accepted that attribution statements about individual weather or climate events are possible," wrote the editors of the first report that explained extreme events in the context of climate change, published in 2012 as a special report of the *Bulletin of the American Meteorological Society*. Since then, an edition of the report has been published each year, examining events that occurred during the previous year and teasing apart the roles of climate change and chaos.

After an extreme weather event, many people ask, what were the chances of that? Unlike most of us, who shake our heads and leave the question unanswered, the scientists who study attribution calculate the odds of the extreme weather event to quantify the influence of climate change on the event. They rely on probabilities. For example, a Texas heat wave is much more likely today than it was four decades ago.

The science of attribution is a new one and there is much to learn, but it's an important step toward understanding how climate change is affecting people on earth. We long to know if extreme events will change in the future, and we want to know if what is happening is normal. Chaos has always played a leading role when it comes to extreme weather, but now climate change is also playing a role. Human-caused climate change had a clear influence on the extreme heat events examined in the 2014 report, but the role of climate change on extreme precipitation events was less clear.

At NOAA's lab in Boulder, Martin Hoerling and colleagues analyzed the Colorado rainfall for the 2014 report, looking into whether the likelihood of extreme rainfall in the region has increased due to climate change. They used modeling to compare the probabilities of extreme rainfall events during the last part of the nineteenth century with recent events and found that the probability of a rainstorm like we saw in 2013 has likely decreased due to climate change. So even if the likelihood of extreme storms has increased worldwide, the probability of this happening in our little corner of the world is actually lower than it used to be. And yet it did happen. Perhaps chaos in the atmosphere outshone climate change.

However, three miles up the road at the National Center for Atmospheric Research, Kevin Trenberth and colleagues were looking at the main source of the moisture that produced the Colorado rainfall: the Pacific Ocean. They found evidence that the storm wouldn't have been as large or disastrous if it were not under the influence of climate change. The surface of the tropical Pacific was unusually warm at the end of the summer, which caused more water to evaporate. The moist air flowed into Colorado, making the atmosphere more humid than ever recorded in September. As climate changes, the sea surface is warming as well, which causes more water to evaporate. Because what goes up into the air must come down, more evaporated water means that there is more water that can precipitate. Even if the probability of a large rainstorm may be decreasing as climate changes in this region, the amount of water in the 2013 storm is related to warmer climate.

■ ■ ■

As a geologist, I prefer the perspective of the sedimentary aftermath to the drama of the storm itself. It was a relief when the rain stopped, the floodwaters subsided, and Adam, Mila, and I could wander a mile or so west for some forensic-style detective work in the sedimentary geology along Fourmile Canyon Creek at the end of September.

With water down to a trickle, the sediments made a nice path. There were occasional stream crossings where water was still flowing fast. Mila jumped into the water and was carried downstream as she doggy paddled furiously. Adam and I mounted a quick rescue operation and scooped her out. The water looked clean at that point, unlike the muddy slurry during the floods. The mud had traveled out of the streambed and onto the streets of the neighborhood, into the basements of homes, or farther out onto the plains. Islands of sediment surrounded the cottonwood trees, which a few weeks before had been located in grassy underbrush. Many of the trees had lost branches but were still alive. They were adapted for this.

In some areas, I could see the cross sections of sediments deposited during the flood — several feet of sand and gravels. At the base of these deposits were larger rocks — rounded cobbles roughly the size of my hiking boots. Heavy rocks like these were all that could stay put during the rush of water. Progressively smaller sediments were stacked in layers above the cobbles, deposited as the water became calmer. This is a typical sequence of sediments deposited by a flood: large cobbles at the base grading up to smaller sediments above.

In other areas, erosion around the bends in the stream had exposed older layers of sediments that lay beneath arid grasses and prickly pear cacti at the creek's edge. The older sediments, newly exposed, had the same grading: upward from large cobbles to smaller sands. They were deposited during a previous storm, evidence that Fourmile Canyon Creek has a history of flash floods. There is a subdivision on top of the sediment layers, a collection of homes built in 1993 that narrowly escaped the waters of this flood.

The line between that housing development and nature is distinct, where bright green turf meets dry prairie grasses. Adam pointed out fake cactus sculptures sitting in a green grass yard. We scratched our feet on real cacti as we stumbled through on the nature side of the line. The single-family homes, each about five thousand square feet and worth about two million dollars, are at the edge of the one hundred–year floodplain. They sit within the five hundred–year floodplain. The line between the natural and built environment is distinct, but it's not a barrier. The wilderness can get in.

In 1969, Gilbert F. White, witnessing people moving into floodplains across America, asked, "How does man adjust to risk and uncertainty in natural systems?" Today people still decide to live in floodplains. How have they adjusted to risk and uncertainty? Do they know they are living within the wild?

I find one answer to White's question by looking into the permitting process for a new apartment complex within the one hundred–year floodplain next to Fourmile Canyon Creek. Instead of adjusting to risk and uncertainty, the developers planned to squelch it with creative engineering. The builders proposed to the city planning board that they would engineer a way out of flood danger. Their plan was to remove the apartments from the floodplain by grading the site and the area along the creek.

They showed the planning board a map of the location as it was — in the high hazard zone of the one hundred–year floodplain. Then they showed the planning board a map of the location as an island of safety entirely surrounded by the floodplain. The plan was approved by FEMA and the City of Boulder.

But many people live in the floodplain without reengineering the wild. When Boulder residents were surveyed about flash flood risk several years before the 2013 flooding, about half of the people surveyed who lived in the one hundred–year or five hundred–year floodplain didn't know that they lived in the floodplain. Others thought their house was in the floodplain, but they were wrong. They just didn't know. It had been decades since there had been a flood. No one who lives in the floodplain had ever seen one.

I asked a resident of Lyons whose subdivision narrowly avoided the floodwaters what he thought about the catastrophe. We were standing next to a scarp the flood had carved into the edge of his neighborhood. His small child was playing in the nearby sea of gravels left by the flood. He told me that if we could now check floods off our natural disaster life list, then he was relieved. This is not uncommon among people who survive disasters. Those brushed by disaster once often feel that they are less likely to encounter another disaster, according to psychologists. The misunderstanding that a one hundred–year flood only happens every one hundred years serves to enhance this feeling. A one hundred–year flood has a 1 percent chance of happening each year, but he was betting that this wouldn't happen again in his lifetime. The cliff beside him quietly betrayed evidence of a previous flood, with large cobbles grading up to sands. These older flood deposits had been topped by grasses and, more recently, homes, including his. The cliff's sediments offered an alternate perspective — one both comforting and unsettling — that it is certain that flash floods have happened before and will happen again, yet it is uncertain when a flood will strike next.

Chapter Eight

Reply Hazy. Try Again.

An oversized plastic billiard ball filled with murky liquid sits in my hands. Inside float twenty possible answers. I close my eyes and silently ask a question. I turn the ball, revealing a little window on its underside, and wait for an answer to emerge through the haze.

This is a Magic 8 Ball, a toy marketed by Mattel. But it is so much more than a toy. The Magic 8 Ball is a tool for prediction. Developed in the 1940s by a couple of guys, one of whom had a fortune-telling mom, the Magic 8 Ball has been providing answers for decades. Sometimes our questions are small ("Should I have another cup of coffee?"), and other times they are larger ("Will I ever find a job?"). Regardless of size, our questions for the 8 Ball are usually about the future.

We rely on predictions of the future. We want to know whether next week's rainstorm will cause a colossal flood. We want to know how much climate will warm in the coming decades and how that will change heat waves, drought, hurricanes, and the way we live in the world. Some aspects of our planet — earthquakes, for example — defy prediction. But weather and climate can be predicted. All we need are highly complex mathematical models. In the case of climate, projections into the future are made with the help of earth system models, which use a headache-inducing volume of calculations to simulate our planet. The world's most powerful computers — roomfuls of whirring metal boxes, wires, and chips called supercomputers — perform the calculations.

Since pondering the future has become something of a hobby for me, and since I cannot afford a supercomputer or an earth system model, I stopped by a local toy store to acquire my very own Magic 8 Ball. Dodging a dozen or so yo-yo artists sending spinning projectiles in all directions, I wandered the shop, looking for the ball. The toy store overflowed with bright boxes and plastic things that made noise. How I thought I'd be able to find an unassuming black ball in that place, I'm not sure. Perhaps the Magic 8 Ball was an outdated relic and no longer sold. After a scan of the chirping, clanking merchandise and cartoon-clad packaging, I gave up and timidly asked the woman behind the counter if they might have a Magic 8 Ball somewhere. She pointed directly behind me, and I turned to see a number of colorful boxes on the shelf over my left shoulder. The toy manufacturer had cloaked the black ball in packaging that camouflaged it among the whimsy.

I'm not interested in the 8 Ball's value as a toy. I am embarking on a serious scientific experiment. I want to know how well this bulb of plastic can predict the future. To assess the Magic 8 Ball's forecasting talents, I will compare its predictions to that of supercomputer models. I will test its ability to predict whether ice will still fill the Arctic by the middle of this century.

Sea ice is the frozen seawater that bobs at the surface of the Arctic Ocean. It waxes and wanes with the seasons, growing all winter and melting all summer, but never completely vanishes. It's the type of ice that you typically see polar bears wandering atop in photos. Scientists who study the future of sea ice project that because our planet is warming, the Arctic will have no sea ice in late summers by the middle of the century. Marika Holland at the National Center for Atmospheric Research and many other research teams work with some of the world's most advanced models of the planet to develop these projections of future sea ice. I want to see what the Magic 8 Ball has to say.

Because this is a scientific study of the 8 Ball's predictive abilities, I shall explain my methods (as is customary in scientific research) so that you, fair reader, may replicate the study if you wish. I will perform multiple trials, asking my 8 Ball the question "Will the Arctic Ocean be ice free by the middle of this century?" five times to see whether there is variability in the ball's answers.

Trial #1: "Reply hazy. Try again."

Research Notes: If I try again, will that be the second trial or an addition to this trial? In either case, the "Reply hazy" answer is a nod to uncertainty, and uncertainty is critical to acknowledge. The 8 Ball must be on the outskirts of what is known.

Trial #2: "Cannot predict now."

Research Notes: Is the 8 Ball on strike? Why can't it give an answer? I suppose this trial is somewhat of a wash. This is why scientists use multiple trials in experiments.

Trial #3: "Signs point to yes."

Research Notes: Our first indication that the 8 Ball and the climate scientists are on the same page, both predicting an ice-free Arctic. Again, there is an allusion to uncertainty here as the answer is not a straightforward "yes."

Trial #4: "Outlook not so good."

Research Notes: If the outlook is not good for an ice-free Arctic, does that mean there will be Arctic ice? Or is the 8 Ball editorializing with its use of the word *good*, implying that the situation will not be good for those who rely on the ice, such as polar bears and the Inuit?

Trial #5: "Concentrate and ask again."

Research Notes: The 8 Ball is avoiding the question.

In conclusion, I'd like to point out that most of the 8 Ball's answers in this thoroughly comprehensive scientific study were not really answers at all. They were on the fence, uncertain, wishy-washy statements that didn't really leave us with anything definitive to go on. Come to think of it, that's what frustrates many people about science.

■ ■ ■

The language of uncertainty — words like *maybe*, or *likely*, or *possibly* — can be especially perplexing to people who are trying to make concrete plans. As people attempt to understand how climate change will affect their lives and the lives of their children, they can get hung up with uncertainty. Uncertain language can make it hard to know whether the current state of our planet is normal, and if it's not normal, whether we should do something about it.

People making decisions about whether to buy a home on a fault line or an eroding coast also want to know what *will* happen instead of what *might* happen. People deciding what fish to order for dinner want to know which species on the menu *will not* go extinct instead of which species *might not* go extinct. We humans want to know what's going to happen, yet scientists keep giving us the same 8 Ball answer of "Reply hazy. Try again."

Recall that there are two main reasons for this language. One reason is that our chaotic planet doesn't allow for complete certainty in many realms, such as how quickly the coast erodes or when an earthquake will happen. The other reason is that scientists need to communicate what is known, what is not known, and what is in between. They need to be clear about whether we know that a fish will invade, whether a storm was affected by climate change, and when, exactly, the Arctic will be ice free. And if we don't know for sure, then it's best to be transparent about how sure we are.

The Intergovernmental Panel on Climate Change (IPCC) develops reports every few years that summarize what we know about climate change and where there is uncertainty. In this case, the uncertainty is mostly due to being on the edge of what's known. Projections of future climate change look at averages and are based on a budget of energy coming to the earth from the sun and energy going from the earth out into space, so they are less affected by chaos than projections of future weather. The climate projections for the future that the IPCC describes are based on models that can handle the complexity of our planet's climate much better than my Magic 8 Ball but cannot predict the future with absolute certainty.

We know more about why climate is changing than we did twenty years ago, which is why the IPCC statements about the cause of recent climate change have become less uncertain and more definitive over time. The 1995 report stated that "the balance of evidence suggests a discernible human influence on global climate," which sounds like an 8 Ball answer to me. The 2001 report was much more certain, stating that "there is new and stronger evidence that most of the warming observed over the last 50 years is attributable to human activities." By 2007, the statement read that "most of the observed increase in global average temperatures since the mid-20th century is *very likely* due to the observed increase in anthropogenic greenhouse gas concentrations." And by 2014, the statement read, "It is *extremely likely* that human influence has been the dominant cause of the observed warming since the mid-20th century." In two decades, we have moved from "evidence suggests" to "extremely likely." There is much less uncertainty about the reason earth's climate is changing.

Moving from less certain to more certain is part of the process of science. In two decades, we have moved the boundary of our understanding about our planet's climate. Scientists are still working on the boundary, but the idea that humans are responsible for climate change is no longer on the outskirts of our understanding.

■ ■ ■

Our understanding of how climate will change in the future is uncertain in large part not because of scientific uncertainty or chaos but because we don't know when humanity will stop using fossil fuels or whether we'll keep polluting the skies with heat-trapping gases. According to the IPCC *Fifth Assessment Report*, published in 2014, average global temperature at the end of the century will likely increase between 0.5°F to 3.1°F (0.3°C to 1.7°C) if we take action quickly, reducing the use of fossil fuels rapidly by 2020. According to experts, this would keep the amount of warming below dangerous levels.

Unfortunately, most of the actions that nations are taking today are not enough to reduce emissions drastically, so we will probably not meet the 2020 goal. Let's say it takes us a couple of decades longer to figure out how to drop greenhouse gas emissions worldwide. If we start decreasing emissions rapidly in 2040, we will be looking at 2°F to 4.7°F (1.1°C to 2.6°C) more warming. But if we give up on reducing emissions entirely and maintain or increase the level of greenhouse gases we are currently pouring into the atmosphere, our planet will likely be 4.7°F to 8.6°F (2.6°C to 4.8°C) warmer at the end of the twenty-first century. That's between three and six times as much warming as we had during the previous century and is above the 2°C threshold often cited as the limit of our ability to adapt. Summers in Chicago would be like summers in Dallas are now. Summers in Detroit would be like summers in Little Rock. New Hampshire would be more like North Carolina if we keep the rate of greenhouse gas emissions high.

■ ■ ■

When my 8 Ball is not giving uncertain answers, its predictions are still not solidly yes or no. You may get an answer like "Don't count on it," which isn't a definitive "no" but is on that side of the bench, or "As I see it, yes," which imparts an almost self-deprecating insecurity.

The 8 Ball's nondefinitive answers can be aggravating. For a while, I carried the thing around in my purse. I haven't needed to make any big decisions lately in my relatively humdrum life, but there are always little decisions. I'm not great with these, so I was ready to welcome any assistance the 8 Ball could provide. After a couple of days of consulting it, I found out that it is, in fact, less decisive than I am. For example, I took it out at Starbucks as I considered my coffee choices. "Grande drip coffee with hazelnut?" I asked it. The 8 Ball hemmed and hawed, holding up the line of under-caffeinated patrons.

When the Magic 8 Ball gives an answer like "Most likely," the questioner is left to interpret just how likely that is. Not so with climate predictions cited in reports of the IPCC. Levels of confidence and likelihood are italicized in the reports because the authors defined those terms. In 2007, the IPCC developed numerical equivalents for words such as *likely* so that people would have an understanding of the odds. Likely events, they decided, are ones that have a 66 to 100 percent probability of happening, while very likely events have a 90 to 100 percent probability of happening and virtually certain events have a greater than 99 percent probability of occurrence.

The 2007 report states that "continued greenhouse gas emissions at or above current rates will cause further warming and induce many changes in the global climate system during the 21st century that would *very likely* be larger than those observed during the 20th century." The report also says, "It is *very likely* that heat waves will be more intense, more frequent and longer lasting in a future warmer climate." The 2014 report concurred and added that it is *virtually certain* that there will be more hot temperature extremes. Given that there is 99 to 100 percent probability, I'd say you could count on it. People who buy lottery tickets would love such odds. However, when psychologist David Budescu and his research team asked people to assign a probability to phrases like "very likely," the survey respondents consistently thought the probability was lower than it actually was.

Unlike my random Magic 8 Ball predictions, the IPCC predictions of future climate are based on earth system models, mock planets made entirely of math. With models, scientists can ask questions about how the earth might change if we add greenhouse gases or if we pave over its surface. They can estimate what the world might be like in the future — what might happen. The calculations take into account the intricate processes of water, carbon, land, air, ocean, and ice — aspects of our planet that affect climate.

But even with all that complexity, a model never perfectly describes what it is representing. Scientists have a saying: "All models are wrong but some models are useful." I've often heard it muttered in response to complaints about the accuracy of scientific models. To make a perfect model — a perfect model of the earth, for example — we would need an exact replica of the earth. That's just not practical. We don't have room to store an extra planet, and it would take an extraordinary amount of funding to maintain both versions of earth. A virtual earth run by a supercomputer is a useful approximation. As long as we understand how it is wrong, we will know how it is useful.

Supercomputer models of the earth are group projects. Hundreds or thousands of researchers each contribute a portion of the equations necessary in order for the model to accurately describe our planet. And because scientists are working together, each researcher needs to communicate what he or she is sure of and what is uncertain. Those areas of uncertainty become areas for future research. And scientists using the model are aware of what it can do and its limitations. For example, if you wanted to project the number of butterflies on earth in the year 2100, you may be dismayed to learn that although earth system models are good at simulating the climate in 2100, they are useless when it comes to simulating future butterfly populations.

How do we know whether a model is doing a good job of predicting the future? One way is to ask a model of earth to predict what happened to the climate during a time when we know what actually happened, for example during the twentieth century. The model — being made of just math and machine — has no idea what happened during the twentieth century, but we do. We were making weather observations around the world each day, which can be averaged to look at trends in global average surface temperature. When models were run to simulate twentieth-century global climate, they pretty accurately showed the amounts and rates of warming that we experienced. Earth system models that can accurately replicate twentieth-century climate are probably doing a good job of predicting twenty-first-century climate.

There is another way to know if predictions are correct. We can wait until the future and see if they were right.

It can't be common for people who predict the future professionally to make a time capsule, but that's what happened when the National Center for Atmospheric Research turned fifty years old, in 2010. Staff contributed suggestions for what to include. Among the 114 items was a model prediction of how much sea ice will still be floating in the Arctic in 2035, the year the time capsule will be opened. A time capsule is an excuse to think about the future and to consider the present as if it were the past. When future NCAR employees, and a few current ones, open the capsule, they will know whether the world is as they predicted.

■ ■ ■

Our projections of the future seem gloomy. Pollution, climate change, overpopulation, extinctions — we live in the age of worry. We're no longer sure it will be fine.

Living in the age of worry has led to the exponential rise in popularity of dystopian novels and films depicting a future where we are facing the most dire consequences of the poor decisions we have made. Perhaps we will have to live in a bubble to avoid pollution. Perhaps food will be so scarce that humans become cannibals or must 3D-print synthetic meat. Perhaps we will search for another planet after we've trashed our own. Yes, they are fictional futures, but they are based on our worst fears in the present: what could happen as our environment decays. British teen Megan Quibell sums up the equation that produces dystopic fiction from real environmental disaster in an April 2015 op-ed in the *Guardian* positing that authors of young adult dystopic fiction "look at all the scary uncertainties we're worrying about in our time and put them into a devastating future, building a (generally terrifying and evil) society around the disaster."

People deal with uncertainty about the future in different ways. Some overreact. Some hyperventilate or buy insurance policies. Others try to knock down the uncertainty as if it were the enemy. A Buddhist may suggest that it's better to get comfortable with uncertainty. That's what I'd suggest too, because when it comes to the science of the environment, we must make decisions despite the uncertainty. But some people use uncertainty to justify inaction, which can be far more dangerous than overreacting.

In a 2001 speech about why he refused to sign the Kyoto Protocol, a global first step in carbon dioxide emissions reductions, George W. Bush said, "no one can say with any certainty what constitutes a dangerous level of warming and, therefore, what level must be avoided."

For people who think that human-induced climate warming of any amount is risky, myself included, Bush's comment was like a doctor saying, "We don't know what size of tumor is dangerous, so we will not do anything about the one that is currently growing inside your body." But for those who see a small amount of warming as benign and a large amount as dangerous, there is a nebulous line somewhere that should not be crossed.

Several years after Bush refused action on climate change because of uncertainty, researchers defined that line as a global average of two degrees Celsius (3.6°F) warmer than the pre-industrial temperature. If we stay below two degrees of warming, humanity will hopefully avoid the most dangerous effects of climate change. We are about halfway there. As of 2016, Earth was 1.2°C (2.2°F) warmer than in 1880, near the start of the Industrial Revolution. It would take an enormous overhaul of technologies and the way we produce energy on earth for us to stay below two degrees. It's unlikely.

There are different perceptions of what the two-degree target means. Some view it as a threshold below which we are safe and above which marks catastrophe. Others see the benefits of limiting climate change to two degrees as outweighing the costs. And now there is evidence that two degrees Celsius might be too much warming. We might be facing catastrophic impacts even if we stay below two degrees, according to James Hansen, director of the Columbia University Climate Science, Awareness, and Solutions Program. In any case, we are going to cross this nebulous line. As the flash flood forecasters said, if you wait until you are certain, you have waited too long.

■ ■ ■

A few years before Bush refused to sign Kyoto, psychology researchers Sylvia Roch and Charles Samuelson were gathering nearly two hundred undergraduate students at Texas A&M University to participate in a study in which they would test how uncertainty impacts environmental decision making. The students arrived at the psychology lab in groups of six. Each was assigned to a computer, and each computer contained a game that simulated a situation where students would harvest natural resources. The natural resources in the game were nondescript. To provide a concrete example, I will describe the scenario in the game as a lake filled with fish. Each student at a computer was harvesting virtual fish from a virtual lake. Meanwhile, virtual people created by the game were also harvesting fish at the same lake. The students were trying to play the game as long as possible and get as many fish as possible.

Some students were provided with information about how many fish were in the lake and their rate of reproduction. Based on this information, they could make sure that they did not take so many that there would be no fish for the next round. But other students were provided with uncertain information about the fish in the lake. These students needed to also make decisions about the number of fish to harvest in each round, despite the uncertainty.

Roch and Samuelson had each of the students fill out a questionnaire that helped identify which students were cooperators and which were non-cooperators. Cooperators are concerned about the good of the whole group. Non-cooperators are not as concerned with the group as they are with themselves. They are individualistic or competitive. Out of the group of students, approximately half were cooperators and half were non-cooperators.

The researchers found that when there was uncertainty about the number of virtual fish, students started harvesting more fish than when the number of fish was known. Then, as the amount of fish began to appear limited, non-cooperators took more while cooperators took less.

"If environmental uncertainty is high, then the major challenge lies with one segment of the population: Resource users with non-cooperative social values," Roch and Samuelson stated in their conclusion.

I think we can say that George W. Bush's decision to not sign the Kyoto Protocol was motivated by non-cooperative social values. But he is not the only one who has used uncertainty to justify inaction. Donald Trump and a number of people in his administration have done the same despite the increased certainty about climate change impacts. One of the most common arguments used by climate deniers (i.e., people who are convinced that climate change is not a threat or is not happening) is that no action is needed because we are not sure what is really happening. Naomi Oreskes and Erik M. Conway make the case in their 2010 book *Merchants of Doubt* that a few individuals staunchly opposed to regulation have acted to create confusion about what we know about climate change, acid rain, tobacco use, and pesticides. One way they have done this is by framing scientific uncertainty as the unknown, breeding doubt even when there is a preponderance of evidence.

What is important to remember is that inaction is an action. Deciding to make no change is a decision. Deciding to not move a lighthouse from an eroding coast is a decision to let it fall. Deciding to not kill lionfish is a decision to let them invade. Deciding to not decrease the amount of carbon dioxide we are putting into the atmosphere is a decision to warm the atmosphere.

If everything were predictable, no one would buy a Magic 8 Ball. We'd know exactly how much earth will warm this century. The problem of climate change would seem without solution if nothing we did could change the outcome. If there is an upside to an uncertain future during this time of environmental consequences, it's that we are not destined to fail at our attempts to stop polluting and warming the earth. We are not destined to live in a dystopia.

The good thing about the uncertainty of human actions to reduce greenhouse gas emissions is that anything could happen. Unlike the rhinos inhaling ash, we have the power to make decisions that will help us be resilient.

Chapter Nine

Space-Age Improbable Possibilities

Probable impossibilities are to be preferred
to improbable possibilities. —Aristotle

I was handed a large ticket with "Capsule Number Six" printed on it. Capsule. I laughed out loud. Other tourists in line did not seem to find their capsule assignment as amusing as I found mine. The term made it sound like we would be getting inside giant Tylenols. I would soon discover that was not far from the truth. Capsule Number Six was a little white half lozenge.

We pilgrims to Saint Louis's masterpiece of modern design, the Gateway Arch, were shepherded into a narrow hallway. We were instructed to turn and face one wall. I did so, plastic ticket in hand. After a bit of a wait, long enough to start me wondering why I was there, the wall in front of us parted with flair, revealing a semicircle of white mushroom-shaped stools inside a small, white domed lozenge. My futuristic capsule had arrived, ready to transport me up a leg of the arch.

Yet this was an old version of the future, a relic of what people long ago thought we might become. A space-age vision, it is a slice of *The Jetsons* and a slice of Woody Allen's *Sleeper* with a sprinkle of Disney. It made me wonder how the idea of the future has been revised in the five decades since the arch and its capsules were made — where we thought we were heading then, and where we think we are heading now.

The architecture of the arch is partly a feat of midcentury technology and partly a feat of design made possible by that technology. We can do

Drawing of the capsule cars in which passengers ride to the top of the Gateway Arch. (Courtesy of the *Saint Louis Post-Dispatch*.)

amazing things with our advances, it says. Architect Eero Saarinen designed the arch, known formally as the Gateway Arch or the Gateway to the West, and construction was completed in 1965. The capsule train was added late in the project, after Saarinen's death in 1961. The arch, a national memorial, is a monument to literal advances in the past — to westward expansion and the pioneers whose covered wagons passed through.

From a distance, it appears to be a narrow rainbow with ends anchored in the flat land next to the Mississippi River. Close up, the arch has a shiny stainless steel exterior. And on the inside, it is fortified with steel and concrete. That interior is where the capsules carried us more than six hundred feet up to the top of the arch.

The walls of my capsule were rounded, which made the space appear deceptively large upon first glance. Ducking though the little sliding doors, I discovered it was tiny. I held my arms out to see if I could touch both sides. I wondered how five of us were going to fit and how long we would have to stay in there. Then I heard the crying child who was entering Capsule Number Five and I took a moment to feel fortunate for my seemingly calm and quiet capsule mates. The doors slid closed with small jerks and jumps like a prop in a grade-school play.

As we trundled toward the top in our train of eight capsules rising inside one of the curved legs, I was giddy with anticipation and could not sit still on my swiveling mushroom. Swivel left. Swivel right. Swivel left again. I couldn't seem to stop no matter how much this annoyed the other passengers. As if this was all ordinary, they perched placidly on their own white mushrooms. I swiveled. Don't they see what this is? I thought. This is not just a monument to the past. It's a vision of the future! Or at least, it used to be.

Attempting to avoid eye contact with my capsule mates, who were fast losing patience, I looked beyond them to the stark whiteness of this odd environment. There was the white domed ceiling, free of ornamentation, with rounded walls and a plain little door. No details to latch onto in this modernist orb. It was a bit like the future, I supposed. You might have a picture of the future that shows broad generalities, but rarely can you see the details of what's to come.

Much as eyes adjusted to the night sky are able to see more stars, I realized on a second sweep of the place that there were details to be seen: fingerprints on white paint, nose prints on the little glass windows in the door, and shoe marks on walls. This vision of the future was showing its age.

Because the current prospects for the future include such environmental gloom, I have become enamored of the glamorous space-age vision of the future. It was a time of limitless possibility. Today's future does not look like that. To see it as the white, orderly world of the capsule is to ignore both nature's chaos and the environmental calamities we have caused. Will our relationship with nature be as strained in the future as it is today? How can we again see the future with optimism, as a place we'd want to head toward, yet not ignore the environmental catastrophe that we are within?

■ ■ ■

At some point, scientific research shifted from predominantly a quest to understand nature and perhaps exploit it (through identifying species and rocks, mapping rivers and mountains, investigating human health, and discovering the chemical makeup of earth) to predominantly a quest to be in control of nature or to at least free ourselves from its constraints. We also started documenting how humans have changed nature. Today there are scientists who seek to understand nature, and long ago there were scientists who sought to control nature, but a shift in the ratios occurred. The Industrial Revolution, starting in the 1800s, may have spurred the shift. It led to technologies that allowed us to live a bit beyond the confines of nature. Evolving technology made it possible to eat food that had never grown from the earth, to have the temperature indoors be significantly different than it is outdoors, and to create buildings (like the Gateway Arch) in shapes that would never have been possible without creative use of steel and concrete.

A case in point: the Gateway Arch, according to some, was designed as an instrument to control the weather, with the power to reroute midwestern storms either toward or away from the city of Saint Louis. That's a myth. The arch is a monument, not a center for storm traffic control. But myths expose our desires and fears, and this myth speaks to both our desire to be in control of nature and our fear of nature.

Perhaps the greatest of existential dilemmas raised in the mid-twentieth century occurred when we developed the ability to leave earth. Human spaceflight brought into question what our ties were to this planet. Just as a relaxing vacation to Bermuda may prompt a vacationer to ask, "Why do I live in Cincinnati?," a trip to the moon may prompt us to ask if we humans really need earth. Do we need nature? What do we need?

Space-age visions of the future were often whimsical — at least the ones made for broad popular consumption. They depicted ways that technology might lead us to utopia. It wasn't necessarily a utopia that I hope to live within, yet I envy those who had these visions. They felt certain about the direction we were headed. They could see the upside to our separation from nature. I cannot.

Case in point: *The Jetsons*. First aired in 1962, this prime-time cartoon sitcom depicted a family of four living in 2062. Then, that was a hundred years into the future. As I write this, it is less than fifty years away. We are more than halfway to *The Jetsons* future.

George Jetson, Jane, his wife, daughter Judy, and his boy Elroy live up high at the Skypad Apartments. Machines cook meatloaf and redecorate rooms at the push of a button. Moving walkways transport the family members through the apartment. The technologies the Jetson family relies on are a major part of each episode. There are many complaints from the cartoon characters about the physical strain caused by button pushing. I suppose a hefty button from the early 1960s was more difficult to push than a modern button. Dialing an entire phone number was a strain on the dialing finger and could take all afternoon. Today, our need to push buttons and dial numbers has been greatly reduced. As a consequence, I no longer know anyone's phone number.

All buildings in this cartoon world were styled like the Seattle Space Needle, but only their tops are depicted, never the ground. I'm not sure this future world includes the planet's surface, which begs the question What's going on down there? Are there cities like Seattle at the base of these buildings? Is there a giant dump? Or have *The Jetsons* futurists allowed it to go wild, with humans living above and beyond the chaos and uncertainty of nature?

Somehow, son Elroy finds Astro the dog and brings him home. To me, dogs seem more connected to nature than humans. My dog sniffs and rolls around in nature on a regular basis. Yet Astro goes for walks on a treadmill that protrudes from the side of the Skypad Apartments, high above the ground. Where does he poop? Somehow the Jetsons' future includes no dog poop.

At some point, my exploration of the old future evolved from casual loafing around and watching cartoons into dedicated research. This often happens. I have an ability to take anything light and fun and analyze it.

The aim of my research is not to examine past visions of the future and discount them as inaccurate. It is instead to look at these visions as one way we saw our relationship with nature. The way I see it, climate change, other environmental calamities, and natural disasters are forcing us to re-define our relationship with nature. Thus, it's helpful to consider where we've been as we try to figure out where we are going.

I hypothesized that visions of the future either would portray humans as vulnerable to nature (wiped out by storms, disease, malnutrition) or would portray nature as vulnerable to us (extinction of species, polluted waters and air) — a dystopia in either case. According to my couch potato research watching old *Jetsons* episodes, neither hypothesis is correct. In the *Jetsons* future, nature is nowhere to be found.

"I figure you must be an outdoor man," says Mr. Spacely, George's boss, when assigning him the role of scout leader in charge of a camping trip. This episode, titled "The Good Little Scouts," in which George and Elroy go camping seems promising. Surely the Jetsons must encounter nature if they are heading into the wild.

But where do they go? The moon. Traveling 238,000 miles from Grand Central Station, George and the kids put on oxygen helmets and step out into Moonhattan, the developed urban center on the moon. They dismiss a yellow moon taxi and hike miles away to a campground. In an adorable combination of past and future, a scout named Arthur goes to the dark side of the moon to change the film in his camera. But the moon wil-derness becomes a scary place when George and a camper get lost. The moon was, and still is, quite wild. Yet it lacks characteristics of earth's nature — such as plants, animals, water, and atmosphere. Or perhaps my definition of nature is too narrow.

"You need a rest, Mrs. Jetson. Get away from all those buttons," advises Jane's doctor in another episode after she nearly has a nervous breakdown. The doctor wears a headband with a reflector on it. "Someplace close to nature," he continues. "One of those dude planets, maybe."

She and a fellow housewife who also needs a break arrive on a dude planet that does not appear to include anything close to nature. The rodeo bulls are mechanical. The horses they ride are mechanical. There is not a living creature aside from the dudes.

When vacationing on the planet of Las Venus, the Jetsons' car is crushed on arrival. But never fear. They give you a brand-new car at the end of your vacation, George explains to Jane. How kind. Is the crushed car recycled into a new car, or is there a giant pile of crushed cars on the other side of Las Venus? These practicalities are not of concern in a cartoon future.

. . .

During the 1964–65 New York World's Fair, which took place two years after *The Jetsons* first aired, more than seventeen million people saw the future. They took a fourteen-minute train ride past elaborate dioramas of future landscapes, moonscapes, and underwater worlds in the Futurama II exhibit by General Motors while a narrator described the possibilities.

I found a tattered guidebook for that world's fair in my local library. I imagined a happy fairgoer toting it about the Queens, New York, fairgrounds over fifty years ago, awed and inspired. Do you think he or she imagined that the little guidebook would wind up in a library two thousand miles away? That I, decades later, would handle it gingerly, treating it with the reverence and awe of a relic?

The guidebook described the New York World's Fair as "a circus and a classroom, a voyage around the world, a look back at the past, a peepshow into the future." I searched the book for the peepshow.

The designers of the exhibits speculated about the future without censuring whimsy. Perhaps it was the influence of Walt Disney that allowed outlandish possibilities to be shared with fair visitors. According to the guidebook, the first demonstration of atomic fusion, a "manmade sun," was showcased in Progressland, presented by General Electric and Walt Disney. In a guidebook illustration, young children look awed by a tabletop nuclear glow. I hope all visitors survived.

"If all the cars that will attend the fair were placed end to end, the traffic jam would stretch around the world 3½ times," proclaims a car advertisement in the guidebook. The idea of this traffic jam would not persuade me to buy a car or visit the world's fair.

The Futurama exhibit at the 1964–65 world's fair showcased the possible future in dioramas, explaining how humans will have tamed wild places such as mountains, space, the bottom of the ocean, jungles, and Antarctica. No timeline was given for these predictions.

"Remember how, back in 1964, man thought [Antarctica] to be frigid wastelands not suitable for habitation? And how they thought the tropics could never be tamed? How wrong they were!" reads a blurb about Futurama.

Gingerly turning pages of the guidebook is the closest I can get to time travel to this past future, since I was not alive then. But wait a minute; thanks to technology, I can attend Futurama. Someone unknown to me converted a GM promotional film of Futurama to digital video and posted it on YouTube. I was able to find it via Google and watch it on my phone while sitting in the city park next to the library on a lovely sunny day. Thank you, technology. Thank you, future! (Note: If you are reading this even one or two years into the future, feel free to take a moment to be amused at my antiquated technologies. Those were the good old days.)

"Technology has found a way to control the wild profusion of this wonder world," says the narrator as the little train of fair attendees enters the rain forest of the future in the video. The process starts with lasers that cut down trees and ends with a smooth coating of pavement. Less trees and more pavement will bring progress to the jungle, the narrator asserts.

Antarctica is peppered with white pods on stilts that look not unlike the Jetsons' futuristic dwellings, except smaller and closer to the ground. As we take a look inside one of the buildings, the narrator announces, "Here is weather central, forecasting to the world the great climate changes born in the Antarctic's never-ending winds." It has the look and feel of early NASA mission control.

The narrator boasts that the ocean contains "food enough to feed seven times the population of the earth." But the oceans of the future will not be just a place of food. They will also be a vacation destination, says the narrator. Unlike a visit to a beach, people in the future will be able to spend a romantic weekend at Hotel Atlantis on the seafloor. The hotel in the diorama looks a lot like the white pods on stilts we saw in the Antarctica diorama. Same architect? I wonder.

. . .

I became absorbed in space-age futures because I couldn't understand them. I can't see the future as a utopia. I am part of a generation that grew up in a world with environmental and sustainability crises. The environmental movement was in full swing during my childhood. I was an infant in 1973 during the first energy crisis, when an oil embargo sent gas prices soaring and the demand for large cars plummeting. The energy crisis that I remember was in the late 1970s. All the parents would park their station wagons in line at the gas station, get out, and grumble that they wished they didn't need this stuff, that our technology was flawed.

In February 1977, President Jimmy Carter spoke to his fellow Americans via television wearing a cardigan sweater. This was a major energy policy speech, and a defining moment in the Carter presidency. His point was that we needed to conserve energy, which is why he donned the cardigan instead of turning up the White House thermostat. As a four-year-old American, I knew none of this. I came to the only possible conclusion: that Mr. Rogers was our president. He, too, would appear on our black-and-white TV and speak directly to his television friends while wearing a cardigan. Perhaps I needed glasses back then, but I found it comforting Mr. President Rogers was asking me to put on a sweater.

The future that I hear about today is one of overpopulation, food and water shortages, energy crises, sea level rise, intense heat waves and severe storms, flooding, sewage, disease — in general, more uncertainty and less control, leaving us vulnerable. It's more a dystopian vision of the future than a space-age utopia. If the Futurama exhibit were made today, it would showcase the mess we have made and the impacts environmental catastrophes will have in the future. A sign at the entrance would recommend that people with heart conditions and anxiety enter at their own risk.

Were people so excited about science and technology during the mid-century space age that they were willing to allow the future of our planet to be guided entirely by technological advances? It occurs to me that some still are. Today people of this mind are known as "techno-optimists." They think that science, technology, and engineering will allow us to get out of any calamity unscathed. On the flip side, there are the "techno-pessimists," known for their ability to itemize the environmental problems we are facing, largely due to our ability to develop technologies that meddle with the planet.

On the surface it might seem that the techno-optimists would be a more fun bunch of people to invite to your cocktail party. Who doesn't want an optimist around? But the techno-optimists need the techno-pessimists. The techno-pessimists help the optimists understand the challenges we are facing. This allows the optimists to dream up technology that might help us avert disaster instead of odd gadgets of limited utility. Without any techno-pessimists, the techno-optimists become out of touch with reality and the challenges we face. Without any techno-optimists around, the techno-pessimists get very gloomy.

I'd invite them both in equal proportions to my cocktail party. They would probably not get along, but after a while, perhaps, they would warm to each other, and hopefully by the wee hours they would have solved at least one looming environmental catastrophe. Of course, it is not going to be that easy. Solutions to the catastrophes will take years of hard work. Today the techno-optimists and techno-pessimists are often fighting about who is right. We could get so much more accomplished if they were working together instead.

For the record, I am part techno-optimist and part techno-pessimist. It really depends which side of the bed I wake up on and what environmental calamity I've just heard about. Maybe we are all some combination of both. If so, we all have the ability to develop solutions and avoid the dystopia. The key is to consider the impacts of any new innovation, not just the intended impacts but the unintended ones as well.

Space-age futures might have been painted as utopias as a reaction to fears of science and technology. As Bill Cotter and Bill Young put it in their book *The 1964–1965 New York World's Fair*, "The country needed to believe that the amazing advances in science and technology would soon lead to a future of limitless promise. The alternative could be the very extinction of humanity."

At that time, science and technology were billing themselves as a method of advancement, but we were also aware that science and technology could work against our own best interests. Two years before the New York World's Fair, and during *The Jetsons* first season, Rachel Carson's book *Silent Spring* made the case that chemical pesticides were harmful to the environment and human health. The nuclear arms race was taking off between the United States and the Soviet Union. We had the ability to wipe out humanity. We realized we had the power to harm the world. We are facing a similar realization today as our greenhouse gases harm the climate, yet I find it challenging to see anything but dystopia ahead.

▪ ▪ ▪

Modular, boxy homes build themselves with the help of robots. Their solar panels collect energy. They reduce waste where they can. On the inside, homebody robots offer breakfast cereal in the morning and vacuum up spills under the table. These expectations for the future, a bit tamer and more practical than those in Futurama, come from *Tomorrow's Home*, a children's picture book published in 1981. It is part of the World of Tomorrow book series. Other titles include *Out Into Space, Our Future Needs, Transport on Earth*, and *Future War and Weapons*. While designed for children, these books addressed questions for people of all ages.

Author Neil Ardley wrote dozens of other books for children, mainly about science topics. Several years after he wrote *Tomorrow's Home*, he coauthored *The Way Things Work* with David Macaulay, which won numerous awards when it was published in the late 1980s and helped get people of all ages to overcome their fear of science.

The illustrator of *Tomorrow's Home* gets no credit. I mention this not because it is relevant to an inquiry into past visions of the future and the role of people in nature, but because I have illustrated children's books and I notice the illustrations. There is a reason these things are called picture books: they are full of pictures.

One thing I appreciate about the illustrations in *Tomorrow's Home*, as compared with the 1960s visions of the future from the world's fair and *The Jetsons*, is that the world of the past appears alongside the world of the future. Futuristic buildings are interspersed with a few castles and other historic structures. (The book was first published in England, where historic buildings and castles are common.) There is even a little green park around a historic building in one of the book illustrations.

The author speculates that computers and robots will allow us more free time. He says that we might spend much of our time exploring the countryside and families will spend more time together. This assertion caused me to shout "Ha!" aloud before checking my email on my iPad and Facebook and Twitter on my phone. Where did the last hour go? What was I doing? Oh, right, back to *Tomorrow's Home*.

A double-page spread about energy efficiency, titled "Waste not, want not," describes self-sufficiency. Today we call the same concept sustainability. This 1980s picture-book vision of the future is somewhat in line with a sustainable world. The ground was considered, natural spaces were preserved, or at least the author acknowledged that they should be preserved. And there was less of a sense of man dominating the planet than there was in Futurama at the New York World's Fair.

"The rancher of the future takes things easy at his remote home. A robot waiter brings him a drink and a terminal linked to the house computer provides entertainment and messages," wrote Ardley. The bronzed rancher wears a Speedo and dark glasses. He reclines on his patio lounge chair, casually accepting a beverage with a swizzle stick from his robot servant. His most rancher-like characteristic, a broad moustache, might have been borrowed from Tom Selleck. Beyond his landscaped terrace and modular home, cows graze and wind turbines spin. Unlike the "dude planet" in *The Jetsons*, this ranch seems somewhat connected with nature, although the rancher is more connected with his lounge chair. At least the cows are real in this future.

We will have varieties of plants that don't need much care, grass that never needs to be cut, and no weeds, the author speculates. "However, it might be difficult to maintain such a garden out in the open," he continues. "Wild plants might invade it, their seeds and spores brought in by birds and insects. So large transparent domes cover this garden of the future. This also means that its owners are able to ignore the climate outside and create a wide range of artificial climates beneath the domes." The illustration shows several domes next to each other — one with people swimming in a warm lake, another with people skating on ice.

Who would do such a thing? I thought. And then several examples came to mind. Every conservatory or greenhouse that houses tropical plants and is located outside the tropics, for instance. The most extreme example I could think of is the indoor ski slope in Dubai, a place known for sandstorms, not snowstorms, where it's over a hundred degrees on an average summer day. The ski area, called Ski Dubai, is located inside a mall. Recently, a troop of Antarctic penguins has been added to Ski Dubai. In an Associated Press story, one of the penguin coordinators, Judy Peterson-Fleming (who happens to also be one of the founders of Denver's aquarium), said, "This is the perfect environment. It's pristine snow. There's no chemicals whatsoever."

■ ■ ■

The Gateway Arch in Saint Louis is an achievement of technology and engineering. Yet even with these advances, the arch, like all human-designed places, is vulnerable to nature. In the past several years it's been suffering from weathering and decay. The stainless steel is stained. The carbon steel is rusted. Maintenance crews mop up water leaks. Alan Weiseman, in his 2007 book *The World without Us*, made the case that our advances in building technology and engineering make us seem unaffected by the forces of nature only as long as we have maintenance workers who are keeping the forces of nature at bay. Maintenance workers are the reason buildings remain standing. Without them, the arch would topple.

We got to the top of the arch and the little doors slid open. "Finally," exclaimed one of my capsule mates as she exited our little pod. Perhaps my swiveling on the mushroom stool had annoyed her. Or perhaps she was just relieved to reach our destination above the Midwest.

I remembered our purpose: to look out from the top. A park ranger (who, amusingly, wore the same uniform as rangers in the wilds of the Rocky Mountains) told people spilling out of capsules that the arch was built as a tribute to westward expansion. Looking through the long, low windowpane of the observation area, I could see a Manifest Destiny of graffiti-covered factory buildings, new lofts with sales-office billboards, parking lots, highways, and fast food reaching to the horizon. This view shattered the stark white future foreshadowed by my capsule journey and brought me back to the messy reality of the present — of climate and environmental change, of battles we fight with nature. The question is, given where we are now in the ruins of our environment, where do we go from here?

Reaction Time

We only find out what people are made of when we watch how they handle a difficult situation. I set out to write this book because I wanted to know what people do when they experience disastrous change on earth, how they treat nature, whether they trust information, whether they are stymied by uncertainty. I wanted to understand why we are not doing enough to stop climate change by looking at our prior experience with change on earth. I learned that we humans have strengths and weaknesses when it comes to dealing with environmental change.

I learned that earth's chaos and uncertainty often leaves us fraught with indecision. It can be hard to know how to act when we are blindsided by rapid change. It can be hard to know when to react when change is gradual and by the time we recognize a problem, it might be too late. That's not unique to climate change. It happens with all sorts of environmental changes.

I also learned that we humans have an amazing capacity to carry on. Late Victorians enjoyed fine dining as their city crumbled, and Cape Cod beachgoers played in the surf as a little house teetered at the edge of a cliff. Pliny the Younger had his Roman nose in a book as earthquakes rumbled and ash darkened the sky. I witnessed women in mermaid costumes swimming with endangered species a mile above the ocean. I learned that when flash floods incapacitate my city, I do laundry.

In uncertain times, when we have no ability to control our situation, focusing on what we can control can be comforting. Plus, we need to eat. We need clean laundry. We need to read and to play on the beach too. While I personally don't need mermaids swimming with sea turtles, some of the kids watching the show really liked it. If we stopped focusing on these small things that we can control, we might be stunned and wide-eyed, like Wile E. Coyote — incapacitated and unable to do anything helpful.

However, our ability to keep calm and carry on also means we avoid fixing problems that we do have the ability to fix. It can divert us from asking: Is this normal? It can lead us to declare: I'm sure it will be fine. Sometimes it's not fine.

We have extraordinary abilities to do nothing — to rationalize inaction. Unlike other animals that must endure or die, our inaction is a choice. In researching stories of disasters, I came across plenty of accounts from people who wished they'd taken swifter action, but I did not come across a single account from anyone who wished they had done nothing. I'm sure we can all think of examples of times when people were too quick to act without having all the information they needed to make a rational decision, but when it comes to natural disasters, and the unnatural variety that we are now generating, every moment of indecision makes us more vulnerable.

Rationalization often involves making up a logical-sounding reason for why something happened and why it is okay when we don't really think that it is okay. It can be used to avoid admitting mistakes or disappointments. For example, let's say you apply for a job. It is down to you and one other candidate. You are excited. Then, the other person gets the job. You say it's for the best, that it would have made you unhappy. Something better is out there. You can come up with many reasons for not being upset. This is rationalization, and in cases when it prevents people from erupting into tears, I'm all for it. But rationalization can also be a powerful way that we fail to deal with climate change. More than a century ago, the developed world invested in systems that relied on burning fossil fuels. Now, we know that our decision had dangerous repercussions for earth's climate. We invent reasons to explain why we chose this path and reasons we are still on it. Rationalization is used to defend the status quo.

Our ability to rationalize will make mitigating climate change more of a challenge. Mitigation will require changing our infrastructure to prevent greenhouse gases from getting into the atmosphere or investing in new technology to pull them out of the atmosphere — enormous undertakings that will not happen if we lack motivation and resist change. Switching from fossil fuels to renewables, for example, requires a huge effort, but there is now evidence that many places are making the effort. Cities including Vancouver, San Diego, San Francisco, Burlington, and Aspen now get most or all of their electricity from renewable sources. Countries such as Iceland, Costa Rica, Paraguay, Norway, and Germany are transitioning to 100 percent renewable energy, and nearly two hundred other countries are increasing renewable energy too. Our time of rationalizing inaction may be coming to an end.

Rationalization plays the same role as denial. People use both denial and rationalization to avoid problems and manage anxiety. These defenses switch on when our actions conflict with our values, according to psychologists.

The book *Climate Change Denial: Heads in the Sand* by Haydn Washington and John Cook outlines many reasons that people deny climate change, either denying that it is happening, denying that it is caused by people, or denying that it is hazardous to our health. In all of these cases, people in denial cherry-pick facts to create alternative ideas. These ideas may help manage anxiety for the denier, but not taking into account all of the evidence means that their alternative ideas are not rooted in the process of science.

In 1968, before the signal of human-caused climate change was detectable, biologist Garrett Hardin wrote, "Natural selection favors the forces of psychological denial. The individual benefits as an individual from his ability to deny the truth even though society as a whole, of which he is a part, suffers."

Hardin was describing the factors that contribute to the tragedy of the commons generally. And yet, I can't help but think of climate change deniers specifically as I read his words, realizing that there may be an evolutionary basis for denying that climate change is a problem. Natural selection did not prepare us to face collective problems.

. . .

Give two people the same information about the risks of climate change. Tell them how sea level rise will flood coasts, how changing storms are causing deluges in some areas and drought in others. Tell them how heat waves are becoming more common and lasting longer, how we are warming the planet faster than it has ever warmed before. No matter what you tell them about the risks we face, the two people will not agree about the extent of the problem.

What I found for the disasters profiled in this book is that a range of perspectives on what constitutes a disaster is the norm. I'm not the only one who is unsure when to pull the handle of the emergency box. Pliny the Elder and Pliny the Younger had different ways of perceiving the risk of volcanic eruption. Some San Franciscans were toasting with wine during the 1906 earthquake and fires while others were fleeing to the hills and considering the frailty of humans as compared with nature. And many of the people at the aquariums I wandered didn't seem to perceive that the oceans were in trouble at all. A range of perspectives is the case for climate change as well.

There are many reasons that we don't all react to environmental threats in the same way. One of those reasons is that we perceive risk differently. Perceiving the risk of climate change is a particular challenge because climate is described through graphs and charts, data and probabilities. This creeping, invisible disaster is often only apparent when it is described with data. The analytical, rational part of our brains steps up to the challenge of interpreting the data, and unfortunately, that is the part of the brain that is least likely to jump into action. It's good at making rational pro and con lists, but is slow to decide what to do.

Despite what the graphs, facts, and data indicate, climate change doesn't feel dangerous to many people. Climate change, even at today's rapid rate, is slow compared to an attacking bear or other imminent threat. It's also difficult to see. It's challenging to perceive the risks when the foe is invisible and slow moving. Humans respond to danger based primarily on instinct and intuition, not on the rational and analytical parts of their brains. If climate change doesn't feel risky, then it doesn't seem imperative that we take action.

There is another reason that different people will perceive different amounts of risk when it comes to climate change. Those who place a large value on nature see that there is so much more at stake as climate warms. Those who do not value nature are less concerned because they see less in harm's way. In the case of moving the lighthouse on Cape Cod, it took local residents to recognize that the value of the lighthouse was more than its ability to be a beacon to mariners. It was of historic value to the community. That recognition spurred action to save the lighthouse. In the case of climate change, people who place a high value on nature see species going extinct at a thousand times the usual extinction rate due to environmental impacts we have caused, including climate change, and realize how much we have to lose. They are more likely to assess the risk as high than those who don't place high value on nature.

Not only do we all perceive risk differently, we also will not make the same decisions when faced with risk. The accounts of individuals enduring disastrous events that I encountered while researching this book reminded me that there is not one way that we as a species react during difficult times. Some will be excited. Others will be concerned. Still others will set the dining table and long for normal. Similarly, we do not all have the same perspective on the climate catastrophe today. Some are terrified. Others are blasé. Some are taking action to stop warming. Others are unsure. And some are vehement that there is not a problem.

Despite the step-by-step instructions provided by In Case of Emergency signs, what people actually do in an emergency will vary. We read too much into why we are not all in agreement about the risks of climate change. Response to other changes on earth is rarely uniform. It follows that our response to climate change would not be uniform either.

Unlike with other disasters, there is an idea that we need everyone's help to stop climate change. However, some people will always be uncertain. Some will not see the risks we face. Others will have non-cooperative social values and resist any change that doesn't benefit themselves. We cannot expect everyone to agree, but that does not mean all hope is lost.

■ ■ ■

In some ways, we live on a chaotic planet that we cannot control, but when it comes to climate change, we can control that. The amount of climate change we are facing in the future is directly related to the choices we make now. Like Wile E. Coyote, once we let go of that stunned expression, we will use our ingenuity to get ourselves out of this mess.

We humans are unique in our ability to cause a problem like climate change. But we are also better equipped to deal with climate change than any other species. The stories in this book demonstrate that we have the ability to take actions to minimize risk. Italians are demanding a better emergency plan for when Vesuvius erupts again. Divers with spears are trying to kill lionfish before they destroy reefs. The Cape Codders did move the lighthouse as the cliff eroded. In Boulder, Colorado, the flood maps are being redrawn to take into account what we learned in the 2013 floods. We are good at picking up the pieces. We have rebuilt San Francisco, designated national seashores to prevent more houses from being built where they will topple into the sea. We know how to mop up the mud after a flood. Unlike the rhinos snuffling ash at the Nebraska watering hole, we have the ability to be resilient.

If we are going to get ourselves out of the climate catastrophe, it will be our abilities for mental time travel that will get us there. We can see the future like no other animal. We have the ability to project what the temperature will be on earth in the decades ahead if we change our energy systems, if we reduce the use of fossil fuels, and if we do not. Our visions of the future may be riddled with uncertainty — we have no idea if we'll each have a flying car, for example — but we can see what lies ahead for global climate.

Our abilities to plan and prepare could serve us well when it comes to climate adaptation. We can plan for changes that allow us to avoid dire impacts of climate change. This includes planning to maintain the freshwater supply during droughts, building drainage for extreme rainfall, and putting air-conditioning in classrooms so that school can be in session even during an extreme heat wave. However there will still be places where proper planning will not allow people to carry on, such as low-lying coastal communities that will need to relocate as glaciers melt, the ocean warms, and sea levels rise.

■ ■ ■

Since 1960, we have added a billion people to earth every twelve to fourteen years. At the current rate, the population of earth in the middle of this century will be about ten billion. With more people on the planet, the lines between wild places and human-built places are increasingly blurred. With more people on the planet, there are more people in harm's way. And there is more risk as climate change breeds stronger storms, drought, and heat waves.

The whimsical space-age visions of the future that I came across placed little to no value on nature. Today we preserve land, protecting it from ourselves. We are told not to step off the trail; otherwise, we might destroy what is wild. All the while we are filling the atmosphere with invisible greenhouses gases. We are metaphorically stepping off the trail all the time.

For decades, we have collectively adopted a "wait and see" mentality when it comes to climate change. The first messages from scientists about climate change decades ago were much more uncertain than what we know today, which is perhaps why we did not jump into action. But the same conclusions that they reached then still hold true today: we are warming the climate, and that is hazardous to our health.

We've been stuck in a quagmire of indecision when it comes to climate change; however, while I was writing this book, I started to see major strides around the world to reduce emissions of heat-trapping greenhouse gases. These are glimmers of hope that we are starting to take action in the face of this disaster. In 2014, for the first time in history, global carbon dioxide emissions plateaued while the economy thrived. China became the world's biggest investor in renewable energy after years of building coal-fired power plants. The European Union pledged to get emissions 40 percent lower than their 1990 levels by 2030. Texas, known for extracting oil and gas, is now home to more wind turbines than any other US state. In 2015, for the first time in history, the nations of the world pledged to sharply reduce emissions, and in 2017 they reaffirmed this commitment despite the United States' wavering. Globally, we are beginning to react to this emergency in meaningful ways.

However, we are not reacting fast enough to stop the warming. The United States has moved from a president who was making progress on climate issues to one who is undoing much of the progress that has been made. During the time that you have spent reading this book, over six and a half billion pounds of carbon dioxide have been released into the atmosphere around the world (based on 2013 emissions and a reading pace of three hundred words per minute). There has been progress, but it has not yet been enough.

"I'm not a big believer in man-made climate change," said President Trump in an interview on March 21, 2016, with the *Washington Post*. It appears that Trump is one of the three in ten people who doesn't see the disaster that is in progress.

People who don't see a disaster will not take action, so helping people understand it is a good idea, but it's also possible that enough of us are shaking off the shock and anxiety and starting to take action to minimize the harm of climate change — enough to make a difference. In a speech to Chinese businesspeople shortly after Trump was elected, Michael Bloomberg, former New York City mayor and current United Nations envoy for cities and climate change, said:

> I am confident that no matter what happens in Washington, no matter what regulations the next administration adopts or rescinds, no matter

what laws the next Congress may pass, we will meet the pledges that the U.S. made in Paris. The reason is simple: Cities, businesses, and citizens will continue reducing emissions because they have concluded — just as China has — that doing so is in their own self-interest.

In the days after June 1, 2017, when Donald Trump pulled the United States out of the Paris Accord, 1,400 U.S. cities, states, and businesses announced that they would meet the pledge despite the national withdrawal.

Our collective ability to change the planet is vast and dangerous, yet it could also be beneficial. It means that we also have the ability to keep the world livable. We have the ability to survive a multitude of other types of change on earth, and we can deal with this too. It's not an insurmountable situation. It will come down to the choices we make as individuals, as nations, and as a global community. But at this point, those choices are still uncertain.

■　■　■

As I researched this book, I was drawn to the stories of individuals facing disasters more than the accounts of communities, regions, or governments. When I came across a San Franciscan who declared his riding outfit to be his most appropriate earthquake attire, Bonaire environmentalists killing invading fish, a toddler babbling to an octopus as the ocean degrades, and a Roman naturalist racing toward an erupting volcano, I realized that it can be the very small decisions that we each make every day that can make a difference during a disaster.

Today, the climate change disaster seems so large and far-reaching that the actions of one small person don't seem like they would make much of a difference. Bike to work instead of driving your car and you will likely see countless other cars on the road spewing carbon dioxide into the atmosphere. Just like the lionfish invading the Caribbean, climate change is a collective action problem. It can be difficult to see how our individual actions can stop it.

However, since we are never going to get all people on the same page, what we can control are our own choices about how to live in this world and whether we will contribute to the problem or help solve it. Those of us who see the disaster need to move ahead and act decisively, even if others can't yet see it.

If you know that climate change jeopardizes things that you value in this world, then do something. Others will follow. When we are not sure what to do, we look to others and see what they are doing. This is called the chameleon effect. If people see everyone else just sitting there doing nothing, then they may not feel compelled to do anything. But if enough people show that they are concerned and take action, others will follow. When you are on your bike and see the cars on the road, don't get disheartened. The drivers of those cars are seeing you too. Some of those drivers might be looking for other ideas about how to get to work. You might not feel like you are making a difference, but you may have a larger impact than you know.

■ ■ ■

Since 1960, American households have tripled the amount of electricity they use at home, and most of that energy comes from fossil fuels such as coal and natural gas. The average American home used about eleven thousand kilowatt-hours (kWh) of electricity per year in 2014 according to the US Energy Information Administration. Also since 1960, we have doubled the amount of gasoline we pump into cars. We have been using more energy. Since most of this energy has come from burning fossil fuels, it increases greenhouse gas emissions.

Most of the actions you can take to help solve climate change as an individual boil down to one thing: reducing the amount of greenhouse gases you are adding to the atmosphere. To do this, you need to figure out ways to reduce the amount of fossil fuels you burn.

Using a carbon footprint calculator to assess how much fossil fuel you are burning and how much carbon dioxide you are producing is a good way to start as you consider how you can reduce. There are several great, free carbon footprint calculators available online. One of my favorites is the CoolClimate Calculator (http://coolclimate.berkeley.edu). It guides you first through calculating the amount of greenhouse gases you are currently adding to the atmosphere, which is a bit depressing. But then it provides a buffet of options for how you can trim that number down, which is quite motivating. The calculator might suggest switching to renewable energy, for example, or unplugging vampire electronics, items like DVD players and TVs that suck energy out of your outlets even when they are turned off. It might prompt you to consider transportation options like telecommuting or riding a bike to work, or think about the carbon footprint of the products you buy, avoiding items enrobed in excess packaging or foods grown halfway around the world. The buffet of solutions is varied enough that there is something for everyone. Some solutions, like unplugging the vampires, cost nothing, while others, like buying an electric car, cost thousands of dollars.

Note that when tallying up your own greenhouse gas emissions, it's likely that you will feel guilty at some point in the process. Guilt is not helpful. I am somewhat envious of the disasters profiled in this book that aren't caused by humans because that allows people to focus on making the situation better without wasting time debating the whodunit. Don't worry about who is at fault. Just do something to help us survive.

Many of the actions that we can take to help stop climate change don't look anything like the actions one might take during a disaster (such as applying pressure to wounds, screaming for help, or digging out from rubble). Many of the individual actions we can take — such as shrinking your electric bill, riding a bike, and avoiding processed foods wrapped in packaging — also improve mental and physical health. So even if each makes only a small impact on global climate, they can have a huge impact on your quality of life.

After years of working to minimize my carbon footprint, I've come to realize that I will never be able to get my greenhouse gas emissions to zero on my own. I'm going to need help. I can get my emissions quite low compared to the American average, but I will continue to add some greenhouses gases to the atmosphere until we have large-scale systems in place in our country and world that allow us to move beyond burning fossil fuels. That is outside the scope of what I can do as an individual. It requires a paradigm shift in our society.

Many solutions to climate change are larger than one's individual actions. They will need to be implemented through new state, local, or national policies. Some criticize asking individuals to focus on their own contributions of greenhouse gases because it ignores big polluting companies and infrastructure. But there is something else that individuals can do. One of the most important actions we can take to solve this problem is to vote for political leaders who are looking for the best ways to solve climate change, transforming our energy infrastructure to run on renewables and funding research on climate adaptation and mitigation strategies.

Although climate change is a global problem, the human population is made up of individuals. Look at a picture of the earth from space and you will not see any of the seven and a half billion individual people on the planet, but they are there, making choices each day about how to live in this world. We need to recognize that not making changes is a decision to make earth a difficult place for us to survive in the future. However, we have control over the consequential side of the risk equation. We can decide what sort of world we want this to be. We can decide how we will survive and thrive. It will be the people who recognize that we are in the midst of a disaster who will make this happen.

Shrinking the Carbon Footprint
of This Book

I sheepishly admit that writing this book released some carbon dioxide into the atmosphere. I did what I could to keep that amount as small as possible, but alas, it is not zero. For two years the computer that I used to write the book ran on solar energy from panels installed on the roof of my former home. For two years it ran on wind energy from the local utility. However, there were six months when it was powered by a natural gas–fired power plant and almost a year when it was powered by a coal-fired power plant. (Note to self: Question the source of the energy that flows from the outlets when you move to a new home.)

When my computer died in 2016, I recycled it and replaced it with a laptop that uses as little energy as possible. I printed numerous drafts of the chapters, enough to cover the floor of my study with paper, which Mila enjoyed napping on. This was recycled paper, and after the drafts were no longer needed, I recycled them. Recycling paper saves both the energy to manufacture new paper and avoids cutting down trees, which soak up much of the carbon dioxide that we are putting into the air.

You may have noticed that there was some travel involved in these stories. I drove to find the ash-covered rhinos in Nebraska. I flew to Vermont for the Wildbranch Writers Workshop that began this book project in earnest and then drove to Cape Cod with Sara and Justin, which inspired an early chapter. I flew to California several times for the chapters about earthquakes and aquariums, and once to Bonaire to find the invading lionfish. I also flew to Baltimore three times to attend the MFA residency at Goucher College for some much-needed guidance, feedback, and inspiration. To make this production of greenhouse gases a little more palatable, I bought carbon offsets, meaning that I gave some money to help plant and care for trees — trees that are hopefully at work today taking some of my carbon dioxide out of the atmosphere.

In Case of Emergency

Look and Listen:
Read the news and learn the science.

One important thing that you can do to help during the climate emergency is to stay informed. Solving climate change is a rapidly evolving area of research. Each day there are new ideas about sources of renewable energy and ways to geoengineer the planet to dampen the warming, new stories of ways that communities and countries are taking action to stop using fossil fuels and be more resilient. Learning what is working in other places on earth can give us ideas for what we can do and what we would like to ask of our elected leaders.

Children's television personality Fred Rogers once said, "When I was a boy and I would see scary things in the news, my mother would say to me, 'Look for the helpers. You will always find people who are helping.' To this day, especially in times of disaster, I remember my mother's words and I am always comforted by realizing that there are still so many helpers — so many caring people in this world." Mr. Rogers was referring to how we can help children during a disaster, but I believe looking for the helpers is comforting for all of us, not just children. It's motivating and inspiring, too.

The websites listed below are some of the sources I turn to for news about how people are researching and solving the climate disaster. The news is not always good, sometimes it is about how people are vulnerable to the problems we have caused, but the articles highlight people who are working to find answers and solutions. They are the helpers that Mr. Rogers mentioned.

- Climate Nexus: http://climatenexus.org
- Climate Central: http://www.climatecentral.org
- Grist — climate news with a dose of humor: http://grist.org
- Yale Climate Connections: http://www.yaleclimateconnections.org

It's also important to keep tabs on this catastrophe since it can be difficult to see on an average day. For information about how climate change is affecting places on earth and the latest numbers about the amount of warming, sea level rise, and sea ice, visit NASA Global Climate Change (http://climate.nasa.gov), NOAA's Climate website (https://www.climate.gov), and the Category 6 Blog from Weather Underground (https://www.wunderground.com/cat6). To learn more about how climate change is affecting regions of the United States, check out the National Climate Assessment (http://nca2014.globalchange.gov).

To learn more about the science of how our planet's climate works, check out *The Thinking Person's Guide to Climate Change* by Robert Henson, a detailed overview of the science of climate for nonscientists. For an explanation of the findings of the Intergovernmental Panel on Climate Change, read *Dire Predictions: Understanding Climate Change* by Michael E. Mann and Lee R. Kump. And if you hear something about climate science that does not seem correct, turn to the Skeptical Science website (https://www.skepticalscience.com), which details common misinformation about climate change and how it compares with real science.

Respond:
Do what you can to quell the catastrophe.

Another important thing that you can do to help during the climate emergency is to take action. While large-scale action of nations and governments is needed, often what you can control is the way you live and the way you vote. Do what you can to stop emitting greenhouse gases into the atmosphere.

- Figure out how to reduce your own greenhouse gas emissions. The CoolClimate Network (http://coolclimate.berkeley.edu) includes an excellent carbon calculator to get you started. Once you calculate how much greenhouse gas you are spewing into the atmosphere, find ways to reduce that amount. Encourage your friends and neighbors to do the same.
- Once you've done what you can, expand outward, considering what your workplace, community, and region needs to do. The US Climate Resilience Toolkit (https://toolkit.climate.gov) developed by NOAA includes tools for individuals, businesses, and communities that want to reduce greenhouse gases, learn how to manage climate-related risks, and improve resilience. Visit the United Nations Climate Change Newsroom (http://newsroom.unfccc.int) to learn about the actions that nations around the world are taking to stop climate change and to gather ideas about what your community or region could do.
- Find out which carbon offsets are the most effective at helping us combat climate change. The Natural Resources Defense Council explains what to look for when purchasing climate offsets (https://www.nrdc.org/stories/should-you-buy-carbon-offsets).
- Get involved with organizations working toward policy change to stop the climate catastrophe, like 350.org (https://350.org), which advocates for keeping carbon dioxide levels in the atmosphere low, and the Union of Concerned Scientists (http://www.ucsusa.org), which advocates for practical environmental action that is based on the latest science research.
- Vote for candidates who are working to stop the disaster. The National Environmental Scorecard from the League of Conservation Voters (http://scorecard.lcv.org) provides objective, factual information about environmental legislation in the United States and the voting records of all members of Congress.
- Most importantly, don't give up.

Further Reading

Albins, M. A., and M. A. Hixon. "Invasive Indo-Pacific *Pterois volitans* Reduce Recruitment of Atlantic Coral-Reef Fishes." *Marine Ecology Progress Series* 367 (2008): 233–238.

Ardley, N. *Tomorrow's Home*. New York: Franklin Watts, 1981.

Association of Zoos and Aquariums. "About the Association of Zoos and Aquariums." https://www.aza.org/about-us.

Ballantyne, R., et al. "Conservation Learning in Wildlife Tourism Settings: Lessons from Research in Zoos and Aquariums." *Environmental Education Research* 13 (2007): 367–383.

Barnes, K. "Europe's Ticking Time Bomb." *Nature*, 473 (2011): 140–141.

BBC News. "Plain English Award for Met Office 'Gobbledygook.'" December 9, 2011. http://www.bbc.com/news/uk-16100112.

Benedict, T. "The Coming of Astro." *The Jetsons*. Directed by J. Barbera and W. Hanna. Aired October 21, 1962.

Black, W. "Dude Planet." *The Jetsons*. Directed by J. Barbera and W. Hanna. Aired February 17, 1963.

Blitzer, B. E. "Las Venus." *The Jetsons*. Directed by J. Barbera and W. Hanna. Aired December 16, 1962.

Bragg, M. A. "Storm-Eroded Dunes Imperil Homes." *Cape Cod Times*, December 28, 2010.

Breining, G. *Super Volcano: The Ticking Time Bomb beneath Yellowstone National Park*. New York: Voyageur Press, 2007.

Bryant, C., and J. Clark. "How Chaos Theory Changed the Universe." *Stuff You Should Know Podcast*, July 19, 2016. http://www.stuffyoushouldknow.com/podcasts/how-chaos-theory-changed-the-universe.htm.

Budescu, D. V., et al. "The Interpretation of IPCC Probabilistic Statements around the World." *Nature Climate Change* 4 (2014): 508–512.

Burness, A. "Boulder's Murky Blue Line Could Get Clarity from Voters in November." *Daily Camera*, August 6, 2016.

Bush, G. W. "President Bush Discusses Global Climate Change," June 11, 2001.

https://georgewbush-whitehouse.archives.gov/news/releases/2001/06/20010611-2.html.

Campbell, T. *The Gateway Arch: A Biography.* New Haven: Yale University Press, 2013.

Cape Cod Commission. "Cape Cod Sea Level Rise." http://www.capecod commission.org/sealevelrise/.

Carson, R., *Silent Spring.* 40th anniversary edition. Boston: Houghton Mifflin Company, 2002.

Carter, J. "Report to the American People on Energy, February 2, 1977." Miller Center, University of Virginia. http://millercenter.org/president/carter /speeches/speech-3396.

Cleetus, R. *Overwhelming Risk: Rethinking Flood Insurance in a World of Rising Seas.* Cambridge, MA: Union of Concerned Scientists, 2014. www.ucsusa.org /floodinsurance.

Corn, J. J., and B. Horrigan. *Yesterday's Tomorrows: Past Visions of the American Future.* Baltimore: Johns Hopkins University Press, 1996.

Cotter, B., and B. Young. *The 1964–1965 New York World's Fair: Creation and Legacy.* Charleston, SC: Arcadia Publishing, 2008.

Cronin, W. *Uncommon Ground: Rethinking the Human Place in Nature.* New York: W. W. Norton and Company, 1996.

Dalai Lama. *The Art of Happiness: A Handbook for Living.* 10th anniversary edition. New York: Riverhead Books, 2009.

de Saint-Exupéry, A. *The Little Prince.* Translated by R. Howard. San Diego: Harcourt, 2000.

Demuth, J. L., J. K. Lazo, and R. E. Morss. "Exploring Variations in People's Sources, Uses, and Perceptions of Weather Forecasts." *Weather, Climate, and Society* (2011): 177–192.

Dregni, E., and J. Dregni. *Follies of Science: 20th Century Visions of Our Fantastic Future.* Denver: Speck Press, 2006.

Environmental Protection Agency. *A Citizens Guide to Radon: The Guide to Protecting Yourself and Your Family from Radon.* EPA Publication Number 402/K-12/002, 2014.

Falk, J., et al. *Why Zoos and Aquariums Matter: Assessing the Impact of a Visit to a Zoo or Aquarium.* Silver Spring, MD: Association of Zoos and Aquariums, 2007.

FEMA. "Flood Map Service Center." https://msc.fema.gov/portal.

Flato, G., et al. "Evaluation of Climate Models." In *Climate Change 2013: The Physical Science Basis. Contribution of Working Group I to the Fifth Assessment*

Report of the Intergovernmental Panel on Climate Change, edited by T. F. Stocker, et al., 741–866. Cambridge, UK, and New York: Cambridge University Press, 2013.

Fradkin, P. L. *The Great Earthquake and Firestorms of 1906: How San Francisco Nearly Destroyed Itself.* Berkeley: University of California Press, 2006.

Frazer, T. K., et al. "Coping with the Lionfish Invasion: Can Targeted Removals Yield Beneficial Effects?" *Reviews in Fisheries Science* 20 (2012): 185–191.

Gambino, M. "Evolution World Tour: Ashfall Fossil Beds, Nebraska." *Smithsonian Magazine*, January 2012.

Gardiner, L. *Catastrophic Colorado! The History and Science of Our Natural Disasters.* Denver: Westcliffe Publishers, 2006.

Gardner, D. *Risk: The Science and Politics of Fear.* London: Virgin Books, 2009.

General Motors. *'64–65 NY World's Fair Futurama Ride Video*, YouTube video, 7:16. March 26, 2008. https://youtu.be/2-5aKoHo5jk.

Goldstone, H. "Cape's Shape Constantly Changing, for Good and Bad." *Cape Cod Times*, August 6, 2011.

Hansen, J., et al. "Ice Melt, Sea Level Rise, and Superstorms: Evidence from Paleoclimate Data, Climate Modeling, and Modern Observations That 2°C Global Warming Could Be Dangerous." *Atmospheric Chemistry and Physics* 16 (2016): 3761–3812.

Hardin, G. "The Tragedy of the Commons." *Science* 162 (1968): 1243–1248.

Henson, R. *The Thinking Person's Guide to Climate Change.* Washington, DC: American Metrological Society, 2014.

Hoerling, M., et al. "Northeast Colorado Extreme Rains Interpreted in a Climate Change Context." *Special Supplement to the Bulletin of the American Meteorological Society* 95 (2014).

Intergovernmental Panel on Climate Change. *Climate Change 2014: Synthesis Report. Contribution of Working Groups I, II, and III to the Fifth Assessment Report of the Intergovernmental Panel on Climate Change.* Edited by Core Writing Team, R. K. Pachauri, and L. A. Meyer. Geneva, Switzerland: IPCC, 2014.

———. *Assessment Reports.* https://www.ipcc.ch/publications_and_data /publications_and_data_reports.shtml \.

Jaeger, C. C., and J. Jaeger. "Three Views on Two Degrees." *Regional Environmental Change* 11(2011): S15–S26.

Kiehl, J. T. *Facing Climate Change: An Integrated Path to the Future.* New York: Columbia University Press, 2015.

Kline, B. *First Along the River: A Brief History of the U.S. Environmental Movement*. 3rd ed. Lanham, MD: Rowman and Littlefield, 2007.

Kolbert, E. *The Sixth Extinction: An Unnatural History*. New York: Henry Holt and Company, 2014.

Leiserowitz, A., et al. *Climate Change in the American Mind: May 2017*. Yale Program on Climate Change Communication. http://climatecommunication.yale.edu/wp-content/uploads/2017/107/Climate-Change-American-Mind-May-2017.pdf.

London, J. "The Story of an Eyewitness." *Collier's*, May 5, 1906.

Lorenz, E. N. *The Essence of Chaos*. Seattle: University of Washington Press, 1994.

Louv, R. *Last Child in the Woods: Saving Our Children from Nature-Deficit Disorder*. Chapel Hill, NC: Algonquin Books, 2008.

Markes, L. "Good Little Scouts." *The Jetsons*. Directed by J. Barbera and W. Hanna. Aired October 28, 1962.

Martinez, J. "Climate Change Concerns Weigh on Cape Home-Buying Decisions." *Boston Globe*, September 13, 2014.

Mastrolorenzo, G., et al. "Lethal Thermal Impact at Periphery of Pyroclastic Surges: Evidences at Pompeii," *PLOS One* 5 (2010). https://doi.org/10.1371/journal.pone.0011127.

McInerney, F. A., and S. L. Wing. "The Paleocene-Eocene Thermal Maximum: A Perturbation of Carbon Cycle, Climate, and Biosphere with Implications for the Future." *Annual Review Earth and Planetary Science* 39 (2011): 489–516.

Morell, V. "Invasive Lionfish Attacks Reefs and Fish as Scientists Scramble." *Science Insider* (April 2010). http://www.sciencemag.org/news/2010/04/invasive-lionfish-attacks-reefs-and-fish-scientists-scramble.

Morris Jr., J. A., et al. "Biology and Ecology of the Invasive Lionfishes, *Pterois miles* and *Pterois volitans*." *Proceedings of the 61st Gulf and Caribbean Fisheries Institute*, November 10–14, 2008.

Morris Jr., J. A., and P. E. Whitfield. "Biology, Ecology, Control, and Management of the Invasive Indo-Pacific Lionfish: An Updated Integrated Assessment." *NOAA/National Ocean Service/Center for Coastal Fisheries and Habitat Research*, NOAA Technical Memorandum NOS NCCOS 99 (2009).

Morss, R. E., et al. "Flash Flood Risks and Warning Decisions: A Mental Models Study of Forecasters, Public Officials, and Media Broadcasters in Boulder, Colorado." *Risk Analysis* 35 (2015): 2009–2028.

Morss, R. E., J. K. Lazo, and J. L. Demuth. "Examining the Use of Weather

Forecasts in Decision Scenarios: Results from a US Survey with Implications for Uncertainty Communication." *Meteorological Applications* 17 (2010): 149–162.

Moser, S. C., and L. Dilling. "Making Climate Hot: Communicating the Urgency and Challenge of Global Climate Change." *Environment: Science and Policy for Sustainable Development* 46 (2004): 32–46.

Mumby, P. J., A. R. Harborne, and D. R. Brumbaugh. "Grouper as a Natural Biocontrol of Invasive Lionfish." *PLoS ONE* 6 (2011). https://doi.org/10.1371/journal.pone.0021510.

National Oceanic and Atmospheric Administration. *Weather Forecast Uncertainty, State of the Science Fact Sheet.* Washington, DC: US Department of Congress, 2009.

National Park Service. "Lighthouse Moves." https://www.nps.gov/caco/learn/historyculture/lighthouse-moves.htm.

National Research Council. *Informing an Effective Response to Climate Change.* Washington, DC: National Academies Press, 2010.

Ocean Project. "America and the Ocean Report 2011." http://theoceanproject.org/reports .

Official Guide New York World's Fair 1964–1965. New York: Time Life Books, 1964.

Oreskes, N., and E. M. Conway. *Merchants of Doubt: How a Handful of Scientists Obscured the Truth on Issues from Tobacco Smoke to Global Warming.* London: Bloomsbury Press, 2010.

Paglia, M. "Big Splash." *Westword*, August 26, 1999.

———. "Going Up and Coming Down." *Westword*, March 13, 2003.

Pahl, S., et al. "Perceptions of Time in Relation to Climate Change." *WIREs Climate Change* 6 (2015): 359.

Peachey, J. A., et al. "How Forecasts Expressing Uncertainty Are Perceived by UK Students." *Weather* 68 (2013): 176–181.

Pellegrino, C. R. *Ghosts of Vesuvius: A New Look at the Last Days of Pompeii, How Towers Fall, and Other Strange Connections.* New York: Harper Perennial, 2005.

Pilkey, O. H. "We Need to Retreat from the Beach." *New York Times*, November 14, 2012.

Pistor, N. J. C., "Gateway Arch Showing Rust and Decay." *Saint Louis Post-Dispatch*, August 22, 2010.

Powell, J. L. *Science and Global Warming, 2014: Methodology.* http://www.jamespowell.org/methodology/newmethodology.html.

Prairie Mountain Publishing. *A Thousand-Year Rain: The Historic Flood of 2013 in Boulder and Larimer Counties.* Battle Ground, WA: Pediment Publishing, 2013.

Quibell, M. "How Dystopia Hammers Home the Reality of Climate Change." *Guardian*, April 20, 2015.

Revkin, A. "Conveying the Climate Story." Presentation, June 2011. Google Science Communication Fellows. https://youtu.be/lU_4OR3hOyo.

Roch, S. G., and C. D. Samuelson. "Effects of Environmental Uncertainty and Social Value Orientation in Resource Dilemmas." *Organizational Behavior and Human Decision Processes* 70 (1997): 221–235.

Rocha, L. A., et al. "Invasive Lionfish Preying on Critically Endangered Reef Fish." *Coral Reefs* 34 (2015): 803–806.

Rose, W. I., C. M. Riley, and S. Dartevelle. "Sizes and Shapes of 10-Ma Distal Fall Pyroclasts in the Ogallala Group, Nebraska." *Journal of Geology* 111 (2003): 115–124.

Rosenthal, E., "Answer for Invasive Species: Put It on a Plate and Eat It." *New York Times*, July 9, 2011.

Scyphers, S. B., et al. "The Role of Citizens in Detecting and Responding to a Rapid Marine Invasion." *Conservation Letters* 8 (2015): 242–250.

Sealey, K. S., et al. "The Invasion of Indo-Pacific Lionfish in the Bahamas: Challenges for a National Response Plan." *Proceedings of the 61st Gulf and Caribbean Fisheries Institute*, November 10–14, 2008.

Slovic, P., et al. "Risk as Analysis and Risk as Feelings: Some Thoughts about Affect, Reason, Risk, and Rationality." *Risk Analysis* 24 (2004): 311–322.

Slovic, P., and E. Peters. "Risk Perception and Affect." *Current Directions in Psychological Science* 15 (2006): 322–325.

Smith, N. S., and K. S. Sealey. "The Lionfish Invasion in the Bahamas: What Do We Know and What to Do About It?" *Proceedings of the 60th Gulf and Caribbean Fisheries Institute*, November 5–9, 2007.

Snider, L. "Boulder's Blue Line Turns 50." *Colorado Daily*, July 21, 2009.

Strahler, A. N. *A Geologist's View of Cape Cod.* Orleans, MA: Parnassus Imprints, 1988.

Suddendorf, T., and J. Busby. "Mental Time Travel in Animals?" *Trends in Cognitive Sciences* 7 (2003): 391–396.

Suddendorf, T., and M. C. Corballis. "The Evolution of Foresight: What Is Mental Time Travel, and Is It Unique to Humans?" *Behavioral and Brain Sciences* 2007: 299–313.

Suplee, H. H. "The Sermon in San Francisco's Stones: How the New City May Be Made Proof against Devastation by Earthquakes and Fire." *Harper's Weekly*, May 26, 1906.

Thoreau, H. D. *Cape Cod*. Boston: Ticknor and Fields, 1865.

Trenberth, K. E., J. T. Fasullo, and T. G. Shepherd. "Attribution of Climate Extreme Events." *Nature Climate Change* (June 22, 2015): 725–730.

UK Met Office. "A Golden Conundrum." *Official Blog of the Met Office News Team*. December 9, 2011. https://blog.metoffice.gov.uk/2011/12/09/a-golden -conundrum/.

United Nations. *The Paris Agreement*. The United Nations Framework Convention on Climate Change, 2015. http://unfccc.int/paris_agreement/items /9485.php.

Virtual Museum of the City of San Francisco. "The Great 1906 Earthquake and Fire." http://www.sfmuseum.org/1906/06.html.

Voorhies, M., et al. *Ashfall Fossil Beds State Historical Park and National Natural Landmark: Present View of an Ancient Past*. Lincoln: University of Nebraska State Museum and Nebraska Game and Parks Commission, 2015.

Walter, C. "Passages to the Sea: The Ocean Journey Aquarium in Denver, Colorado." *Sunset Magazine*, September 1999.

Washington, H., and J. Cook. *Climate Change Denial: Head in the Sand*. Abingdon, UK: Taylor and Francis, 2011.

Weiseman, A. *The World without Us*. New York: St. Martin's Press, 2008.

White, G. F. *Strategies of American Water Management*. Ann Arbor: University of Michigan Press, 1969.

Whitfield, P. E., et al. "Biological Invasion of the Indo-Pacific Lionfish *Pterois volitans* along the Atlantic Coast of North America." *Marine Ecology Progress Series* 235 (2002): 289–297.

Wilderness Act, 16 U.S.C. 1131–1136 (1964). http://wilderness.nps.gov/document /wildernessAct.pdf.

Wilkinson, P. *Pompeii: The Last Day*. London: BBC Worldwide, 2004.

Williams, J. *The USA Today Weather Book*. New York: Knopf Doubleday Publishing Group, 1997.

Wong, A. N., "A Brief History of the Magic 8 Ball." *Mental Floss*, August 24, 2015.

Index

COMPLETE GUIDE TO

VITAMINS
MINERALS & SUPPLEMENTS

H. Winter Griffith, M.D.

FISH
BOOKS

Publishers: *Fred W. Fisher, Helen V. Fisher
& Howard W. Fisher*
Chief Editor: *Judith Schuler*
Editors: *Veronica Durie, Joyce Bush*
Art Director: *Josh Young*
Cover photo: *Ray Manley Studios*

Published by Fisher Books
3499 N. Campbell Avenue, Suite 909
Tucson, Arizona 85719
(602) 325-5263

CIP DATA
Library of Congress Cataloging-in-Publication Data
Griffith, H. Winter, 1926
   Complete guide to vitamins, minerals
   & supplements.
   Bibliography: p.
   Includes index.
   1. Vitamins—Handbooks, manuals, etc. 2.
Minerals in human nutrition—Handbooks, manuals,
etc. 3. Dietary supplements—Handbooks, manuals,
etc. I. Harrison, Gail. II. Levinson, Dan.
QP771.G75 1987 613.2'8 87-25156

ISBN 1-55561-006-4

©1988 Fisher Books

Printed in U.S.A.
Printing 10 9 8 7 6

Notice: The information in this book is true
and complete to the best of our knowledge.
This book is intended only as an informative
guide for those wishing to know more about
vitamins, minerals, supplements and
medicinal herbs. This book is not intended to
replace, countermand or conflict with the
advice given to you by your physician. He or
she knows your history, symptoms, signs,
allergies, general health and the many other
variables that challenge his/her judgment in
caring for you as a patient. The information in
this book is general and is offered with no
guarantees on the part of the author or Fisher
Books. The author and publisher disclaim all
liability in connection with use of this book.

Contents

Technical Consultants

Gail Harrison, Ph.D., Professor
Department of Family Community Medicine
University of Arizona College of Medicine

Dan Levinson, M.D.,
Adjunct Associate Professor
Department of Family Community Medicine
University of Arizona College of Medicine

About the Author

H. Winter Griffith, M.D., received his medical degree from Emory University in 1953 and spent more than 20 years in private practice. At Florida State University, he established a basic medical-science program and also directed the family-practice residency program at Tallahassee Memorial Hospital. After moving to the southwest, he became associate professor of Family and Community Medicine at the University of Arizona College of Medicine, where he is currently an adjunct professor. At present, he devotes most of his time to writing medical-information books for non-medical readers. He has published several popular books and continues to pursue this avenue of interest.

Dedication

To each of you who wishes to be informed enough to become the most important member of your own health-care team.

Vitamins, Minerals and Supplements

Everyone consumes vitamins, minerals or supplements in some form. Most people rely on diet to supply all they need. Many take pills, tonics, capsules or injections to meet their needs. Of those who supplement their diet with products from non-food sources, most take the amounts recommended by knowledgeable professionals. The vitamins, minerals and supplements they take may help them feel better. Without vitamins and minerals from some source, life cannot endure.

The American Institute of Nutrition and The American Society for Clinical Nutrition recently issued an official statement on vitamin and mineral supplements. This statement was developed jointly with the American Dietetic Association and the National Council Against Health Fraud. The American Medical Association's Council on Scientific Affairs reviewed this statement and found it to be consistent with its official statement on dietary supplements. The statement reads:

"Healthy children and adults should obtain adequate nutrient intakes from dietary sources. Meeting nutrient needs by choosing a variety of foods in moderation, rather than by supplementation, reduces the potential risk for both nutrient deficiencies and nutrient excesses. Individual recommendations regarding supplements and diets should come from physicians and registered dietitians."

Supplement usage may be indicated in various circumstances. Some of these situations are listed below.

- Women with excessive menstrual bleeding may need iron supplements.
- Pregnant or breast-feeding women have an increased need of certain nutrients, especially iron, folic acid and calcium.
- People with very low calorie intakes frequently consume diets that do not meet their needs for all nutrients.
- Some vegetarians may not receive adequate calcium, iron, zinc and vitamin B-12.
- Newborns are commonly given a single dose of vitamin K to prevent abnormal bleeding. (This is done under the direction of a physician.)
- Certain disorders or diseases and some medications interfere with nutrient intake, digestion, absorption, metabolism or excretion. This changes vitamin and mineral requirements.

Nutrients are potentially toxic when ingested in sufficiently large amounts. Safe intake levels vary widely from nutrient to nutrient and may vary with the age and health of the individual. In addition, high-dosage vitamin and mineral supplements can interfere with the normal metabolism of other nutrients and with the therapeutic effects of certain drugs.

The *Recommended Dietary Allowance* (RDA) represents the best currently available assessment of safe and adequate intakes. It serves as the basis for the recommended daily allowances shown on many product labels. There are no demonstrated benefits of self-supplementation beyond these allowances. When no RDA has been established, a daily Estimated Safe Intake may be listed.

Every health professional wants consumers to take proper nutrients and supplements if they need them. But some people abuse these essential substances by taking megadoses—doses 10 to 20 times the recommended amount or more.

Some people believe if one pill is good, 20 pills must be better. They also believe vitamins, minerals and supplements aren't medicine. They are wrong!

Vitamins, minerals and supplements *are* medicine because with some doses in some people, they can cause a change in the body's physiology or internal anatomy. Many of these substances are not regulated by the federal Food and Drug Administration (FDA), but that doesn't mean they can't cause harmful as well as beneficial effects. As the information in this book points out, these substances cause side effects, adverse reactions, interactions with other drugs and unexpected problems. Some people should *not* take certain things because of a unique situation, such as pregnancy or age. Others need to be aware that medical conditions, such as heart problems or various disease conditions, can be an indication not to take certain substances.

When you make the decision about whether to supplement your diet with vitamins and minerals, there are important things to remember. Too much can be harmful—don't overdose or take megadoses! For example, megadoses of vitamin A can cause bone pain and hypertension. It can also cause birth defects in babies if a pregnant woman takes megadoses. Long-term excess vitamin E can cause a low sperm count, degeneration of testicles and sterility. Vitamin C can also have toxic effects, including gout and perhaps kidney stones. Megadoses of this vitamin can interfere with the white blood cells' ability to kill bacteria. This can make infections worse, rather than clearing them up.

Many conditions may also rule out taking some substances. Always be alert to any side effects or interactions you may experience that could put your health in jeopardy.

It's up to you to be a "smart" consumer of the vitamins, minerals and supplements your body may need. Get them from the food you eat when you can—supplement with available products when you must.

Vitamins

Vitamins are chemical compounds necessary for growth, health, normal metabolism and physical well-being. Some vitamins are essential parts of *enzymes*—the chemical molecules that catalyze or facilitate the completion of chemical reactions. Other vitamins form essential parts of *hormones*—the chemical substances that promote and protect body health and reproduction. If you're in good health, you need vitamins only in small amounts. They can be found in sufficient quantities in the foods you eat. This assumes you eat a normal, well-balanced diet of foods grown in a nutritionally adequate soil. Traditionally, vitamins have been divided into two categories: *fat soluble* and *water soluble*.

Fat-soluble vitamins can be stored in the body. If you take excessive amounts of fat-soluble vitamins, they accumulate to provide needed amounts at a later time. That's the good news. The bad news is, if you take *excessive* amounts of fat-soluble vitamins, toxic levels can accumulate in storage areas such as the liver. Too much of any fat-soluble vitamin can lead to potentially dangerous, long-term physical problems.

Water-soluble vitamins cannot be stored in the body to any great extent. The daily amount you need must be provided by what you eat or drink each day or two.

Under some circumstances, it may be difficult or impossible for you to obtain and assimilate enough vitamins simply by eating your customary diet. The amount of vitamins you need, such as during illness or following surgery, may be increased. A vitamin supplement may be necessary. People with special needs for supplements or others at risk of vitamin deficiency are identified and discussed in detail later in this section. See page 9.

Taking vitamin supplements cannot take the place of good nutrition. Vitamins do not provide energy. Your body needs other substances besides vitamins for adequate nutrition, including carbohydrates, fats, proteins and minerals. Vitamins cannot help maintain a healthy body except in the presence of other nutrients, mainly from food and minerals.

Detailed charts are provided for many important, necessary vitamins including:

- Vitamin A
- Vitamin C
- Vitamin D
- Vitamin E
- All the B vitamins

Minerals

Minerals are inorganic chemical elements not attached to a carbon atom. They participate in many biochemical and physiological processes necessary for optimum growth, development and health. There is a clear and important distinction between the terms *mineral* and *trace element*. If the body requires more than 100 milligrams of a mineral each day, the substance is labeled *mineral*. If the body requires less than 100 milligrams of a mineral each day, the substance is labeled *trace element*.

Many minerals are essential parts of enzymes. They also participate actively in regulating many physiological functions, including transporting oxygen to each of the body's 60-trillion cells, providing the stimulus for muscles to contract and in many ways guaranteeing normal function of the central nervous system. Minerals are required for growth, maintenance, repair and health of tissues and bones.

Most minerals are widely distributed in foods. *Severe* mineral deficiency is unusual in the Western world. Of all essential minerals, only a few may be deficient in a typical diet. Even so, there are exceptions. Iron deficiency is common in infants, children and pregnant women. Zinc and copper deficiencies occur fairly frequently.

Detailed charts are provided for many minerals, including:

- Calcium
- Chloride
- Magnesium
- Phosphorous
- Potassium
- Sodium
- Sulfur

In addition, detailed charts are provided for trace elements, including:

- Chromium
- Fluorine
- Iodine
- Iron
- Manganese
- Molybdenum
- Selenium
- Vanadium
- Zinc

Nickel, tin, silicon and arsenic are also considered essential. Charts are *not* included because all available information about these trace elements comes from studies done on animals, not human experiments or experience.

Multivitamin/Mineral Preparations

Some foods contain all the nutrients you need. For healthy people who are past the growing stage and not over 55, food is the best, most-reliable source of nutrients—if you eat a well-balanced, nutritious diet every day. Not many people fit all these parameters.

If you or your children need supplementation, it is probably better to take one of the commercially available multivitamin/mineral preparations rather than attempt to augment your food intake with separate products containing only one or two substances. Commercial over-the-counter products usually have a good balance of nutrients. Taking separate products may lead to an imbalance of nutrients, which can lead to an overabundance of one substance at the expense of decreased absorption or effectiveness of another. The cost is also much less if you take a combination

3

product rather than separate products.

There are exceptions to this rule. For example, iron and folic-acid needs during pregnancy should be met with a single product.

Most major pharmacentical manufacturers supply widely advertised combination products. The brand names are too numerous to list and change constantly. Your pharmacist or doctor should be able to recommend a good source for a superior multivitamin/ mineral preparation.

If you study vitamins and minerals, you may find you need supplements for one reason or another. I hope this book provides you with enough information to choose wisely or be able to ask the right questions to find out what is best for you.

Supplements

Supplements are chemical substances that are neither vitamins nor minerals, but they have received notice as nutritional supplements. Many supplements have proven effects in the body but may not yet have proved safe and effective when taken in pill or capsule form to supplement normal food intake. Speculated benefits and claims frequently go beyond what can be proved at present. These include anti-aging properties and claims that substances create and preserve health.

People separate into two distinct groups almost immediately when talk turns to supplementation. On one hand, the traditional medical establishment (of which I am part—partially renegade— but still a part) usually cries, "Eat a well-balanced diet, and you'll get all the carbohydrates, fat, fiber, protein, vitamins, minerals and micronutrients you need."

But hard data now available about our "normal, well-balanced" diet shows we are overfed and undernourished. The majority of experts in the medical field and in nutrition now agree. We consume too many calories, too much fat, too little fiber, too much refined sugar, too much sodium and not enough unrefined carbohydrates. So insisting a normal, well-balanced diet is all we need is a concept that is in deep trouble.

On the other hand, some view every new supplement or every new promising piece of information about the existing supplements as a miracle that will cure our ills if the overly conservative medical establishment and the FDA will get out of the way. Advertisers are quite successful with this group because many people are easily persuaded if they take a product, they will be healthier, live longer and look and feel sexier, slimmer and smarter.

Not much is written that takes a middle ground. I believe this position represents the true status of human nutrition at present. But this book *does* take a middle ground! No personal opinions are expressed, only the consensus of the majority of experts, presented as impartially as possible.

Selected supplements discussed in this book include amino acids, nucleic acids and other supplements. Detailed charts are provided for many supplements. Amino acids detailed in charts include:
- Arginine
- L-cysteine
- L-lysine
- Methionine
- Phenylalanine

Nucleic acids detailed in charts include:
- Adenosine
- DNA/RNA
- Inosine
- Orotate
- Taurine
- Tryptophan
- Tyrosine

Other supplements detailed in charts include:
- Coenzyme Q
- Dietary fiber
- Gamma-linolenic acid (evening primrose oil)
- Inositol
- L-carnitine
- Lecithin
- Omega-3 fatty acids (EPA/Max EPA)
- Superoxide dismutase
- Wheatgrass

How to Use the Vitamin-Minerals-Supplements Section of This Book

Information in this book is organized in condensed, easy-to-read charts. Each vitamin or mineral is described in a multipage format as shown in the sample charts on the following pages. Vitamins and minerals are arranged alphabetically by the most frequently recognized name—usually a generic name instead of a brand name.

Each name appears at the top of the chart. For example, vitamin B-12 is frequently called *cyanocobalamin*. Both names are given when there are two or more names. In addition, many substances are sold by brand names rather than generic names. Commonly available commercial brand names are listed, beginning on page 477.

To learn more about any vitamin, mineral or supplement, you need only one name. Look in the Index for any name you have, page 495. The Index provides a page number for the information you seek about that vitamin, mineral, supplement or herb.

The book is divided into five main section—vitamins, minerals, nucleic acids and amino acids, other supplements and medicinal herbs. Information on the medicinal-herbs section and charts begins on page 255. Names of substances are listed alphabetically within each section. Chart design is the same for every substance in the first three sections. When you become familiar with the chart, you can quickly find the information you seek.

On the next few pages, each numbered section is explained. This information will help you read and understand the charts that begin on page 22.

1-Generic name

Each chart is titled by the *generic name*. Sometimes there are two or more generic names. Each entry is titled by the most common one. Each name is listed on the top line or under the *Basic Information* section. If a substance has two or more generic names, the Index includes a reference for each name.

A product container may show a generic name, a brand name or both. Any name listed on the label should be listed in the Index. If the container has no name, ask the pharmacist or health-store attendant for the name.

2-Brand names

A brand name is frequently shorter and easier to remember than a generic name. Brand names are keyed to the generic name and can be found on page 477. The most-common ones are listed. New brands appear frequently, and brands are sometimes taken off the market. No list can reflect instantaneous changes. Inclusion of a brand name does *not* imply recommendation or endorsement. Exclusion does *not* imply a brand name is less effective or less safe than those listed.

Most vitamins, minerals and supplements contain inert, or inactive, fillers or solvents. Manufacturers choose inert ingredients that preserve the product without interfering with the action of the active ingredients. Inert ingredients are sometimes listed on the label of a product and may represent substances to which you are allergic. Read *all* labels carefully. Avoid taking any products with ingredients you are allergic to or that cause adverse reactions.

Occasionally a tablet or capsule contains a small amount of sodium or sugar. Liquids frequently contain alcohol. If you're on a diet or take medication that restricts any substance, ask your doctor,

Guide to Vitamins, Minerals, Supplements Charts

To find information about a specific vitamin, mineral, amino acid, nucleic acid or supplement, look in the easy-to-read charts starting on page 22. Charts like the samples shown below and on the opposite page appear alphabetically by generic name.

A *generic* name is the official chemical name. A substance listed by generic name may

1—Riboflavin (Vitamin B-2)

Basic Information

2 —— Brand names, see page 477.
3 —— Available from natural sources? Yes
4 —— Available from synthetic sources? Yes
5 —— Prescription required? No
6 —— Fat-soluble or water-soluble: Water-soluble

 Natural Sources

Almonds
Brewer's yeast
Cheese
Chicken
Organ meats (beef, kidney)
Wheat germ

 Reasons to Use

• Aids in release of energy from food.
• Maintains healthy mucous membranes lining respiratory, digestive, circulatory and excretory tracts when used in conjunction with vitamin A.
• Preserves integrity of nervous system, skin, eyes.
• Promotes normal growth and development.
• Aids in treating infections, stomach problems, burns, alcoholism, liver disease.

 Unproved Speculated Benefits

• Cures various eye diseases.
• Treats skin disorders.
• Prevents cancer.
• Increases body growth during normal developmental stages.
• Helps overcome infertility.
• Prevents stress.
• Stimulates hair growth in bald men.
• Improves vision.

 Who Needs Additional Amounts?

• Anyone with inadequate caloric or nutritional dietary intake or increased nutritional requirements.
• Pregnant or breast-feeding women.
• Those who abuse alcohol or other drugs.
• People with a chronic wasting illness, excess stress for long periods or who have recently undergone surgery.
• Athletes and workers who participate in vigorous physical activities.

• Those with a portion of the gastrointestinal tract surgically removed.
• People with recent severe burns or injuries.
• Those who rely almost exclusively on processed foods for their daily diet.
• Women taking oral contraceptives or estrogen.

 Deficiency Symptoms ——11

• Cracks and sores in corners of mouth
• Inflammation of tongue and lips
• Eyes overly sensitive to light and easily tired
• Itching and scaling of skin around nose, mouth, scrotum, forehead, ears, scalp
• Trembling
• Dizziness
• Insomnia
• Slow learning
• Itching, burning and reddening of eyes
• Damage to cornea of eye.

 Unproved Speculated Symptoms ——12

• Mild anemia
• Mild lethargy
• Acne
• Migraine headaches
• Muscle cramps

 Lab Tests to Detect Deficiency ——13

• Serum riboflavin
• Erythrocyte riboflavin
• Glutathione reductase

 Dosage and Usage Information

Recommended Dietary Allowance (RDA):
Estimate of adequate daily intake by the Food —14
and Nutrition Board of the National Research
Council, 1980. See Glossary.

have many brand names. Brand names for the United States and Canada are in a list starting on page 477.

Chart design and information presentation is similar for vitamins, minerals, amino acids, nucleic acids and supplements. On the next few pages I explain each of the sections, using the numbers on the charts as a reference. All charts are organized similarly, making it easy to make comparisons.

Riboflavin (Vitamin B-2)

Age	RDA
0–6 months	0.4mg
6–12 months	0.6mg
1–3 years	0.8mg
4–6 years	1.0mg
7–10 years	1.4mg
Males	
11–14 years	1.6mg
15–22 years	1.7mg
23–50 years	1.6mg
51+ years	1.4mg
Females	
11–22 years	1.3mg
23+ years	1.2mg
Pregnant	+0.3mg
Lactating	+0.5mg

24 Storage:
- Store in cool, dry place away from direct light, but don't freeze.
- Store safely out of reach of children.
- Don't store in bathroom medicine cabinet. Heat and moisture may change action of vitamin.

25 Others:
- Unlikely to cause toxic symptoms in healthy people with normal kidney function.

15 What this vitamin does:
- Acts as component in two co-enzymes (flavin mononucleotide and flavin adenine dinucleotide) needed for normal tissue respiration.
- Activates pyridoxine.

16 Miscellaneous information:
- A balanced diet prevents deficiency without supplements.
- Large doses may produce dark-yellow urine.
- Processing food may decrease quantity of vitamin B-2.
- Mixing with baking soda destroys riboflavin.

17 Available as:
- Tablets: Swallow whole with full glass of liquid. Don't chew or crush. Take with or immediately after food to decrease stomach irritation.
- A constituent of many multivitamin/mineral preparations.

Warnings and Precautions

18 Don't take if you:
- Are allergic to any B vitamin.
- Have chronic kidney failure.

19 Consult your doctor if you are:
- Pregnant or planning a pregnancy.

20 Over age 55:
- Need for vitamin B-2 is greater.

21 Pregnancy:
- Don't take megadoses.

22 Breast-feeding:
- Don't take megadoses.

23 Effect on lab tests:
- Urinary catecholamine concentration may show false elevation.
- Urobilongen determinations (Ehrlich's) may produce false-positive results.

Overdose/Toxicity

26 Signs and symptoms:
Dark urine, nausea, vomiting.

What to do:
27 For symptoms of overdosage: Discontinue vitamin, and consult doctor. Also see *Adverse Reactions or Side Effects* section below.
For accidental overdosage (such as child taking entire bottle): Dial 911 (emergency), 0 for operator or your nearest Poison Control Center.

28 Adverse Reactions or Side Effects

Reaction or effect	What to do
Yellow urine, with large doses	No action necessary

29 Interaction with Medicine, Vitamins or Minerals

Interacts with	Combined effect
Anti-depressants (tricyclic)	Decreases B-2 effect.
Phenothiazines	Decreases B-2 effect.
Probenecid	Decreases B-2 effect.

30 Interaction with Other Substances

Tobacco decreases absorption. Smokers may require supplemental vitamin B-2.

Alcohol prevents uptake and absorption of vitamin B-2.

pharmacist or retail merchant to suggest another form.

3-Available from natural sources?
4-Available from synthetic sources?

Many vitamins, minerals and supplements are advertised as "natural," implying the product is derived from natural sources as opposed to synthetic sources. By definition, minerals are basic chemical substances that can't be manufactured (or synthesized) from other substances. However, many vitamins and supplements are derived from both sources.

This is confusing to many consumers. Many manufacturers have done everything possible to take financial advantage of that confusion. Advertisers claim natural sources are good and synthetic sources are bad. The truth is, natural and synthetic versions of the same chemical are identical!

Don't pay extra money for *natural* vitamins or supplements. They all have the same effect on your body. The *synthetic* version may even be purer or less contaminated with extraneous materials such as insecticides and fertilizers.

5-Prescription required?

Most vitamins, minerals and supplements are available without prescription. Some formulas with higher dosages to treat specific diseases require a prescription from your doctor. "Yes" means your doctor must prescribe. "No" means you can buy this product without prescription. The information about a generic product is the same, whether it requires a prescription or not. If generic ingredients are the same, non-prescription products have the same uses, dangers, warnings, precautions, side effects and interactions with other substances that prescription products do.

6-Fat soluble or water soluble?

This section applies *only* to vitamins. Fat-soluble vitamins can accumulate in the body and might cause toxic effects in excessive doses, either in a single day or in small, periodic excesses over a long time. Water-soluble vitamins do not accumulate to any great extent in the body. Except under unusual circumstances, the body readily eliminates excess water-soluble accumulation in the body. The dangers of water-soluble vitamins generally depend on the effects of excessive dosages taken over a relatively short period.

7-Natural sources

This is a list of the food and beverage sources from which vitamins, minerals and supplements may be obtained. No attempt has been made to rank them according to the richest sources. They are listed alphabetically. If you want more information about natural sources, many reference works are available at your local library.

8-Reasons to use

This section consists of *proved benefits*, including body functions the substance maintains or improves. It also lists disease processes and malfunctions the substance cures or improves. These proved benefits have withstood the scrutiny of scientifically controlled studies with results published in medical literature. This medical literature is subjected to review by top authorities in many fields before the material can be published in respected scientific journals.

9-Unproved speculated benefits

Some authors and many newspaper, magazine and television advertisers make unjustified, sometimes outrageous, claims for products.

This list contains claims that have *not* withstood the same scientific scrutiny the

8

Reasons to use section has passed. These claims may be as accurate and as effective as the proved claims. But they haven't been proved with well-controlled studies. Such studies can take years to complete and may be very expensive. Until such studies have been completed, the claims must be listed as unproved.

10-Who needs additional amounts?

People listed in this category are most likely to need significant care to regain or maintain normal health or who are less likely to meet their requirements through diet alone. A summary of groups follows, with a list of reasons why the risk is greater.

Anyone with inadequate dietary intake or increased nutritional needs—Included in this group are people whose energy needs are less than 1,200 calories a day. Fewer than 1,200 calories a day for energy requirements almost never provides enough vitamins and minerals, so supplements are needed. Those most likely to have inadequate dietary intake include:

ès People of small stature or body build who eat only minimal nutrients per day to maintain current weight.

ès Elderly people with greatly decreased daily activities. This applies particularly to aging women.

ès People who have had limbs amputated.

ès People with reduced physical activity because of activity-limiting disease, such as coronary-artery disease, intermittent lameness, angina pectoris.

ès Fad dieters with a dietary imbalance and inadequacy.

ès People with eating disorders such as anorexia nervosa and bulimia.

ès Vegetarians.

Older people (55 and over)—People in this age group may have inadequate dietary intake because of difficulty obtaining an adequate diet, or because of disability and depression.

Pregnancy—Pregnant women uniformly need supplementation of folic acid and iron. Sometimes they need other supplements as well. Pregnant women need to increase dietary intake so total body weight increases from 12 to 30 pounds during pregnancy. Many women do not consume enough calories to allow this weight gain and therefore develop a nutritional deficiency. This causes a need for supplementation with a well-rounded, well-balanced preparation containing vitamins and minerals as well as the need for separate folic-acid and iron supplementation.

Ask your doctor for recommendations on specific brand names of acceptable multivitamin/mineral preparations. Also seek advice about folic acid and iron.

Breast-feeding women—Breast-feeding women who are healthy and active need to continue supplementation. Make sure you get enough iron. Talk to your doctor about your concerns.

Most authorities suggest iron and folic-acid supplements for pregnant and breast-feeding women should be taken as separate products. Iron occasionally causes gastrointestinal side effects that are so uncomfortable to some women that they discontinue the supplements.

Another important nutritional factor with breast-feeding is the need for extra fluids. Fluid deficiency can be as disabling as a nutritional deficiency. Drink at least eight 8-ounces glasses of water a day.

People who abuse alcohol and other drugs—People who consume too much alcohol are likely to develop nutritional deficiencies. Much of the daily caloric intake of these people is the alcohol they consume; it is deficient in nutritional substances. In addition, there is also poor absorption of food and increased excretion of nutrients because of diarrhea and fluid loss. When the excessive alcohol consumption stops, the nutritional deficiency can be treated with good food and supplements for a while, if liver disease has not already occurred.

9

Abuse of other drugs frequently leads to decreased appetite and decreased interest in food. Addicts need supplements of both vitamins and minerals.

People with a chronic wasting illness—This group includes people with malignant disease, chronic malabsorption, hyperthyroidism, chronic obstructive pulmonary disease, congestive heart failure, cystic fibrosis and other illnesses. Nutritional risk is increased because these people have greatly increased caloric and nutritional requirements that are difficult to satisfy with food.

People who have recently undergone surgery—Surgery can cause a relative deficiency, even if a person is well nourished before surgery. People who have undergone surgery on the gastrointestinal tract are particularly likely to develop deficiences during the post-operative period. Supplementation is very helpful. Vitamins and minerals are frequently administered intravenously until the patient can eat. After that, most people benefit from vitamin and mineral supplements for several weeks post-operatively.

People with a portion of the gastrointestinal tract removed—These people are likely to develop deficiencies because important nutrient-absorbing parts of the gastrointestinal tract may be absent from the body. A good multivitamin/mineral preparation usually prevents signs and symptoms of deficiences. Vitamin B-12 must be supplemented for life (usually by injection) for *all* people with a significant portion of the stomach removed.

People who must take medicines—Many medications can cause a deficiency of vitamins and minerals. Specific drugs are listed in the separate profiles for each substance in this book. In general, laxatives, antacids, medicines to treat epilepsy, oral contraceptives and several other medications can cause a special need for supplementation for adequate vitamin and mineral absorption.

People who have recently sustained severe injuries or severe burns—The nutritional requirement for these people is greatly increased. Faster healing and recovery can be aided by adequate supplementation. Ask your doctor for specific advice.

11-Deficiency symptoms

Contains a list of *proved symptoms*. These symptoms of deficiency have withstood the scrutiny of scientifically controlled studies with results published in medical literature.

12-Unproved speculated symptoms

Contains deficiency symptoms that have *not* withstood the scrutiny of scientifically controlled studies. Results have not been published in medical literature.

13-Lab tests to detect deficiency

Sometimes clinical features—medical history, signs and symptoms as interpreted by a competent professional—are all that are required to make an accurate diagnosis of deficiency. At other times, although clinical features may suggest a specific diagnosis, objective proof by a specific laboratory test adds confidence. As much data as can be collected is desirable before committing to a prolonged, sometimes expensive, sometimes hazardous, course of treatment. When lab tests are readily available and reasonable in cost, doctors can treat their patients with greater confidence than is possible without laboratory confirmation of the diagnosis. This section lists many of those studies.

Note: Analysis of hair samples to detect deficiencies of minerals and trace elements, while easily available commercially, cannot be regarded as a valid test. Minerals and trace elements appear in shampoos, hair-care products and generally in the environment. In addition, when nutrition is poor for any reason, hair growth actually slows—causing greater concentration of minerals in the hair. This greater concentration gives falsely high values. Hair tests are entirely without value except for experimental purposes.

14-Recommended dietary allowance (RDA)

RDA is an estimate of amounts of a nutrient required daily by people with the highest requirements in the general healthy population. Estimates are made by the Food and Nutrition Board of the National Academy of Sciences, which began publishing recommendations in the 1940s. Recommendations are updated periodically to reflect changing opinions of the majority of experts. Because knowledge of human requirements is always changing, there is continuing controversy over the optimal levels of intake.

Many dietitians and nutrition experts believe the RDAs do not ensure optimum health. The RDAs represent the only official guide to safety. They have been carefully calculated and, at the very least, are a good reference point.

It is probably impossible to get all the recommended nutrients from today's diet because the foods it includes are highly processed and refined.

For some nutrients, no RDA has been established. An Estimated Safe Intake (per day) may be included.

15-What this substance does

Includes a brief discussion of the part each substance plays in chemical reactions or combinations that affect growth, development and health maintenance.

16-Miscellaneous information

Information in this section doesn't fit readily into other information blocks on the charts. Some information includes:

  🙵 Cooking tips to preserve the substance during food preparation.

  🙵 Time lapse before changes can be expected.

  🙵 Information of special interest.

17-Available as

Different available forms of the vitamin, mineral or supplement are discussed. These include tablets, powders, capsules, injections and oral forms.

18-Don't take if you have

Lists circumstances when use of this vitamin, mineral or supplement may not be safe. In formal medical literature, these circumstances are called *absolute contraindications.*

19-Consult your doctor if you have

Lists conditions under which a vitamin, mineral or supplement should be used with caution. In formal medical literature, these circumstances are frequently listed as *relative contraindications.* Using this product under these circumstances may require special consideration on your part and your doctor's. The rule is—*the potential benefit must outweigh the possible risk!*

20-Over age 55

As a person ages, physical changes occur that require special consideration when using vitamins, minerals and supplements. Liver and kidney function usually decreases, metabolism slows and other changes take place. These are expected and must be considered.

Most chemical substances introduced into the body are metabolized or excreted at a rate that depends on kidney and liver functions. In the aging population, smaller doses or longer intervals between doses may be necessary to prevent an unhealthy concentration of vitamins, minerals or supplements. These principles are exactly the same for therapeutic medicines and drugs. Toxic effects, severe side effects and adverse reactions occur more frequently and may cause more serious problems in this age group.

21-Pregnancy

Pregnancy creates an increased need for optimal nutrition, which may be difficult to maintain without using some supplemental vitamins and minerals. What you take depends on your age, your present state of nutrition, your state of health and other factors. Work with your doctor to determine what supplements you will need and how much. Don't take *any* substance without consulting your doctor first!

22-Breast-feeding

Lactating mothers require sound nutrition. Follow your doctor's recommendations about diet, vitamins, minerals and supplements during this time. Don't be reluctant to ask questions and challenge your doctor regarding these important topics. But don't take *any* substance without consulting your doctor first!

23-Effect on lab tests

This section lists lab studies that may be affected when you take vitamins, minerals or supplements. Possible effects include causing a false-positive or false-negative test, resulting in a low result or high result when your actual physical state is the opposite. In general, some tests can be performed accurately only after discontinuing vitamins, minerals or supplements for a few days before the test is scheduled.

24-Storage

This serves as a reminder to keep these substances safely away from children. It also discusses how and where to store vitamins, minerals and supplements.

25-Others

Special warnings and precautions appear here if they don't fit any other specific information block. This section may contain information about the best time to take the substance, instructions about mixing or diluting or anything else that is important about this substance.

26-Overdose signs and symptoms

Symptoms listed are the ones most likely to develop with accidental or deliberate overdose. An overdosed person may not show all symptoms listed and may experience other symptoms not listed. Sometimes signs and symptoms are identical with ones listed as side effects or adverse reactions. The difference is intensity and severity. You must be the judge. Consult a doctor or poison control center if you are in doubt.

27-What to do

If you suspect an overdose, whether symptoms are apparent or not, follow instructions in this section. Expanded instructions for overdose or *anaphylaxis*—severe, life-threatening allergic reaction—appear in the Glossary. See page 485.

28-Adverse reactions or side effects

Adverse reactions or side effects are symptoms that may occur when you ingest any substance, whether it is food, medicine, vitamin, mineral, herb or supplement. These are effects on the body other than the desired effect for which you take them.

The term *side effect* may include an expected, perhaps unavoidable, effect of a vitamin, mineral or supplement. For example, various forms of niacin may cause dramatic dizziness and flushing of the face and neck in the blush zone in almost everyone who takes a high enough dose. These symptoms are harmless, although sometimes uncomfortable, and have nothing to do with the intended use or therapeutic effect of niacin.

The term *adverse effect* is more significant. These effects can cause hazards that outweigh benefits.

29-Interaction with medicine, vitamins or minerals

Vitamins, minerals, supplements, herbs and various medicines may interact in your body with other vitamins, minerals, supplements, herbs and medicines. It doesn't matter if they are prescription or non-prescription, natural or synthetic.

Interactions affect absorption, elimination or distribution of the substances that interact with each other. Sometimes they are beneficial, but at other times they are deadly. You may not be able to determine from the chart whether an interaction is good or bad. Don't guess! Ask your doctor or pharmacist—some interactions can kill!

30-Interaction with other substances

This list includes possible interactions with food, beverages, tobacco, cocaine, alcohol and other substances you may ingest.

Medicinal Herbs

You may be wondering why a conservative, traditional doctor would write about medicinal herbs. Many medical professionals ignore their existence. There are several reasons why these substances are included in this book.

A popular backlash currently exists against traditional medicine as it is practiced today. The medical profession has brought some negative feelings upon itself. Part of this backlash takes the form of returning to "natural" medicine—specifically to any of the 2,500 herbs that have been used throughout history for medicinal purposes. People self-prescribe these plant materials and believe they are saving time and money by not consulting a traditional physician. Because medicinal herbs are natural and unregulated, many people believe they are without hazards. This is not true!

For centuries, people have collected herbs to use for medicinal purposes. Very little of the experience has been written down. It has been passed down verbally instead. Most uses for herbs in the past and most of the reasons people use them today are probably without scientific foundation. Yet some of mankind's most useful medicines, such as digitalis, rauwolfia (used for mental illness and hypertension), cromlyn (used for preventing asthma attacks) and curare (a muscle relaxant), have all come from herbal "folk remedies."

Many medicinal herbs have pharmacological properties that we know are useful. But at the same time they may be harmful or toxic. Medicinal herbs are available in many forms. Most have *not* been scrutinized for safety and effectiveness by the FDA.

People have turned to medicinal herbs, believing they are "natural," safe, effective and wonderful. However, experience has taught us *any* effective medicine can also have uncomfortable side effects, adverse reactions and dangerous possible toxicity, just as many pharmaceuticals do.

Active ingredients of medicinal herbs vary greatly, whether you personally collect plant drugs or buy them. Variable factors include:

- Conditions under which the plant was grown (soil conditions, temperature, season).
- Degree of maturity of the plant when it was collected.
- Type of drying process.
- Type and duration of storage.

In conventional medicine, these variables are controlled by manufacturing procedures or government tests or assays to standardize the amount of the active principal and therefore the predictable safety and efficacy of the material. None of these safeguards exist for medicinal herbs.

The Placebo

The *placebo effect* has long been held as an advantage of using medicinal herbs. Many scientists and researchers claim most herbs do not really help people—it's the placebo effect of using these herbs that really heals. The word *placebo* comes from a Latin predecessor meaning "to please" or "to serve." Under a strict interpretation of the term as it is now used, a placebo medication has no pharmacologically or biologically active ingredients. Another interpretation asserts amounts *commonly* used could not affect the body, but *large amounts* of the same substance may.

For centuries, healers have helped people who were ill, no matter what the illness. Many ancient healers used remedies that have no pharmacological effects in the body. But these remedies were not always useless. They frequently proved to be very effective.

Modern studies conclusively prove *all* remedies help relieve symptoms in *some* people. In the early 1900s, many patients and physicians believed placebo therapy was quackery. Today, we know this to be untrue. Placebos *can* mimic the effect of almost any active drug. Placebo effects are real, although we cannot scientifically determine why.

How does the placebo effect work? We don't know for certain, but there are different theories.

Endorphins—chemicals normally present in the brain—can be activated by exercise, stress, mental exercises and imaging. Once endorphins have been activated, they kill pain the same way narcotics kill pain. Placebo treatment can trigger the production of hormones in the body, such as cortisone and adrenalin. This can affect the way we behave, the way we feel, the way we think. If the placebo can cause production of these chemicals, this may relieve symptoms of many disorders.

Harder to explain is the part that "power of suggestion" may play in the effectiveness of any remedy, whether it is a powerful drug, a supplement, an herb or a placebo. The gentle touch of the healer, the taste and smell of the product, the packaging, the cost—all are factors that have been studied and found to play a part in the placebo effect.

Understanding Common Terms

When you read about medicinal herbs, some of the following terms are used repeatedly. They refer to ways in which medicinal herbs can be useful.

Compress—Cloth is soaked in a cool liquid form of an herb, wrung out and applied directly to skin.

Decoction—Herb is boiled 10 to 15 minutes, then allowed to steep.

Extract—Solution resulting from soaking herb in cold water for 24 hours.

Fomentation—Cloth is soaked in a hot liquid form of an herb, wrung out and applied directly to skin.

Infusions—Tea is prepared by steeping herb in hot water. Infusions can be made from any part of a plant.

Ointment—Powdered form of an herb is mixed with any soft-based salve, such as lanolin, wax or lard.

Poultice—Herb applied to a moistened cloth, then applied directly to skin.

Powder—Useful part of herb is ground into a powder.

Syrup—Herb is added to brown sugar dissolved in boiling water, then boiled and strained.

Tincture—Powdered herb is added to a 50-50 solution of alcohol and water.

Points to Remember

Precautions apply to herbal medications. Read the checklist on page 19. Also keep in mind:

ã Children under age 2 should *not* be given herbal medications.

ã Pregnant and lactating women should avoid herbal medicines because of potential damage to the fetus or breast-feeding child.

ã Collecting medicinal herbs for yourself is unwise, unless you have received a great deal of training. Correctly identifying plants and knowing how to select, preserve and use them properly requires a great deal of knowledge and judgment.

The medicinal-herb section of the book, page 255, contains profiles of the herbs most generally available and most frequently used in the United States and Canada. An extensive toxicity list, beginning on page 458, follows the charts. This list contains the names and possible toxic effects of over 350 medicinal herbs. Some herbs in this list do not have charts in this book. They are included for your reference.

Guide to Medicinal-Herbs Charts

The medicinal-herb information in this book is organized into condensed, easy-to-read charts. Each medicinal herb is described on a 1-page chart, as shown in the sample chart below. Charts are arranged alphabetically by the most-common herbal name. If you cannot find a name, look for alternate names in the Index or ask your herbal-medication retailer.

Boneset (Richweed, White Snakeroot, Ague Weed) —A

B— **Basic Information**

Biological name (genus and species):
Eupatorium perfoliatum, E. rugosum

C— *Parts used for medicinal purposes:*
Leaves
Petals/flower

D— *Chemicals this herb contains:*
Eupatroin
Resin (See Glossary)
Sugar
Tremetrol
Volatile oils (See Glossary)
Wax (See Glossary)

E— **Known Effects**

• Irritates gastrointestinal tract.
• Can produce "milk sickness" in humans, an acute disease characterized by trembling, vomiting and severe abdominal pain. It is caused by eating dairy products or beef from cattle poisoned by eating boneset.
• Increases perspiration.
• Causes vomiting.

F— **Miscellaneous information:**
• Tremetrol can accumulate slowly in animal bodies and cause toxic symptoms. It may do the same in humans.

G— **Unproved Speculated Benefits**

• Decreases blood sugar.
• Treats malaria.
• Treats fever.

Warnings and Precautions

H— **Don't take if you:**
• Are pregnant, think you may be pregnant or plan pregnancy in the near future.
• Have any chronic disease of the gastrointestinal tract, such as stomach or duodenal ulcers, esophageal reflux (reflux esophagitis), ulcerative colitis, spastic colitis, diverticulosis, diverticulitis.

I— **Consult your doctor if you:**
• Take this herb for any medical problem that doesn't improve in 2 weeks. There may be safer, more-effective treatments.

• Take any medicinal drugs or herbs including aspirin, laxatives, cold and cough remedies, antacids, vitamins, minerals, amino acids, supplements, other prescription or non-prescription drugs.

J— **Pregnancy:**
• Dangers outweigh any possible benefits. Don't use.

K— **Breast-feeding:**
• Dangers outweigh any possible benefits. Don't use.

L— **Infants and children:**
• Treating infants and children under 2 with any herbal preparation is hazardous.

M— **Others:**
• Dangers outweigh any possible benefits. Don't use.

N— **Storage:**
• Keep cool and dry, but don't freeze. Store safely away from children.

O— **Safe dosage:**
• At present no "safe" dosage has been established.

P— **Toxicity**

Comparative-toxicity rating not available from standard references.

For symptoms of toxicity: See below.

Q— **Adverse Reactions, Side Effects or Overdose Symptoms**

Signs and symptoms:	What to do:
Breathing difficulties	Seek emergency treatment.
Coma	Seek emergency treatment.
Drooling	Discontinue. Call doctor when convenient.
Muscle trembling	Discontinue. Call doctor immediately.
Nausea	Discontinue. Call doctor immediately.
Stiffness	Discontinue. Call doctor when convenient.
Vomiting	Discontinue. Call doctor immediately.
Weakness	Discontinue. Call doctor when convenient.

A-Popular name

Each chart is titled by the most popular name. Sometimes there may be two or more names. Alternate names are shown in parentheses. The Index contains a reference to each name listed. Popular names may vary in different parts of the world.

B-Biological name (genus and species)

Identifies the medicinal herb by genus and species. These Latin names are commonly used by biologists and plant scientists. They are included to help you make a positive identification.

C-Parts used for medicinal purposes

Describes what parts of the herb are used to supply the expected effects. Roots, leaves, bark and flowers are commonly used portions of the plant. Sometimes the entire plant is used.

D-Chemicals this herb contains

Chemicals and family names of chemically related groups are listed. Chemically related groups include saponins, tannins, volatile oils and others.

E-Known effects

Expected effects of these chemicals are the identified chemical actions of the medicinal herb being discussed. These effects have been identified and validated by scientists and researchers through various studies. Some effects may be beneficial; others are harmful.

F-Miscellaneous information

Contains information that doesn't fit into other information blocks on the chart.

G-Unproved speculated benefits

List of symptoms or medical problems this drug has been reported to treat or improve. These claims may be accurate, but they haven't been proved with well-controlled studies.

H-Don't take if you

Lists circumstances under which the use of this herb may not be safe. In formal medical literature, these circumstances are listed as *absolute contraindications*.

I-Consult your doctor if you

Lists conditions in which this herb should be used with caution. In formal medical literature, these circumstances are called *relative contraindications*. Using an herb under these circumstances may require special consideration by you and your doctor. The rule to follow is— *the potential benefit must outweigh the possible risk.*

J-Pregnancy

As more is learned about effective medications, including herbal medications, the more health-care workers fear the possible effects of any medicinal product on an unborn child. This fear holds for *all* chemicals that cause changes in the body. The fact herbal medicines occur naturally does not free them from possibly causing harm. *The best rule to follow is don't take anything during pregnancy if you can avoid it!*

K-Breast-feeding

Although a breast-feeding newborn infant is not as likely to be harmed as an unborn fetus, caution should be observed. If you take a medicine or an herb during the time you breast-feed, do so *only* under professional supervision.

L-Infants and children

Treating infants and children under 2 years old with any herbal medication or preparation is hazardous. Dosages, uses and effects of an herb cannot be gauged easily with a young child. Do not use medicinal herbs to treat a problem your child may have without first discussing it thoroughly with your doctor.

M-Others

Warnings and precautions appear here if they don't fit into other categories.

N-Storage

This serves as a reminder to keep these substances safely away from children. It also discusses how and where to store medicinal herbs.

O-Safe dosage

Safe dosages have *not* been documented by procedures outlined by the FDA. It is impossible to list a "safe" dosage and have it carry any significance. People who have had experience with herbs are usually qualified to predict safe doses if they know the person's age, past medical history and some important facts about his or her current health.

Many reputable distributors of herb products have recommendations for ranges of safety, but these may vary a great deal from manufacturer to manufacturer, according to age and purity of the product. The most important fact to understand is the more you ingest of a medicinal herb over a long period of time, the more likely a toxic reaction will occur. Most available herbs are safe when taken in small doses for short periods of time. Never fall into the trap of thinking "if a little is good, more is better."

P-Toxicity

Includes a general, average toxicity rating for each medicinal herb.

Q-Adverse reactions, side effects or overdose symptoms

Adverse reactions or side effects are symptoms that may occur when you ingest any substance, whether it is food, medicine, vitamin, mineral, herb or supplement. These are effects on the body other than the desired effect for which you take them.

The term *adverse effect* means the effects can cause hazards that outweigh benefits.

The term *side effect* may include an expected, perhaps unavoidable, effect of a vitamin, mineral, supplement or medicinal herb. For example, a side effect of horseradish may be nausea. This symptom is harmless although sometimes uncomfortable and has nothing to do with the intended use.

If you suspect an overdose, whether symptoms are present or not, follow instructions in this section.

Warning

Whether you use medicinal herbs or not is your decision. If you choose to use them, be sure you take them with knowledge and understanding of what they are. Know the supplier, and be sure you know the possible dangers. And consider that self-medication with medicinal herbs may prevent you from receiving better help from more effective medications that have withstood critical scientific investigations.

Checklist for Safer Use of Vitamins, Minerals, Supplements & Medicinal Herbs

The most important caution regarding all vitamins, minerals, supplements and medicinal herbs deals with the amount you take. Despite many popular articles in magazines and newspapers and reports on television, large doses of some of these substances can be hazardous to your health. Don't believe sensational advertisements and take large doses or megadoses. The belief "if a little does good, a lot will do much more" has no place in rational thinking regarding products to protect your health. Stay within safe-dose ranges!

1. Learn all you can about the vitamins, minerals, supplements and medicinal herbs *before* you take them. Information sources include this book, books from your public library, your doctor or your pharmacist.

2. Don't take vitamins, minerals, supplements or medicinal herbs prescribed for someone else, even if your symptoms are the same. At the same time, keep prescription items to yourself. They may be harmful to someone else.

3. Tell your doctor or health-care professional about any symptoms you experience that you suspect may be caused by anything you take.

4. Take vitamins, minerals, supplements and medicinal herbs in good light after you have identified the contents of the container. If you wear glasses, put them on to check and recheck labels.

5. Don't keep medicine by your bedside. You may unknowingly repeat a dose when you are half-asleep or confused.

6. Know the names of all the substances you take.

7. Read labels on medications you take.

If information in incomplete, ask your pharmacist for more details.

8. If they are in liquid form, shake vitamins, minerals, supplements and medicinal herbs before you take them.

9. Store all vitamins, minerals, supplements and medicinal herbs in cool places away from sunlight and moisture. Bathroom medicine cabinets are usually unacceptable because it's too warm and too humid there.

10. If a vitamin, mineral, supplement or medicinal herb requires refrigeration, don't freeze!

11. Obtain a standard measuring spoon from your pharmacy for liquid vitamins and a graduated dropper to use for liquid preparations for infants and children.

12. Follow manufacturer's or doctor's suggestions regarding diet instructions. Some products work better on a full stomach. Others work best on an empty stomach. Some products work best when you follow a special diet. For example, a low-salt diet enhances effectiveness of any product expected to lower blood pressure.

13. Avoid any substance you know you are allergic to.

14. If you become pregnant while taking any vitamin, mineral, supplement or medicinal herb, tell your physician and discontinue taking it until you have discussed it with him or her. Try to remember the exact dose and the length of time you have taken the substance.

15. Tell your health-care worker about vitamins, minerals, supplements, medicinal herbs and other substances you take, even if you

bought them without a prescription. During an illness or prior to surgery, this information is *crucial*. Even mention antacids, laxatives, tonics and over-the-counter preparations. Many people believe these products are completely safe and forget to inform doctors, nurses or pharmacists they are using them.

16. Regard all vitamins, minerals, supplements and medicinal herbs as potentially harmful to children. Store them safely away from their reach.

Store any substances that may be harmful out of the reach of children.

17. Alcohol, marijuana, cocaine, other mood-altering drugs and tobacco can cause life-threatening interactions when mixed with some vitamins, minerals, supplements and medicinal herbs. They can also prevent treatment from being effective or delay your return to good health. Common sense dictates you avoid them, particularly during an illness.

Vitamins

Vitamins are chemical compounds necessary for normal growth, health, metabolism and physical well-being. They provide essential parts of enzymes—the chemical molecules that catalyze or facilitate the completion of chemical reactions. Vitamins also form an essential part of many hormones—the chemical substances that promote and protect body health and reproduction.

If you're in good health, vitamins are needed only in small amounts. They are usually found in sufficient quantities in the foods you eat.

Vitamin A (Beta-carotene, Retinol)

Basic Information

Beta-carotene is a previtamin-A compound found in plants. The body converts beta-carotene to vitamin A.
Brand names, see page 477.
Available from natural sources? Yes
Available from synthetic sources? Yes
Prescription required? No
Fat-soluble or water-soluble: Fat-soluble

Natural Sources

Apricots, fresh	Liver
Asparagus	Mustard greens
Broccoli	Pumpkin
Cantaloupe	Spinach
Carrots, sliced	Squash, winter
Endive, raw	Sweet potatoes
Kale	Watermelon
Leaf lettuce	

Reasons to Use

- Aids in treatment of many eye disorders, including prevention of night blindness and formation of visual purple in the eye.
- Promotes bone growth, teeth development, reproduction.
- Helps form and maintain healthy skin, hair, mucous membranes.
- Builds body's resistance to respiratory infections.
- Helps treat acne, impetigo, boils, carbuncles, open ulcers when applied externally.

Unproved Speculated Benefits

- Helps control glaucoma.
- Buffers against cancer.
- Guards against effects of pollution and smog.
- Cushions against stress.
- Speeds healing.
- Helps in removal of age spots.
- Fights infections.
- Fights skin diseases.
- Shortens duration of some illnesses.

Who Needs Additional Amounts?

- Anyone with inadequate caloric or nutritional dietary intake or increased nutritional requirements.
- Pregnant or breast-feeding women.
- Those who abuse alcohol or other drugs.
- People with a chronic wasting illness, excess stress for long periods or who have recently undergone surgery.
- Those with a portion of the gastrointestinal tract surgically removed.
- People with recent severe burns or injuries.

Deficiency Symptoms

- Night blindness
- Lack of tear secretion
- Changes in eyes with eventual blindness if deficiency is severe and untreated
- Susceptibility to respiratory infection
- Dry, rough skin
- Changes in mucous membranes
- Weight loss
- Poor bone growth
- Weak tooth enamel
- Diarrhea
- Slow growth

Unproved Speculated Symptoms

- Bone thickening
- Kidney stones
- Diarrhea
- Birth defects
- Reduced production of steroid hormones

Lab Tests to Detect Deficiency

Many months of deficiency required before lab studies reflect deficiency.
- Plasma vitamin A and plasma carotene
- Dark-adaptation test
- Electronystagmogram
- Electroretinogram

Vitamin A (Beta-carotene, Retinol)

Dosage and Usage Information

Recommended Dietary Allowance (RDA):
Estimate of adequate daily intake by the Food and Nutrition Board of the National Research Council, 1980. See Glossary.
RDA for vitamin A is expressed in retinol equivalents (RE). One RE = 1mcg retinol or 6mcg beta-carotene. IU = International units.

Age	Retinol Equivalents	International Units
0–6 months	420RE	2,100IU
6–12 months	400RE	2,100IU
1–3 years	400RE	2,000IU
4–6 years	500RE	2,500IU
7–10 years	700RE	3,300IU
Males		
11+ years	1,000RE	5,000IU
Females		
11+ years	800RE	4,000IU
Pregnant	+200RE	+1,000IU
Lactating	+400RE	+2,000IU

What this vitamin does:
• Essential for normal function of retina. Combines with red pigment of retina (opsin) to form rhodopsin, which is necessary for sight in partial darkness.
• May act as co-factor in enzyme systems.
• Necessary for growth of bone, testicular function, ovarian function, embryonic development, regulation of growth, differentiation of tissues.

Miscellaneous information:
• Many months of a vitamin-A-deficient diet are required before symptoms develop. Average person has a 2-year supply of vitamin A stored in the liver.
• Steroids are produced by the adrenal gland and are part of the natural response to stress and immune function. Failure to make these important hormones leaves immune system in a less-than-ideal state.

Available as:
• Extended-release capsules or tablets: Swallow whole with full glass of liquid. Don't chew or crush. Take with or immediately after food to decrease stomach irritation.
• Oral solution: Dilute in at least 1/2 glass water or other liquid. Take with meals or 1 to 1-1/2 hours after meals unless otherwise directed by your doctor.
• A constituent of many multivitamin/mineral preparations.
• Some forms available by generic name.

Warnings and Precautions

Don't take if you:
• Are allergic to any preparation containing vitamin A.

Consult your doctor if you have:
• Cystic fibrosis
• Diabetes
• Intestinal disease with diarrhea
• Kidney disease
• Liver disease
• Overactive thyroid function
• Disease of the pancreas

Over age 55:
• More likely to be malnourished and need supplement.
• Dosage must be taken carefully to avoid possible toxicity.

Pregnancy:
• Daily doses exceeding 6,000IU can produce growth retardation and urinary-tract malformations of fetus.
• Don't take megadoses.

Breast-feeding:
• Don't take megadoses.

Effect on lab tests:
• With chronic vitamin-A toxicity, lab tests show *increased* blood glucose, blood-urea nitrogen, serum calcium, serum cholesterol, serum triglycerides.
• Poor results on dark-adaptation test (See Glossary)
• Poor results on electronystagmogram (See Glossary)
• Poor results on electroretinogram (See Glossary)

Storage:
• Store in cool, dry place away from direct light, but don't freeze.
• Store safely out of reach of children.
• Don't store in bathroom medicine cabinet. Heat and moisture may change action of vitamin.

Others:
• Children are more sensitive to vitamin A and are more likely to develop toxicity with dosages exceeding the RDA.
• Toxicity is slowly reversible on withdrawal of vitamin A but may persist for several weeks.

Vitamin A (Beta-carotene, Retinol), Continued

Overdose/Toxicity

Signs and symptoms:
Bleeding from gums or sore mouth, bulging soft spot on head in babies, sometimes hydrocephaly ("water on brain"), confusion or unusual excitement, diarrhea, dizziness, double vision, headache, irritability, dry skin, hair loss, peeling skin on lips, palms and in other areas, seizures, vomiting, enlarged spleen and liver. *Note:* Toxicity symptoms usually appear about 6 hours after ingestion of overdoses of vitamin A. Symptoms may also develop gradually if overdose is milder and over a long period of time.

What to do:
For symptoms of overdosage: Discontinue vitamin, and consult doctor. Also see *Adverse Reactions or Side Effects* section below.
For accidental overdosage (such as child taking entire bottle): Dial 911 (emergency), 0 for operator or your nearest Poison Control Center.

Adverse Reactions or Side Effects

Reaction or effect	What to do
Abdominal pain	Discontinue. Call doctor immediately.
Appetite loss	Discontinue. Call doctor when convenient.
Bone or joint pain	Discontinue. Call doctor immediately.
Discomfort, tiredness or weakness	Discontinue. Call doctor when convenient.
Drying or cracking of skin or lips	Discontinue. Call doctor immediately.
Fever	Discontinue. Call doctor immediately.
Hair loss	Discontinue. Call doctor immediately.
Headache	Discontinue. Call doctor when convenient.
In children, premature closure of epiphyses (the end parts of bones where growth occurs from birth through adolescence)	Discontinue. Call doctor immediately.
Increase in frequency of urination	Discontinue. Call doctor when convenient.
Increased sensitivity of skin to sunlight	Discontinue. Call doctor when convenient.
Irritability	Discontinue. Call doctor immediately.
Vomiting	Seek emergency treatment.
Yellow-orange patches on soles of feet, palms of hands or skin around nose and lips	Seek emergency treatment.

Interaction with Medicine, Vitamins or Minerals

Interacts with	Combined effect
Antacids	Decreases absorption of vitamin A and fat-soluble vitamins D, E, K.
Anti-coagulants	Increases likelihood of spontaneous or hidden bleeding.
Cholestyramine, colestipol	Decreases absorption of vitamin A.
Mineral oil, neomycin, sucralfate, isioretinoin	Increases likelihood of vitamin-A toxicity.
Oral contraceptives	Increases vitamin-A concentrations.
Vitamin E	Normal amount facilitates absorption, storage in liver and utilization of vitamin A. Excessive dosage may deplete vitamin-A stores in liver.

Interaction with Other Substances

Tobacco decreases absorption. Smokers may need supplementary vitamin A.

Chronic alcoholism interferes with the body's ability to transport and use vitamin A.

Ascorbic Acid (Vitamin C)

Basic Information

Brand names, see page 477.
Available from natural sources? Yes
Available from synthetic sources? Yes
Prescription required? No
Fat-soluble or water-soluble: Water-soluble

Natural Sources

Black currants	Orange juice
Broccoli	Oranges
Brussels sprouts	Papayas
Cabbage	Potatoes
Collards	Rose hips
Grapefruit	Spinach
Green peppers	Strawberries
Guava	Sweet and hot peppers
Kale	Tangerines
Lemons	Tomatoes
Mangos	Watercress

Reasons to Use

- Promotes healthy capillaries, gums, teeth.
- Aids iron absorption.
- Helps heal wounds and broken bones.
- Prevents and treats scurvy.
- Treats anemia, especially for iron-deficiency anemia.
- Treats urinary-tract infections.
- Helps form collagen in connective tissue.
- Increases iron absorption from intestines.
- Contributes to hemoglobin and red-blood-cell production in bone marrow.
- Blocks production of nitrosamines.

Unproved Speculated Benefits

- Prevents or cures the common cold and other infections.
- Cures some forms of cancer.
- Reduces cholesterol.
- Protects against heart disease.
- Prevents blood clots.
- Prevents allergies.
- Prevents or cures poisoning from various substances.
- Cures arthritis, skin ulcers, hay fever.
- Reduces rectal polyps.
- Alleviates mental illness.
- Relieves herpes infections of eyes and genitals.
- Prevents periodontal disease.
- Detoxifies those who abuse alcohol and drugs.
- Heals bed sores.
- Retards aging.

Who Needs Additional Amounts?

- Anyone with inadequate caloric or nutritional dietary intake or increased nutritional requirements.
- Older people (over 55 years).
- Pregnant or breast-feeding women.
- Those who abuse alcohol or other drugs.
- People with a chronic wasting illness, acute illness with fever, hyperthyroidism, tuberculosis, cold exposure.
- Anyone who experiences excess stress for long periods or who has recently undergone surgery.
- Athletes and workers who participate in vigorous physical activities.
- Those with a portion of the gastrointestinal tract surgically removed.
- People with recent severe burns or injuries.
- Those receiving kidney dialysis.
- Infants on unfortified formulas.

Deficiency Symptoms

- Scurvy: muscle weakness, swollen gums, loss of teeth, tiredness, depression, bleeding under skin, bleeding gums
- Shortness of breath
- Digestive difficulties
- Easy bruising
- Swollen or painful joints
- Nosebleeds
- Anemia: weakness, tiredness, paleness
- Frequent infections
- Slow healing of wounds

Unproved Speculated Symptoms

- Blood-vessel weakness

Lab Tests to Detect Deficiency

- Vitamin-C levels in blood plasma.
- Measurement of ascorbic-acid level in white-blood cells. (Expensive and used mostly for experimental purposes.)

⟫▸

Ascorbic Acid (Vitamin C), Continued

 ## Dosage and Usage Information

Recommended Dietary Allowance (RDA):
Estimate of adequate daily intake by the Food and Nutrition Board of the National Research Council, 1980. See Glossary.

Age	RDA
0–12 months	35mg
1–10 years	45mg
11–14 years	50mg
15+ years	60mg
Pregnant	+20mg
Lactating	+40mg

What this vitamin does:
- Necessary for collagen formation and tissue repair.
- Participates in oxidation-reduction reactions.
- Needed for metabolism of phenylalanine, tyrosine, folic acid, iron.
- Helps utilization of carbohydrates, synthesis of fats and proteins, preservation of integrity of blood-vessel walls.
- Strengthens blood vessels.

Miscellaneous information:
Food preparation tips to conserve vitamin C:
- Eat food raw or minimally cooked.
- Shorten cooking time by putting vegetables in very small amounts of water.
- Avoid prolonged standing of food at room temperature.
- Avoid overexposure of food to air and light.
- Avoid soaking vegetables.

Available as:
- Tablets: Swallow whole with full glass of liquid. Don't chew or crush. Take with meals or 1 to 1-1/2 hours after meals unless otherwise directed by your doctor.
- Extended-release capsules or tablets: Swallow whole with full glass of liquid. Don't chew or crush. Take with or immediately after food to decrease stomach irritation.
- Oral solution: Dilute in at least 1/2 glass water or other liquid. Take with meals or 1 to 1-1/2 hours after meals unless otherwise directed by your doctor.
- Injectable forms are administered by doctor or nurse.
- Chewable tablets: Chew well before swallowing.
- Effervescent tablets: Allow to dissolve completely in liquid before swallowing.
- A constituent of many multivitamin/mineral preparations.

 ## Warnings and Precautions

Don't take if you:
- Are allergic to vitamin C.

Consult your doctor if you have:
- Gout.
- Kidney stones.
- Sickle-cell anemia.

Over age 55:
- Needs are greater.
- Side effects are more likely.
- If you take 1,000mg a day or more, drink *at least 2 quarts* of water.

Pregnancy:
- Requires vitamin-C supplements because of demands made by bone development, teeth and connective-tissue formation of fetus. Consult doctor to ensure correct dose.
- If mother takes megadoses, newborn may develop deficiency symptoms after birth.
- Don't take megadoses.

Breast-feeding:
- Requires vitamin-C supplementation to support rapid growth of child. Consult doctor to ensure correct dose.
- Don't take megadoses.

Effect on lab tests:
With megadoses (10 times recommended RDA):
- Blood in stool. Large doses may cause false-negative test results.
- LDH and SGOT (See Glossary)
- Glucose in urine. Depends on method used.
- Serum bilirubin. False low level.
- Urinary pH. False low level.

Storage:
- Store in cool, dry place away from direct light, but don't freeze.
- Store safely out of reach of children.
- Don't store in bathroom medicine cabinet. Heat and moisture may change action of vitamin.

Others:
- Very high doses may cause kidney stones, although reported studies do not confirm this.

Overdose/Toxicity

Signs and symptoms:
Flushed face, headache, increased urination, lower-abdominal cramps, mild diarrhea, nausea, vomiting for oral forms. Dizziness and faintness (if given by injection).

What to do:
For symptoms of overdosage: Discontinue vitamin and consult doctor. Also see *Adverse Reactions or Side Effects* section below.
For accidental overdosage (such as child taking entire bottle): Dial 911 (emergency), 0 for operator or your nearest Poison Control Center.

Adverse Reactions or Side Effects

Reaction or effect	What to do
Anemia	Discontinue. Call doctor immediately.
Flushed face	Discontinue. Call doctor when convenient.
Headache	Discontinue. Call doctor when convenient.
Increased urination	Discontinue. Call doctor when convenient.
Lower abdominal cramps	Seek emergency treatment.
Mild diarrhea	Discontinue. Call doctor when convenient.
Nausea	Seek emergency treatment.
Vomiting	Seek emergency treatment.

Interaction with Medicine, Vitamins or Minerals

Interacts with	Combined effect
Aminosalicylic acid (PAS for tuberculosis)	Increases chance of formation of drug crystals in urine. Large doses of vitamin C must be taken to produce this effect.
Anti-cholinergics	Decreases anti-cholinergic effect.
Anti-coagulants (oral)	Decreases anti-coagulant effect.
Aspirin	Decreases vitamin-C effect.
Barbiturates	Decreases vitamin-C effect. Increases barbiturate effect.
Calcium	Assists in absorption of calcium.
Copper	Decreases absorption of copper. Large doses of vitamin C must be taken to produce this effect.
Iron supplements	Increases iron effect.
Mineral oil	Decreases vitamin-C effect.
Oral contraceptives	Decreases vitamin-C effect.
Quinidine	Decreases quinidine effect.
Salicylates	Decreases vitamin-C effect.
Sulfa drugs	Decreases vitamin-C effect. May cause kidney stones.
Tetracyclines	Decreases vitamin-C effect.

Interaction with Other Substances

Tobacco decreases absorption. Smokers may require supplemental vitamin C.

Alcohol can be more rapidly broken down in body with large doses of vitamin C.

Vitamin B-12

Basic Information

Vitamin B-12 is also called cyanocobalamin.
Brand names, see page 477.
Available from natural sources? Yes
Available from synthetic sources? Yes
Prescription required? Yes, for high doses and injectable forms
Fat-soluble or water-soluble: Water-soluble

Natural Sources

Beef	Liverwurst
Beef liver	Mackerel
Blue cheese	Milk
Clams	Milk products
Eggs	Sardines
Flounder	Snapper
Herring	Swiss cheese

Note: Vitamin B-12 is not found in vegetables.

Reasons to Use

- Promotes normal growth and development.
- Treats some types of nerve damage.
- Treats pernicious anemia.
- Treats and prevents vitamin B-12 deficiencies in people who have had a portion of the gastrointestinal tract surgically removed.
- Prevents vitamin-B12 deficiency in vegan vegetarians and persons with absorption diseases.

Unproved Speculated Benefits

- Helps mental and nervous disorders.
- Improves resistance to infection and disease.
- Increases appetite.
- Promotes growth of someone who has smaller-than-average stature.
- Improves memory and the ability to learn.
- Increases energy.

Who Needs Additional Amounts?

- Strict vegetarians.
- Anyone with inadequate caloric or nutritional dietary intake or increased nutritional requirements.
- Those who abuse alcohol or other drugs.
- People with a chronic wasting illness, excess stress for long periods or who have recently undergone surgery.

- Those with a portion of the gastrointestinal tract surgically removed.
- People with recent severe burns or injuries.

Deficiency Symptoms

- Pernicious anemia, with the following symptoms:
  Fatigue, profound
  Weakness, especially in arms and legs
  Sore tongue
  Nausea, appetite loss, weight loss
  Bleeding gums
  Numbness and tingling in hands and feet
  Difficulty maintaining balance
  Pale lips, pale tongue, pale gums
  Yellow eyes and skin
  Shortness of breath
  Depression
  Confusion and dementia
  Headache
  Poor memory

Unproved Speculated Symptoms

- Aging
- Allergies
- Eye problems
- Slow growth
- Skin problems
- Easy fatigue
- Mental symptoms
- Sterility
- Thyroid disorders
- Menstrual disorders
- Delusions and hallucinations

Lab Tests to Detect Deficiency

- Serum vitamin B-12, a radioactive study usually performed with serum-folic-acid test, called the *Schilling Test*
- Reticulocyte count

Dosage and Usage Information

Recommended Dietary Allowance (RDA):
Estimate of adequate daily intake by the Food and Nutrition Board of the National Research Council, 1980. See Glossary.

⟫➤

Age	RDA
0–6 months	0.5mcg
6–12 months	1.5mcg
1–3 years	2mcg
4–6 years	2.5mcg
7–10 years	3mcg
11+ years	3mcg
Pregnant	+1mcg
Lactating	+1mcg

What this vitamin does:
• Acts as co-enzyme for normal DNA synthesis.
• Promotes normal fat and carbohydrate metabolism and protein syntheses.
• Promotes growth, cell development, blood-cell development, manufacture of covering to nerve cells, maintenance of normal function of nervous system.

Miscellaneous information:
• There is a very low incidence of toxicity of vitamin B-12, even with large amounts up to 1,000mcg/day.

Available as:
• Oral and injectable forms. Oral forms are used only as diet supplement. Only people with portions of the gastrointestinal tract removed surgically or those with pernicious anemia require injections.
• Tablets: Swallow whole with full glass of liquid. Don't chew or crush. Take with meals or 1 to 1-1/2 hours after meals unless otherwise directed by your doctor.
• Extended-release capsules or tablets: Swallow whole with full glass of liquid. Don't chew or crush. Take with or immediately after food to decrease stomach irritation.
• Injectable forms are administered by doctor or nurse.
• A constituent of many multivitamin/mineral preparations.

Warnings and Precautions

Don't take if you:
• Are allergic to B-12 given by injection. Allergy to injections produces itching, redness, swelling and rarely blood-pressure drop with loss of consciousness.
• Have Leber's disease.

Consult your doctor if you have:
• Gout.

Over age 55:
• No problems expected.

Pregnancy:
• No problems expected.
• Don't take megadoses.

Breast-feeding:
• No problems expected.
• Don't take megadoses.

Effect on lab tests:
• Tests for serum potassium may show precipitous drop (hypokalemia) during 48 hours after beginning treatment for anemia.

Storage:
• Store in cool, dry place away from direct light, but don't freeze. Liquid forms should be refrigerated.
• Store safely out of reach of children.
• Don't store in bathroom medicine cabinet. Heat and moisture may change action of vitamin.

Others:
• The injectable form is the only effective form to treat pernicious anemia or people with portions of the gastrointestinal tract surgically removed. These individuals do not absorb oral forms.

Overdose/Toxicity

Signs and symptoms:
• If taken with large doses of vitamin C, vitamin B-12 may cause nosebleed, ear bleeding, dry mouth.

What to do:
For symptoms of overdosage: Discontinue vitamin, and consult doctor. Also see *Adverse Reactions or Side Effects* section below.
For accidental overdosage (such as child taking entire bottle): Dial 911 (emergency), 0 for operator or your nearest Poison Control Center.
»→

Vitamin B-12, Continued

 Adverse Reactions or Side Effects

Reaction or effect	What to do
Diarrhea (rare)	Discontinue. Call doctor immediately.
Itching skin after injections (rare)	Seek emergency treatment.

 Interaction with Medicine, Vitamins or Minerals

Interacts with	Combined effect
Aminosalicylates	Reduces absorption of vitamin B-12.
Antibiotics	May cause false-low test results for vitamin B-12.
Ascorbic acid (vitamin C)	Large doses may destroy vitamin B-12. Separate doses by at least 1 hour.
Chloramphenicol	May prevent therapeutic response when vitamin B-12 is used to treat anemia.
Cholestyramine	Reduces absorption of vitamin B-12.
Colchicine	Reduces absorption of vitamin B-12.
Folic acid	Large doses decrease vitamin B-12 concentrations in blood.
Neomycin (oral forms only)	Reduces absorption of vitamin B-12.
Potassium in extended-release forms	Reduces absorption of vitamin B-12. May increase need for vitamin B-12.

 Interaction with Other Substances

Tobacco decreases absorption. Smokers may require supplemental vitamin B-12.

Alcohol in excessive amounts for long periods may lead to vitamin B-12 deficiency.

Biotin (Vitamin H)

Basic Information

Biotin is also called vitamin H.
Available from natural sources? Yes
Available from synthetic sources? No
Prescription required? No
Fat-soluble or water-soluble: Water-soluble

Natural Sources

Brewer's yeast
Brown rice
Bulgur wheat
Butter
Calves' liver
Cashew nuts
Cheese
Chicken
Eggs
Green peas
Lentils
Mackerel
Meats
Milk
Oats
Peanuts
Soybeans
Split peas
Sunflower seeds
Tuna
Walnuts

Reasons to Use

• Helps formation of fatty acids.
• Facilitates metabolism of amino acids and carbohydrates.
• Promotes normal health of sweat glands, nerve tissue, bone marrow, male sex glands, blood cells, skin, hair.

Unproved Speculated Benefits

• Cures baldness.
• Alleviates muscle pain.
• Cures dermatitis.
• Alleviates depression.

Who Needs Additional Amounts?

• Anyone with inadequate caloric or nutritional dietary intake or increased nutritional requirements.
• People who consume huge quantities of raw eggs, which contain a compound that inhibits biotin. Cooking eggs destroys this compound and eliminates the problem.

Deficiency Symptoms

Babies:
• Dry scaling on scalp and face
Adults:
• Fatigue
• Depression
• Sleepiness
• Nausea
• Loss of appetite
• Muscular pains
• Loss of muscular reflexes
• Tongue becomes smooth and pale
• Hair loss
• Blood-cholesterol levels increase
• Anemia
• Skin disorders

Unproved Speculated Symptoms

• Sudden infant death syndrome (SIDS)

Lab Tests to Detect Deficiency

• None available, except for experimental purposes.

Dosage and Usage Information

Recommended Dietary Allowance (RDA):
No RDA has been established. Estimated safe intake given below:

Age	Estimated Safe Intake
0–6 months	35mg/day
6–12 months	50mg/day
1–3 years	65mg/day
4–6 years	85mg/day
7–10 years	120mg/day
11+ years	100–200mg/day

What this vitamin does:
• Biotin is necessary for normal growth, development and health.

Miscellaneous information:
• Intestinal bacteria produce all the biotin the body needs, so there is no substantial evidence that normal, healthy adults need dietary supplements of biotin.

⟫➤

Biotin (Vitamin H), Continued

Available as:
- Tablets or capsules: Swallow whole with full glass of liquid. Don't chew or crush. Take with or immediately after food to decrease stomach irritation.
- A constituent of many multivitamin/mineral preparations.

 Warnings and Precautions

Don't take if you:
- No specific precautions.

Consult your doctor if you have:
- No specific precautions.

Over age 55:
- No specific precautions.

Pregnancy:
- No specific precautions.
- Don't take megadoses.

Breast-feeding:
- No specific precautions.
- Don't take megadoses.

Effect on lab tests:
- None expected.

Storage:
- Store in cool, dry place away from direct light, but don't freeze.
- Store safely out of reach of children.
- Don't store in bathroom medicine cabinet. Heat and moisture may change action of vitamin.

 Overdose/Toxicity

Signs and symptoms:
Supplements in amounts suggested by manufacturers on the label are non-toxic.

What to do:
For symptoms of overdosage: Discontinue vitamin, and consult doctor.
For accidental overdosage (such as child taking entire bottle): Dial 911 (emergency), 0 for operator or your nearest Poison Control Center.

 Adverse Reactions or Side Effects

None expected.

 Interaction with Medicine, Vitamins or Minerals

Interacts with	Combined effect
Antibiotics (broad spectrum)	Destroys "friendly" bacteria in intestines that produce biotin. This can lead to significant biotin deficiency.
Sulfonamides	Destroys "friendly" bacteria in intestines that produce biotin. This can lead to significant biotin deficiency.

 Interaction with Other Substances

Tobacco decreases absorption. Smokers may require supplemental biotin.

Foods
- Eating large quantities of *raw egg whites* may cause biotin deficiency. Egg whites contain *avidin*, which prevents biotin from being absorbed into the body.

Calcifidiol

Basic Information

Calcifidiol is a form of vitamin D. *It is also called* ergocalciferol.
Available from natural sources? Yes
Available from synthetic sources? Yes
Prescription required? Yes
Fat-soluble or water-soluble: Fat-soluble

Natural Sources

Cod-liver oil	Salmon
Halibut-liver oil	Sardines
Herring	Sunlight
Mackerel	Vitamin-D-fortified milk

Reasons to Use

- Calcifidiol is a form of vitamin D primarily used as an additional medicine to treat hypocalcemia (not enough calcium in blood). It is also frequently prescribed to treat bone disease in people undergoing renal dialysis and those with hypoparathyroidism.
- Regulates growth, hardening and repair of bone by controlling absorption of calcium and phosphorus from small intestine.
- Prevents rickets.
- Treats post-operative muscle contractions.
- Works with calcium to control bone formation.
- Promotes normal growth and development of infants and children, particularly bones and teeth.

Unproved Speculated Benefits

- Cures arthritis.
- Prevents colon cancer.
- Treats aging symptoms.
- Treats acne.
- Treats alcoholism.
- Treats herpes simplex and herpes zoster.
- Treats cystic fibrosis.

Who Needs Additional Amounts?

- Children who live in sunshine-deficient areas.
- Anyone with inadequate caloric or nutritional dietary intake or increased nutritional requirements.
- Older people (over 55 years), especially women after menopause.
- Pregnant or breast-feeding women.
- Those who abuse alcohol or other drugs.
- People with a chronic wasting illness, excess stress for long periods or those who have recently undergone surgery.
- Those with a portion of the gastrointestinal tract surgically removed.
- People with recent severe burns or injuries.

Deficiency Symptoms

- Rickets (a childhood deficiency disease): bent, bowed legs, malformations of joints or bones, late tooth development, weak muscles, listlessness.
- Osteomalacia (adult rickets): muscle weakness and spasm, brittle, easily broken bones, pain in ribs, lower spine, pelvis and legs.

Unproved Speculated Symptoms

- Muscle diseases (myopathies)

Lab Tests to Detect Deficiency

- Reduced levels of vitamin D forms in blood.
- Decreased serum phosphate, decreased calcium, increased alkaline phosphatase, urinary hydroxyproline, PTH levels.
- Bone X-ray.

Dosage and Usage Information

Recommended Dietary Allowance (RDA):
Estimate of adequate daily intake by the Food and Nutrition Board of the National Research Council, 1980. See Glossary.

Age	RDA
0–6 months	10mcg
6–12 months	10mcg
1–10 years	10mcg
Males	
11–18 years	10mcg
19–22 years	7.5mcg
23+ years	5mcg
Females	
11–18 years	10mcg
19–22 years	7.5mcg
23+ years	5mcg
Pregnant	+5mcg
Lactating	+5mcg

➤

Calcifidiol, Continued

What this vitamin does:
- Absorbs and uses calcium and phosphorous to make bone.
- Essential for normal growth and development.

Miscellaneous information:
- Take at the same time every day.
- Put liquid vitamin D directly into mouth or mix with cereal, fruit juice or food.

Available as:
- Extended-release capsules or tablets: Swallow whole with full glass of liquid. Don't chew or crush. Take with or immediately after food to decrease stomach irritation.
- Oral solution: Dilute in at least 1/2 glass water or other liquid. Take with meals or 1 to 1-1/2 hours after meals unless otherwise directed by your doctor.
- Some forms available by generic name.

Warnings and Precautions

Don't take if you:
- Are allergic to vitamin D, ergocalciferol or any vitamin-D derivative.

Consult your doctor if you have:
- Any plans to become pregnant while taking vitamin D.
- Epilepsy.
- Heart or blood-vessel disease.
- Kidney, liver, pancreatic disease.
- Chronic diarrhea.
- Intestinal problems.
- Sarcoidosis.

Over age 55:
- Adverse reactions and side effects are more likely. Supplements are often necessary.

Pregnancy:
- Taking too much during pregnancy may cause abnormalities in fetus. Consult doctor before taking supplement to ensure correct dosage.
- Don't take megadoses.

Breast-feeding:
- Important for you to receive correct amount so enough vitamin D is available for normal growth and development of baby. Consult doctor about supplements.
- Don't take megadoses.

Effect on lab tests:
- May decrease serum alkaline phosphatase.
- May increase levels of calcium, cholesterol and phosphate in test results.
- May increase level of magnesium in test results.
- May increase amounts of calcium and phosphorous in urine.

Storage:
- Store in cool, dry place away from direct light, but don't freeze. Avoid overexposure to air.
- Store safely out of reach of children.
- Don't store in bathroom medicine cabinet. Heat and moisture may change action of vitamin.

Others:
- Absence of sunlight prevents natural formation of vitamin D by skin. Sunshine provides sufficient amounts of vitamin D for people who live in sunny climates. Those who live in northern areas with fewer days of sunshine and extended periods of cloud cover and darkness must depend on dietary sources for vitamin D.
- Avoid megadoses.

Overdose/Toxicity

Signs and symptoms:
High blood pressure, irregular heartbeat, nausea, weight loss, seizures, abdominal pain, appetite loss, mental and physical-growth retardation, premature hardening of arteries, kidney damage.

What to do:
For symptoms of overdosage: Discontinue vitamin, and consult doctor. Also see *Adverse Reactions or Side Effects* section below.

For accidental overdosage (such as child taking entire bottle): Dial 911 (emergency), 0 for operator or your nearest Poison Control Center.

For toxic symptoms: Discontinue vitamin, and seek immediate medical help. Hospitalization may be necessary.

Adverse Reactions or Side Effects

Reaction or effect	What to do
Appetite loss	Discontinue. Call doctor when convenient.
Constipation	Discontinue. Call doctor when convenient.
Diarrhea	Discontinue. Call doctor immediately.
Dry mouth	Discontinue. Call doctor when convenient.
Headache	Discontinue. Call doctor immediately.
Increased thirst	Discontinue. Call doctor when convenient.
Mental confusion	Discontinue. Call doctor immediately.
Metallic taste	Discontinue. Call doctor when convenient.
Nausea	Discontinue. Call doctor immediately.
Unusual tiredness	Discontinue. Call doctor when convenient.
Vomiting	Discontinue. Call doctor immediately.

Interaction with Medicine, Vitamins or Minerals

Interacts with	Combined effect
Antacids with aluminum	Decreases absorption of vitamin D and fat-soluble vitamins A, D, E, K.
Antacids with magnesium	May cause too much magnesium in blood, especially for people with kidney failure.
Anti-convulsants	May reduce effect of vitamin D from natural sources and require supplements to prevent loss of strength in bones.
Barbiturates	May reduce effect of vitamin D from natural sources and require supplements to prevent loss of strength in bones.
Calcitonin	Reduces effect of calcitonin when treating hypercalcemia.
Calcium (high doses)	Increases risk of hypercalcemia.
Cholestyramine	Impairs absorption of vitamin D. May need supplements.
Colestipol	Impairs absorption of vitamin D. May need supplements.
Digitalis preparations	Increases risk of heartbeat irregularities.
Diuretics, thiazide	Increases risk of hypercalcemia.
Hydantoin	May reduce effect of vitamin D from natural sources and require supplements to prevent loss of strength in bones.
Mineral oil	Increases absorption of vitamin D. May need supplements.
Phosphorous-containing medicines	Increases risk of too much phosphorous in blood.
Primidone	May reduce effect of vitamin D from natural sources and require supplements to prevent loss of strength in bones.
Vitamin-D derivatives, such as calciferol, calcitrol, dihydrotachysterol, ergocalciferol	Additive effects may increase potential for toxicity.

Interaction with Other Substances

Chronic alcoholism depletes liver stores of vitamin D.

Calcitrol

Basic Information

Calcitrol is a form of vitamin D. *It is also called* ergocalciferol.
Available from natural sources? Yes
Available from synthetic sources? Yes
Prescription required? Yes
Fat-soluble or water-soluble: Fat-soluble

Natural Sources

Cod-liver oil	Salmon
Halibut-liver oil	Sardines
Herring	Sunlight
Mackerel	Vitamin-D-fortified milk

Reasons to Use

- Calcitrol is a form of vitamin D primarily used to treat hypocalcemia (not enough calcium in blood). It is also frequently prescribed to treat bone disease in patients undergoing renal dialysis, and patients with hypoparathyroidism.
- Regulates growth, hardening and repair of bone by controlling absorption of calcium and phosphorus from small intestine.
- Prevents rickets.
- Treats post-operative muscle contractions.
- Works with calcium to control bone formation.
- Promotes normal growth and development of infants and children, particularly bones and teeth.

Unproved Speculated Benefits

- Cures arthritis.
- Prevents colon cancer.
- Treats aging symptoms.
- Treats acne.
- Treats alcoholism.
- Treats herpes simplex and herpes zoster.
- Treats cystic fibrosis.

Who Needs Additional Amounts?

- Children who live in sunshine-deficient areas.
- Anyone with inadequate caloric or nutritional dietary intake or increased nutritional requirements.
- Older people (over 55 years), especially women after menopause.
- Pregnant or breast-feeding women.
- Those who abuse alcohol or other drugs.

- People with a chronic wasting illness, excess stress for long periods or those who have recently undergone surgery.
- Those with a portion of the gastrointestinal tract surgically removed.
- People with recent severe burns or injuries.

Deficiency Symptoms

- Rickets (a childhood deficiency disase): bent, bowed legs, malformations of joints or bones, late tooth development, weak muscles, listlessness.
- Osteomalacia (adult rickets): muscle weakness and spasm, brittle, easily broken bones, pain in ribs, lower spine, pelvis and legs,

Unproved Speculated Symptoms

- Muscle diseases (myopathies)

Lab Tests to Detect Deficiency

- Reduced levels of vitamin D forms in blood.
- Decreased serum phosphate, decreased calcium, increased alkaline phosphatase, urinary hydroxyproline, PTH levels.
- Bone X-ray.

Dosage and Usage Information

Recommended Dietary Allowance (RDA):
Estimate of adequate daily intake by the Food and Nutrition Board of the National Research Council, 1980. See Glossary.

Age	RDA
0–6 months	10mcg
6–12 months	10mcg
1–10 years	10mcg
Males	
11–18 years	10mcg
19–22 years	7.5mcg
23+ years	5mcg
Females	
11–18 years	10mcg
19–22 years	7.5mcg
23+ years	5mcg
Pregnant	+5mcg
Lactating	+5mcg

»»➤

What this vitamin does:
- Absorbs and uses calcium and phosphorous to make bone.
- Essential for normal growth and development.

Miscellaneous information:
- Take at the same time every day.
- Put liquid vitamin D directly into mouth or mix with cereal, fruit juice or food.

Available as:
- Extended-release capsules or tablets: Swallow whole with full glass of liquid. Don't chew or crush. Take with or immediately after food to decrease stomach irritation.
- Oral solution: Dilute in at least 1/2 glass water or other liquid. Take with meals or 1 to 1-1/2 hours after meals unless otherwise directed by your doctor.
- Some forms available by generic name.

Warnings and Precautions

Don't take if you:
- Are allergic to vitamin D, ergocalciferol or any vitamin-D derivative.

Consult your doctor if you have:
- Any plans to become pregnant while taking vitamin D.
- Epilepsy.
- Heart or blood-vessel disease.
- Kidney, liver, pancreatic disease.
- Chronic diarrhea.
- Intestinal problems.
- Sarcoidosis.

Over age 55:
- Adverse reactions and side effects are more likely. Supplements are often necessary.

Pregnancy:
- Taking too much during pregnancy may cause abnormalities in fetus. Consult doctor before taking supplement to ensure correct dosage.
- Don't take megadoses.

Breast-feeding:
- Important for you to receive correct amount so enough vitamin D is available for normal growth and development of baby. Consult doctor about supplements.
- Don't take megadoses.

Effect on lab tests:
- May decrease serum alkaline phosphatase.
- May increase levels of calcium, cholesterol and phosphate in test results.
- May increase level of magnesium in test results.
- May increase amounts of calcium and phosphorous in urine.

Storage:
- Store in cool, dry place away from direct light, but don't freeze. Avoid overexposure to air.
- Store safely out of reach of children.
- Don't store in bathroom medicine cabinet. Heat and moisture may change action of vitamin.

Others:
- Absence of sunlight prevents natural formation of vitamin D by skin. Sunshine provides sufficient amounts of vitamin D for people who live in sunny climates. Those who live in northern areas with fewer days of sunshine and extended periods of cloud cover and darkness must depend on dietary sources for vitamin D.
- Avoid megadoses.

Overdose/Toxicity

Signs and symptoms:
High blood pressure, irregular heartbeat, nausea, weight loss, seizures, abdominal pain, appetite loss, mental and physical-growth retardation, premature hardening of arteries, kidney damage.

What to do:
For symptoms of overdosage: Discontinue vitamin, and consult doctor. Also see *Adverse Reactions or Side Effects* section below.
For accidental overdosage (such as child taking entire bottle): Dial 911 (emergency), 0 for operator or your nearest Poison Control Center.
For toxic symptoms: Discontinue vitamin, and seek immediate medical help. Hospitalization may be necessary.

》➤

Calcitrol, Continued

 ## Adverse Reactions or Side Effects

Reaction or effect	What to do
Appetite loss	Discontinue. Call doctor when convenient.
Constipation	Discontinue. Call doctor when convenient.
Diarrhea	Discontinue. Call doctor immediately.
Dry mouth	Discontinue. Call doctor when convenient.
Headache	Discontinue. Call doctor immediately.
Increased thirst	Discontinue. Call doctor when convenient.
Mental confusion	Discontinue. Call doctor immediately.
Metallic taste	Discontinue. Call doctor when convenient.
Nausea	Discontinue. Call doctor immediately.
Unusual tiredness	Discontinue. Call doctor when convenient.
Vomiting	Discontinue. Call doctor immediately.

 ## Interaction with Medicine, Vitamins or Minerals

Interacts with	Combined effect
Antacids with aluminum	Decreases absorption of vitamin D and fat-soluble vitamins A, D, E, K.
Antacids with magnesium	May cause too much magnesium in blood, especially for people with kidney failure.
Anti-convulsants	May reduce effect of vitamin D from natural sources and require supplements to prevent loss of strength in bones.
Barbiturates	May reduce effect of vitamin D from natural sources and require supplements to prevent loss of strength in bones.
Calcitonin	Reduces effect of calcitonin when treating hypercalcemia.
Calcium (high doses)	Increases risk of hypercalcemia.
Cholestyramine	Impairs absorption of vitamin D. May need supplements.
Colestipol	Impairs absorption of vitamin D. May need supplements.
Digitalis preparations	Increases risk of heartbeat irregularities.
Diuretics, thiazide	Increases risk of hypercalcemia.
Hydantoin	May reduce effect of vitamin D from natural sources and require supplements to prevent loss of strength in bones.
Mineral oil	Increases absorption of vitamin D. May need supplements.
Phosphorous-containing medicines	Increases risk of too much phosphorous in blood.
Primidone	May reduce effect of vitamin D from natural sources and require supplements to prevent loss of strength in bones.
Vitamin-D derivatives, such as calciferol, calcitrol, dihydrotachysterol, ergocalciferol	Additive effects may increase potential for toxicity.

 ## Interaction with Other Substances

Chronic alcoholism depletes liver stores of vitamin D.

Children's Multivitamin With Fluoride

Basic Information

Brand names, see page 477.
Available from natural sources? No
Available from synthetic sources? Yes
Prescription required? Yes
Fat-soluble or water-soluble: Fat-soluble and
water-soluble

Natural Sources

These are all manufactured products.

Reasons to Use

• Prevents vitamin deficiency of essential fat-soluble and water-soluble vitamins when the daily diet doesn't include enough of these vitamins needed for good health.
• Prevents dental caries in children who live in areas where naturally occurring fluoride in drinking water is inadequate.

Unproved Speculated Benefits

• Prevents dental caries in adults.
• Prevents dental plaque bacteria from causing damage to normal teeth.

Who Needs Additional Amounts?

• Anyone with inadequate caloric or nutritional dietary intake or increased nutritional requirements.

Deficiency Symptoms

• Frequent dental caries (cavities in teeth)
• Failure to grow and develop normally

Unproved Speculated Symptoms

• Dental plaque in adolescents and adults

Lab Tests to Detect Deficiency

• None readily available to test for multiple-vitamin deficiency.
• Dental X-rays detect dental caries and suggest fluoride deficiency.

Dosage and Usage Information

Recommended Dietary Allowance (RDA):
No RDA has been established for multiple vitamins. See individual vitamin charts for recommendations for vitamins A, D, C.

What this vitamin does:
• Fluoride becomes incorporated into bone and teeth, promotes remineralization of decalcified enamel and *may* interfere with growth and development of bacteria that cause dental plaque.

Miscellaneous information:
• Fluorides dissolve easily and can be absorbed easily from the stomach and intestines.
• Fluoride applications to teeth and fluoride toothpaste and mouthwash also help prevent caries.

Available as:
• Oral solution: Dilute in at least 1/2 glass water or other liquid. Take after brushing teeth, immediately before bedtime.
• Chewable tablets: Chew well before swallowing. Take after brushing teeth, immediately before bedtime.

Warnings and Precautions

Don't take if you:
• Are allergic to vitamins A, D, C or fluoride.
• Have evidence of dental fluorosis (dark-brown stains on teeth).

Consult your doctor if you have:
• Ever lived in an area where fluoride in drinking water is excessive.
• Hypothyroidism.

Over age 55:
• Probably not useful.

Pregnancy:
• Take only under medical supervision.

Breast-feeding:
• Take only under medical supervision.

Children's Multivitamin With Fluoride, Continued

Effect on lab tests:
- Falsely decreases serum acid phosphatase
- Falsely increases SGOT (serum aspartate aminotransferase)
- Falsely decreases PBI (protein-bound iodine)
- Decreases serum calcium

Storage:
- Store in cool, dry place away from direct light, but don't freeze.
- Store safely out of reach of children.
- Don't store in bathroom medicine cabinet. Heat and moisture may change action of combination.

Overdose/Toxicity

Signs and symptoms:
Black, tarry stools, bloody vomit, diarrhea, drowsiness, shallow breathing, abdominal cramping, pain, increased salivation

What to do:
For symptoms of overdosage: Discontinue vitamin, and consult doctor. Also see *Adverse Reactions or Side Effects* section below.
For accidental overdosage (such as child taking entire bottle): Dial 911 (emergency), 0 for operator or your nearest Poison Control Center.

Adverse Reactions or Side Effects

Reaction or effect	What to do
Aching pain in bones	Discontinue. Call doctor immediately.
Appetite loss	Discontinue. Call doctor when convenient.
Joint stiffness	Discontinue. Call doctor when convenient.
Mouth sores	Discontinue. Call doctor immediately.
Skin rash	Discontinue. Call doctor immediately.
Weight loss, large	Discontinue. Call doctor immediately.
White, brown or black mottled discoloration of teeth	Discontinue. Call doctor when convenient.

Interaction with Medicine, Vitamins or Minerals

Interacts with	Combined effect
Anti-coagulants	Increases bleeding
Aluminum hydroxide	Decreases absorption of vitamins and fluoride
Iron supplements	Decreases effect of iron
Vitamin-D preparations	Increases possibility of toxic effects of vitamin D

Interaction with Other Substances

Milk and **milk products** may decrease absorption of fluoride. Take fluoride at least 2 hours before or after drinking milk.

Basic Information

Vitamin D is also called cholecalciferol.
Brand names, see page 477.
Available from natural sources? Yes
Available from synthetic sources? Yes
Prescription required? No
Fat-soluble or water-soluble: Fat-soluble

Natural Sources

Cod-liver oil	Salmon
Halibut-liver oil	Sardines
Herring	Sunlight
Mackerel	Vitamin-D-fortified milk

Reasons to Use

- Regulates growth, hardening and repair of bone by controlling absorption of calcium and phosphorus from small intestine.
- Prevents rickets.
- Treats hypocalcemia (low blood calcium) in kidney disease.
- Treats post-operative muscle contractions.
- Works with calcium to control bone formation.
- Promotes normal growth and development of infants and children, particularly bones and teeth.

Unproved Speculated Benefits

- Cures arthritis.
- Prevents colon cancer.
- Treats aging symptoms.
- Treats acne.
- Treats alcoholism.
- Treats herpes simplex and herpes zoster.
- Treats cystic fibrosis.

Who Needs Additional Amounts?

- Children who live in sunshine-deficient areas.
- Anyone with inadequate caloric or nutritional dietary intake or increased nutritional requirements.
- Older people (over 55 years), especially women after menopause.
- Pregnant or breast-feeding women.
- Those who abuse alcohol or other drugs.
- People with a chronic wasting illness, excess stress for long periods or who have recently undergone surgery.
- Those with a portion of the gastrointestinal tract surgically removed.
- People with recent severe burns or injuries.

Deficiency Symptoms

- Rickets (a childhood deficiency disease): bent, bowed legs, malformations of joints or bones, late tooth development, weak muscles, listlessness.
- Osteomalacia (adult rickets): pain in ribs, lower spine, pelvis and legs, muscle weakness and spasm, brittle, easily broken bones.

Unproved Speculated Symptoms

- Muscle diseases (myopathies)

Lab Tests to Detect Deficiency

- Reduced levels of vitamin D forms in blood.
- Decreased serum phosphate, decreased calcium, increased alkaline phosphatase, urinary hydroxyproline, PTH levels.
- Bone X-ray.

Dosage and Usage Information

Recommended Dietary Allowance (RDA):
Estimate of adequate daily intake by the Food and Nutrition Board of the National Research Council, 1980. See Glossary.

Age	RDA
0–6 months	10mcg
6–12 months	10mcg
1–10 years	10mcg
Males	
11–18 years	10mcg
19–22 years	7.5mcg
23+ years	5mcg
Females	
11–18 years	10mcg
19–22 years	7.5mcg
23+ years	5mcg
Pregnant	+5mcg
Lactating	+5mcg

What this vitamin does:
- Absorbs and uses calcium and phosphorous to make bone.
- Essential for normal growth and development.

»▸

Vitamin D, Continued

Miscellaneous information:
- Take at the same time every day.
- Put liquid vitamin D directly into mouth or mix with cereal, fruit juice or food.

Available as:
- Extended-release capsules or tablets: Swallow whole with full glass of liquid. Don't chew or crush. Take with or immediately after food to decrease stomach irritation.
- Oral solution: Dilute in at least 1/2 glass water or other liquid. Take with meals or 1 to 1-1/2 hours after meals unless otherwise directed by your doctor.
- A constituent of many multivitamin/mineral preparations.
- Some forms available by generic name.

Warnings and Precautions

Don't take if you:
- Are allergic to vitamin D, ergocalciferol or any vitamin-D derivative.

Consult your doctor if you have:
- Any plans to become pregnant while taking vitamin D.
- Epilepsy.
- Heart or blood-vessel disease.
- Kidney, liver, pancreatic disease.
- Chronic diarrhea.
- Intestinal problems.
- Sarcoidosis.

Over age 55:
- Adverse reactions and side effects are more likely. Supplements are often necessary.

Pregnancy:
- Taking too much during pregnancy may cause abnormalities in fetus. Consult doctor before taking supplement to ensure correct dosage.
- Don't take megadoses.

Breast-feeding:
- Important for you to receive correct amount so enough vitamin D is available for normal growth and development of baby. Consult doctor about supplements.
- Don't take megadoses.

Effect on lab tests:
- May decrease serum alkaline phosphatase.
- May increase levels of calcium, cholesterol and phosphate in test results.
- May increase level of magnesium in test results.
- May increase amounts of calcium and phosphorous in urine.

Storage:
- Store in cool, dry place away from direct light, but don't freeze. Avoid overexposure to air.
- Store safely out of reach of children.
- Don't store in bathroom medicine cabinet. Heat and moisture may change action of vitamin.

Others:
- Absence of sunlight prevents natural formation of vitamin D by skin. Sunshine provides sufficient amounts of vitamin D for people who live in sunny climates. Those who live in northern areas with fewer days of sunshine and extended periods of cloud cover and darkness must depend on dietary sources for vitamin D.
- Avoid megadoses.

Overdose/Toxicity

Signs and symptoms:
High blood pressure, irregular heartbeat, nausea, weight loss, seizures, abdominal pain, appetite loss, mental and physical-growth retardation, premature hardening of arteries, kidney damage.

What to do:
For symptoms of overdosage: Discontinue vitamin, and consult doctor. Also see *Adverse Reactions or Side Effects* section below.

For accidental overdosage (such as child taking entire bottle): Dial 911 (emergency), 0 for operator or your nearest Poison Control Center.

For toxic symptoms: Discontinue vitamin and seek immediate medical help. Hospitalization may be necessary.

Adverse Reactions or Side Effects

Reaction or effect	What to do
Appetite loss	Discontinue. Call doctor when convenient.
Constipation	Discontinue. Call doctor when convenient.
Diarrhea	Discontinue. Call doctor immediately.
Dry mouth	Discontinue. Call doctor when convenient.
Headache	Discontinue. Call doctor immediately.
Increased thirst	Discontinue. Call doctor when convenient.
Mental confusion	Discontinue. Call doctor immediately.
Metallic taste	Discontinue. Call doctor when convenient.
Nausea	Discontinue. Call doctor immediately.
Unusual tiredness	Discontinue. Call doctor when convenient.
Vomiting	Discontinue. Call doctor immediately.

Interaction with Medicine, Vitamins or Minerals

Interacts with	Combined effect
Antacids with aluminum	Decreases absorption of vitamin D and fat-soluble vitamins A, D, E, K.
Antacids with magnesium	May cause too much magnesium in blood, especially for people with kidney failure.
Anti-convulsants	May reduce effect of vitamin D from natural sources and require supplements to prevent loss of strength in bones.
Barbiturates	May reduce effect of vitamin D from natural sources and require supplements to prevent loss of strength in bones.
Calcitonin	Reduces effect of calcitonin when treating hypercalcemia.
Calcium (high doses)	Increases risk of hypercalcemia.
Cholestyramine	Impairs absorption of vitamin D. May need supplements.
Colestipol	Impairs absorption of vitamin D. May need supplements.
Digitalis preparations	Increases risk of heartbeat irregularities.
Diuretics, thiazide	Increases risk of hypercalcemia.
Hydantoin	May reduce effect of vitamin D from natural sources and require supplements to prevent loss of strength in bones.
Mineral oil	Increases absorption of vitamin D. May need supplements.
Phosphorous-containing medicines	Increases risk of too much phosphorous in blood.
Primidone	May reduce effect of vitamin D from natural sources and require supplements to prevent loss of strength in bones.
Vitamin-D derivatives, such as calciferol, calcitrol, dihydrotachysterol, ergocalciferol	Additive effects may increase potential for toxicity.

Interaction with Other Substances

Chronic alcoholism depletes liver stores of vitamin D.

Dihydrotachysterol

Basic Information

Dihydrotachysterol is a form of vitamin D. *It is also called* ergocalciferol.
Available from natural sources? Yes
Available from synthetic sources? Yes
Prescription required? Yes
Fat-soluble or water-soluble: Fat-soluble

 ## Natural Sources

Cod-liver oil	Salmon
Halibut-liver oil	Sardines
Herring	Sunlight
Mackerel	Vitamin-D-fortified milk

 ## Reasons to Use

• Dihydrotachysterol is primarily used to treat hypocalcemia (not enough calcium in blood) in people with chronic kidney failure or hypoparathyroidism.
• Regulates growth, hardening and repair of bone by controlling absorption of calcium and phosphorus from small intestine.
• Prevents rickets.
• Treats post-operative muscle contractions.
• Works with calcium to control bone formation.
• Promotes normal growth and development of infants and children, particularly bones and teeth.

 ## Unproved Speculated Benefits

• Cures arthritis.
• Prevents colon cancer.
• Treats aging symptoms.
• Treats acne.
• Treats alcoholism.
• Treats herpes simplex and herpes zoster.
• Treats cystic fibrosis.

 ## Who Needs Additional Amounts?

• Children who live in sunshine-deficient areas.
• Anyone with inadequate caloric or nutritional dietary intake or increased nutritional requirements.
• Older people (over 55 years), especially women after menopause.
• Pregnant or breast-feeding women.
• Those who abuse alcohol or other drugs.

• People with a chronic wasting illness, excess stress for long periods or those who have recently undergone surgery.
• Those with a portion of the gastrointestinal tract surgically removed.
• People with recent severe burns or injuries.

 ## Deficiency Symptoms

• Rickets (a childhood deficiency disease): bent, bowed legs, malformations of joints or bones, late tooth development, weak muscles, listlessness.
• Osteomalacia (adult rickets): muscle weakness and spasm, brittle, easily broken bones, pain in ribs, lower spine, pelvis, legs.

 ## Unproved Speculated Symptoms

• Muscle diseases (myopathies)

 ## Lab Tests to Detect Deficiency

• Reduced levels of vitamin D forms in blood.
• Decreased serum phosphate, decreased calcium, increased alkaline phosphatase, urinary hydroxyproline, PTH levels.
• Bone X-ray.

 ## Dosage and Usage Information

Recommended Dietary Allowance (RDA):
Estimate of adequate daily intake by the Food and Nutrition Board of the National Research Council, 1980. See Glossary.

Age	RDA
0–6 months	10mcg
6–12 months	10mcg
1–10 years	10mcg
Males	
11–18 years	10mcg
19–22 years	7.5mcg
23+ years	5mcg
Females	
11–18 years	10mcg
19–22 years	7.5mcg
23+ years	5mcg
Pregnant	+5mcg
Lactating	+5mcg

»➤

Dihydrotachysterol

What this vitamin does:
- Absorbs and uses calcium and phosphorous to make bone.
- Essential for normal growth and development.

Miscellaneous information:
- Take at the same time every day.
- Put liquid vitamin D directly into mouth or mix with cereal, fruit juice or food.

Available as:
- Extended-release capsules or tablets: Swallow whole with full glass of liquid. Don't chew or crush. Take with or immediately after food to decrease stomach irritation.
- Oral solution: Dilute in at least 1/2 glass water or other liquid. Take with meals or 1 to 1-1/2 hours after meals unless otherwise directed by your doctor.
- A constituent of many multivitamin/mineral preparations.
- Some forms available by generic name.

 ## Warnings and Precautions

Don't take if you:
- Are allergic to vitamin D, ergocalciferol or any vitamin-D derivative.

Consult your doctor if you have:
- Any plans to become pregnant while taking vitamin D.
- Epilepsy.
- Heart or blood-vessel disease.
- Kidney, liver, pancreatic disease.
- Chronic diarrhea.
- Intestinal problems.
- Sarcoidosis.

Over age 55:
- Adverse reactions and side effects are more likely. Supplements are often necessary.

Pregnancy:
- Taking too much during pregnancy may cause abnormalities in fetus. Consult doctor before taking supplement to ensure correct dosage.
- Don't take megadoses.

Breast-feeding:
- Important for you to receive correct amount so enough vitamin D is available for normal growth and development of baby. Consult doctor about supplements.
- Don't take megadoses.

Effect on lab tests:
- May decrease serum alkaline phosphatase.
- May increase levels of calcium, cholesterol and phosphate in test results.
- May increase level of magnesium in test results.
- May increase amounts of calcium and phosphorous in urine.

Storage:
- Store in cool, dry place away from direct light, but don't freeze. Avoid overexposure to air.
- Store safely out of reach of children.
- Don't store in bathroom medicine cabinet. Heat and moisture may change action of vitamin.

Others:
- Absence of sunlight prevents natural formation of vitamin D by skin. Sunshine provides sufficient amounts of vitamin D for people who live in sunny climates. Those who live in northern areas with fewer days of sunshine and extended periods of cloud cover and darkness must depend on dietary sources for vitamin D.
- Avoid megadoses.

 ## Overdose/Toxicity

Signs and symptoms:
High blood pressure, irregular heartbeat, nausea, weight loss, seizures, abdominal pain, appetite loss, mental and physical-growth retardation, premature hardening of arteries, kidney damage.

What to do:
For symptoms of overdosage: Discontinue vitamin, and consult doctor. Also see *Adverse Reactions or Side Effects* section below.

For accidental overdosage (such as child taking entire bottle): Dial 911 (emergency), 0 for operator or your nearest Poison Control Center.

For toxic symptoms: Discontinue vitamin, and seek immediate medical help. Hospitalization may be necessary.

———»▸

Dihydrotachysterol, Continued

 Adverse Reactions or Side Effects

Reaction or effect	What to do
Appetite loss	Discontinue. Call doctor when convenient.
Constipation	Discontinue. Call doctor when convenient.
Diarrhea	Discontinue. Call doctor immediately.
Dry mouth	Discontinue. Call doctor when convenient.
Headache	Discontinue. Call doctor immediately.
Increased thirst	Discontinue. Call doctor when convenient.
Mental confusion	Discontinue. Call doctor immediately.
Metallic taste	Discontinue. Call doctor when convenient.
Nausea	Discontinue. Call doctor immediately.
Unusual tiredness	Discontinue. Call doctor when convenient.
Vomiting	Discontinue. Call doctor immediately.

 Interaction with Medicine, Vitamins or Minerals

Interacts with	Combined effect
Antacids with aluminum	Decreases absorption of vitamin D and fat-soluble vitamins A, D, E, K.
Antacids with magnesium	May cause too much magnesium in blood, especially for people with kidney failure.
Anti-convulsants	May reduce effect of vitamin D from natural sources and require supplements to prevent loss of strength in bones.
Barbiturates	May reduce effect of vitamin D from natural sources and require supplements to prevent loss of strength in bones.
Calcitonin	Reduces effect of calcitonin when treating hypercalcemia.
Calcium (high doses)	Increases risk of hypercalcemia.
Cholestyramine	Impairs absorption of vitamin D. May need supplements.
Colestipol	Impairs absorption of vitamin D. May need supplements.
Digitalis preparations	Increases risk of heartbeat irregularities.
Diuretics, thiazide	Increases risk of hypercalcemia.
Hydantoin	May reduce effect of vitamin D from natural sources and require supplements to prevent loss of strength in bones.
Mineral oil	Increases absorption of vitamin D. May need supplements.
Phosphorous-containing medicines	Increases risk of too much phosphorous in blood.
Primidone	May reduce effect of vitamin D from natural sources and require supplements to prevent loss of strength in bones.
Vitamin-D derivatives, such as calciferol, calcitrol, dihydrotachysterol, ergocalciferol	Additive effects may increase potential for toxicity.

 Interaction with Other Substances

Chronic alcoholism depletes liver stores of vitamin D.

Basic Information

Vitamin E is also called alpha-tocopherol.
Brand names, see page 477.
Available from natural sources? Yes
Available from synthetic sources? Yes
Prescription required? Yes, for injectable forms
Fat-soluble or water-soluble: Fat-soluble

 ## Natural Sources

Almonds	Peanut oil
Apricot oil	Safflower nuts
Corn oil	Sunflower seeds
Cottonseed oil	Walnuts
Hazelnuts (filberts)	Wheat germ
Margarine	Whole-wheat flour

 ## Reasons to Use

- Promotes normal growth and development.
- Treats and prevents vitamin-E deficiency in premature or low-birth-weight infants.
- Prevents oxidation of free radicals in body.
- Acts as anti-blood clotting agent.
- Protects tissue against oxidation.
- Promotes normal red-blood-cell formation.

 ## Unproved Speculated Benefits

- Treats fibrocystic disease of breast.
- Treats circulatory problems of lower extremities.
- Treats sickle-cell anemia.
- Treats lung toxicity from air pollution.
- Prevents or alleviates coronary-artery heart disease.
- Enhances sexual performance.
- Improves muscle strength and stamina.
- Heals burns and wounds.
- Retards aging.
- Prevents hair loss.
- Prevents abortion.
- Treats menopause.
- Helps overcome infertility.
- Treats bee stings, liver spots on hands, bursitis, diaper rash.
- Prevents and treats cancer.
- Decreases scarring.
- Improves athletic performance.
- Treats muscular dystrophy, heart degeneration, anemia.
- Treats acne.
- Prevents eye problems and lung problems in low-birth-weight or premature infants.

 ## Who Needs Additional Amounts?

- Anyone with inadequate caloric or nutritional dietary intake or increased nutritional requirements.
- Older people (over 55 years).
- Those who abuse alcohol or other drugs.
- People who have a chronic wasting illness, excess stress for long periods or those who have recently undergone surgery.
- Those with part of the gastrointestinal tract surgically removed.
- People with recent severe burns or injuries.
- People with hyperthyroidism.

 ## Deficiency Symptoms

Premature infants and children:
- Irritability
- Edema
- Hemolytic anemia

Adults:
- Lack of vitality
- Lethargy
- Apathy
- Inability to concentrate
- Irritability
- Disinterest in physical activity
- Decreased sexual performance
- Muscle weakness

 ## Unproved Speculated Symptoms

- Indigestion
- Low libido and impotence
- Premature aging
- Chest pain

 ## Lab Tests to Detect Deficiency

- Blood tocopherol level
- Excess creatine in urine to indicate muscle breakdown
- Red-blood-cell fragility test

»▶

Vitamin E, Continued

Dosage and Usage Information

Estimate of adequate daily intake by the Food and Nutrition Board of the National Research Council, 1980. See Glossary.

Age	RDA
0–12 months	3–4mg
1–7 years	5–7mg
11–18 years	8mg
Males	
18+ years	10mg
Females	
18+ years	8mg
Pregnant	+ 2mg
Lactating	+ 3mg

Note: 1mg vitamin E equals 1IU. Labels may list as mg or IU.

What this vitamin does:
• Prevents a chemical reaction called *oxidation*. Excessive oxidation can sometimes cause harmful effects.
• Acts as a co-factor in several enzyme systems.

Miscellaneous information:
• Take at same time every day.
• Vitamin E is a constituent of many skin ointments, salves and creams. Claims for beneficial effects have not been confirmed, but topical application probably does not cause harm.
• May require several weeks of treatment before symptoms caused by deficiency will improve.
• Freezing may destroy vitamin E.
• Extreme heat causes vitamin E to break down. Avoid deep-fat frying foods that are natural sources of vitamin E.
• Vitamin E functions as an anti-oxidant, prevents enzyme action of peroxidase on unsaturated bonds of cell membranes and protects red blood cells from disintegrating.

Available as:
• Tablets or capsules: Swallow whole with full glass of liquid. Don't chew or crush. Take with or immediately after food to decrease stomach irritation.
• Drops: Dilute dose in beverage before swallowing, or squirt directly into mouth.
• A constituent of many multivitamin/mineral preparations.

Warnings and Precautions

Don't take if you:
• Are allergic to vitamin E.

Consult your doctor if you have:
• Iron-deficiency anemia.
• Bleeding or clotting problems.
• Cystic fibrosis.
• Intestinal problems.
• Liver disease.
• Overactive thyroid.

Over age 55:
• No problems expected.

Pregnancy:
• No problems expected, except with megadoses.

Breast-feeding:
• No problems expected.
• Don't take megadoses.

Effect on lab tests:
• Serum cholesterol and serum triglycerides may register *high* if you take large doses of vitamin E.

Storage:
• Store in cool, dry area away from direct light, but don't freeze.
• Store safely out of reach of children.
• Don't store in bathroom medicine cabinet. Heat and moisture may change action of the vitamin.

Others:
• Beware of megadoses.

Overdose/Toxicity

Signs and symptoms:
High doses deplete vitamin-A stores in body. Very high doses (over 800 mg/day) causes tendency to bleed, altered immunity, impaired sex functions, increased risk of blood clots, altered metabolism of thyroid, pituitary and adrenal hormones.

What to do:
For other symptoms of toxicity: Discontinue vitamin, and consult doctor. Also see *Adverse Reactions or Side Effects* section below.
For accidental overdosage (such as a child taking entire bottle): Dial 911 (emergency), 0 for operator or your nearest Poison Control Center.

➤➤

Adverse Reactions or Side Effects

Reaction or Effect	What to do
Abdominal pain	Discontinue. Call doctor immediately.
Breast enlargement	Discontinue. Call doctor when convenient.
Diarrhea	Discontinue. Call doctor immediately.
Dizziness	Discontinue. Call doctor when convenient.
Flu-like symptoms	Discontinue. Call doctor immediately.
Headache	Discontinue. Call doctor when convenient.
Nausea	Discontinue. Call doctor immediately.
Tiredness or weakness	Discontinue. Call doctor when convenient.
Vision blurred	Discontinue. Call doctor immediately.

Interaction with Medicine, Vitamins or Minerals

Interacts with	Combined effect
Antacids	Decreases vitamin-E absorption.
Anti-coagulants, coumarin- or indandione-type	May increase spontaneous or hidden bleeding.
Cholestyramine	May decrease absorption of vitamin E.
Colestipol	May decrease absorption of vitamin E.
Iron supplements	Decreases effect of iron supplement in people with iron-deficiency anemia. Decreases vitamin-E effect in healthy people.
Mineral oil	May decrease absorption of vitamin E.
Sucralfate	May decrease absorption of vitamin E.
Vitamin A	Facilitates absorption, storage and utilization of vitamin A. Reduces potential toxicity of vitamin A. Excessive doses of vitamin E causes vitamin-A depletion.

Interaction with Other substances

Tobacco decreases absorption. Smokers may require supplemental vitamin E.

Chronic alcoholism depletes vitamin-E stores in liver.

Folic Acid (Vitamin B-9)

Basic Information

Folic acid is also called folate, pteroyglutamic acid, folacin.
Brand names, see page 477.
Available from natural sources? Yes
Available from synthetic sources? Yes
Prescription required? Yes, for injectable forms
Fat-soluble or water-soluble? Water-soluble

Natural Sources

Barley	Lentils
Beans	Orange juice
Brewer's yeast	Oranges
Calves' liver	Peas
Endive	Rice
Fruits	Soybeans
Garbanzo beans	Split peas
(chickpeas)	Sprouts
Green, leafy	Wheat
vegetables	Wheat germ

Reasons to Use

- Promotes normal red-blood-cell formation.
- Maintains nervous system, intestinal tract, sex organs, white blood cells, normal patterns of growth.
- Regulates embryonic and fetal development of nerve cells.
- Promotes normal growth and development.
- Treats anemias due to folic-acid deficiency occurring from alcoholism, liver disease, hemolytic anemia, sprue, pregnancy, breast-feeding, oral-contraceptive use.

Unproved Speculated Benefits

- Prevents mental problems.
- Acts as a natural analgesic or pain killer.

Who Needs Additional Amounts?

- Anyone with inadequate caloric or nutritional dietary intake or increased nutritional requirements.
- Older people (over 55 years).
- Pregnant or breast-feeding women.
- Women who use oral contraceptives.
- Those who abuse alcohol or other drugs

- People with a chronic wasting illness, excess stress for long periods or those who have recently undergone surgery.
- Those with a portion of the gastrointestinal tract surgically removed.
- People with recent severe burns or injuries.
- Young infants not receiving breast milk or fortified commercial formula.
- Extremely ill people who must be fed intravenously or by naso-gastric tube.

Deficiency Symptoms

- Hemolytic and megaloblastic anemia in which red blood cells are large and uneven in size, have a shorter life span or are likely to have cell membranes rupture
- Irritability
- Weakness
- Lack of energy
- Sleeping difficulties
- Paleness
- Sore red tongue
- Mild mental symptoms, such as forgetfulness and confusion
- Diarrhea

Unproved Speculated Symptoms

- Depression
- Cervical dysplasia
- Psychosis

Lab Tests to Detect Deficiency

- Serum folic acid
- Blood cells showing macrocytic anemia coupled with normal levels of B-12 in blood

Dosage and Usage Information

Recommended Dietary Allowance (RDA):
Estimate of adequate daily intake by the Food and Nutrition Board of the National Research Council, 1980. See Glossary.   ⫸

Folic Acid (Vitamin B-9)

Age	RDA
0–6 months	30mcg
6–12 months	45mcg
1–3 years	100mcg
4–6 years	200mcg
7–10 years	300mcg
11+ years	400mcg
Pregnant	+400mcg
Lactating	+100mcg

What this vitamin does:
- Acts as co-enzyme for normal DNA synthesis.
- Functions as part of co-enzyme in amino acid and nucleoprotein synthesis.
- Promotes normal red-blood-cell formation.

Miscellaneous information:
- Cooking vegetables causes loss of some folic-acid content.

Available as:
- Tablets: Swallow whole with full glass of liquid. Don't chew or crush. Take with meals or 1 to 1-1/2 hours after meals unless otherwise directed by your doctor.

Note: Folic acid is sometimes omitted from multivitamin/mineral preparations. Check labels.

Warnings and Precautions

Don't take if you:
- Have pernicious anemia. Folic acid will make the blood appear normal, but neurological problems may progress and be irreversible.
- Take anti-convulsant medication.

Consult your doctor if you have:
- Anemia

Over age 55:
- No problems expected.

Pregnancy:
- No problems expected.
- Don't take megadoses.

Breast-feeding:
- No problems expected.
- Don't take megadoses.

Effect on lab tests:
- May cause false-low results in tests for vitamin B-12.

Storage:
- Store in cool, dry place away from direct light, but don't freeze.
- Store safely out of reach of children.
- Don't store in bathroom medicine cabinet. Heat and moisture may change action of vitamin.

Others:
- Renal dialysis reduces blood folic acid. Patients on dialysis should increase RDA by 300%.

Overdose/Toxicity

Signs and symptoms:
Prolonged use of high doses can produce damaging folacin crystals in the kidney. Doses over 1,500mcg/day can cause appetite loss, nausea, flatulence, abdominal distension, may obscure existence of pernicious anemia.

What to do:
For symptoms of overdosage: Discontinue vitamin, and consult doctor. Also see *Adverse Reactions or Side Effects* section below.
For accidental overdosage (such as child taking entire bottle): Dial 911 (emergency), 0 for operator or your nearest Poison Control Center.

—»→

VITAMIN

Folic Acid <small>(Vitamin B-9), Continued</small>

Adverse Reactions or Side Effects

Reaction or effect	What to do
Bright-yellow urine (always)	Nothing.
Diarrhea	Discontinue. Call doctor immediately.
Fever	Discontinue. Call doctor immediately.
Skin rash	Discontinue. Call doctor when convenient.

Interaction with Medicine, Vitamins or Minerals

Interacts with	Combined effect
Analgesics	Decreases effect of folic acid.
Antibiotics	May cause false-low results in tests for serum-folic acid.
Anti-convulsants	Decreases effect of folic acid and anti-convulsant.
Chloramphenicol	Produces folic-acid deficiency.
Cortisone drugs	Decreases effect of folic acid.
Methotrexate	Decreases effect of folic acid.
Oral contraceptives	Decreases effect of folic acid. Those who take oral contraceptives require additional folic acid.
Phenytoin	Decrease phenytoin effect. Patients taking phenytoin should avoid taking folic acid.
Pyrimethamine	Decreases effect of folic acid and interferes with effectiveness of pyrimethamine. Avoid this combintation.
Quinine	Decreases effect of folic acid.
Sulfasalazine and other sulfa drugs	Decreases effect of folic acid.
Trimethoprim	Decreases effect of folic acid.
Trimterene	Decreases effect of folic acid.

Interaction with Other Substances

Tobacco decreases absorption. Smokers may require supplemental folic acid.

Alcohol abuse makes deficiency more likely. Alcoholism is the principal cause of folic-acid deficiency.

Basic Information

Menadiol is one form of Vitamin K.
Brand names, see page 477.
Available from natural sources? Yes
Available from synthetic sources? Yes
Prescription required? Yes
Fat-soluble or water-soluble: Fat-soluble.
Menadiol sodium diphosphate is water-soluble.

Natural Sources

Alfalfa	Green tea
Brussels sprouts	Oats
Cabbage	Soybeans
Camembert cheese	Spinach
Cauliflower	Turnip greens
Cheddar cheese	

Reasons to Use

- Promotes normal growth and development.
- Prevents hemorrhagic disease of the newborn.
- Prevents abnormal bleeding, particularly in those with chronic intestinal disease or those taking anti-coagulant medicines. Vitamin K is normally manufactured in the intestinal tract by "friendly" bacteria. If bacteria are destroyed or damaged by disease or antibiotics, vitamin-K deficiency may develop.
- Treats bleeding disorders due to vitamin-K deficiency.

Unproved Speculated Benefits

- None.

Who Needs Additional Amounts?

- Anyone with inadequate caloric or nutritional dietary intake or increased nutritional requirements.
- Those with a portion of the gastrointestinal tract surgically removed.
- People with recent severe burns or injuries.
- Premature newborns.
- Those with recent severe burns or injuries.
- Anyone taking antibiotics that may destroy normal "friendly" bacteria in the intestinal tract.
- People who do not have enough bile to absorb fats. Replacement must be given by injection.

Deficiency Symptoms

Infants:
- Failure to grow and develop normally.
- Hemorrhagic disease of the newborn characterized by vomiting blood and bleeding from intestine, umbilical cord, circumcision site. Symptoms begin 2 or 3 days after birth.

Adults:
- Abnormal blood clotting that can lead to nosebleeds, blood in urine, stomach bleeding, bleeding from capillaries or skin causing spontaneous black-and-blue marks, prolonged clotting time (a laboratory test).

Unproved Speculated Symptoms

- Excessive diarrhea.

Lab Tests to Detect Deficiency

- Prothrombin time.
- Serum prothrombin.
- Serum vitamin K.

Dosage and Usage Information

Recommended Dietary Allowance (RDA):
No RDA has been established. Adequate and safe range is 2mcg/kg body weight per day. Estimated Safe Intake/day is given below.

Age	Estimated Safe Intake
0–6 months	12mcg
6–12 months	10–20mcg
1–3 years	15–30mcg
4–6 years	20–40mcg
7–10 years	30–60mcg
11–17 years	50–100mcg
18+ years	70–140mcg

What this vitamin does:
- Promotes production of active prothrombin (factor II), proconvertin (factor VII) and other clotting factors. These are all necessary for normal blood clotting.

Miscellaneous information:
- Very little vitamin K is lost from processing or cooking foods.
- When a severe bleeding disorder exists due to a vitamin-K deficiency, fresh whole blood may be needed during severe bleeding episodes.
- There is a significant delay before vitamin K becomes effective when given by injection.

Menadiol (Vitamin K), Continued

Available as:
- Tablets: Swallow whole with full glass of liquid. Don't chew or crush. Take with meals or 1 to 1-1/2 hours after meals unless otherwise directed by your doctor.
- Injectable forms are administered by doctor or nurse.

Note: Vitamin K is not usually included in most multivitamin/mineral preparations.

Warnings and Precautions

Don't take if you:
- Are allergic to vitamin K.
- Have a G6PD deficiency. See Glossary.
- Have liver disease.

Consult your doctor if you have:
- Cystic fibrosis.
- Had prolonged diarrhea.
- Had prolonged intestinal problems.
- Taken any other medicines.
- Plans for surgery (including dental surgery) in the near future.

Over age 55:
- No problems expected.

Pregnancy:
- No studies available in humans. Avoid if possible.
- Don't take megadoses.

Breast-feeding:
- Don't take megadoses.

Effect on lab tests:
- Changes prothrombin times.

Storage:
- Store in cool, dry place away from direct light, but don't freeze.
- Store safely out of reach of children.
- Don't store in bathroom medicine cabinet. Heat and moisture may change action of vitamin.

Others:
- Avoid overdosage. Vitamin K is a fat-soluble vitamin. Excess intake can lead to impaired liver function.
- Tell any dentist or doctor who plans surgery that you take vitamin K.

Overdose/Toxicity

Signs and symptoms:
In Infants: Brain damage.
In All: Large doses may impair liver function.

What to do:
For symptoms of overdosage: Discontinue vitamin, and consult doctor. Also see *Adverse Reactions or Side Effects* section below.
For accidental overdosage (such as child taking entire bottle): Dial 911 (emergency), 0 for operator or your nearest Poison Control Center.

Adverse Reactions or Side Effects

Reaction or effect	What to do
Hemolytic anemia in infants	Seek emergency treatment.
Hyperbilirubinemia (too much bilirubin in the blood) in newborns or infants given too much vitamin K	Seek emergency treatment.
Jaundice (yellow skin and eyes) resulting from hyperbilirubinemia	Seek emergency treatment.
Allergic reactions, including:	
Face flushing	Discontinue. Call doctor immediately.
Gastrointestinal upset	Discontinue. Call doctor immediately.
Rash	Discontinue. Call doctor immediately.
Redness, pain or swelling at injection site	Discontinue. Call doctor immediately.
Skin itching	Seek emergency treatment.

≫➤

Interaction with Medicine, Vitamins or Minerals

Interacts with	Combined effect
Anti-coagulants (oral)	Decreases anti-coagulant effect.
Antibiotics, broad spectrum	Causes vitamin-K deficiency.
Cholestyramine	Decreases vitamin-K effect.
Colestipol	Decreases vitamin-K effect.
Coumarin (isolated from sweet clover)	Decreases vitamin-K effect.
Mineral oil (long term)	Causes vitamin-K deficiency.
Primaquine	Increases potential for toxic side effects.
Quinidine	Causes vitamin-K deficiency.
Salicylates	Increases need for vitamin K when administered over long time.
Sucralfate	Decreases vitamin-K effect.
Sulfa drugs	Causes vitamin-K deficiency.

Interaction with Other Substances

None known

V I T A M I N

Niacin (Vitamin B-3)

Basic Information

Niacin is also called Vitamin B-3.
Brand names, see page 477.
Available from natural sources? Yes
Available from synthetic sources? Yes
Prescription required? Yes, for high doses used
for cholesterol reduction
Fat-soluble or water-soluble: Water-soluble

Natural Sources

Beef liver	Salmon
Brewer's yeast	Sunflower seeds
Chicken, white meat	Swordfish
Halibut	Tuna
Peanuts	Turkey
Pork	Veal

Reasons to Use

- Maintains normal function of skin, nerves, digestive system.
- Reduces cholesterol and triglycerides in blood.
- Corrects niacin deficiency.
- Dilates blood vessels.
- Treats vertigo (dizziness) and ringing in ears.
- Prevents premenstrual headache.
- Treats pellagra.

Unproved Speculated Benefits

- Prevents heart attacks.
- Treats or prevents motion sickness.
- Alleviates mental illness, notably schizophrenia.
- Cures depression.
- Prevents migraine headaches.
- Improves poor digestion.
- Protects against pollutants and toxins.
- Treats leprosy.
- Stimulates sex drive.

Who Needs Additional Amounts?

- Anyone with inadequate caloric or nutritional dietary intake or increased nutritional requirements.
- Older people (over 55 years).
- Pregnant or breast-feeding women.
- Those who abuse alcohol or other drugs.

- People with a chronic wasting illness including malignancies, pancreatic insufficiency, cirrhosis of the liver, sprue.
- Anyone who experiences excess stress for long periods or who has recently undergone surgery.
- Athletes and workers who participate in vigorous physical activities.
- Those with a portion of the gastrointestinal tract surgically removed.
- People with recent severe burns or injuries.
- Those with diabetes.
- Infants born with errors of metabolism (congenital disorders due to chromosome abnormalities).
- Anyone with hyperthyroidism.

Deficiency Symptoms

Early Symptoms
- Muscle weakness
- General fatigue
- Loss of appetite
- Headaches
- Swollen, red tongue
- Skin lesions, including rashes, dry scaly skin, wrinkles, coarse skin texture
- Nausea and vomiting
- Dermatitis
- Diarrhea
- Irritability
- Dizziness

Late Symptoms of severe deficiency called *pellagra:*
- Dementia
- Death

Unproved Speculated Symptoms

- Acne
- Poor circulation
- Mental problems

Lab Tests to Detect Deficiency

- Urinary N-1 methylnicotinamide
- Urinary 2—pyrindone/N-1 methylnicotinamide. Test results not always conclusive.
- Abnormal-liver-function studies

Dosage and Usage Information

Recommended Dietary Allowance (RDA):
Estimate of adequate daily intake by the Food and Nutrition Board of the National Research Council, 1980. See Glossary.

Age	RDA
0–6 months	6mg
6–12 months	8mg
1–3 years	9mg
4–6 years	11mg
7–10 years	16mg
Males	
11–18 years	18mg
19–22 years	19mg
23–50 years	18mg
50+ years	16mg
Females	
11–14 years	15mg
15–22 years	14mg
23+ years	13mg
Pregnant	+2mg
Lactating	+4mg

What this vitamin does:
- Aids in release of energy from foods.
- Helps synthesize DNA.
- Becomes component of two co-enzymes (NAD and NADP), which are both necessary for utilization of fats, tissue respiration and production of sugars.

Miscellaneous information:
- The body manufactures niacin from tryptophan, an amino acid.

Available as:
- Tablets or capsules: Swallow whole with full glass of liquid. Don't chew or crush. Take with meals or 1 to 1-1/2 hours after meals unless otherwise directed by your doctor.
- Extended-release capsules or tablets: Swallow whole with full glass of liquid. Don't chew or crush. Take with or immediately after food to decrease stomach irritation.
- Oral solution: Dilute in at least 1/2 glass water or other liquid. Take with meals or 1 to 1-1/2 hours after meals unless otherwise directed by your doctor.
- Injectable forms are administered by doctor or nurse.
- A constituent of many multivitamin/mineral preparations.
- Some forms available by generic name.

Warnings and Precautions

Don't take if you:
- Are allergic to niacin or any niacin-containing vitamin mixtures.
- Have impaired liver function.
- Have an active peptic ulcer.

Consult your doctor if you have:
- Diabetes.
- Gout.
- Gallbladder or liver disease.

Over age 55:
- Response to drug cannot be predicted. Dose must be individualized.

Pregnancy:
- Risk to fetus with high doses outweighs benefits. Do not use.

Breast-feeding:
- Studies are inconclusive. Consult doctor about supplements.
- Don't take megadoses.

Effect on lab tests:
- Urinary catecholamine concentration may falsely elevate results.
- Urine glucose (using Benedict's reagent) may produce false-positive reactions.
- Falsely elevates blood sugar.
- Falsely increases growth-hormone level in blood.
- Falsely elevates blood uric acid with large daily doses.

Storage:
- Store in cool, dry place away from direct light, but don't freeze.
- Store safely out of reach of children.
- Don't store in bathroom medicine cabinet. Heat and moisture may change action of vitamin.

Others:
- High dosages over long periods may cause liver damage or aggravate a stomach ulcer.

Overdose/Toxicity

Signs and symptoms:
Body flush, nausea, vomiting, abdominal cramps, diarrhea, weakness, lightheadedness, headache, fainting, sweating, high blood sugar, high uric acid, heart-rhythm disturbances, jaundice.

➤➤

Niacin (Vitamin B-3), Continued

What to do:

For symptoms of overdosage: Discontinue vitamin, and consult doctor. Also see *Adverse Reactions or Side Effects* section below.

For accidental overdosage (such as child taking entire bottle): Dial 911 (emergency), 0 for operator or your nearest Poison Control Center.

 Adverse Reactions or Side Effects

Reaction or effect	What to do
Abdominal pain	Discontinue. Call doctor immediately.
Diarrhea	Discontinue. Call doctor when convenient.
Faintness	Discontinue. Call doctor immediately.
Headache	Discontinue. Call doctor when convenient.
"Hot" feeling, with skin flushed in blush zone (always)	Nothing.
Jaundice (yellow skin and eyes)	Discontinue. Call doctor immediately.
Nausea or vomiting	Discontinue. Call doctor immediately.
Skin dryness	Discontinue. Call doctor when convenient.
Vomiting	Discontinue. Call doctor immediately.

 Interaction with Medicine, Vitamins or Minerals

Interacts with	Combined effect
Anti-diabetics	Decreases anti-diabetic effect.
Beta-adrenergic blockers	Lowers blood pressure to extremely low level.
Chenodiol	Decreases chenodiol effect.
Guanethidine	Increases guanethidine effect.
Isoniazid	Decreases niacin effect.
Mecamylamine	Lowers blood pressure to extremely low level.
Pargyline	Lowers blood pressure to extremely low level.

 Interaction with Other Substances

Tobacco decreases absorption. Smokers may require supplemental niacin.

Alcohol may cause extremely low blood pressure. Use caution.

Niacinamide

Basic Information

Niacinamide is a form of vitamin B-3 and sometimes called nicotinamide.
Brand names, see page 477.
Available from natural sources? Yes
Available from synthetic sources? Yes
Prescription required? Yes, for high doses used for cholesterol reduction
Fat-soluble or water-soluble: Water-soluble

Natural Sources

Beef liver	Salmon
Brewer's yeast	Sunflower seeds
Chicken	Swordfish
Halibut	Tuna
Peanuts	Turkey
Pork	Veal

Reasons to Use

- Maintains normal function of skin, nerves, digestive system.
- Corrects niacin deficiency.
- Dilates blood vessels.
- Treats dizziness and ringing in ears.
- Prevents premenstrual headache.
- Treats pellagra.

Unproved Speculated Benefits

- Prevents heart attacks.
- Treats or prevents motion sickness.
- Alleviates mental illness, notably schizophrenia.
- Cures depression.
- Prevents migraine or headaches.
- Improves digestion.
- Protects against pollutants and toxins.
- Treats leprosy.
- Stimulates sex drive.

Who Needs Additional Amounts?

- Anyone with inadequate caloric or dietary intake or increased nutritional requirements.
- Those who abuse alcohol or other drugs.
- People with a chronic wasting illness, including malignancies, pancreatic insufficiency, cirrhosis of the liver, sprue.

- Anyone who experiences excess stress for long periods, or who has recently undergone surgery.
- Athletes and workers who participate in vigorous physical activities.
- Those with a portion of the gastrointestinal tract surgically removed.
- People with recent severe burns or injuries.
- Those with diabetes.
- Anyone with hyperthyroidism.

Deficiency Symptoms

Early symptoms:
- Muscular weakness
- General fatigue
- Irritability
- Dizziness
- Loss of appetite
- Headaches
- Swollen, red tongue
- Skin lesions, including rashes, dry scaly skin in areas exposed to sunlight, wrinkles, coarse skin texture
- Nausea and vomiting

Late symptoms of severe deficiency called *pellagra*:
- Dementia
- Death

Unproved Speculated Symptoms

- Acne
- Poor circulation
- Mental problems

Lab Tests to Detect Deficiency

- Urinary N-1 methylnicotinamide
- Urinary 2—pyrindone/N-1 methylnicotinamide; test results not always conclusive
- Liver-function studies

VITAMIN

>>>

Niacinamide, Continued

Dosage and Usage Information

Recommended Dietary Allowance (RDA):
Estimate of adequate daily intake by the Food and Nutrition Board of the National Research Council, 1980. See Glossary.

Age	RDA
0–6 months	6mg
6–12 months	8mg
1–3 years	9mg
4–6 years	11mg
7–10 years	16mg
Males	
11–18 years	18mg
19–22 years	19mg
23–50 years	18mg
50+ years	16mg
Females	
11–14 years	15mg
15–22 years	14mg
23+ years	13mg
Pregnant	+2mg
Lactating	+4mg

What this vitamin does:
- Aids in release of energy from foods.
- Helps synthesis of DNA.
- Becomes a component of two co-enzymes (NAD and NADP), which are necessary for utilization of fats, tissue respiration, production of sugars.

Miscellaneous information:
- The body manufactures niacinamide from tryptophan, an amino acid.

Available as:
- Tablets or capsules: Swallow whole with full glass of liquid. Don't chew or crush. Take with meals or 1 to 1-1/2 hours after meals unless otherwise directed by your doctor.
- Extended-release capsules or tablets: Swallow whole with full glass of liquid. Don't chew or crush. Take with or immediately after food to decrease stomach irritation.
- Oral solution: Dilute in at least 1/2 glass water or other liquid. Take with meals or 1 to 1-1/2 hours after meals unless otherwise directed by your doctor.
- Injectable forms are administered by doctor or nurse.
- Some forms available by generic name.

Warnings and Precautions

Don't take if you:
- Are allergic to niacin or any niacin-containing vitamin mixtures.

- Have impaired liver function.
- Have an active peptic ulcer.

Consult your doctor if you have:
- Diabetes.
- Gout.
- Gallbladder or liver disease.

Over age 55:
- Response to drug cannot be predicted. Dose must be individualized.

Pregnancy:
- Risk with high doses to unborn child outweighs benefits. Don't use.

Breast-feeding:
- Studies inconclusive. Consult doctor about supplements.
- Don't take megadoses.

Effect on lab tests:
- Urinary catecholamine concentration may show falsely elevate results.
- Urine glucose (using Benedict's reagent) may produce false-positive reactions.
- Falsely elevates blood sugar.
- Falsely increases growth-hormone level in blood.
- Falsely elevates blood-uric acid with large daily doses.

Storage:
- Store in cool, dry place away from direct light, but don't freeze.
- Store safely out of reach of children.
- Don't store in bathroom medicine cabinet. Heat and moisture may change action of vitamin.

Others:
- High doses over long periods may cause liver damage or aggravate a stomach ulcer.

Overdose/Toxicity

Signs and symptoms:
Body flush, nausea, vomiting, abdominal cramps, diarrhea, weakness, lightheadedness, fainting, sweating, headache, high blood sugar, high uric acid, heart-rhythm disturbances, jaundice.

What to do:
For symptoms of overdosage: Discontinue vitamin, and consult doctor. Also see *Adverse Reactions or Side Effects* section below.
For accidental overdosage (such as child taking entire bottle): Dial 911 (emergency), 0 for operator or your nearest Poison Control Center. »➤

Adverse Reactions or Side Effects

Reaction or effect	What to do
Abdominal pain	Discontinue. Call doctor immediately.
Diarrhea	Discontinue. Call doctor immediately.
Faintness	Discontinue. Call doctor immediately.
Headache	Discontinue. Call doctor when convenient.
"Hot" feeling, with skin flushed in blush zone (always)	Nothing.
Jaundice (yellow skin and eyes)	Seek emergency treatment.
Nausea	Discontinue. Call doctor immediately.
Skin dryness	Discontinue. Call doctor immediately.
Vomiting	Discontinue. Call doctor immediately.

Interaction with Medicine, Vitamins or Minerals

Interacts with	Combined effect
Anti-diabetics	Decreases anti-diabetic effect.
Beta-adrenergic blockers	Lowers blood pressure to extremely low level.
Chenodiol	Decreases chenodiol effect.
Guanethidine	Increases guanethidine effect.
Isoniazid	Decreases niacin effect.
Mecamylamine	Lowers blood pressure to extremely low level.
Pargyline	Lowers blood pressure to extremely low level.

Interaction with Other Substances

Tobacco decreases absorption. Smokers may require supplemental niacin.

Alcohol may cause excessively low blood pressure. Use caution.

Pantothenic Acid (Vitamin B-5)

Basic Information

Available from natural sources? Yes
Available from synthetic sources? Yes
Prescription required? Yes, for injectable forms
Fat-soluble-or water-soluble: Water-soluble

Natural Sources

Blue cheese	Meats, all kinds
Brewer's yeast	Peanuts
Corn	Peas
Eggs	Soybeans
Lentils	Sunflower seeds
Liver	Wheat germ
Lobster	Whole-grain products

Reasons to Use

• Promotes normal growth and development.
• Aids in release of energy from foods.
• Helps synthesis of numerous body materials.

Unproved Speculated Benefits

• Stimulates wound healing.
• Alleviates stress.
• Restores gray hair to normal hair color.
• Prevents hair from turning gray.
• Cures allergies.
• Treats alcoholism, liver cirrhosis.
• Treats constipation.
• Treats fatigue.
• Treats stomach ulcers.
• Retards aging.

Who Needs Additional Amounts?

• Anyone with inadequate caloric or nutritional dietary intake or increased nutritional requirements.
• Older people (over 55 years).
• Pregnant or breast-feeding women.
• Those who abuse alcohol or other drugs.
• People with a chronic wasting illness, excess stress for long periods or who have recently undergone surgery.
• Athletes and workers who participate in vigorous physical activities.

• People with a portion of the gastrointestinal tract surgically removed.
• People with recent severe burns or injuries.

Deficiency Symptoms

None proved for pantothenic acid alone. However, lack of one B vitamin usually means lack of other B nutrients. Pantothenic acid is usually given with other B vitamins if there are symptoms of *any* vitamin-B deficiency, including excessive fatigue, sleep disturbances, loss of appetite, nausea.

Unproved Speculated Symptoms

• Nerve damage
• Breathing problems
• Skin problems
• Gray hair
• Arthritis
• Allergies
• Birth defects
• Mental fatigue
• Headaches
• Sleep disturbances
• Muscle spasms, cramps

Lab Tests to Detect Deficiency

• Methods are limited and expensive. Tests are used only for research at present. Methods are available to measure blood levels and levels in 24-hour urine collections.

Dosage and Usage Information

Recommended Dietary Allowance (RDA):
No RDA has been established. Estimated safe intake given below.

Age	Estimated Safe Intake
0–6 months	2mg/day
6 months–3 years	3mg/day
4–6 years	3–4mg/day
7–9 years	4–5mg/day
10+ years	4–7mg/day

Pregnancy and lactation may increase the need by one-third.

What this vitamin does:
• Acts as co-enzyme in energy metabolism of carbohydrates, protein and fat.

Available as:
- Tablets: Swallow whole with full glass of liquid. Don't chew or crush. Take with meals or 1 to 1-1/2 hours after meals unless otherwise directed by your doctor.
- A constituent of many multivitamin/mineral preparations.
- Pantothenic acid is also sold as dexpanthenol (panthoderm), a lotion or cream applied to burns, cuts or abrasions. It relieves itching and soothes the wound.

Warnings and Precautions

Don't take if you:
- Are allergic to pantothenic acid.
- Are taking levodopa for Parkinson's disease.

Consult your doctor if you have:
- Hemophilia.

Over age 55:
- No problems expected.

Pregnancy:
- Don't exceed recommended dose.

Breast-feeding:
- Don't exceed recommended dose.

Effect on lab tests:
- None expected.

Storage:
- Store in cool, dry place away from direct light, but don't freeze.
- Store safely out of reach of children.
- Don't store in bathroom medicine cabinet. Heat and moisture may change action of vitamin.

Others:
- Avoid megadoses.
- Don't exceed recommended doses if you take pantothenic acid without medical supervision.

Overdose/Toxicity

Signs and symptoms:
Diarrhea and water retention with ingestion of megadoses—over 10 to 20 grams/day (10,000mg-20,000mg). This dose is not life-threatening.

What to do:
For symptoms of overdosage: Discontinue vitamin, and consult doctor.
For accidental overdosage (such as child taking entire bottle): Dial 911 (emergency), 0 for operator or your nearest Poison Control Center.

Adverse Reactions or Side Effects

Reaction or effect	What to do
None expected with normal intake	Call doctor if you suspect new symptoms are caused by taking pantothenic acid.

Interaction with Medicine, Vitamins or Minerals

Interacts with	Combined effect
Levodopa	Small amounts of pantothenic acid nullify levodopa's effect. Carbidopa-levodopa combination is not affected by this interaction.

Interaction with Other Substances

Tobacco decreases absorption. Smokers may require supplemental vitamin B-5.

VITAMIN

Phytonadione (Vitamin K)

Basic Information

Phytonadione is one form of vitamin K.
Brand names, see page 477.
Available from natural sources? Yes
Available from synthetic sources? Yes
Prescription required? Yes
Fat-soluble or water-soluble: Fat-soluble

Natural Sources

Alfalfa	Green tea
Brussels sprouts	Oats
Cabbage	Soybeans
Camembert cheese	Spinach
Cauliflower	Turnip greens
Cheddar cheese	

Reasons to Use

- Promotes normal growth and development.
- Prevents hemorrhagic disease of the newborn.
- Prevents abnormal bleeding, particularly in those with chronic intestinal disease or those taking anti-coagulant medicines. Vitamin K is normally manufactured in the intestinal tract by "friendly" bacteria. If bacteria are destroyed or damaged by disease or antibiotics, vitamin-K deficiency may develop.
- Treats bleeding disorders due to vitamin K deficiency.

Unproved Speculated Benefits

- None.

Who Needs Additional Amounts?

- Anyone with inadequate caloric or nutritional dietary intake or increased nutritional requirements.
- Those with a portion of the gastrointestinal tract surgically removed.
- Premature newborns.
- People with recent severe burns or injuries.
- Anyone taking antibiotics that may destroy normal "friendly" bacteria in the intestinal tract.
- People who do not have enough bile to absorb fats. Replacement must be given by injection.

Deficiency Symptoms

Infants:
- Failure to grow and develop normally.
- Hemorrhagic disease of the newborn characterized by vomiting blood and bleeding from the intestine, umbilical cord, circumcision site. Symptoms begin 2 or 3 days after birth.

Adults:
- Abnormal blood clotting that can lead to nosebleeds, blood in urine, stomach bleeding, bleeding from capillaries or skin causing spontaneous black-and-blue marks, prolonged clotting time (a laboratory test).

Unproved Speculated Symptoms

- Excessive diarrhea.

Lab Tests to Detect Deficiency

- Prothrombin time
- Serum prothrombin
- Serum vitamin K

Dosage and Usage Information

Recommended Dietary Allowance (RDA):
No RDA has been established. Adequate and safe range is 2mcg/kg body weight per day. Estimated safe intake per day given below.

Age	Estimated Safe Intake
0–6 months	12mcg/day
6 months–12 months	10–20mcg/day
1–3 years	15–30mcg/day
4–6 years	20–40mcg/day
7–10 years	30–60mcg/day
11–17 years	50–100mcg/day
18+ years	70–140mcg/day

What this vitamin does:
- Promotes production of active prothrombin (factor II), proconvertin (factor VII) and other clotting factors. All are necessary for normal blood clotting.

Miscellaneous information:
- Very little vitamin K is lost from processing or cooking foods.
- When a severe bleeding disorder exists due to a vitamin-K deficiency, fresh whole blood may be needed during severe bleeding episodes.
- There is a significant delay before vitamin K becomes effective when given by injection.

Phytonadione (Vitamin K)

Available as:
- Tablets: Swallow whole with full glass of liquid. Don't chew or crush. Take with meals or 1 to 1-1/2 hours after meals unless otherwise directed by your doctor.
- Injectable forms are administered by doctor or nurse.

Note: Vitamin K is not usually included in most multivitamin/mineral preparations.

Warnings and Precautions

Don't take if you:
- Are allergic to vitamin K.
- Have a G6PD deficiency. See Glossary.
- Have liver disease.

Consult your doctor if you have:
- Cystic fibrosis.
- Had prolonged diarrhea.
- Prolonged intestinal problems.
- Taken any other medicines.
- Plans for surgery (including dental surgery) in the near future.

Over age 55:
- No problems expected.

Pregnancy:
- No studies available in humans. Avoid if possible.
- Don't take megadoses.

Breast-feeding:
- Don't take megadoses.

Effect on lab tests:
- Changes prothrombin times.

Storage:
- Store in cool, dry place away from direct light, but don't freeze.
- Store safely out of reach of children.
- Don't store in bathroom medicine cabinet. Heat and moisture may change action of vitamin.

Others:
- Avoid overdosage. Vitamin K is a fat-soluble vitamin. Excess intake can lead to impaired liver function.
- Tell any dentist or doctor who plans surgery that you take vitamin K.

Overdose/Toxicity

Signs and symptoms:
In Infants: Brain damage
In All: Large doses may impair liver function.

What to do:
For symptoms of overdosage: Discontinue vitamin, and consult doctor. Also see *Adverse Reactions or Side Effects* section below.

For accidental overdosage (such as child taking entire bottle): Dial 911 (emergency), 0 for operator or your nearest Poison Control Center.

Adverse Reactions or Side Effects

Reaction or effect	What to do
Hemolytic anemia in infants	Seek emergency treatment.
Hyperbilirubinemia (too much bilirubin in the blood) in newborns or infants given too much vitamin K	Seek emergency treatment.
Jaundice (yellow skin and eyes) resulting from hyperbilirubinemia	Seek emergency treatment.
Allergic reactions, including:	
Face flushing	Discontinue. Call doctor immediately.
Gastrointestinal upset	Discontinue. Call doctor immediately.
Rash	Discontinue. Call doctor immediately.
Redness, pain or swelling at injection site	Discontinue. Call doctor immediately.
Skin itching	Seek emergency treatment.

»→

Phytonadione (Vitamin K), Continued

 Interaction with Medicine, Vitamins or Minerals

Interacts with	Combined effect
Anti-coagulants (oral)	Decreases anti-coagulant effect.
Antibiotics, broad spectrum	Causes vitamin-K deficiency.
Cholestyramine	Decreases vitamin-K effect.
Colestipol	Decreases vitamin-K effect.
Coumarin (isolated from sweet clover)	Decreases vitamin-K effect.
Mineral oil (long term)	Causes vitamin-K deficiency.
Primaquine	Increases potential for toxic side effects.
Quinidine	Causes vitamin-K deficiency.
Salicylates	Increases need for vitamin K when administered over long time.
Sucralfate	Decreases vitamin-K effect.
Sulfa drugs	Causes vitamin-K deficiency.

 Interaction with Other Substances

None known

Basic Information

Pyridoxine is also called pyridoxal phosphate.
Brand names, see page 477.
Available from natural sources? Yes
Available from synthetic sources? Yes
Prescription required? No
Fat-soluble-or water-soluble: Water-soluble

Natural Sources

Avocados	Rice
Bananas	Salmon
Bran	Shrimp
Brewer's yeast	Soybeans
Carrots	Sunflower seeds
Flour, whole-wheat	Tuna
Hazelnuts (filberts)	Wheat germ
Lentils	

Reasons to Use

- Participates actively in many chemical reactions of proteins and amino acids.
- Helps normal function of brain.
- Promotes normal red-blood-cell formation.
- Maintains chemical balance among body fluids.
- Regulates excretion of water.
- Helps in energy production and resistance to stress.
- Acts as co-enzyme in carbohydrate, protein and fat metabolism.
- Treats some forms of anemia.
- Treats cycloserine and isoniazid poisoning.

Unproved Speculated Benefits

- Treats or prevents depression when used with oral contraceptives.
- Treats premenstrual syndrome.
- Reduces breast milk in nursing mothers with congested breasts.
- Relieves morning sickness.
- Helps arthritis.
- Cures migraines.
- Relieves nausea.
- Acts as a tranquilizer.
- Relieves nervous and muscle disorders.
- Prevents tooth decay.
- Lowers blood cholesterol.
- Retards aging.
- Treats diabetes.
- Treats mental retardation.
- Improves vision.

- Helps weight-reduction efforts.
- Helps infertility.
- Cures carpal-tunnel syndrome.

Who Needs Additional Amounts?

- Anyone with inadequate caloric or nutritional dietary intake or increased nutritional requirements.
- Older people (over 55 years).
- Pregnant or breast-feeding women.
- Those who abuse alcohol or other drugs.
- People with a chronic wasting illness, excess stress for long periods or who have recently undergone surgery.
- Those with a portion of the gastrointestinal tract surgically removed.
- People with recent severe burns or injuries.
- Women taking oral contraceptives or estrogen.

Deficiency Symptoms

Symptoms of vitamin B-6 deficiency are non-specific and hard to reproduce experimentally.
- Weakness
- Mental confusion
- Irritability
- Nervousness
- Insomnia
- Poor coordination walking
- Hyperactivity
- Abnormal electroencephalogram
- Anemia
- Skin lesions
- Discoloration of tongue
- Muscle twitching
- Kidney stones

Unproved Speculated Symptoms

- Depression
- Diabetes

Lab Tests to Detect Deficiency

- Pyridoxine level in blood
- Xanthurenic-acid level in urine

»▶

Pyridoxine (Vitamin B-6), Continued

Dosage and Usage Information

Recommended Dietary Allowance (RDA):
Estimate of adequate daily intake by the Food and Nutrition Board of the National Research Council, 1980. See Glossary.

Age	RDA
0–6 months	0.3mg
6–12 months	0.6mg
1–3 years	0.9mg
4–6 years	1.3mg
7–10 years	1.8mg
Males	
11+ years	2.2mg
Females	
11+ years	2.0mg
Pregnant	+0.6mg
Lactating	+0.5mg

What this vitamin does:
- Acts as co-enzyme for metabolic functions affecting protein, carbohydrates and fat utilization.
- Promotes conversion of tryptophan to niacin or serotonin.

Miscellaneous information:
- Avoid cooking foods that contain vitamin B-6 in large amounts of water.
- Freezing vegetables results in a 30 to 56% reduction of vitamin B-6.
- Canning vegetables results in a 57 to 77% reduction of vitamin B-6.

Available as:
- Tablets: Swallow whole with full glass of liquid. Don't chew or crush. Take with meals or 1 to 1-1/2 hours after meals unless otherwise directed by your doctor.
- Extended-release capsules or tablets: Swallow whole with full glass of liquid. Don't chew or crush. Take with or immediately after food to decrease stomach irritation.
- A constituent of many multivitamin/mineral preparations.

Warnings and Precautions

Don't take if you:
- You are allergic to vitamin B-6.

Consult your doctor if you have:
- Been under severe stress with illness, burns, an accident, recent surgery.
- Intestinal problems.
- Liver disease.
- Overactive thyroid.
- Parkinson's disease.

Over age 55:
- More likely to have marginal deficiency.

Pregnancy:
- Don't take megadoses.

Breast-feeding:
- Megadoses can cause dangerous side effects in the infant.

Effect on lab tests:
- May produce false-positive results in urobilinogen determinations using Ehrlich's reagent.

Storage:
- Store in cool, dry place away from direct light, but don't freeze.
- Store safely out of reach of children.
- Don't store in bathroom medicine cabinet. Heat and moisture may change action of vitamin.

Others:
- Regular B-6 supplements are recommended if you take chloramphenicol, cycloserine, ethionamide, hydralazine, immunosuppressants, isoniazid or penicillamine. These decrease pyridoxine absorption and can cause anemia or tingling and numbness in hands and feet.
- Don't crush, break or chew tablets before swallowing.

Overdose/Toxicity

Signs and symptoms:
Clumsiness, numbness in hands and feet.

What to do:
For symptoms of overdosage: Discontinue vitamin, and consult doctor. Also see *Adverse Reactions or Side Effects* section below.
For accidental overdosage (such as child taking entire bottle): Dial 911 (emergency), 0 for operator or your nearest Poison Control Center.

➣➤

Adverse Reactions or Side Effects

Reaction or effect	What to do
Doses of 200mg/day can produce dependency, requiring need to continue to take high doses (undesirable).	Discontinue megadoses gradually.
Large doses (2 to 6 grams of pyridoxine/ day) taken for several months are reported to cause severe sensory neuropathy (see Glossary) with unsteady gait, numb feet and hands, clumsiness.	Discontinue megadoses. Call doctor immediately.
Causes depression when taken with oral contraceptive pills.	Discontinue pyridoxine. Call doctor when convenient.

 Interaction with Medicine, Vitamins or Minerals

Interacts with	Combined effect
Chloramphenicol, cycloserine, ethionamide, hydralazine, isoniazid, penicillamine, immuno-suppressants, such as adrenocorticoids, azathioprine, chlorambucil, ACTH, cyclophosphamide, cyclosporine, mercaptopurine	May increase excretion of pyridoxine and cause anemia or peripheral neuritis, which includes pain, numbness and coldness in feet and fingertips. If you take these medicines, you may need increased pyridoxine. Consult your doctor.
Estrogen or oral contraceptives	Increases requirements of pyridoxine. Also causes depression.
Levodopa	Prevents levodopa from controlling symptoms of Parkinson's disease. This problem does not occur with carbidopa-levodopa combination.
Phenytoin	Large doses of B-6 hasten break-down of phenytoin.

 Interaction with Other Substances

Tobacco decreases absorption. Smokers may require supplemental vitamin B-6.

Riboflavin (Vitamin B-2)

Basic Information

Brand names, see page 477.
Available from natural sources? Yes
Available from synthetic sources? Yes
Prescription required? No
Fat-soluble or water-soluble: Water-soluble

 Natural Sources

Almonds
Brewer's yeast
Cheese
Chicken
Organ meats (beef, kidney)
Wheat germ

 Reasons to Use

- Aids in release of energy from food.
- Maintains healthy mucous membranes lining respiratory, digestive, circulatory and excretory tracts when used in conjunction with vitamin A.
- Preserves integrity of nervous system, skin, eyes.
- Promotes normal growth and development.
- Aids in treating infections, stomach problems, burns, alcoholism, liver disease.

 Unproved Speculated Benefits

- Cures various eye diseases.
- Treats skin disorders.
- Prevents cancer.
- Increases body growth during normal developmental stages.
- Helps overcome infertility.
- Prevents stress.
- Stimulates hair growth in bald men.
- Improves vision.

 Who Needs Additional Amounts?

- Anyone with inadequate caloric or nutritional dietary intake or increased nutritional requirements.
- Pregnant or breast-feeding women.
- Those who abuse alcohol or other drugs.
- People with a chronic wasting illness, excess stress for long periods or who have recently undergone surgery.
- Athletes and workers who participate in vigorous physical activities.

- Those with a portion of the gastrointestinal tract surgically removed.
- People with recent severe burns or injuries.
- Those who rely almost exclusively on processed foods for their daily diet.
- Women taking oral contraceptives or estrogen.

 Deficiency Symptoms

- Cracks and sores in corners of mouth
- Inflammation of tongue and lips
- Eyes overly sensitive to light and easily tired
- Itching and scaling of skin around nose, mouth, scrotum, forehead, ears, scalp
- Trembling
- Dizziness
- Insomnia
- Slow learning
- Itching, burning and reddening of eyes
- Damage to cornea of eye.

 Unproved Speculated Symptoms

- Mild anemia
- Mild lethargy
- Acne
- Migraine headaches
- Muscle cramps

 Lab Tests to Detect Deficiency

- Serum riboflavin
- Erythrocyte riboflavin
- Glutathione reductase

 Dosage and Usage Information

Recommended Dietary Allowance (RDA):
Estimate of adequate daily intake by the Food and Nutrition Board of the National Research Council, 1980. See Glossary.　　　　⟫▶

Riboflavin (Vitamin B-2)

Age	RDA
0–6 months	0.4mg
6–12 months	0.6mg
1–3 years	0.8mg
4–6 years	1.0mg
7–10 years	1.4mg
Males	
11–14 years	1.6mg
15–22 years	1.7mg
23–50 years	1.6mg
51+ years	1.4mg
Females	
11–22 years	1.3mg
23+ years	1.2mg
Pregnant	+0.3mg
Lactating	+0.5mg

What this vitamin does:
- Acts as component in two co-enzymes (flavin mononucleotide and flavin adenine dinucleotide) needed for normal tissue respiration.
- Activates pyridoxine.

Miscellaneous information:
- A balanced diet prevents deficiency without supplements.
- Large doses may produce dark-yellow urine.
- Processing food may decrease quantity of vitamin B-2.
- Mixing with baking soda destroys riboflavin.

Available as:
- Tablets: Swallow whole with full glass of liquid. Don't chew or crush. Take with or immediately after food to decrease stomach irritation.
- A constituent of many multivitamin/mineral preparations.

Warnings and Precautions

Don't take if you:
- Are allergic to any B vitamin.
- Have chronic kidney failure.

Consult your doctor if you are:
- Pregnant or planning a pregnancy.

Over age 55:
- Need for vitamin B-2 is greater.

Pregnancy:
- Don't take megadoses.

Breast-feeding:
- Don't take megadoses.

Effect on lab tests:
- Urinary catecholamine concentration may show false elevation.
- Urobilongen determinations (Ehrlich's) may produce false-positive results.

Storage:
- Store in cool, dry place away from direct light, but don't freeze.
- Store safely out of reach of children.
- Don't store in bathroom medicine cabinet. Heat and moisture may change action of vitamin.

Others:
- Unlikely to cause toxic symptoms in healthy people with normal kidney function.

Overdose/Toxicity

Signs and symptoms:
Dark urine, nausea, vomiting.

What to do:
For symptoms of overdosage: Discontinue vitamin, and consult doctor. Also see *Adverse Reactions or Side Effects* section below.
For accidental overdosage (such as child taking entire bottle): Dial 911 (emergency), 0 for operator or your nearest Poison Control Center.

Adverse Reactions or Side Effects

Reaction or effect	What to do
Yellow urine, with large doses	No action necessary

Interaction with Medicine, Vitamins or Minerals

Interacts with	Combined effect
Anti-depressants (tricyclic)	Decreases B-2 effect.
Phenothiazines	Decreases B-2 effect.
Probenecid	Decreases B-2 effect.

Interaction with Other Substances

Tobacco decreases absorption. Smokers may require supplemental vitamin B-2.

Alcohol prevents uptake and absorption of vitamin B-2.

Thiamine (Vitamin B-1)

Basic Information

Brand names, see page 477.
Available from natural sources? Yes
Available from synthetic sources? Yes
Prescription required? Yes, for injectable forms
Fat-soluble or water-soluble: Water-soluble

Natural Sources

Beef kidney	Navy beans, dried
Beef liver	Pork
Brewer's yeast	Rice bran
Flour,	Rice, brown, raw
rye and whole-wheat	Salmon steak
Garbanzo beans	Soybeans, dried
(chickpeas), dried	Sunflower seeds, dried
Kidney beans, dried	Wheat germ
	Whole-grain products

Reasons to Use

- Keeps mucous membranes healthy.
- Maintains normal function of nervous system, muscles, heart.
- Aids in treatment of herpes zoster.
- Promotes normal growth and development.
- Treats beriberi (thiamine-deficiency disease).
- Replaces deficiency caused by alcoholism, cirrhosis, overactive thyroid, infection, breast-feeding, absorption diseases, pregnancy, prolonged diarrhea, burns.

Unproved Speculated Benefits

- Cures depression.
- Prevents fatigue.
- Is used as an insect repellent. If you take large amounts of thiamine by mouth, insects are repelled by unpleasant taste and odor of thiamine in perspiration.
- Treats motion sickness.
- Decreases pain.
- Improves appetite, digestion, mental alertness.

Who Needs Additional Amounts?

- People who abuse alcohol or other drugs. Alcoholics need more thiamine. Thiamine accelerates metabolism, using extra carbohydrates and calories from alcohol.
- Anyone with inadequate caloric or nutritional dietary intake or increased nutritional requirements.

- Older people (over 55 years).
- Pregnant or breast-feeding women.
- People with a chronic wasting illness, especially diabetes, excess stress for long periods or who have recently undergone surgery.
- People with a portion of the gastrointestinal tract surgically removed.
- Those with recent severe burns or injuries.
- People with liver disease, overactive thyroid, prolonged diarrhea.

Deficiency Symptoms

Normal deficiency:
- Loss of appetite
- Fatigue
- Nausea
- Vomiting
- Mental problems, such as rolling of eyeballs, depression, memory loss, difficulty concentrating and dealing with details, personality changes, rapid heartbeat
- Gastrointestinal disorders
- Muscles become tender and atrophied

Gross deficiency:
- Leads eventually to beriberi, which is rare, except in severely ill alcoholics.
- Pain or tingling in arms or legs
- Decreased reflex activity
- Fluid accumulation in arms and legs
- Heart enlargement
- Constipation
- Nausea
- Vomiting

Unproved Speculated Symptoms

- Gastric hydrochloric acid lower than normal
- Nerve problems
- Skin problems
- Ulcerative colitis

Lab Tests to Detect Deficiency

- Transketolase function study on red blood cells
- Pyruvic-acid blood level
- 24-hour urine collection

»▶

Dosage and Usage Information

Recommended Dietary Allowance (RDA):
Estimate of adequate daily intake by the Food and Nutrition Board of the National Research Council, 1980. See Glossary.

Age	RDA
0–6 months	0.3mg
6–12 months	0.5mg
1–3 years	0.7mg
4–6 years	0.9mg
7–10 years	1.2mg
Males	
11–18 years	1.4mg
19–22 years	1.5mg
23–50 years	1.4mg
51+ years	1.2mg
Females	
11–22 years	1.1mg
23+ years	1.0mg
Pregnant	+0.4mg
Lactating	+0.5mg

What this vitamin does:
- Functions in combination with adenosine triphosphate to form co-enzyme necessary for converting carbohydrate into energy in muscles and nervous system.

Miscellaneous information:
- Cook foods in minimum amount of water or steam.
- Avoid high cooking temperatures and long heat exposure.
- Avoid using baking soda when you take thiamine unless it is used as a leavening agent in baked products.
- Thiamine is stable when frozen and stored.
- A balanced diet should provide enough thiamine for healthy people to make supplementation unnecessary. Best dietary sources of thiamine are whole-grain cereals and meat.
- Take at same time every day.
- If you forget a dose, take it when you remember it. Return to regular schedule.

Available as:
- Tablets: Swallow whole with full glass of liquid. Don't chew or crush. Take with meals or 1 to 1-1/2 hours after meals unless otherwise directed by your doctor.
- Liquid: Dilute in at least 1/2 glass of water or other liquid. Take with meals or 1 to 1-1/2 hours after meals unless otherwise directed by your doctor.
- Injectable forms are administered by doctor or nurse.

Warnings and Precautions

Don't take if you:
- Are allergic to any B vitamin.

Consult your doctor if you have:
- Liver or kidney disease.

Over age 55:
- No problems expected.

Pregnancy:
- Consult doctor about supplements.
- Don't take megadoses.

Breast-feeding:
- No problems expected. Consult doctor about supplements.
- Don't take megadoses.

Effect on lab tests:
- Interferes with results of serum theophylline.
- May produce false-positive results in tests for uric acid or urobilinogen.

Storage:
- Store in cool, dry place away from direct light, but don't freeze.
- Store safely out of reach of children.
- Don't store in bathroom medicine cabinet. Heat and moisture may change action of vitamin.

Others:
- Most excess thiamine is excreted in urine if kidney function is normal.

Overdose/Toxicity

Signs and symptoms:
Occasionally large doses of vitamin B-1 have caused hypersensitive reactions resembling anaphylactic shock. Several-hundred milligrams may cause drowsiness in some people.

What to do:
For symptoms of overdosage: Discontinue vitamin, and consult doctor. Also see *Adverse Reactions or Side Effects* section below.
For accidental overdosage (such as child taking entire bottle): Dial 911 (emergency), 0 for operator or your nearest Poison Control Center.

Thiamine (Vitamin B-1), Continued

 Adverse Reactions or Side Effects

Reaction or effect	What to do
Skin rash or itching (rare)	Discontinue. Call doctor immediately.
Wheezing (more likely after intravenous dose)	Seek emergency treatment.

 Interaction with Medicine, Vitamins or Minerals

Interacts with	Combined effect
Drugs used to relax muscles during surgery	Produces excessive muscle relaxation. Tell your doctor before surgery if you are taking supplements.

 Interaction with Other Substances

Tobacco decreases absorption. Smokers may require supplemental vitamin B-1.

Alcohol reduces intestinal absorption of vitamin B-1, which is necessary to metabolize alcohol.

Beverages
• Carbonates and citrates (additives listed on many beverage labels) decrease thiamine effect.

Foods
• Carbonates and citrates (additives listed on many food labels) decrease thiamine effect.

Minerals

Many minerals are essential parts of enzymes. They also actively participate in regulating many physiological functions. These include transporting oxygen to each of the body's cells, providing sparks to make muscles contract and participating in many ways to guarantee normal function of the central nervous system. Minerals are required for growth, maintenance, repair and health of tissues and bones.

Most minerals (zinc is an exception) are widely distributed in foods. *Severe* mineral deficiency is unusual in the United States and Canada. A typical diet will only be deficient in a few essential minerals. Even so, there are exceptions. Iron deficiency is common in infants, children and pregnant women. Zinc and copper deficiencies occur frequently.

Alumina/Magnesia/Calcium Carbonate

Basic Information

Provides additional magnesium. Other substances in title are not supplied in large enough amounts for use by your body.
Brand names, see page 477.
Available from natural sources? Yes
Available from synthetic sources? No
Prescription required? Yes, for some forms

Natural Sources

Almonds	Leafy, green	Snails
Bluefish	vegetables	Soybeans
Carp	Mackerel	Sunflower
Cod	Molasses	seeds
Flounder	Nuts	Swordfish
Halibut	Ocean perch	Wheat germ
Herring	Shrimp	

Reasons to Use

- Aids bone growth.
- Aids function of nerves and muscles, including regulation of normal heart rhythm.
- Keeps metabolism steady.
- Conducts nerve impulses.
- Works as laxative in large doses.
- Acts as antacid in small doses.
- Strengthens tooth enamel.

Unproved Speculated Benefits

- Cures alcoholism.
- Cures kidney stones.
- Alleviates heart disease.
- Helps control bad breath and body odor.
- Helps contain lead poisoning.

Who Needs Additional Amounts?

- Anyone with inadequate caloric or dietary intake or increased nutritional requirements.
- Those who abuse alcohol or other drugs.
- People with a chronic wasting illness or who have recently undergone surgery.
- Those with recent severe burns or injuries.

Deficiency Symptoms

Following symptoms occur rarely:
- Muscle contractions
- Convulsions
- Confusion and delirium
- Irritability
- Nervousness
- Skin problems
- Hardening of soft tissues

Unproved Speculated Symptoms

- Sudden heart failure

Lab Tests to Detect Deficiency

- Serum magnesium

Dosage and Usage Information

Recommended Dietary Allowance (RDA):
Estimate of adequate daily intake by the Food and Nutrition Board of the National Research Council, 1980 (See Glossary).

Age	RDA
0–6 months	50mg
6–12 months	70mg
1–3 years	150mg
4–6 years	200mg
7–10 years	250mg
Males	
11–14 years	350mg
15–18 years	400mg
18+ years	350mg
Females	
11+ years	300mg
Pregnant	+150mg
Lactating	+150mg

What this mineral does:
- Activates essential enzymes.
- Affects metabolism of proteins and nucleic acids.
- Helps transport sodium and potassium across cell membranes.
- Influences calcium levels inside cells.

Available as:
- Oral suspension: Dilute in at least 1/2 glass of water or other liquid. Take with meals or 1 to 1-1/2 hours after meals unless otherwise directed by your doctor.
- Chewable tablets: Chew well before swallowing.

Alumina/Magnesia/Calcium Carbonate

Warnings and Precautions

Don't take if you have:
• Kidney failure.
• Heart block (unless you have a pacemaker).
• An ileostomy.

Consult your doctor if you have:
• Chronic constipation, colitis, diarrhea.
• Symptoms of appendicitis.
• Stomach or intestinal bleeding.

Over age 55:
• Adverse reactions/side effects more likely.

Pregnancy:
• Risk to fetus outweighs benefits. Don't use.

Breast-feeding:
• Avoid magnesium in large quantities.
• If you must take temporarily, discontinue breast-feeding. Consult doctor for advice on maintaining milk supply.
• Don't take megadoses.

Effect on lab tests:
• Makes test for stomach-acid secretion inaccurate.
• May increase or decrease serum-phosphate concentrations.
• May decrease serum and urine pH.

Storage:
• Store in cool, dry place away from direct light, but don't freeze.
• Store safely out of reach of children.
• Don't store in bathroom medicine cabinet. Heat and moisture may change action of mineral.

Others:
• Chronic kidney disease frequently causes body to retain excess magnesium.
• Adverse reactions, side effects and interactions with medicines, vitamins or minerals occur *only rarely* when too much magnesium is taken for too long or if you have kidney disease.

Overdose/Toxicity

Signs and symptoms:
Severe nausea and vomiting, extremely low blood pressure, extreme muscle weakness, difficulty breathing, heartbeat irregularity.

What to do:
For symptoms of overdosage: Discontinue mineral, and consult doctor immediately. Also see *Adverse Reactions or Side Effects* section below.
For accidental overdosage (such as child taking entire bottle): Dial 911 (emergency), 0 for operator or your nearest Poison Control Center.

Adverse Reactions or Side Effects

Reaction or effect	What to do
Abdominal pain	Discontinue. Call doctor immediately.
Appetite loss	Discontinue. Call doctor when convenient.
Diarrhea	Discontinue. Call doctor immediately.
Irregular heartbeat	Seek emergency treatment.
Mood changes or mental changes	Discontinue. Call doctor when convenient.
Nausea	Discontinue. Call doctor immediately.
Tiredness or weakness	Discontinue. Call doctor when convenient.
Urination discomfort	Discontinue. Call doctor when convenient.
Vomiting	Discontinue. Call doctor immediately.

Interaction with Medicine, Vitamins or Minerals

Interacts with	Combined effect
Cellulose sodium phosphate	Decreases magnesium effect. Take 1 or more hours apart.
Fat-soluble vitamins (A, E, K)	Decreases absorption of mineral.
Ketoconazole	Reduces absorption of ketoconazole. Take 2 hours apart.
Mecamylamine	May slow urinary excretion of mecamylamine. Avoid combination.
Tetracycline	Decreases absorption of tetracycline.
Vitamin D	May raise magnesium level too high.

Interaction with Other Substances

None known

Alumina/Magnesia/Simethicone

Basic Information

Provides additional magnesium. Other substances in title are not supplied in large enough amounts for use by your body.
Brand names, see page 477.
Available from natural sources? Yes
Available from synthetic sources? No
Prescription required? Yes, for some forms

Natural Sources

Almonds	Leafy, green	Snails
Bluefish	vegetables	Soybeans
Carp	Mackerel	Sunflower
Cod	Molasses	seeds
Flounder	Nuts	Swordfish
Halibut	Ocean perch	Wheat germ
Herring	Shrimp	

Reasons to Use

- Aids bone growth.
- Aids function of nerves and muscles, including regulation of normal heart rhythm.
- Keeps metabolism steady.
- Conducts nerve impulses.
- Works as laxative in large doses.
- Acts as antacid in small doses.
- Strengthens tooth enamel.

Unproved Speculated Benefits

- Cures alcoholism.
- Cures kidney stones.
- Alleviates heart disease.
- Helps control bad breath and body odor.
- Helps contain lead poisoning.

Who Needs Additional Amounts?

- Anyone with inadequate caloric or dietary intake or increased nutritional requirements.
- Those who abuse alcohol or other drugs.
- People with a chronic wasting illness or who have recently undergone surgery.
- Those with recent severe burns or injuries.

Deficiency Symptoms

Following symptoms occur rarely:
- Muscle contractions
- Convulsions
- Confusion and delirium
- Irritability
- Nervousness
- Skin problems
- Hardening of soft tissues

Unproved Speculated Symptoms

- Sudden heart failure

Lab Tests to Detect Deficiency

- Serum magnesium

Dosage and Usage Information

Recommended Dietary Allowance (RDA):
Estimate of adequate daily intake by the Food and Nutrition Board of the National Research Council, 1980 (See Glossary).

Age	RDA
0–6 months	50mg
6–12 months	70mg
1–3 years	150mg
4–6 years	200mg
7–10 years	250mg
Males	
11–14 years	350mg
15–18 years	400mg
18+ years	350mg
Females	
11+ years	300mg
Pregnant	+150mg
Lactating	+150mg

What this mineral does:
- Activates essential enzymes.
- Affects metabolism of proteins and nucleic acids.
- Helps transport sodium and potassium across cell membranes.
- Influences calcium levels inside cells.

Available as:
- Oral suspension: Dilute in at least 1/2 glass of water or other liquid. Take with meals or 1 to 1-1/2 hours after meals unless otherwise directed by your doctor.
- Chewable tablets: Chew well before swallowing.

Alumina/Magnesia/Simethicone

Warnings and Precautions

Don't take if you have:
• Kidney failure.
• Heart block (unless you have a pacemaker).
• An ileostomy.

Consult your doctor if you have:
• Chronic constipation, colitis, diarrhea.
• Symptoms of appendicitis.
• Stomach or intestinal bleeding.

Over age 55:
• Adverse reactions/side effects more likely.

Pregnancy:
• Risk to fetus outweighs benefits. Don't use.

Breast-feeding:
• Avoid magnesium in large quantities.
• If you must take temporarily, discontinue breast-feeding. Consult doctor for advice on maintaining milk supply.
• Don't take megadoses.

Effect on lab tests:
• Makes test for stomach-acid secretion inaccurate.
• May increase or decrease serum-phosphate concentrations.
• May decrease serum and urine pH.

Storage:
• Store in cool, dry place away from direct light, but don't freeze.
• Store safely out of reach of children.
• Don't store in bathroom medicine cabinet. Heat and moisture may change action of mineral.

Others:
• Chronic kidney disease frequently causes body to retain excess magnesium.
• Adverse reactions, side effects and interactions with medicines, vitamins or minerals occur *only rarely* when too much magnesium is taken for too long or if you have kidney disease.

Overdose/Toxicity

Signs and symptoms:
Severe nausea and vomiting, extremely low blood pressure, extreme muscle weakness, difficulty breathing, heartbeat irregularity.

What to do:
For symptoms of overdosage: Discontinue mineral, and consult doctor immediately. Also see *Adverse Reactions or Side Effects* section below.
For accidental overdosage (such as child taking entire bottle): Dial 911 (emergency), 0 for operator or your nearest Poison Control Center.

Adverse Reactions or Side Effects

Reaction or effect	What to do
Abdominal pain	Discontinue. Call doctor immediately.
Appetite loss	Discontinue. Call doctor when convenient.
Diarrhea	Discontinue. Call doctor immediately.
Irregular heartbeat	Seek emergency treatment.
Mood changes or mental changes	Discontinue. Call doctor when convenient.
Nausea	Discontinue. Call doctor immediately.
Tiredness or weakness	Discontinue. Call doctor when convenient.
Urination discomfort	Discontinue. Call doctor when convenient.
Vomiting	Discontinue. Call doctor immediately.

Interaction with Medicine, Vitamins or Minerals

Interacts with	Combined effect
Cellulose sodium phosphate	Decreases magnesium effect. Take 1 or more hours apart.
Fat-soluble vitamins (A, E, K)	Decreases absorption of mineral.
Ketoconazole	Reduces absorption of ketoconazole. Take 2 hours apart.
Mecamylamine	May slow urinary excretion of mecamylamine. Avoid combination.
Tetracycline	Decreases absorption of tetracycline.
Vitamin D	May raise magnesium level too high.

Interaction with Other Substances

None known

Alumina/Magnesium Carbonate

Basic Information

Provides additional magnesium. Other substances in title are not supplied in large enough amounts for use by your body.
Brand names, see page 477.
Available from natural sources? Yes
Available from synthetic sources? No
Prescription required? Yes, for some forms

Natural Sources

Almonds	Leafy, green	Snails
Bluefish	vegetables	Soybeans
Carp	Mackerel	Sunflower
Cod	Molasses	seeds
Flounder	Nuts	Swordfish
Halibut	Ocean perch	Wheat germ
Herring	Shrimp	

Reasons to Use

- Aids bone growth.
- Aids function of nerves and muscles, including regulation of normal heart rhythm.
- Keeps metabolism steady.
- Conducts nerve impulses.
- Works as laxative in large doses.
- Acts as antacid in small doses.
- Strengthens tooth enamel.

Unproved Speculated Benefits

- Cures alcoholism.
- Cures kidney stones.
- Alleviates heart disease.
- Helps control bad breath and body odor.
- Helps contain lead poisoning.

Who Needs Additional Amounts?

- Anyone with inadequate caloric or dietary intake or increased nutritional requirements.
- Those who abuse alcohol or other drugs.
- People with a chronic wasting illness or who have recently undergone surgery.
- Those with recent severe burns or injuries.

Deficiency Symptoms

Following symptoms occur rarely:
- Muscle contractions
- Convulsions
- Confusion and delirium
- Irritability
- Nervousness
- Skin problems
- Hardening of soft tissues

Unproved Speculated Symptoms

- Sudden heart failure

Lab Tests to Detect Deficiency

- Serum magnesium

Dosage and Usage Information

Recommended Dietary Allowance (RDA):
Estimate of adequate daily intake by the Food and Nutrition Board of the National Research Council, 1980 (See Glossary).

Age	RDA
0–6 months	50mg
6–12 months	70mg
1–3 years	150mg
4–6 years	200mg
7–10 years	250mg
Males	
11–14 years	350mg
15–18 years	400mg
18+ years	350mg
Females	
11+ years	300mg
Pregnant	+150mg
Lactating	+150mg

What this mineral does:
- Activates essential enzymes.
- Affects metabolism of proteins and nucleic acids.
- Helps transport sodium and potassium across cell membranes.
- Influences calcium levels inside cells.

Available as:
- Oral suspension: Dilute in at least 1/2 glass of water or other liquid. Take with meals or 1 to 1-1/2 hours after meals unless otherwise directed by your doctor.
- Chewable tablets: Chew well before swallowing.

Alumina/Magnesium Carbonate

Warnings and Precautions

Don't take if you have:
• Kidney failure.
• Heart block (unless you have a pacemaker).
• An ileostomy.

Consult your doctor if you have:
• Chronic constipation, colitis, diarrhea.
• Symptoms of appendicitis.
• Stomach or intestinal bleeding.

Over age 55:
• Adverse reactions/side effects more likely.

Pregnancy:
• Risk to fetus outweighs benefits. Don't use.

Breast-feeding:
• Avoid magnesium in large quantities.
• If you must take temporarily, discontinue breast-feeding. Consult doctor for advice on maintaining milk supply.
• Don't take megadoses.

Effect on lab tests:
• Makes test for stomach-acid secretion inaccurate.
• May increase or decrease serum-phosphate concentrations.
• May decrease serum and urine pH.

Storage:
• Store in cool, dry place away from direct light, but don't freeze.
• Store safely out of reach of children.
• Don't store in bathroom medicine cabinet. Heat and moisture may change action of mineral.

Others:
• Chronic kidney disease frequently causes body to retain excess magnesium.
• Adverse reactions, side effects and interactions with medicines, vitamins or minerals occur *only rarely* when too much magnesium is taken for too long or if you have kidney disease.

Overdose/Toxicity

Signs and symptoms:
Severe nausea and vomiting, extremely low blood pressure, extreme muscle weakness, difficulty breathing, heartbeat irregularity.

What to do:
For symptoms of overdosage: Discontinue mineral, and consult doctor immediately. Also see *Adverse Reactions or Side Effects* section below.
For accidental overdosage (such as child taking entire bottle): Dial 911 (emergency), 0 for operator or your nearest Poison Control Center.

Adverse Reactions or Side Effects

Reaction or effect	What to do
Abdominal pain	Discontinue. Call doctor immediately.
Appetite loss	Discontinue. Call doctor when convenient.
Diarrhea	Discontinue. Call doctor immediately.
Irregular heartbeat	Seek emergency treatment.
Mood changes or mental changes	Discontinue. Call doctor when convenient.
Nausea	Discontinue. Call doctor immediately.
Tiredness or weakness	Discontinue. Call doctor when convenient.
Urination discomfort	Discontinue. Call doctor when convenient.
Vomiting	Discontinue. Call doctor immediately.

Interaction with Medicine, Vitamins or Minerals

Interacts with	Combined effect
Cellulose sodium phosphate	Decreases magnesium effect. Take 1 or more hours apart.
Fat-soluble vitamins (A, E, K)	Decreases absorption of mineral.
Ketoconazole	Reduces absorption of ketoconazole. Take 2 hours apart.
Mecamylamine	May slow urinary excretion of mecamylamine. Avoid combination.
Tetracycline	Decreases absorption of tetracycline.
Vitamin D	May raise magnesium level too high.

Interaction with Other Substances

None known

Alumina/Magnesium Trisilicate

Basic Information
Provides additional magnesium. Other substances in title are not supplied in large enough amounts for use by your body. Brand names, see page 477.
Available from natural sources? Yes
Available from synthetic sources? No
Prescription required? Yes, for some forms

Natural Sources

Almonds	Leafy, green	Snails
Bluefish	vegetables	Soybeans
Carp	Mackerel	Sunflower
Cod	Molasses	seeds
Flounder	Nuts	Swordfish
Halibut	Ocean perch	Wheat germ
Herring	Shrimp	

Reasons to Use

- Aids bone growth.
- Aids function of nerves and muscles, including regulation of normal heart rhythm.
- Keeps metabolism steady.
- Conducts nerve impulses.
- Works as laxative in large doses.
- Acts as antacid in small doses.
- Strengthens tooth enamel.

Unproved Speculated Benefits

- Cures alcoholism.
- Cures kidney stones.
- Alleviates heart disease.
- Helps control bad breath and body odor.
- Helps contain lead poisoning.

Who Needs Additional Amounts?

- Anyone with inadequate caloric or dietary intake or increased nutritional requirements.
- Those who abuse alcohol or other drugs.
- People with a chronic wasting illness or who have recently undergone surgery.
- Those with recent severe burns or injuries.

Deficiency Symptoms

Following symptoms occur rarely:
- Muscle contractions
- Convulsions
- Confusion and delirium
- Irritability
- Nervousness
- Skin problems
- Hardening of soft tissues

Unproved Speculated Symptoms

- Sudden heart failure

Lab Tests to Detect Deficiency

- Serum magnesium

Dosage and Usage Information

Recommended Dietary Allowance (RDA):
Estimate of adequate daily intake by the Food and Nutrition Board of the National Research Council, 1980 (See Glossary).

Age	RDA
0–6 months	50mg
6–12 months	70mg
1–3 years	150mg
4–6 years	200mg
7–10 years	250mg
Males	
11–14 years	350mg
15–18 years	400mg
18+ years	350mg
Females	
11+ years	300mg
Pregnant	+150mg
Lactating	+150mg

What this mineral does:
- Activates essential enzymes.
- Affects metabolism of proteins and nucleic acids.
- Helps transport sodium and potassium across cell membranes.
- Influences calcium levels inside cells.

Available as:
- Chewable tablets: Chew well before swallowing.

➤➤➤

Alumina/Magnesium Trisilicate

Warnings and Precautions

Don't take if you have:
- Kidney failure.
- Heart block (unless you have a pacemaker).
- An ileostomy.

Consult your doctor if you have:
- Chronic constipation, colitis, diarrhea.
- Symptoms of appendicitis.
- Stomach or intestinal bleeding.

Over age 55:
- Adverse reactions/side effects more likely.

Pregnancy:
- Risk to fetus outweighs benefits. Don't use.

Breast-feeding:
- Avoid magnesium in large quantities.
- If you must take temporarily, discontinue breast-feeding. Consult doctor for advice on maintaining milk supply.
- Don't take megadoses.

Effect on lab tests:
- Makes test for stomach-acid secretion inaccurate.
- May increase or decrease serum-phosphate concentrations.
- May decrease serum and urine pH.

Storage:
- Store in cool, dry place away from direct light, but don't freeze.
- Store safely out of reach of children.
- Don't store in bathroom medicine cabinet. Heat and moisture may change action of mineral.

Others:
- Chronic kidney disease frequently causes body to retain excess magnesium.
- Adverse reactions, side effects and interactions with medicines, vitamins or minerals occur *only rarely* when too much magnesium is taken for too long or if you have kidney disease.

Overdose/Toxicity

Signs and symptoms:
Severe nausea and vomiting, extremely low blood pressure, extreme muscle weakness, difficulty breathing, heartbeat irregularity.

What to do:
For symptoms of overdosage: Discontinue mineral, and consult doctor immediately. Also see *Adverse Reactions or Side Effects* section below.
For accidental overdosage (such as child taking entire bottle): Dial 911 (emergency), 0 for operator or your nearest Poison Control Center.

Adverse Reactions or Side Effects

Reaction or effect	What to do
Abdominal pain	Discontinue. Call doctor immediately.
Appetite loss	Discontinue. Call doctor when convenient.
Diarrhea	Discontinue. Call doctor immediately.
Irregular heartbeat	Seek emergency treatment.
Mood changes or mental changes	Discontinue. Call doctor when convenient.
Nausea	Discontinue. Call doctor immediately.
Tiredness or weakness	Discontinue. Call doctor when convenient.
Urination discomfort	Discontinue. Call doctor when convenient.
Vomiting	Discontinue. Call doctor immediately.

Interaction with Medicine, Vitamins or Minerals

Interacts with	Combined effect
Cellulose sodium phosphate	Decreases magnesium effect. Take 1 or more hours apart.
Fat-soluble vitamins (A, E, K)	Decreases absorption of mineral.
Ketoconazole	Reduces absorption of ketoconazole. Take 2 hours apart.
Mecamylamine	May slow urinary excretion of mecamylamine. Avoid combination.
Tetracycline	Decreases absorption of tetracycline.
Vitamin D	May raise magnesium level too high.

Interaction with Other Substances

None known

Alumina/Magnesium Trisilicate/Sodium Bicarbonate

Basic Information

Provides additional magnesium. Other substances in title are not supplied in large enough amounts for use by your body.
Brand names, see page 477.
Available from natural sources? Yes
Available from synthetic sources? No
Prescription required? Yes, for some forms

Natural Sources

Almonds	Leafy, green	Snails
Bluefish	vegetables	Soybeans
Carp	Mackerel	Sunflower
Cod	Molasses	seeds
Flounder	Nuts	Swordfish
Halibut	Ocean perch	Wheat germ
Herring	Shrimp	

Reasons to Use

- Aids bone growth.
- Aids function of nerves and muscles, including regulation of normal heart rhythm.
- Keeps metabolism steady.
- Conducts nerve impulses.
- Works as laxative in large doses.
- Acts as antacid in small doses.
- Strengthens tooth enamel.

Unproved Speculated Benefits

- Cures alcoholism.
- Cures kidney stones.
- Alleviates heart disease.
- Helps control bad breath and body odor.
- Helps contain lead poisoning.

Who Needs Additional Amounts?

- Anyone with inadequate caloric or dietary intake or increased nutritional requirements.
- Those who abuse alcohol or other drugs.
- People with a chronic wasting illness or who have recently undergone surgery.
- Those with recent severe burns or injuries.

Deficiency Symptoms

Following symptoms occur rarely:
- Muscle contractions
- Convulsions
- Confusion and delirium
- Irritability
- Nervousness
- Skin problems
- Hardening of soft tissues

Unproved Speculated Symptoms

- Sudden heart failure

Lab Tests to Detect Deficiency

- Serum magnesium

Dosage and Usage Information

Recommended Dietary Allowance (RDA):
Estimate of adequate daily intake by the Food and Nutrition Board of the National Research Council, 1980 (See Glossary).

Age	RDA
0–6 months	50mg
6–12 months	70mg
1–3 years	150mg
4–6 years	200mg
7–10 years	250mg
Males	
11–14 years	350mg
15–18 years	400mg
18+ years	350mg
Females	
11+ years	300mg
Pregnant	+150mg
Lactating	+150mg

What this mineral does:
- Activates essential enzymes.
- Affects metabolism of proteins and nucleic acids.
- Helps transport sodium and potassium across cell membranes.
- Influences calcium levels inside cells.

Available as:
- Chewable tablets: Chew well before swallowing.

»▶

Alumina/Magnesium Trisilicate/Sodium Bicarbonate

Warnings and Precautions

Don't take if you have:
• Kidney failure.
• Heart block (unless you have a pacemaker).
• An ileostomy.

Consult your doctor if you have:
• Chronic constipation, colitis, diarrhea.
• Symptoms of appendicitis.
• Stomach or intestinal bleeding.

Over age 55:
• Adverse reactions/side effects more likely.

Pregnancy:
• Risk to fetus outweighs benefits. Don't use.

Breast-feeding:
• Avoid magnesium in large quantities.
• If you must take temporarily, discontinue breast-feeding. Consult doctor for advice on maintaining milk supply.
• Don't take megadoses.

Effect on lab tests:
• Makes test for stomach-acid secretion inaccurate.
• May increase or decrease serum-phosphate concentrations.
• May decrease serum and urine pH.

Storage:
• Store in cool, dry place away from direct light, but don't freeze.
• Store safely out of reach of children.
• Don't store in bathroom medicine cabinet. Heat and moisture may change action of mineral.

Others:
• Chronic kidney disease frequently causes body to retain excess magnesium.
• Adverse reactions, side effects and interactions with medicines, vitamins or minerals occur *only rarely* when too much magnesium is taken for too long or if you have kidney disease.

Overdose/Toxicity

Signs and symptoms:
Severe nausea and vomiting, extremely low blood pressure, extreme muscle weakness, difficulty breathing, heartbeat irregularity.

What to do:
For symptoms of overdosage: Discontinue mineral, and consult doctor immediately. Also see *Adverse Reactions or Side Effects* section below.
For accidental overdosage (such as child taking entire bottle): Dial 911 (emergency), 0 for operator or your nearest Poison Control Center.

Adverse Reactions or Side Effects

Reaction or effect	What to do
Abdominal pain	Discontinue. Call doctor immediately.
Appetite loss	Discontinue. Call doctor when convenient.
Diarrhea	Discontinue. Call doctor immediately.
Irregular heartbeat	Seek emergency treatment.
Mood changes or mental changes	Discontinue. Call doctor when convenient.
Nausea	Discontinue. Call doctor immediately.
Tiredness or weakness	Discontinue. Call doctor when convenient.
Urination discomfort	Discontinue. Call doctor when convenient.
Vomiting	Discontinue. Call doctor immediately.

Interaction with Medicine, Vitamins or Minerals

Interacts with	Combined effect
Cellulose sodium phosphate	Decreases magnesium effect. Take 1 or more hours apart.
Fat-soluble vitamins (A, E, K)	Decreases absorption of mineral.
Ketoconazole	Reduces absorption of ketoconazole. Take 2 hours apart.
Mecamylamine	May slow urinary excretion of mecamylamine. Avoid combination.
Tetracycline	Decreases absorption of tetracycline.
Vitamin D	May raise magnesium level too high.

Interaction with Other Substances

None known

85

Calcium Carbonate

Basic Information
Brand names, see page 477.
Available from natural sources? Yes
Available from synthetic sources? Yes
Prescription required? Some forms, yes;
others, no

Natural Sources

Almonds	Salmon, canned
Brazil nuts	Sardines, canned
Caviar	Shrimp
Cheese	Soybeans
Kelp	Tofu
Milk	Turnip greens
Milk products	Yogurt
Molasses	

Reasons to Use

- Helps prevent osteoporosis in older people.
- Treats calcium depletion in people with hypoparathyroidism, osteomalacia, rickets.
- Treats low-calcium levels in people taking anti-convulsant medication.
- Treats tetany (severe muscle spasms) caused by insect bites, sensitivity reactions, cardiac arrest, lead poisoning.
- Is used as an antidote to magnesium poisoning.
- Prevents muscle cramps in some people.
- Promotes normal growth and development.
- Builds bones and teeth
- Maintains bone density and strength.
- Buffers acid in stomach and acts as antacid.
- Helps regulate heartbeat, blood clotting, muscle contraction.
- Treats neonatal hypocalcemia.
- Promotes storage and release of some body hormones.
- Promotes use of amino acids.
- Lowers phosphate concentrations in people with chronic kidney disease.

Unproved Speculated Benefits

- Helps prevent insomnia and anxiety (acts as a natural tranquilizer).
- Helps prevent hypertension.
- Treats allergies.
- Decreases likelihood of hardening of arteries.
- Treats leg cramps.
- Treats diabetes.
- Treats throat spasms.

Who Needs Additional Amounts?

- Anyone with inadequate caloric or dietary intake or increased nutritional requirements or who does not like or consume milk products.
- People allergic to milk and milk products or who don't tolerate them well.
- Older people (over 55 years), particularly women.
- Women throughout adult life, especially during pregnancy and lactation, but not limited to these times.
- Those who abuse alcohol or other drugs.
- People with a chronic wasting illness, excess stress for long periods or who have recently undergone surgery.
- Those with a portion of the gastrointestinal tract surgically removed.
- People with recent severe burns or injuries.

Deficiency Symptoms

- Osteoporosis (late symptoms): frequent fractures in spine and other bones, deformed spinal column with humps, loss of height
- Osteomalacia: frequent fractures
- Muscle contractions
- Convulsive seizures
- Muscle cramps
- Low backache

Unproved Speculated Symptoms

- Uncontrollable temper outbursts

Lab Tests to Detect Deficiency

- 24-hour urine collection to measure calcium levels (Sulkowitch)
- Serum-calcium levels
- Imaging procedures to scan for bone density (more reliable than above tests)

Dosage and Usage Information

Recommended Dietary Allowance (RDA):
Estimate of adequate daily intake by the Food and Nutrition Board of the National Research Council, 1980 (See Glossary).
Many doctors and nutritionists recommend women take more calcium than quoted by the ⟫⟫

RDA. They recommend 1,000 milligrams per day for premenopausal women and 1,500 milligrams per day for post-menopausal women and elderly men.

Age	RDA
0–6 months	360mg
6–12 months	540mg
1–10 years	800mg
11–18 years	1,000mg
18+ years	800mg
Pregnant	+400mg
Lactating	+400mg

Different types of calcium supplements contain more available calcium (also called *elemental calcium*) than others. To provide 1,000mg of available calcium, you must take:

  4 tablets/day of 625mg calcium carbonate
  4 tablets/day of 650mg calcium carbonate
  4 tablets/day of 750mg calcium carbonate
  3 tablets/day of 835mg calcium carbonate
  2 tablets/day of 1,250mg calcium carbonate
  2 tablets/day of 1,500mg calcium carbonate

Check contents of product you choose to determine how many tablets are needed to provide the amount of calcium you require.

What this mineral does:
• Participates in metabolic functions necessary for normal activity of nervous, muscular, skeletal systems.
• Plays important role in normal heart function, kidney function, blood clotting, blood-vessel integrity.
• Helps utilization of vitamin B-12.

Miscellaneous information:
• Bones serve as storage site for calcium in the body. There is a constant interchange between calcium in bone and the bloodstream.
• Foods rich in calcium (or supplements) help maintain the balance between bone needs and blood needs.
• Don't discard outer parts of vegetables during food preparation.
• Exercise, a balanced diet, calcium from natural sources or supplements and estrogens are important in treating and preventing osteoporosis.

Available as:
• Tablets: Swallow whole with full glass of liquid. Don't chew or crush. Take with meals or 1 to 1-1/2 hours after meals unless otherwise directed by your doctor.
• Chewable tablets: Chew well before swallowing.

Warnings and Precautions

Don't take if you:
• Are allergic to calcium or antacids.
• Have kidney stones.
• Have a high blood-calcium level.
• Have sarcoidosis.

Consult your doctor if you have:
• Kidney disease.
• Chronic constipation, colitis, diarrhea.
• Stomach or intestinal bleeding.
• Irregular heartbeat.

Over age 55:
• Adverse reactions and side effects are more likely.
• Diarrhea or constipation are particularly likely.

Pregnancy:
• May need extra calcium. Consult doctor about supplements.
• Don't take megadoses.

Breast-feeding:
• Drug passes into milk. Consult doctor about supplements.
• Don't take megadoses.

Effect on lab tests:
• Serum-amylase and serum 11-hydroxycorticosteroid concentrations can be increased.
• Decreases serum-phosphate concentration with excessive, prolonged use.

Storage:
• Store in cool, dry area away from direct light, but don't freeze.
• Store safely out of reach of children.
• Don't store in bathroom medicine cabinet. Heat and moisture may change action of mineral.

Others:
• Dolomite or bone meal are probably *unsafe* sources of calcium because they contain lead.
• Avoid taking calcium within 1 or 2 hours of meals or ingestion of other medicines, if possible.
• Some calcium carbonate is derived from oyster shells. Calcium carbonate derived from this source is *not* recommended!

⟩⟩▸

Calcium Carbonate, Continued

Overdose/Toxicity

Signs and symptoms:
Confusion, high blood pressure, increased sensitivity of eyes and skin to light, increased thirst, slow or irregular heartbeat, depression, bone or muscle pain, nausea, vomiting, skin itching, skin rash, increased urination.

What to do:
For symptoms of overdosage: Discontinue mineral, and consult doctor immediately. Also see *Adverse Reactions or Side Effects* section below.
For accidental overdosage (such as child taking entire bottle): Dial 911 (emergency), 0 for operator or your nearest Poison Control Center.

Adverse Reactions or Side Effects

Reaction or effect	What to do
Early signs of too much calcium in blood:	
Appetite loss	Discontinue. Call doctor when convenient.
Constipation	Discontinue. Call doctor when convenient.
Drowsiness	Discontinue. Call doctor immediately.
Dry mouth	Discontinue. Call doctor when convenient.
Headache	Discontinue. Call doctor when convenient.
Metallic taste	Discontinue. Call doctor when convenient.
Tiredness or weakness	Discontinue. Call doctor immediately.
Late signs of too much calcium in blood:	
Confusion	Discontinue. Call doctor immediately.
Depression	Discontinue. Call doctor when convenient.
High blood pressure	Discontinue. Call doctor immediately.
Increased thirst	Discontinue. Call doctor when convenient.
Increased urination	Discontinue. Call doctor when convenient.
Muscle or bone pain	Discontinue. Call doctor immediately.
Nausea	Discontinue. Call doctor immediately.
Skin rash	Discontinue. Call doctor immediately.
Slow or irregular heartbeat	Seek emergency treatment.
Vomiting	Discontinue. Call doctor immediately.

Interaction with Medicine, Vitamins or Minerals

Interacts with	Combined effect
Digitalis preparations	Heartbeat irregularities.
Iron supplements	Decreases absorption of iron unless vitamin C is taken at same time.
Magnesium-containing medications or supplements	Increases blood level of both.
Oral contraceptives and estrogens	May increase calcium absorption.
Potassium supplements	Increases chance of heartbeat irregularities.
Tetracyclines (oral)	Decreases absorption of tetracycline.
Vitamin A (megadoses)	Stimulates bone loss.
Vitamin D (megadoses)	Excessively increases absorption of calcium supplements.

Interaction with Other Substances

Tobacco decreases absorption.

Alcohol decreases absorption.

Beverages
• Tea decreases absorption.
• Coffee decreases absorption.
• Don't take calcium with milk or other dairy products so your body can absorb the most calcium from food *and* calcium supplement.

Foods
• Avoid eating spinach, rhubarb, bran, whole-grain cereals, fresh fruits or fresh vegetables at same time you take calcium. They may prevent efficient absorption.

Calcium Carbonate/Magnesia

Basic Information

Provides additional magnesium. Other substances in title are not supplied in large enough amounts for use by your body.
Brand names, see page 477.
Available from natural sources? Yes
Available from synthetic sources? No
Prescription required? Yes, for some forms

Natural Sources

Almonds	Leafy, green	Snails
Bluefish	vegetables	Soybeans
Carp	Mackerel	Sunflower
Cod	Molasses	seeds
Flounder	Nuts	Swordfish
Halibut	Ocean perch	Wheat germ
Herring	Shrimp	

Reasons to Use

- Aids bone growth.
- Aids function of nerves and muscles, including regulation of normal heart rhythm.
- Keeps metabolism steady.
- Conducts nerve impulses.
- Works as laxative in large doses.
- Acts as antacid in small doses.
- Strengthens tooth enamel.

Unproved Speculated Benefits

- Cures alcoholism.
- Cures kidney stones.
- Alleviates heart disease.
- Helps control bad breath and body odor.
- Helps contain lead poisoning.

Who Needs Additional Amounts?

- Anyone with inadequate caloric or dietary intake or increased nutritional requirements.
- Those who abuse alcohol or other drugs.
- People with a chronic wasting illness or who have recently undergone surgery.
- Those with recent severe burns or injuries.

Deficiency Symptoms

Following symptoms occur rarely:
- Muscle contractions
- Convulsions
- Confusion and delirium
- Irritability
- Nervousness
- Skin problems
- Hardening of soft tissues

Unproved Speculated Symptoms

- Sudden heart failure

Lab Tests to Detect Deficiency

- Serum magnesium

Dosage and Usage Information

Recommended Dietary Allowance (RDA):
Estimate of adequate daily intake by the Food and Nutrition Board of the National Research Council, 1980 (See Glossary).

Age	RDA
0–6 months	50mg
6–12 months	70mg
1–3 years	150mg
4–6 years	200mg
7–10 years	250mg
Males	
11–14 years	350mg
15–18 years	400mg
18+ years	350mg
Females	
11+ years	300mg
Pregnant	+150mg
Lactating	+150mg

What this mineral does:
- Activates essential enzymes.
- Affects metabolism of proteins and nucleic acids.
- Helps transport sodium and potassium across cell membranes.
- Influences calcium levels inside cells.

Available as:
- Chewable tablets: Chew well before swallowing.

➤➤➤

Calcium Carbonate/Magnesia, Continued

 Warnings and Precautions

Don't take if you have:
• Kidney failure.
• Heart block (unless you have a pacemaker).
• An ileostomy.

Consult your doctor if you have:
• Chronic constipation, colitis, diarrhea.
• Symptoms of appendicitis.
• Stomach or intestinal bleeding.

Over age 55:
• Adverse reactions/side effects more likely.

Pregnancy:
• Risk to fetus outweighs benefits. Don't use.

Breast-feeding:
• Avoid magnesium in large quantities.
• If you must take temporarily, discontinue breast-feeding. Consult doctor for advice on maintaining milk supply.
• Don't take megadoses.

Effect on lab tests:
• Makes test for stomach-acid secretion inaccurate.
• May increase or decrease serum-phosphate concentrations.
• May decrease serum and urine pH.

Storage:
• Store in cool, dry place away from direct light, but don't freeze.
• Store safely out of reach of children.
• Don't store in bathroom medicine cabinet. Heat and moisture may change action of mineral.

Others:
• Chronic kidney disease frequently causes body to retain excess magnesium.
• Adverse reactions, side effects and interactions with medicines, vitamins or minerals occur *only rarely* when too much magnesium is taken for too long or if you have kidney disease.

 Overdose/Toxicity

Signs and symptoms:
Severe nausea and vomiting, extremely low blood pressure, extreme muscle weakness, difficulty breathing, heartbeat irregularity.

What to do:
For symptoms of overdosage: Discontinue mineral, and consult doctor immediately. Also see *Adverse Reactions or Side Effects* section below.
For accidental overdosage (such as child taking entire bottle): Dial 911 (emergency), 0 for operator or your nearest Poison Control Center.

 Adverse Reactions or Side Effects

Reaction or effect	What to do
Abdominal pain	Discontinue. Call doctor immediately.
Appetite loss	Discontinue. Call doctor when convenient.
Diarrhea	Discontinue. Call doctor immediately.
Irregular heartbeat	Seek emergency treatment.
Mood changes or mental changes	Discontinue. Call doctor when convenient.
Nausea	Discontinue. Call doctor immediately.
Tiredness or weakness	Discontinue. Call doctor when convenient.
Urination discomfort	Discontinue. Call doctor when convenient.
Vomiting	Discontinue. Call doctor immediately.

 Interaction with Medicine, Vitamins or Minerals

Interacts with	Combined effect
Cellulose sodium phosphate	Decreases magnesium effect. Take 1 or more hours apart.
Fat-soluble vitamins (A, E, K)	Decreases absorption of mineral.
Ketoconazole	Reduces absorption of ketoconazole. Take 2 hours apart.
Mecamylamine	May slow urinary excretion of mecamylamine. Avoid combination.
Tetracycline	Decreases absorption of tetracycline.
Vitamin D	May raise magnesium level too high.

 Interaction with Other Substances

None known

Calcium Carbonate/Magnesia/Simethicone

Basic Information

Provides additional magnesium. Other substances in title are not supplied in large enough amounts for use by your body.
Brand names, see page 477.
Available from natural sources? Yes
Available from synthetic sources? No
Prescription required? Yes, for some forms

Natural Sources

Almonds	Leafy, green	Snails
Bluefish	vegetables	Soybeans
Carp	Mackerel	Sunflower
Cod	Molasses	seeds
Flounder	Nuts	Swordfish
Halibut	Ocean perch	Wheat germ
Herring	Shrimp	

Reasons to Use

- Aids bone growth.
- Aids function of nerves and muscles, including regulation of normal heart rhythm.
- Keeps metabolism steady.
- Conducts nerve impulses.
- Works as laxative in large doses.
- Acts as antacid in small doses.
- Strengthens tooth enamel.

Unproved Speculated Benefits

- Cures alcoholism.
- Cures kidney stones.
- Alleviates heart disease.
- Helps control bad breath and body odor.
- Helps contain lead poisoning.

Who Needs Additional Amounts?

- Anyone with inadequate caloric or dietary intake or increased nutritional requirements.
- Those who abuse alcohol or other drugs.
- People with a chronic wasting illness or who have recently undergone surgery.
- Those with recent severe burns or injuries.

Deficiency Symptoms

Following symptoms occur rarely:
- Muscle contractions
- Convulsions
- Confusion and delirium
- Irritability
- Nervousness
- Skin problems
- Hardening of soft tissues

Unproved Speculated Symptoms

- Sudden heart failure

Lab Tests to Detect Deficiency

- Serum magnesium

Dosage and Usage Information

Recommended Dietary Allowance (RDA):
Estimate of adequate daily intake by the Food and Nutrition Board of the National Research Council, 1980 (See Glossary).

Age	RDA
0–6 months	50mg
6–12 months	70mg
1–3 years	150mg
4–6 years	200mg
7–10 years	250mg
Males	
11–14 years	350mg
15–18 years	400mg
18+ years	350mg
Females	
11+ years	300mg
Pregnant	+150mg
Lactating	+150mg

What this mineral does:
- Activates essential enzymes.
- Affects metabolism of proteins and nucleic acids.
- Helps transport sodium and potassium across cell membranes.
- Influences calcium levels inside cells.

Available as:
- Tablets: Swallow whole with full glass of liquid. Don't chew or crush. Take with meals or 1 to 1-1/2 hours after meals unless otherwise directed by your doctor.

»→

Calcium Carbonate/Magnesia/Simethicone, Continued

 Warnings and Precautions

Don't take if you have:
- Kidney failure.
- Heart block (unless you have a pacemaker).
- An ileostomy.

Consult your doctor if you have:
- Chronic constipation, colitis, diarrhea.
- Symptoms of appendicitis.
- Stomach or intestinal bleeding.

Over age 55:
- Adverse reactions/side effects more likely.

Pregnancy:
- Risk to fetus outweighs benefits. Don't use.

Breast-feeding:
- Avoid magnesium in large quantities.
- If you must take temporarily, discontinue breast-feeding. Consult doctor for advice on maintaining milk supply.
- Don't take megadoses.

Effect on lab tests:
- Makes test for stomach-acid secretion inaccurate.
- May increase or decrease serum-phosphate concentrations.
- May decrease serum and urine pH.

Storage:
- Store in cool, dry place away from direct light, but don't freeze.
- Store safely out of reach of children.
- Don't store in bathroom medicine cabinet. Heat and moisture may change action of mineral.

Others:
- Chronic kidney disease frequently causes body to retain excess magnesium.
- Adverse reactions, side effects and interactions with medicines, vitamins or minerals occur *only rarely* when too much magnesium is taken for too long or if you have kidney disease.

 Overdose/Toxicity

Signs and symptoms:
Severe nausea and vomiting, extremely low blood pressure, extreme muscle weakness, difficulty breathing, heartbeat irregularity.

What to do:
For symptoms of overdosage: Discontinue mineral, and consult doctor immediately. Also see *Adverse Reactions or Side Effects* section below.

For accidental overdosage (such as child taking entire bottle): Dial 911 (emergency), 0 for operator or your nearest Poison Control Center.

 Adverse Reactions or Side Effects

Reaction or effect	What to do
Abdominal pain	Discontinue. Call doctor immediately.
Appetite loss	Discontinue. Call doctor when convenient.
Diarrhea	Discontinue. Call doctor immediately.
Irregular heartbeat	Seek emergency treatment.
Mood changes or mental changes	Discontinue. Call doctor when convenient.
Nausea	Discontinue. Call doctor immediately.
Tiredness or weakness	Discontinue. Call doctor when convenient.
Urination discomfort	Discontinue. Call doctor when convenient.
Vomiting	Discontinue. Call doctor immediately.

 Interaction with Medicine, Vitamins or Minerals

Interacts with	Combined effect
Cellulose sodium phosphate	Decreases magnesium effect. Take 1 or more hours apart.
Fat-soluble vitamins (A, E, K)	Decreases absorption of mineral.
Ketoconazole	Reduces absorption of ketoconazole. Take 2 hours apart.
Mecamylamine	May slow urinary excretion of mecamylamine. Avoid combination.
Tetracycline	Decreases absorption of tetracycline.
Vitamin D	May raise magnesium level too high.

 Interaction with Other Substances

None known

Basic Information

Brand names, see page 477.
Available from natural sources? Yes
Available from synthetic sources? Yes
Prescription required? Some forms, yes;
others, no

Natural Sources

Almonds	Salmon, canned
Brazil nuts	Sardines, canned
Caviar	Shrimp
Cheese	Soybeans
Kelp	Tofu
Milk	Turnip greens
Milk products	Yogurt
Molasses	

Reasons to Use

- Helps prevent osteoporosis in older people.
- Treats calcium depletion in people with hypoparathyroidism, osteomalacia, rickets.
- Treats low-calcium levels in people taking anti-convulsant medication.
- Treats tetany (severe muscle spasms) caused by insect bites, sensitivity reactions, cardiac arrest, lead poisoning.
- Is used as an antidote to magnesium poisoning.
- Prevents muscle cramps in some people.
- Promotes normal growth and development.
- Builds bones and teeth.
- Maintains bone density and strength.
- Buffers acid in stomach and acts as antacid.
- Helps regulate heartbeat, blood clotting, muscle contraction.
- Treats neonatal hypocalcemia.
- Promotes storage and release of some body hormones.
- Promotes use of amino acids.
- Lowers phosphate concentrations in people with chronic kidney disease.

Unproved Speculated Benefits

- Helps prevent insomnia and anxiety (acts as a natural tranquilizer).
- Helps prevent hypertension.
- Treats allergies.
- Decreases likelihood of hardening of arteries.
- Treats leg cramps.
- Treats diabetes.
- Treats throat spasms.

Who Needs Additional Amounts?

- Anyone with inadequate caloric or nutritional dietary intake or increased nutritional requirements or who does not like or consume milk products.
- People allergic to milk and milk products or who don't tolerate them well.
- Older people (over 55 years), particularly women.
- Women throughout adult life, especially during pregnancy and lactation, but not limited to these times.
- Those who abuse alcohol or other drugs.
- People who have a chronic wasting illness, excess stress for long periods or who have recently undergone surgery.
- Those with a portion of the gastrointestinal tract surgically removed.
- People with recent severe burns or injuries.

Deficiency Symptoms

- Osteoporosis (late symptoms): frequent fractures in spine and other bones, deformed spinal column with humps, loss of height
- Osteomalacia: frequent fractures
- Muscle contractions
- Convulsive seizures
- Muscle cramps
- Low backache

Unproved Speculated Symptoms

- Uncontrollable temper outbursts

Lab Tests to Detect Deficiency

- 24-hour urine collection to measure calcium levels (Sulkowitch)
- Serum-calcium levels
- Imaging procedures to scan for bone density (more reliable than above tests)

》➤

MINERAL

Calcium Citrate, Continued

Dosage and Usage Information

Recommended Dietary Allowance (RDA):

Estimate of adequate daily intake by the Food and Nutrition Board of the National Research Council, 1980 (See Glossary).

Many doctors and nutritionists recommend women take more calcium than quoted by the RDA. They recommend 1,000 milligrams per day for premenopausal women and 1,500 milligrams per day for post-menopausal women and elderly men.

Age	RDA
0–6 months	360mg
6–12 months	540mg
1–10 years	800mg
11–18 years	1,000mg
18+ years	800mg
Pregnant	+400mg
Lactating	+400mg

Different types of calcium supplements contain more available calcium (also called *elemental calcium*) than others. To provide 1,000mg of available calcium, you must take 5 tablets/day of 950mg calcium citrate. Take in divided doses after meals. Check contents of product to determine how many tablets are needed to provide the amount of calcium you require.

What this mineral does:

- Participates in metabolic functions necessary for normal activity of nervous, muscular, skeletal systems.
- Plays important role in normal heart function, kidney function, blood clotting, blood-vessel integrity.
- Helps utilization of vitamin B-12.

Miscellaneous information:

- Bones serve as storage site for calcium in the body. There is a constant interchange between calcium in bone and the bloodstream.
- Foods rich in calcium (or supplements) help maintain the balance between bone needs and blood needs.

Available as:

- Tablets: Swallow whole with full glass of liquid. Don't chew or crush. Take with meals or 1 to 1-1/2 hours after meals unless otherwise directed by your doctor.

Warnings and Precautions

Don't take if you:

- Are allergic to calcium or antacids.
- Have kidney stones.
- Have a high blood-calcium level.
- Have sarcoidosis.

Consult your doctor if you have:

- Kidney disease.
- Chronic constipation, colitis, diarrhea.
- Stomach or intestinal bleeding.
- Irregular heartbeat.

Over age 55:

- Adverse reactions and side effects are more likely.
- Diarrhea or constipation are particularly likely.

Pregnancy:

- May need extra calcium. Consult doctor about supplements.
- Don't take megadoses.

Breast-feeding:

- Drug passes into milk. Consult doctor about supplements.
- Don't take megadoses.

Effect on lab tests:

- Serum-amylase and serum 11-hydroxycorticosteroid concentrations can be increased.
- Decreases serum-phosphate concentration with excessive, prolonged use.

Storage:

- Store in cool, dry area away from direct light, but don't freeze.
- Store safely out of reach of children.
- Don't store in bathroom medicine cabinet. Heat and moisture may change action of mineral.

Others:

- Dolomite or bone meal are probably *unsafe* sources of calcium because they contain lead.
- Avoid taking calcium within 1 or 2 hours of meals or ingestion of other medicines, if possible.

Overdose/Toxicity

Signs and symptoms:

Confusion, high blood pressure, increased sensitivity of eyes and skin to light, increased thirst, slow or irregular heartbeat, depression, bone or muscle pain, nausea, vomiting, skin itching, skin rash, increased urination.

What to do:

For symptoms of overdosage: Discontinue mineral, and consult doctor immediately. Also see *Adverse Reactions or Side Effects* section below.

For accidental overdosage (such as child taking entire bottle): Dial 911 (emergency), 0 for operator or your nearest Poison Control Center.

Adverse Reactions or Side Effects

Reaction or effect	What to do
Early signs of too much calcium in blood:	
Appetite loss	Discontinue. Call doctor when convenient.
Constipation	Discontinue. Call doctor when convenient.
Drowsiness	Discontinue. Call doctor immediately.
Dry mouth	Discontinue. Call doctor when convenient.
Headache	Discontinue. Call doctor when convenient.
Metallic taste	Discontinue. Call doctor when convenient.
Tiredness or weakness	Discontinue. Call doctor immediately.
Late signs of too much calcium in blood:	
Confusion	Discontinue. Call doctor immediately.
Depression	Discontinue. Call doctor when convenient.
High blood pressure	Discontinue. Call doctor immediately.
Increased thirst	Discontinue. Call doctor when convenient.
Increased urination	Discontinue. Call doctor when convenient.
Muscle or bone pain	Discontinue. Call doctor immediately.
Nausea	Discontinue. Call doctor immediately.
Skin rash	Discontinue. Call doctor immediately.
Slow or irregular heartbeat	Seek emergency treatment.
Vomiting	Discontinue. Call doctor immediately.

 Interaction with Medicine, Vitamins or Minerals

Interacts with	Combined effect
Digitalis preparations	Heartbeat irregularities.
Iron supplements	Decreases absorption of iron unless vitamin C is taken at same time.
Magnesium-containing medications or supplements	Increases blood level of both.
Oral contraceptives and estrogens	May increase calcium absorption.
Potassium supplements	Increases chance of heartbeat irregularities.
Tetracyclines (oral)	Decreases absorption of tetracycline.
Vitamin A (megadoses)	Stimulates bone loss.
Vitamin D (megadoses)	Excessively increases absorption of calcium supplements.

 Interaction with Other Substances

Tobacco decreases absorption.

Alcohol decreases absorption.

Beverages
- Tea decreases absorption.
- Coffee decreases absorption.
- Don't take calcium with milk or other dairy products so your body can absorb the most calcium from food *and* calcium supplement.

Foods
- Avoid eating spinach, rhubarb, bran, whole-grain cereals, fresh fruits or fresh vegetables at same time you take calcium. They may prevent efficient absorption.

MINERAL

Calcium Glubionate

Basic Information
Brand names, see page 477.
Available from natural sources? Yes
Available from synthetic sources? Yes
Prescription required? Some forms, yes;
others, no

 ## Natural Sources

Almonds	Salmon, canned
Brazil nuts	Sardines, canned
Caviar	Shrimp
Cheese	Soybeans
Kelp	Tofu
Milk	Turnip greens
Milk products	Yogurt
Molasses	

 ## Reasons to Use

- Helps prevent osteoporosis in older people.
- Treats calcium depletion in people with hypoparathyroidism, osteomalacia, rickets.
- Treats low-calcium levels in people taking anti-convulsant medication.
- Treats tetany (severe muscle spasms) caused by insect bites, sensitivity reactions, cardiac arrest, lead poisoning.
- Is used as an antidote to magnesium poisoning.
- Prevents muscle cramps in some people.
- Promotes normal growth and development.
- Builds bones and teeth.
- Maintains bone density and strength.
- Buffers acid in stomach and acts as antacid.
- Helps regulate heartbeat, blood clotting, muscle contraction.
- Treats neonatal hypocalcemia.
- Promotes storage and release of some body hormones.
- Promotes use of amino acids.
- Lowers phosphate concentrations in people with chronic kidney disease.

 ## Unproved Speculated Benefits

- Helps prevent insomnia and anxiety (acts as a natural tranquilizer).
- Helps prevent hypertension.
- Treats allergies.
- Decreases likelihood of hardening of arteries.
- Treats leg cramps.
- Treats diabetes.
- Treats throat spasms.

 ## Who Needs Additional Amounts?

- Anyone with inadequate caloric or nutritional dietary intake or increased nutritional requirements or who does not like or consume milk products.
- People allergic to milk and milk products or who don't tolerate them well.
- Older people (over 55 years), particularly women.
- Women throughout adult life, especially during pregnancy and lactation, but not limited to these times.
- Those who abuse alcohol or other drugs.
- People who have a chronic wasting illness, excess stress for long periods or who have recently undergone surgery.
- Those with a portion of the gastrointestinal tract surgically removed.
- People with recent severe burns or injuries.

 ## Deficiency Symptoms

- Osteoporosis (late symptoms): frequent fractures in spine and other bones, deformed spinal column with humps, loss of height
- Osteomalacia: frequent fractures
- Muscle contractions
- Convulsive seizures
- Muscle cramps
- Low backache

 ## Unproved Speculated Symptoms

- Uncontrollable temper outbursts

 ## Lab Tests to Detect Deficiency

- 24-hour urine collection to measure calcium levels (Sulkowitch)
- Serum-calcium levels
- Imaging procedures to scan for bone density (more reliable than above tests)

 ## Dosage and Usage Information

Recommended Dietary Allowance (RDA):
Estimate of adequate daily intake by the Food and Nutrition Board of the National Research Council, 1980 (See Glossary).
Many doctors and nutritionists recommend »➤

women take more calcium than quoted by the RDA. They recommend 1,000 milligrams per day for premenopausal women and 1,500 milligrams per day for post-menopausal women and elderly men.

Age	RDA
0–6 months	360mg
6–12 months	540mg
1–10 years	800mg
11–18 years	1,000mg
18+ years	800mg
Pregnant	+400mg
Lactating	+400mg

Different types of calcium supplements contain more available calcium (also called *elemental calcium*) than others. To provide 1,000mg of available calcium, you must take 12 teaspoons a day of calcium glubionate. Take in divided doses after meals.

What this mineral does:
- Participates in metabolic functions necessary for normal activity of nervous, muscular, skeletal systems.
- Plays important role in normal heart function, kidney function, blood clotting, blood-vessel integrity.
- Helps utilization of vitamin B-12.

Miscellaneous information:
- Bones serve as storage site for calcium in the body. There is a constant interchange between calcium in bone and the bloodstream.
- Foods rich in calcium (or supplements) help maintain the balance between bone needs and blood needs.
- Don't discard outer parts of vegetables during food preparation.
- Exercise, a balanced diet, calcium from natural sources or supplements and estrogens are important in treating and preventing osteoporosis.

Available as:
- Syrup: Dilute in at least 1/2 glass of water or other liquid. Take with meals or 1 to 1-1/2 hours after meals unless otherwise directed by your doctor.

Warnings and Precautions

Don't take if you:
- Are allergic to calcium or antacids.
- Have kidney stones.
- Have a high blood-calcium level.
- Have sarcoidosis.

Consult your doctor if you have:
- Kidney disease.
- Chronic constipation, colitis, diarrhea.
- Stomach or intestinal bleeding.
- Irregular heartbeat.

Over age 55:
- Adverse reactions and side effects are more likely.
- Diarrhea or constipation are particularly likely.

Pregnancy:
- May need extra calcium. Consult doctor about supplement.
- Don't take megadoses.

Breast-feeding:
- Drug passes into milk. Consult doctor about need for supplements.
- Don't take megadoses.

Effect on lab tests:
- Serum-amylase and serum 11-hydroxycorticosteroid concentrations can be increased.
- Decrease serum-phosphate concentration decreased by excessive, prolonged use.

Storage:
- Store in cool, dry area away from direct light, but don't freeze.
- Store safely out of reach of children.
- Don't store in bathroom medicine cabinet. Heat and moisture may change action of mineral.

Others:
- Dolomite or bone meal are probably *unsafe* sources of calcium because they contain lead.
- Avoid taking calcium within 1 or 2 hours of meals or ingestion of other medicines, if possible.

Overdose/Toxicity

Signs and symptoms:
Confusion, high blood pressure, increased sensitivity of eyes and skin to light, increased thirst, slow or irregular heartbeat, depression, bone or muscle pain, nausea, vomiting, skin itching, skin rash, increased urination.

What to do:
For symptoms of overdosage: Discontinue mineral, and consult doctor immediately. Also see *Adverse Reactions or Side Effects* section below.
For accidental overdosage (such as child taking entire bottle): Dial 911 (emergency), 0 for operator or your nearest Poison Control Center.

Calcium Glubionate, Continued

 Adverse Reactions or Side Effects

Reaction or effect	What to do
Early signs of too much calcium in blood:	
Appetite loss	Discontinue. Call doctor when convenient.
Constipation	Discontinue. Call doctor when convenient.
Drowsiness	Discontinue. Call doctor immediately.
Dry mouth	Discontinue. Call doctor when convenient.
Headache	Discontinue. Call doctor when convenient.
Metallic taste	Discontinue. Call doctor when convenient.
Tiredness or weakness	Discontinue. Call doctor immediately.
Late signs of too much calcium in blood:	
Confusion	Discontinue. Call doctor immediately.
Depression	Discontinue. Call doctor when convenient.
High blood pressure	Discontinue. Call doctor immediately.
Increased thirst	Discontinue. Call doctor when convenient.
Increased urination	Discontinue. Call doctor when convenient.
Muscle or bone pain	Discontinue. Call doctor immediately.
Nausea	Discontinue. Call doctor immediately.
Skin rash	Discontinue. Call doctor immediately.
Slow or irregular heartbeat	Seek emergency treatment.
Vomiting	Discontinue. Call doctor immediately.

 Interaction with Medicine, Vitamins or Minerals

Interacts with	Combined effect
Digitalis preparations	Heartbeat irregularities.
Iron supplements	Decreases absorption of iron unless vitamin C is taken at same time.
Magnesium-containing medications or supplements	Increases blood level of both.
Oral contraceptives and estrogens	May increase calcium absorption.
Potassium supplements	Increases chance of heartbeat irregularities.
Tetracyclines (oral)	Decreases absorption of tetracycline.
Vitamin A (megadoses)	Stimulates bone loss.
Vitamin D (megadoses)	Excessively increases absorption of calcium supplements.

 Interaction with Other Substances

Tobacco decreases absorption.

Alcohol decreases absorption.

Beverages
- Tea decreases absorption.
- Coffee decreases absorption.
- Don't take calcium with milk or other dairy products so your body can absorb the most calcium from food *and* calcium supplement.

Foods
- Avoid eating spinach, rhubarb, bran, whole-grain cereals, fresh fruits or fresh vegetables at same time you take calcium. They may prevent efficient absorption.

Basic Information

Brand names, see page 477.
Available from natural sources? Yes
Available from synthetic sources? Yes
Prescription required? Some forms, yes;
others, no

Natural Sources

Almonds	Salmon, canned
Brazil nuts	Sardines, canned
Caviar	Shrimp
Cheese	Soybeans
Kelp	Tofu
Milk	Turnip greens
Milk products	Yogurt
Molasses	

Reasons to Use

- Helps prevent osteoporosis in older people.
- Treats calcium depletion in people with hypoparathyroidism, osteomalacia, rickets.
- Treats low-calcium levels in people taking anti-convulsant medication.
- Treats tetany (severe muscle spasms) caused by insect bites, sensitivity reactions, cardiac arrest, lead poisoning.
- Is used as an antidote to magnesium poisoning.
- Prevents muscle cramps in some people.
- Promotes normal growth and development.
- Builds bones and teeth.
- Maintains bone density and strength.
- Buffers acid in stomach and acts as antacid.
- Helps regulate heartbeat, blood clotting, muscle contraction.
- Treats neonatal hypocalcemia.
- Promotes storage and release of some body hormones.
- Promotes use of amino acids.
- Lowers phosphate concentrations in people with chronic kidney disease.

Unproved Speculated Benefits

- Helps prevent insomnia and anxiety (acts as a natural tranquilizer).
- Helps prevent hypertension.
- Treats allergies.
- Decreases likelihood of hardening of arteries.
- Treats leg cramps.
- Treats diabetes.
- Treats throat spasms.

Who Needs Additional Amounts?

- Anyone with inadequate caloric or nutritional dietary intake or increased nutritional requirements or who does not like or consume milk products.
- People allergic to milk and milk products or who don't tolerate them well.
- Older people (over 55 years), particularly women.
- Women throughout adult life, especially during pregnancy and lactation, but not limited to these times.
- Those who abuse alcohol or other drugs.
- People who have a chronic wasting illness, excess stress for long periods or who have recently undergone surgery.
- Those with a portion of the gastrointestinal tract surgically removed.
- People with recent severe burns or injuries.

Deficiency Symptoms

- Osteoporosis (late symptoms): frequent fractures in spine and other bones, deformed spinal column with humps, loss of height
- Osteomalacia: frequent fractures
- Muscle contractions
- Convulsive seizures
- Muscle cramps
- Low backache

Unproved Speculated Symptoms

- Uncontrollable temper outbursts

Lab Tests to Detect Deficiency

- 24-hour urine collection to measure calcium levels (Sulkowitch)
- Serum-calcium levels
- Imaging procedures to scan for bone density (more reliable than above tests)

⟫▶

MINERAL

Calcium Gluconate, Continued

Dosage and Usage Information

Recommended Dietary Allowance (RDA):
Estimate of adequate daily intake by the Food and Nutrition Board of the National Research Council, 1980 (See Glossary).
Many doctors and nutritionists recommend women need more calcium than quoted by the RDA. They recommend 1,000 milligrams per day for premenopausal women and 1,500 milligrams per day for post-menopausal women and elderly men.

Age	RDA
0–6 months	360mg
6–12 months	540mg
1–10 years	800mg
11–18 years	1,000mg
18+ years	800mg
Pregnant	+400mg
Lactating	+400mg

Different types of calcium supplements contain more available calcium (also called *elemental calcium*) than others. To provide 1,000mg of available calcium, you must take:
  22 tablets/day of 500mg calcium gluconate
  17 tablets/day of 650mg calcium gluconate
  11 tablets/day of 1,000mg calcium gluconate
Take in divided doses after meals. Check contents of product you choose to determine how many tablets are needed to provide the amount of calcium you require.

What this mineral does:
- Participates in metabolic functions necessary for normal activity of nervous, muscular, skeletal systems.
- Plays important role in normal heart function, kidney function, blood clotting, blood-vessel integrity.
- Helps utilization of vitamin B-12.

Miscellaneous information:
- Bones serve as storage site for calcium in the body. There is a constant interchange between calcium in bone and the bloodstream.
- Foods rich in calcium (or supplements) help maintain the balance between bone needs and blood needs.
- Don't discard outer parts of vegetables during food preparations.
- Exercise, a balanced diet, calcium from natural sources or supplements and estrogens are important in treating and preventing osteoporosis.

Available as:
- Chewable tablets: Chew well before swallowing.
- Injectable forms are administered by doctor or nurse.

Warnings and Precautions

Don't take if you:
- Are allergic to calcium or antacids.
- Have kidney stones.
- Have a high blood-calcium level.
- Have sarcoidosis.

Consult your doctor if you have:
- Kidney disease.
- Chronic constipation, colitis, diarrhea.
- Stomach or intestinal bleeding.
- Irregular heartbeat.

Over age 55:
- Adverse reactions and side effects are more likely.
- Diarrhea or constipation are particularly likely.

Pregnancy:
- May need extra calcium. Consult doctor about supplements.
- Don't take megadoses.

Breast-feeding:
- Drug passes into milk. Consult doctor about need for supplements.
- Don't take megadoses.

Effect on lab tests:
- Serum-amylase and serum 11-hydroxycorticosteroid concentrations can be increased.
- Decreases serum-phosphate concentration with excessive, prolonged use.

Storage:
- Store in cool, dry area away from direct light, but don't freeze.
- Store safely out of reach of children.
- Don't store in bathroom medicine cabinet. Heat and moisture may change action of mineral.

Others:
- Dolomite or bone meal are probably *unsafe* sources of calcium because they contain lead.
- Avoid taking calcium within 1 or 2 hours of meals or ingestion of other medicines, if possible.

Overdose/Toxicity

Signs and symptoms:
Confusion, high blood pressure, increased sensitivity of eyes and skin to light, increased thirst, slow or irregular heartbeat, depression, bone or muscle pain, nausea, vomiting, skin itching, skin rash, increased urination.

》▶

What to do:

For symptoms of overdosage: Discontinue mineral, and consult doctor immediately. Also see *Adverse Reactions or Side Effects* section below.

For accidental overdosage (such as child taking entire bottle): Dial 911 (emergency), 0 for operator or your nearest Poison Control Center.

Adverse Reactions or Side Effects

Reaction or effect	What to do
Early signs of too much calcium in blood:	
Appetite loss	Discontinue. Call doctor when convenient.
Constipation	Discontinue. Call doctor when convenient.
Drowsiness	Discontinue. Call doctor immediately.
Dry mouth	Discontinue. Call doctor when convenient.
Headache	Discontinue. Call doctor when convenient.
Metallic taste	Discontinue. Call doctor when convenient.
Tiredness or weakness	Discontinue. Call doctor immediately.
Late signs of too much calcium in blood:	
Confusion	Discontinue. Call doctor immediately.
Depression	Discontinue. Call doctor when convenient.
High blood pressure	Discontinue. Call doctor immediately.
Increased thirst	Discontinue. Call doctor when convenient.
Increased urination	Discontinue. Call doctor when convenient.
Muscle or bone pain	Discontinue. Call doctor immediately.
Nausea	Discontinue. Call doctor immediately.
Skin rash	Discontinue. Call doctor immediately.
Slow or irregular heartbeat	Seek emergency treatment.
Vomiting	Discontinue. Call doctor immediately.

Interaction with Medicine, Vitamins or Minerals

Interacts with	Combined effect
Digitalis preparations	Heartbeat irregularities.
Iron supplements	Decreases absorption of iron unless vitamin C is taken at same time.
Magnesium-containing medications or supplements	Increases blood level of both.
Oral contraceptives and estrogens	May increase calcium absorption.
Potassium supplements	Increases chance of heartbeat irregularities.
Tetracyclines (oral)	Decreases absorption of tetracycline.
Vitamin A (megadoses)	Stimulates bone loss.
Vitamin D (megadoses)	Excessively increases absorption of calcium supplements

Interaction with Other Substances

Tobacco decreases absorption.

Alcohol decreases absorption.

Beverages
- Tea decreases absorption.
- Coffee decreases absorption.
- Don't take calcium with milk or other dairy products so your body can absorb the most calcium from food *and* calcium supplement.

Foods
- Avoid eating spinach, rhubarb, bran, whole-grain cereals, fresh fruits or fresh vegetables at same time you take calcium. They may prevent efficient absorption.

MINERAL

Calcium Lactate

Basic Information
Brand names, see page 477.
Available from natural sources? Yes
Available from synthetic sources? Yes
Prescription required? Some forms, yes;
others, no

Natural Sources

Almonds	Salmon, canned
Brazil nuts	Sardines, canned
Caviar	Shrimp
Cheese	Soybeans
Kelp	Tofu
Milk	Turnip greens
Milk products	Yogurt
Molasses	

Reasons to Use

- Helps prevent osteoporosis in older people.
- Treats calcium depletion in people with hypoparathyroidism, osteomalacia, rickets.
- Treats low-calcium levels in people taking anti-convulsant medication.
- Treats tetany (severe muscle spasms) caused by insect bites, sensitivity reactions, cardiac arrest, lead poisoning.
- Is used as an antidote to magnesium poisoning.
- Prevents muscle cramps in some people.
- Promotes normal growth and development.
- Builds bones and teeth.
- Maintains bone density and strength.
- Buffers acid in stomach and acts as antacid.
- Helps regulate heartbeat, blood clotting, muscle contraction.
- Treats neonatal hypocalcemia.
- Promotes storage and release of some body hormones.
- Promotes use of amino acids.
- Lowers phosphate concentrations in people with chronic kidney disease.

Unproved Speculated Benefits

- Helps prevent insomnia and anxiety (acts as a natural tranquilizer).
- Helps prevent hypertension.
- Treats allergies.
- Decreases likelihood of hardening of arteries.
- Treats leg cramps.
- Treats diabetes.
- Treats throat spasms.

Who Needs Additional Amounts?

- Anyone with inadequate caloric or nutritional dietary intake or increased nutritional requirements or who does not like or consume milk products.
- People allergic to milk and milk products or who don't tolerate them well.
- Older people (over 55 years), particularly women.
- Women throughout adult life, especially during pregnancy and lactation, but not limited to these times.
- Those who abuse alcohol or other drugs.
- People who have a chronic wasting illness, excess stress for long periods or who have recently undergone surgery.
- Those with a portion of the gastrointestinal tract surgically removed.
- People with recent severe burns or injuries.

Deficiency Symptoms

- Osteoporosis (late symptoms): frequent fractures in spine and other bones, deformed spinal column with humps, loss of height
- Osteomalacia: frequent fractures
- Muscle contractions
- Convulsive seizures
- Muscle cramps
- Low backache

Unproved Speculated Symptoms

- Uncontrollable temper outbursts

Lab Tests to Detect Deficiency

- 24-hour urine collection to measure calcium levels (Sulkowitch)
- Serum-calcium levels
- Imaging procedures to scan for bone density (more reliable than above tests)

Dosage and Usage Information

Recommended Dietary Allowance (RDA):
Estimate of adequate daily intake by the Food and Nutrition Board of the National Research Council, 1980 (See Glossary).
Many doctors and nutritionists recommend ⟫▶

women take more calcium than quoted by the RDA. They recommend 1,000 milligrams per day for premenopausal women and 1,500 milligrams per day for post-menopausal women and elderly men.

Age	RDA
0–6 months	360mg
6–12 months	540mg
1–10 years	800mg
11–18 years	1,000mg
18+ years	800mg
Pregnant	+400mg
Lactating	+400mg

Different types of calcium supplements contain more available calcium (also called *elemental calcium*) than others. To provide 1,000mg of available calcium, you must take:

   24 tablets/day of 325mg calcium lactate
   12 tablets/day of 650mg calcium lactate

Take in divided doses after meals. Check contents of product you choose to determine how many tablets are needed to provide the amount of calcium you require.

What this mineral does:
- Participates in metabolic functions necessary for normal activity of nervous, muscular, skeletal systems.
- Plays important role in normal heart function, kidney function, blood clotting, blood-vessel integrity.
- Helps utilization of vitamin B-12.

Miscellaneous information:
- Bones serve as storage site for calcium in the body. There is a constant interchange between calcium in bone and the bloodstream.
- Foods rich in calcium (or supplements) help maintain the balance between bone needs and blood needs.
- Don't discard outer parts of vegetables during food preparations.
- Exercise, a balanced diet, calcium from natural sources or supplements and estrogens are important in treating and preventing osteoporosis.

Available as:
- Tablets: Swallow whole with full glass of liquid. Don't chew or crush. Take with meals or 1 to 1-1/2 hours after meals unless otherwise directed by your doctor.

Warnings and Precautions

Don't take if you:
- Are allergic to calcium or antacids.
- Have kidney stones.
- Have a high blood-calcium level.
- Have sarcoidosis.

Consult your doctor if you have:
- Kidney disease.
- Chronic constipation, colitis, diarrhea.
- Stomach or intestinal bleeding.
- Irregular heartbeat.

Over age 55:
- Adverse reactions and side effects are more likely.
- Diarrhea or constipation are particularly likely.

Pregnancy:
- May need extra calcium. Consult doctor about supplements.
- Don't take megadoses.

Breast-feeding:
- Drug passes into milk. Consult doctor about need for supplements.
- Don't take megadoses.

Effect on lab tests:
- Serum-amylase and serum 11-hydroxycorticosteroid concentrations can be increased.
- Decreases serum-phosphate concentration with excessive, prolonged use.

Storage:
- Store in cool, dry area away from direct light, but don't freeze.
- Store safely out of reach of children.
- Don't store in bathroom medicine cabinet. Heat and moisture may change action of mineral.

Others:
- Dolomite or bone meal are probably *unsafe* sources of calcium because they contain lead.
- Avoid taking calcium within 1 or 2 hours of meals or ingestion of other medicines, if possible.

Overdose/Toxicity

Signs and symptoms:
Confusion, high blood pressure, increased sensitivity of eyes and skin to light, increased thirst, slow or irregular heartbeat, depression, bone or muscle pain, nausea, vomiting, skin itching, skin rash, increased urination.

What to do:
For symptoms of overdosage: Discontinue mineral, and consult doctor immediately. Also see *Adverse Reactions or Side Effects* section below.
For accidental overdosage (such as child taking entire bottle): Dial 911 (emergency), 0 for operator or your nearest Poison Control Center.

Calcium Lactate, Continued

 Adverse Reactions or Side Effects

Reaction or effect	What to do
Early signs of too much calcium in blood:	
Appetite loss	Discontinue. Call doctor when convenient.
Constipation	Discontinue. Call doctor when convenient.
Drowsiness	Discontinue. Call doctor immediately.
Dry mouth	Discontinue. Call doctor when convenient.
Headache	Discontinue. Call doctor when convenient.
Metallic taste	Discontinue. Call doctor when convenient.
Tiredness or weakness	Discontinue. Call doctor immediately.
Late signs of too much calcium in blood:	
Confusion	Discontinue. Call doctor immediately.
Depression	Discontinue. Call doctor when convenient.
High blood pressure	Discontinue. Call doctor immediately.
Increased thirst	Discontinue. Call doctor when convenient.
Increased urination	Discontinue. Call doctor when convenient.
Muscle or bone pain	Discontinue. Call doctor immediately.
Nausea	Discontinue. Call doctor immediately.
Skin rash	Discontinue. Call doctor immediately.
Slow or irregular heartbeat	Seek emergency treatment.
Vomiting	Discontinue. Call doctor immediately.

 Interaction with Medicine, Vitamins or Minerals

Interacts with	Combined effect
Digitalis preparations	Heartbeat irregularities.
Iron supplements	Decreases absorption of iron unless vitamin C is taken at same time.
Magnesium-containing medications or supplements	Increases blood level of both.
Oral contraceptives and estrogens	May increase calcium absorption.
Potassium supplements	Increases chance of heartbeat irregularities.
Tetracyclines (oral)	Decreases absorption of tetracycline.
Vitamin A (megadoses)	Stimulates bone loss.
Vitamin D (megadoses)	Excessively increases absorption of calcium supplements.

 Interaction with Other Substances

Tobacco decreases absorption.

Alcohol decreases absorption.

Beverages
- Tea decreases absorption.
- Coffee decreases absorption.
- Don't take calcium with milk or other dairy products so your body can absorb the most calcium from food *and* calcium supplement.

Foods
- Avoid eating spinach, rhubarb, bran, whole-grain cereals, fresh fruits or fresh vegetables at same time you take calcium. They may prevent efficient absorption.

Calcium/Magnesium Carbonate

Basic Information

Provides additional magnesium. Other substances in title are not supplied in large enough amounts for use by your body.
Brand names, see page 477.
Available from natural sources? Yes
Available from synthetic sources? No
Prescription required? Yes, for some forms

Natural Sources

Almonds	Leafy, green	Snails
Bluefish	vegetables	Soybeans
Carp	Mackerel	Sunflower
Cod	Molasses	seeds
Flounder	Nuts	Swordfish
Halibut	Ocean perch	Wheat germ
Herring	Shrimp	

Reasons to Use

- Aids bone growth.
- Aids function of nerves and muscles, including regulation of normal heart rhythm.
- Keeps metabolism steady.
- Conducts nerve impulses.
- Works as laxative in large doses.
- Acts as antacid in small doses.
- Strengthens tooth enamel.

Unproved Speculated Benefits

- Cures alcoholism.
- Cures kidney stones.
- Alleviates heart disease.
- Helps control bad breath and body odor.
- Helps contain lead poisoning.

Who Needs Additional Amounts?

- Anyone with inadequate caloric or dietary intake or increased nutritional requirements.
- Those who abuse alcohol or other drugs.
- People with a chronic wasting illness or who have recently undergone surgery.
- Those with recent severe burns or injuries.

Deficiency Symptoms

Following symptoms occur rarely:
- Muscle contractions
- Convulsions
- Confusion and delirium
- Irritability
- Nervousness
- Skin problems
- Hardening of soft tissues

Unproved Speculated Symptoms

- Sudden heart failure

Lab Tests to Detect Deficiency

- Serum magnesium

Dosage and Usage Information

Recommended Dietary Allowance (RDA):
Estimate of adequate daily intake by the Food and Nutrition Board of the National Research Council, 1980 (See Glossary).

Age	RDA
0–6 months	50mg
6–12 months	70mg
1–3 years	150mg
4–6 years	200mg
7–10 years	250mg
Males	
11–14 years	350mg
15–18 years	400mg
18+ years	350mg
Females	
11+ years	300mg
Pregnant	+150mg
Lactating	+150mg

What this mineral does:
- Activates essential enzymes.
- Affects metabolism of proteins and nucleic acids.
- Helps transport sodium and potassium across cell membranes.
- Influences calcium levels inside cells.

Available as:
- Tablets: Swallow whole with full glass of liquid. Don't chew or crush. Take with meals or 1 to 1-1/2 hours after meals unless otherwise directed by your doctor.
- Liquid: Dilute in at least 1/2 glass of water or other liquid. Take with meals or 1 to 1-1/2 hours after meals unless otherwise directed by your doctor.
- Chewable tablets: Chew well before swallowing.

》▸

Calcium/Magnesium Carbonate

 Warnings and Precautions

Don't take if you have:
• Kidney failure.
• Heart block (unless you have a pacemaker).
• An ileostomy.

Consult your doctor if you have:
• Chronic constipation, colitis, diarrhea.
• Symptoms of appendicitis.
• Stomach or intestinal bleeding.

Over age 55:
• Adverse reactions/side effects more likely.

Pregnancy:
• Risk to fetus outweighs benefits. Don't use.

Breast-feeding:
• Avoid magnesium in large quantities.
• If you must take temporarily, discontinue breast-feeding. Consult doctor for advice on maintaining milk supply.
• Don't take megadoses.

Effect on lab tests:
• Makes test for stomach-acid secretion inaccurate.
• May increase or decrease serum-phosphate concentrations.
• May decrease serum and urine pH.

Storage:
• Store in cool, dry place away from direct light, but don't freeze.
• Store safely out of reach of children.
• Don't store in bathroom medicine cabinet. Heat and moisture may change action of mineral.

Others:
• Chronic kidney disease frequently causes body to retain excess magnesium.
• Adverse reactions, side effects and interactions with medicines, vitamins or minerals occur *only rarely* when too much magnesium is taken for too long or if you have kidney disease.

 Overdose/Toxicity

Signs and symptoms:
Severe nausea and vomiting, extremely low blood pressure, extreme muscle weakness, difficulty breathing, heartbeat irregularity.

What to do:
For symptoms of overdosage: Discontinue mineral, and consult doctor immediately. Also see *Adverse Reactions or Side Effects*.
For accidental overdosage (such as child taking entire bottle): Dial 911 (emergency), 0 for operator or your nearest Poison Control Center.

 Adverse Reactions or Side Effects

Reaction or effect	What to do
Abdominal pain	Discontinue. Call doctor immediately.
Appetite loss	Discontinue. Call doctor when convenient.
Diarrhea	Discontinue. Call doctor immediately.
Irregular heartbeat	Seek emergency treatment.
Mood changes or mental changes	Discontinue. Call doctor when convenient.
Nausea	Discontinue. Call doctor immediately.
Tiredness or weakness	Discontinue. Call doctor when convenient.
Urination discomfort	Discontinue. Call doctor when convenient.
Vomiting	Discontinue. Call doctor immediately.

 Interaction with Medicine, Vitamins or Minerals

Interacts with	Combined effect
Cellulose sodium phosphate	Decreases magnesium effect. Take 1 or more hours apart.
Fat-soluble vitamins (A, E, K)	Decreases absorption of mineral.
Ketoconazole	Reduces absorption of ketoconazole. Take 2 hours apart.
Mecamylamine	May slow urinary excretion of mecamylamine. Avoid combination.
Tetracycline	Decreases absorption of tetracycline.
Vitamin D	May raise magnesium level too high.

 Interaction with Other Substances

None known

Calcium/Magnesium Carbonate/Magnesium Oxide

Basic Information

Provides additional magnesium. Other substances in title are not supplied in large enough amounts for use by your body.
Brand names, see page 477.
Available from natural sources? Yes
Available from synthetic sources? No
Prescription required? Yes, for some forms

Natural Sources

Almonds	Leafy, green	Snails
Bluefish	vegetables	Soybeans
Carp	Mackerel	Sunflower
Cod	Molasses	seeds
Flounder	Nuts	Swordfish
Halibut	Ocean perch	Wheat germ
Herring	Shrimp	

Reasons to Use

- Aids bone growth.
- Aids function of nerves and muscles, including regulation of normal heart rhythm.
- Keeps metabolism steady.
- Conducts nerve impulses.
- Works as laxative in large doses.
- Acts as antacid in small doses.
- Strengthens tooth enamel.

Unproved Speculated Benefits

- Cures alcoholism.
- Cures kidney stones.
- Alleviates heart disease.
- Helps control bad breath and body odor.
- Helps contain lead poisoning.

Who Needs Additional Amounts?

- Anyone with inadequate caloric or dietary intake or increased nutritional requirements.
- Those who abuse alcohol or other drugs.
- People with a chronic wasting illness or who have recently undergone surgery.
- Those with recent severe burns or injuries.

Deficiency Symptoms

Following symptoms occur rarely:
- Muscle contractions
- Convulsions
- Confusion and delirium
- Irritability
- Nervousness
- Skin problems
- Hardening of soft tissues

Unproved Speculated Symptoms

- Sudden heart failure

Lab Tests to Detect Deficiency

- Serum magnesium

Dosage and Usage Information

Recommended Dietary Allowance (RDA):
Estimate of adequate daily intake by the Food and Nutrition Board of the National Research Council, 1980 (See Glossary).

Age	RDA
0–6 months	50mg
6–12 months	70mg
1–3 years	150mg
4–6 years	200mg
7–10 years	250mg
Males	
11–14 years	350mg
15–18 years	400mg
18+ years	350mg
Females	
11+ years	300mg
Pregnant	+150mg
Lactating	+150mg

What this mineral does:
- Activates essential enzymes.
- Affects metabolism of proteins and nucleic acids.
- Helps transport sodium and potassium across cell membranes.
- Influences calcium levels inside cells.

Available as:
- Tablets: Swallow whole with full glass of liquid. Don't chew or crush. Take with meals or 1 to 1-1/2 hours after meals unless otherwise directed by your doctor.

➤➤➤

MINERAL

Calcium/Magnesium Carbonate/Magnesium Oxide

 Warnings and Precautions

Don't take if you have:
- Kidney failure.
- Heart block (unless you have a pacemaker).
- An ileostomy.

Consult your doctor if you have:
- Chronic constipation, colitis, diarrhea.
- Symptoms of appendicitis.
- Stomach or intestinal bleeding.

Over age 55:
- Adverse reactions/side effects more likely.

Pregnancy:
- Risk to fetus outweighs benefits. Don't use.

Breast-feeding:
- Avoid magnesium in large quantities.
- If you must take temporarily, discontinue breast-feeding. Consult doctor for advice on maintaining milk supply.
- Don't take megadoses.

Effect on lab tests:
- Makes test for stomach-acid secretion inaccurate.
- May increase or decrease serum-phosphate concentrations.
- May decrease serum and urine pH.

Storage:
- Store in cool, dry place away from direct light, but don't freeze.
- Store safely out of reach of children.
- Don't store in bathroom medicine cabinet. Heat and moisture may change action of mineral.

Others:
- Chronic kidney disease frequently causes body to retain excess magnesium.
- Adverse reactions, side effects and interactions with medicines, vitamins or minerals occur *only rarely* when too much magnesium is taken for too long or if you have kidney disease.

 Overdose/Toxicity

Signs and symptoms:
Severe nausea and vomiting, extremely low blood pressure, extreme muscle weakness, difficulty breathing, heartbeat irregularity.

What to do:
For symptoms of overdosage: Discontinue mineral, and consult doctor immediately. Also see *Adverse Reactions or Side Effects*.
For accidental overdosage (such as child taking entire bottle): Dial 911 (emergency), 0 for operator or your nearest Poison Control Center.

 Adverse Reactions or Side Effects

Reaction or effect	What to do
Abdominal pain	Discontinue. Call doctor immediately.
Appetite loss	Discontinue. Call doctor when convenient.
Diarrhea	Discontinue. Call doctor immediately.
Irregular heartbeat	Seek emergency treatment.
Mood changes or mental changes	Discontinue. Call doctor when convenient.
Nausea	Discontinue. Call doctor immediately.
Tiredness or weakness	Discontinue. Call doctor when convenient.
Urination discomfort	Discontinue. Call doctor when convenient.
Vomiting	Discontinue. Call doctor immediately.

 Interaction with Medicine, Vitamins or Minerals

Interacts with	Combined effect
Cellulose sodium phosphate	Decreases magnesium effect. Take 1 or more hours apart.
Fat-soluble vitamins (A, E, K)	Decreases absorption of mineral.
Ketoconazole	Reduces absorption of ketoconazole. Take 2 hours apart.
Mecamylamine	May slow urinary excretion of mecamylamine. Avoid combination.
Tetracycline	Decreases absorption of tetracycline.
Vitamin D	May raise magnesium level too high.

 Interaction with Other Substances

None known

Calcium Phosphate

Basic Information

This form of calcium is also called tribasic calcium phosphate *or* dibasic calcium phosphate.
Brand names, see page 477.
Available from natural sources? Yes
Available from synthetic sources? Yes
Prescription required? Some forms, yes; others, no

Natural Sources

Almonds	Salmon, canned
Brazil nuts	Sardines, canned
Caviar	Shrimp
Cheese	Soybeans
Kelp	Tofu
Milk	Turnip greens
Milk products	Yogurt
Molasses	

Reasons to Use

- Helps prevent osteoporosis in older people.
- Treats calcium depletion in people with hypoparathyroidism, osteomalacia, rickets.
- Treats low-calcium levels in people taking anti-convulsant medication.
- Treats tetany (severe muscle spasms) caused by insect bites, sensitivity reactions, cardiac arrest, lead poisoning.
- Is used as an antidote to magnesium poisoning.
- Prevents muscle cramps in some people.
- Promotes normal growth and development.
- Builds bones and teeth.
- Maintains bone density and strength.
- Buffers acid in stomach and acts as antacid.
- Helps regulate heartbeat, blood clotting, muscle contraction.
- Treats neonatal hypocalcemia.
- Promotes storage and release of some body hormones.
- Promotes use of amino acids.
- Lowers phosphate concentrations in people with chronic kidney disease.

Unproved Speculated Benefits

- Helps prevent insomnia and anxiety (acts as a natural tranquilizer).
- Helps prevent hypertension.
- Treats allergies.
- Decreases likelihood of hardening of arteries.

- Treats leg cramps.
- Treats diabetes.
- Treats throat spasms.

Who Needs Additional Amounts?

- Anyone with inadequate caloric or nutritional dietary intake or increased nutritional requirements or who does not like or consume milk products.
- People allergic to milk and milk products or who don't tolerate them well.
- Older people (over 55 years), particularly women.
- Women throughout adult life, especially during pregnancy and lactation, but not limited to these times.
- Those who abuse alcohol or other drugs.
- People who have a chronic wasting illness, excess stress for long periods or who have recently undergone surgery.
- Those with a portion of the gastrointestinal tract surgically removed.
- People with recent severe burns or injuries.

Deficiency Symptoms

- Osteoporosis (late symptoms): frequent fractures in spine and other bones, deformed spinal column with humps, loss of height
- Osteomalacia: frequent fractures
- Muscle contractions
- Convulsive seizures
- Muscle cramps
- Low backache

Unproved Speculated Symptoms

- Uncontrollable temper outbursts

Lab Tests to Detect Deficiency

- 24-hour urine collection to measure calcium levels (Sulkowitch)
- Serum-calcium levels
- Imaging procedures to scan for bone density (more reliable than above tests)

Calcium Phosphate, Continued

 Dosage and Usage Information

Recommended Dietary Allowance (RDA):
Estimate of adequate daily intake by the Food and Nutrition Board of the National Research Council, 1980 (See Glossary).
Many doctors and nutritionists recommend women take more calcium than quoted by the RDA. They recommend 1,000 milligrams per day for premenopausal women and 1,500 milligrams per day for post-menopausal women and elderly men.

Age	RDA
0–6 months	360mg
6–12 months	540mg
1–10 years	800mg
11–18 years	1,000mg
18+ years	800mg
Pregnant	+400mg
Lactating	+400mg

Different types of calcium supplements contain more available calcium (also called *elemental calcium*) than others. To provide 1,000mg of available calcium, you must take:
4 tablets/day of 800mg calcium phosphate
2 tablets/day of 1,600mg calcium phosphate
Take in divided doses after meals. Check contents of product you choose to determine how many tablets are needed to provide the amount of calcium you require.

What this mineral does:
• Participates in metabolic functions necessary for normal activity of nervous, muscular, skeletal systems.
• Plays important role in normal heart function, kidney function, blood clotting, blood-vessel integrity.
• Helps utilization of vitamin B-12.

Miscellaneous information:
• Bones serve as storage site for calcium in the body. There is a constant interchange between calcium in bone and the bloodstream.
• Foods rich in calcium (or supplements) help maintain the balance between bone needs and blood needs.
• Don't discard outer parts of vegetables during food preparation.
• Exercise, a balanced diet, calcium from natural sources or supplements and estrogens are important in treating and preventing osteoporosis.

Available as:
• Tablets: Swallow whole with full glass of liquid. Don't chew or crush. Take with meals or 1 to 1-1/2 hours after meals unless otherwise directed by your doctor.

 Warnings and Precautions

Don't take if you:
• Are allergic to calcium or antacids.
• Have kidney stones.
• Have a high blood-calcium level.
• Have sarcoidosis.

Consult your doctor if you have:
• Kidney disease.
• Chronic constipation, colitis, diarrhea.
• Stomach or intestinal bleeding.
• Irregular heartbeat.

Over age 55:
• Adverse reactions and side effects are more likely.
• Diarrhea or constipation are particularly likely.

Pregnancy:
• May need extra calcium. Consult doctor about supplements.
• Don't take megadoses.

Breast-feeding:
• Drug passes into milk. Consult doctor about need for supplements.
• Don't take megadoses.

Effect on lab tests:
• Serum-amylase and serum 11-hydroxycorticosteroid concentrations can be increased.
• Decreases serum-phosphate concentration with excessive, prolonged use.

Storage:
• Store in cool, dry area away from direct light, but don't freeze.
• Store safely out of reach of children.
• Don't store in bathroom medicine cabinet. Heat and moisture may change action of mineral.

Others:
• Dolomite or bone meal are probably *unsafe* sources of calcium because they contain lead.
• Avoid taking calcium within 1 or 2 hours of meals or ingestion of other medicines, if possible.

Overdose/Toxicity

Signs and symptoms:
Confusion, high blood pressure, increased sensitivity of eyes and skin to light, increased thirst, slow or irregular heartbeat, depression, bone or muscle pain, nausea, vomiting, skin itching, skin rash, increased urination.

What to do:
For symptoms of overdosage: Discontinue mineral, and consult doctor immediately. Also see *Adverse Reactions or Side Effects* section below.
For accidental overdosage (such as child taking entire bottle): Dial 911 (emergency), 0 for operator or your nearest Poison Control Center.

Adverse Reactions or Side Effects

Reaction or effect	What to do
Early signs of too much calcium in blood:	
Appetite loss	Discontinue. Call doctor when convenient.
Constipation	Discontinue. Call doctor when convenient.
Drowsiness	Discontinue. Call doctor immediately.
Dry mouth	Discontinue. Call doctor when convenient.
Headache	Discontinue. Call doctor when convenient.
Metallic taste	Discontinue. Call doctor when convenient.
Tiredness or weakness	Discontinue. Call doctor immediately.
Late signs of too much calcium in blood:	
Confusion	Discontinue. Call doctor immediately.
Depression	Discontinue. Call doctor when convenient.
High blood pressure	Discontinue. Call doctor immediately.
Increased thirst	Discontinue. Call doctor when convenient.
Increased urination	Discontinue. Call doctor when convenient.
Muscle or bone pain	Discontinue. Call doctor immediately.
Nausea	Discontinue. Call doctor immediately.
Skin rash	Discontinue. Call doctor immediately.
Slow or irregular heartbeat	Seek emergency treatment.
Vomiting	Discontinue. Call doctor immediately.

Interaction with Medicine, Vitamins or Minerals

Interacts with	Combined effect
Digitalis preparations	Heartbeat irregularities.
Iron supplements	Decreases absorption of iron unless vitamin C is taken at same time.
Magnesium-containing medications or supplements	Increases blood level of both.
Oral contraceptives and estrogens	May increase calcium absorption.
Potassium supplements	Increases chance of heartbeat irregularities.
Tetracyclines (oral)	Decreases absorption of tetracycline.
Vitamin A (megadoses)	Stimulates bone loss.
Vitamin D (megadoses)	Excessively increases absorption of calcium supplements.

Interaction with Other Substances

Tobacco decreases absorption.

Alcohol decreases absorption.

Beverages
• Tea decreases absorption.
• Coffee decreases absorption.
• Don't take calcium with milk or other dairy products so your body can absorb the most calcium from food *and* calcium supplement.

Foods
• Avoid eating spinach, rhubarb, bran, whole-grain cereals, fresh fruits or fresh vegetables at same time you take calcium. They may prevent efficient absorption.

MINERAL

Charcoal, Activated

Basic Information

Brand names, see page 477.
Available from natural sources? No
Available from synthetic sources? Yes
Prescription required? No

Natural Sources

Not available from natural sources

Reasons to Use

- Helps prevent poison from being absorbed from stomach and intestines.
- Treats poisonings from medication.
- Helps absorb gas in intestinal tract.

Unproved Speculated Benefits

- Removes bacteria and contaminants from food.
- Acts as a hangover cure.
- Treats infections.
- Treats hiccups.
- Treats diarrhea.

Who Needs Additional Amounts?

- People who have ingested poisonous substances.

Deficiency Symptoms

- None

Unproved Speculated Symptoms

- None

Lab Tests to Detect Deficiency

- None available, except for experimental purposes.

Dosage and Usage Information

Recommended Dietary Allowance (RDA):
No RDA has been established.

What this mineral does:
- Charcoal can adsorb almost anything it contacts. An *adsorbent* is a substance that attaches things to its surface rather than absorbing them to itself. Activated charcoal is the *only* adsorbent presently recognized by an FDA panel of experts for the treatment of ingested poison.

Miscellaneous information:
- Activated charcoal is used for treatment of potential toxicity from ingesting poisonous substances.
- Charcoal used for medicinal purposes is treated with steam under high pressure and high temperature.
- Charcoal will *not* help remove from the intestinal tract the following potential poisons—cyanide, caustic alkalis, ethyl alcohol (as in whiskey and beer), methyl alcohol, iron supplements, mineral acids.

Available as:
- Powder: Dissolve powder in cold water or juice. Take with meals or 1 to 1-1/2 hours after meals unless otherwise directed by your doctor.
- Tablets: Swallow whole with full glass of liquid. Don't chew or crush. Take with meals or 1 to 1-1/2 hours after meals unless otherwise directed by your doctor.
- Capsules: Swallow whole with full glass of liquid. Don't chew or crush. Take with meals or 1 to 1-1/2 hours after meals unless otherwise directed by your doctor.

Warnings and Precautions

Don't take if:
- Poison ingested is lye (or other strong alkali), strong acid (such as sulfuric acid), cyanide, iron, ethyl alcohol or methyl alcohol. Charcoal will *not* prevent these poisons from causing ill effects.

Consult your doctor if you have:
- Taken charcoal as an antidote for poison.

Over age 55:
- No problems expected.

Pregnancy:
- No problems expected.
- Don't take megadoses.

➠➤

Breast-feeding:
• No problems expected.
• Don't take megadoses.

Effect on lab tests:
• None expected.

Storage:
• Store in cool, dry place away from direct light, but don't freeze.
• Store safely out of reach of children.
• Don't store in bathroom medicine cabinet. Heat and moisture may change action of mineral.

Others:
• If used with syrup of ipecac, wait until vomiting has stopped before giving charcoal.

Overdose/Toxicity

Signs and symptoms:
None expected. Charcoal is not absorbed from intestines into bloodstream.

What to do:
Overdose unlikely to threaten life. If person takes much larger amount than prescribed, call doctor, Poison-Control Center or hospital emergency room for instructions

Adverse Reactions or Side Effects

Reaction or effect	What to do
Black bowel movements	Normal, expected side effect. No action necessary.

Interaction with Medicine, Vitamins or Minerals

Interacts with	Combined effect
Acetylcysteine, oral (an antidote to acetaminophen overdose)	Nullifies effectiveness of acetylcysteine.
Any medication taken at same time	May decrease absorption of medicine.

Interaction with Other Substances

Foods
• **Ice cream** decreases charcoal effect.
• **Sherbet** decreases charcoal effect.

MINERAL

Chloride

Basic Information
Available from natural sources? Yes
Available from synthetic sources? Yes
Prescription required? No

Natural Sources

Salt substitutes (potassium chloride)
Sea salt
Table salt (sodium chloride)
Found in combination with other molecules

Reasons to Use

• Regulates body's electrolyte balance.
• Regulates body's acid-base balance.

Unproved Speculated Benefits

• None known

Who Needs Additional Amounts?

• Anyone with inadequate caloric or nutritional dietary intake or increased nutritional requirements.
• Older people (over 55 years).
• Those who abuse alcohol or other drugs.
• People with a chronic wasting illness, excess stress for long periods or who have recently undergone surgery.
• Athletes and workers who participate in vigorous physical activities.
• Those with a portion of the gastrointestinal tract surgically removed.
• People with recent severe burns or injuries.

Deficiency Symptoms

• Continuous vomiting
• When chloride is intentionally neglected in infant-formula preparations, infant develops metabolic alkalosis, hypovolemia and significant urinary loss. Psychomotor defects, memory loss and growth retardation also occur.
• Upsets balance of acids and bases in body fluids (rare)

• Nausea
• Vomiting
• Confusion
• Weakness
• Coma

Unproved Speculated Symptoms

• None

Lab Tests to Detect Deficiency

• Serum chloride

Dosage and Usage Information

Recommended Dietary Allowance (RDA):
Estimate of adequate daily intake by the Food and Nutrition Board of the National Research Council, 1980 (See Glossary).

Age	RDA
0–6 months	0.275–0.7g
6–12 months	0.4–1.2g
1–3 years	0.5–1.5g
4–6 years	0.7–2.1g
7–10 years	0.925–2.775g
11–17 years	1.4–4.2g
18+ years	1.75–5.1g

What this mineral does:
• Chloride is a constituent of acid in the stomach (hydrochloric acid).
• Interacts with sodium, potassium and carbon dioxide to maintain acid-base balance in body cells and fluids. It is crucial to normal health.
• Concentrations of sodium, potassium, carbon dioxide and chlorine are controlled by mechanisms inside each body cell.

Miscellaneous information:
• Healthy people do not have to make any special efforts to maintain sufficient chloride.
• Eating a balanced diet supplies all daily needs.
• Extremely ill patients, with acid-base imbalance, require hospitalization, frequent laboratory studies and skillful professional care.

Available as:
• Sodium-chloride (salt) tablets. These may cause stomach distress and overload on kidneys.
• A constituent of many multivitamin/mineral preparations.

»»

Warnings and Precautions

Don't take if you:
• No known contraindications.

Consult your doctor if you have:
• No known contraindications.

Over age 55:
• No special problems expected.

Pregnancy:
• No special problems expected.
• Don't take megadoses.

Breast-feeding:
• No special problems expected.
• Don't take megadoses.

Effect on lab tests:
• No special problems expected.

Storage:
• Store in cool, dry place away from direct light, but don't freeze.
• Store safely out of reach of children.
• Don't store in bathroom medicine cabinet. Heat and moisture may change action of mineral.

Overdose/Toxicity

Signs and symptoms:
Upset balance of acids and bases in body fluids can occur with "too-much-chloride" or with "too-little-chloride." Symptoms of either include weakness, confusion, coma.

What to do:
For symptoms of overdosage: Discontinue mineral, and consult doctor.
For accidental overdosage (such as child taking a large amount): Dial 911 (emergency), 0 for operator or your nearest Poison Control Center.

Adverse Reactions or Side Effects

None expected

Interaction with Medicine, Vitamins or Minerals

Interacts with	Combined effect
Chlorine	Maintains normal acid-base balance in body.
Potassium	Maintains normal acid-base balance in body.
Sodium	Maintains normal acid-base balance in body.

Interaction with Other Substances

None known

MINERAL

Chromium

Basic Information
Available from natural sources? Yes
Available from synthetic sources? No
Prescription required? No

Natural Sources

Beef	Fish and seafood
Brewer's yeast	Fresh fruit
Calves' liver	Oysters
Chicken	Potatoes, with skin
Dairy products	Whole-grain products
Eggs	

Reasons to Use

• Promotes glucose metabolism.
• Helps insulin regulate blood sugar.
• Decreases insulin requirements and improves glucose tolerance of some people with maturity-onset diabetes.

Unproved Speculated Benefits

• Relieves atherosclerosis and diabetes.
• Facilitates binding of insulin to cell membrane.

Who Needs Additional Amounts?

• Anyone with inadequate caloric or dietary intake or increased nutritional requirements.
• Those who abuse alcohol or other drugs.
• People with a chronic wasting illness or who have recently undergone surgery.
• Those with a portion of the gastrointestinal tract surgically removed.
• People with recent severe burns or injuries.

Deficiency Symptoms

• Reduced tissue sensitivity to glucose, similar to diabetes
• Disturbances of glucose, fat and protein metabolism
• Symptoms exhibited by people with maturity-onset diabetes, such as overweight, fatigue, excess thirst, increased appetite, frequent urination, decreased resistance to infection, urinary-tract infections and yeast infections of the skin, mouth and vagina

Unproved Speculated Symptoms

• None

Lab Tests to Detect Deficiency

• Serum chromium
• Hair analysis is *not* a reliable test for deficiency or toxicity

Dosage and Usage Information

Recommended Dietary Allowance (RDA):
No RDA has been established. Estimated safe range of intake per day given below.

Age	Estimated Safe Intake
0–6 months	0.01–0.04mg
6–12 months	0.02–0.06mg
1–3 years	0.02–0.08mg
4–6 years	0.03–0.12mg
7+ years	0.05–0.20mg

What this mineral does:
• Aids transport of amino acids to liver and heart cells.
• Enhances effect of insulin in glucose utilization.

Miscellaneous information:
• Chromium toxicity can result from industrial overexposure, such as tanning, electroplating, steel making, abrasives manufacturing, cement manufacturing, diesel-locomotive repairs, furniture polishing, fur processing, glass making, jewelry making, metal cleaning, oil drilling, photography, textile dyeing, wood-preservative manufacturing.
• Nutritional science has yet to determine exact amounts of chromium in most foods. Less than 1% of dietary chromium is absorbed.

Available as:
• A constituent of many multivitamin/mineral preparations.

》▶

Warnings and Precautions

Don't take if you:
• Work in an environment that has high concentrations of chromium.

Consult your doctor if you have:
• Diabetes.
• Lung disease.
• Liver disease.
• Kidney disease.

Over age 55:
• No special needs if you eat a balanced diet.

Pregnancy:
• Avoid during pregnancy.

Breast-feeding:
• Avoid during breast-feeding.

Effect on lab tests:
• Diagnostic tests, such as red-blood-cell-survival studies, performed after radioactive hexavalent chromium is used for 3 months may cause falsely elevated levels in blood.

Storage:
• Store in cool, dry place away from direct light, but don't freeze.
• Store safely out of reach of children.
• Don't store in bathroom medicine cabinet. Heat and moisture may change action of mineral.

Overdose/Toxicity

Signs and symptoms:
Dietary form has very low toxicity. Long-term exposure to chromium may lead to skin problems, perforation of nasal septum, lung cancer, liver, impairment, kidney impairment.

What to do:
For symptoms of overdosage: Discontinue mineral, and consult doctor.
For accidental overdosage (such as child taking entire bottle): Dial 911 (emergency), 0 for operator or your nearest Poison Control Center.

Adverse Reactions or Side Effects

None expected

Interaction with Medicine, Vitamins or Minerals

Interacts with	Combined effect
Insulin	May decrease amount of insulin needed to treat diabetes.

Interaction with Other Substances

Sugar is partially destroyed by chromium.

MINERAL

Cobalt

Basic Information

Available from natural sources? Yes
Available from synthetic sources? Yes
Prescription required? No

Natural Sources

Beet greens	Lettuce
Buckwheat	Liver
Cabbage	Milk
Clams	Oysters
Figs	Spinach
Kidney	Watercress

Note: Small amounts in diet satisfy requirements, except under unusual circumstances.

Reasons to Use

- Promotes normal red-blood-cell formation.
- Acts as substitute for manganese in activation of several enzymes.
- Replaces zinc in some enzymes.

Unproved Speculated Benefits

- Treats anemia that does not respond to other treatment.
- Prevents and treats pernicious anemia.

Who Needs Additional Amounts?

Supplements are difficult to locate, so adequate food sources become more important.
- People with recent severe burns or injuries.
- Those with anorexia nervosa or bulimia.
- Vegetarians.

Deficiency Symptoms

- Pernicious anemia, with the following symptoms:
  Weakness, especially in arms and legs
  Sore tongue
  Nausea, appetite loss, weight loss
  Bleeding gums
  Numbness and tingling in hands and feet
  Difficulty maintaining balance
  Pale lips, pale tongue, pale gums
  Yellow eyes and skin
  Shortness of breath
  Depression
  Confusion and dementia
  Headache
  Poor memory

Unproved Speculated Symptoms

- None

Lab Tests to Detect Deficiency

- Concentration in human plasma
- Measured in bioassay as part of vitamin B-12

Dosage and Usage Information

Recommended Dietary Allowance (RDA):
No RDA has been established.

What this mineral does:
- Acts as a catalyst in complex reactions to form vitamin B-12.

Miscellaneous information:
- This is a trace element stored mainly in the liver.
- Deficiency is extremely rare.
- Cobalt is a necessary ingredient to manufacture vitamin B-12 in the body. A deficiency of cobalt may lead to a deficiency of vitamin B-12 and therefore to pernicious anemia.

Available as:
- Capsules: Swallow whole with full glass of liquid. Don't chew or crush. Take with meals or 1 to 1-1/2 hours after meals unless otherwise directed by your doctor.
- A constituent of many multivitamin/mineral preparations.

Warnings and Precautions

Don't take if you:
- Are healthy and eat a nutritious balanced diet.

Consult your doctor if you:
- No problems expected.

Over age 55:
- Eat a balanced diet to prevent deficiency.

Pregnancy:
- No problems expected, except with megadoses.
- Don't take megadoses.

Breast-feeding:
- No problems expected, except with megadoses.
- Don't take megadoses.

Effect on lab tests:
- None expected.

Storage:
- Store in cool, dry place away from direct light, but don't freeze.
- Store safely out of reach of children.
- Don't store in bathroom medicine cabinet. Heat and moisture may change action of mineral.

 ## Overdose/Toxicity

Signs and symptoms:
- In megadoses, 20-30mg per day, cobalt can produce polycythemia, enlargement of thyroid gland and enlargement of the heart leading to congestive heart failure (See Glossary).
- Cobalt toxicity can cause thyroid overgrowth in infants.

What to do:
For symptoms of overdosage: Discontinue mineral, and consult doctor. Also see *Adverse Reactions or Side Effects* section below.

For accidental overdosage (such as child taking entire bottle): Dial 911 (emergency), 0 for operator or your nearest Poison Control Center.

 ## Adverse Reactions or Side Effects

Reaction or effect	What to do
With megadoses:	
Polycythemia	Discontinue. Call doctor immediately.
Enlargement of thyroid gland	Discontinue. Call doctor immediately.
Enlargement of heart	Discontinue. Call doctor immediately.

 ## Interaction with Medicine, Vitamins or Minerals

Interacts with	Combined effect
Colchicine	May cause inaccurate laboratory studies of cobalt or vitamin B-12.
Neomycin	May cause inaccurate laboratory studies of cobalt or vitamin B-12.
Para-aminosalicylic acid	May cause inaccurate laboratory studies of cobalt or vitamin B-12.
Phenytoin	May cause inaccurate laboratory studies of cobalt or vitamin B-12.

 ## Interaction with Other Substances

Some beer contains cobalt as a stabilizer. People who consume large quantities of **cobalt-stabilized beer** over long periods may develop cobalt toxicity leading to cardiomyopathy and congestive heart failure.

MINERAL

Copper

Basic Information

Available from natural sources? Yes
Available from synthetic sources? No
Prescription required? No

Natural Sources

Barley	Mushrooms
Brazil nuts	Mussels
Cashew nuts	Oats
Hazelnuts (filberts)	Oysters
Honey	Peanuts
Lentils	Salmon
Molasses, black-strap	Walnuts
	Wheat germ

Reasons to Use

- Promotes normal red-blood-cell formation.
- Acts as a catalyst in storage and release of iron to form hemoglobin for red blood cells.
- Assists in production of several enzymes involved in respiration.
- Promotes connective-tissue formation and central-nervous-system function.
- Is used as a nutritional supplement for anyone receiving prolonged feedings through veins or tubes into the stomach.

Unproved Speculated Benefits

- Stimulates hair growth in bald men.
- Treats anemia.
- Protects against cancer.
- Protects against cardiovascular disease.
- Reduces inflammation.
- Helps arthritis.

Who Needs Additional Amounts?

- Anyone with inadequate caloric or dietary intake or increased nutritional requirements.
- Older people (over 55 years).
- Pregnant or breast-feeding women.
- Those who abuse alcohol or other drugs.
- People with a chronic wasting illness, particularly those with chronic diarrhea, malabsorption disorders, kidney disease.
- Anyone who experiences excess stress for long periods or who has recently undergone surgery.

- Those with a portion of the gastrointestinal tract surgically removed.
- People with recent severe burns or injuries.
- Malnourished children whose diet consists of milk without supplements.
- People who receive intravenous nourishment for long periods of time.

Deficiency Symptoms

- Anemia
- Low white-blood-cell count associated with reduced resistance to infection
- Faulty collagen formation
- Bone demineralization

Unproved Speculated Symptoms

- Arthritis
- Cancer
- Heart disease
- Baldness
- Anemia

Lab Tests to Detect Deficiency

- Plasma copper levels
- Urine copper levels in 24-hour collection

Dosage and Usage Information

Recommended Dietary Allowance (RDA):
No RDA has been established. Estimated safe intake given below.

Age	Estimated Safe Intake
0–6 months	0.5–0.7mg/day
6–12 months	0.7–1.0mg/day
1–3 years	1.0–1.5mg/day
4–6 years	1.5–2.0mg/day
7–10 years	2.0–2.5mg/day
11+ years	2.0–3.0mg/day

What this mineral does:
- Copper is an essential component of a number of proteins and enzymes, including lysyl, hydroxylase, dopamine beta-hydroxylase.

Miscellaneous information:
- Plasma-copper levels may *increase* in people with rheumatoid arthritis, pregnancy, cirrhosis of the liver, myocardial infarction (heart attack), schizophrenia, tumors, severe infections. »→

- Processed foods may reduce normal copper absorption.
- Plasma-copper levels *decrease* with hypothyroidism, dysproteinuria of infancy, kwashiorkor, sprue, nephrosis.
- Hair analysis may be used as a measure of copper nutrition. (An unreliable test.)
- Most nutritionists recommend a balanced diet rather than extra supplementation that could upset the body's delicate mineral balance.

Available as:
- Tablets: Swallow whole with full glass of liquid. Don't chew or crush. Take with meals or 1 to 1-1/2 hours after meals unless otherwise directed by your doctor.
- A constituent of many multivitamin/mineral preparations.

Warnings and Precautions

Don't take if you:
- Have hepatolenticular degeneration (Wilson's disease).

Consult your doctor if you:
- Are considering taking a copper supplement.

Over age 55:
- No special considerations.

Pregnancy:
- Increased plasma copper levels are noted during pregnancy. Significance of this to human health is unknown at present.
- Don't take megadoses.

Breast-feeding:
- No information available at present.
- Don't take megadoses.

Effect on lab tests:
- Cobalt, iron, nickel and oral contraceptives with estrogens can cause false-positive or elevated copper values.

Storage:
- Store in cool, dry place away from direct light, but don't freeze.
- Store safely out of reach of children.
- Don't store in bathroom medicine cabinet. Heat and moisture may change action of mineral.

Overdose/Toxicity

Signs and symptoms:
Nausea, vomiting, muscle aches, abdominal pain, anemia.

What to do:
For symptoms of overdosage: Discontinue mineral, and consult doctor.
For accidental overdosage (such as child taking entire bottle): Dial 911 (emergency), 0 for operator or your nearest Poison Control Center.

Adverse Reactions or Side Effects

None expected

Interaction with Medicine, Vitamins or Minerals

Interacts with	Combined effect
Cadmium	Can interfere with copper absorption and utilization.
Fiber	Can interfere with copper absorption and utilization.
Molybdenum	Maintains appropriate ratio of copper to molybdenum in body. If you have excessive amounts of copper, your molybdenum level drops. If you have excessive amounts of molybdenum, your copper level drops.
Oral contraceptives	Increases copper level. Significance unknown at present.
Phytates (cereals, vegetables)	Can interfere with copper absorption and utilization.
Vitamin C	Decreases absorption of copper. Large doses of vitamin C must be taken to produce this effect.
Zinc	Can interfere with copper absorption and utilization.

Interaction with Other Substances

None known

MINERAL

Dihydroxyaluminum Aminoacetate/ Magnesia/Alumina

Basic Information

Provides additional magnesium. Other substances in title are not supplied in large enough amounts for use by your body. Brand names, see page 477.
Available from natural sources? Yes
Available from synthetic sources? No
Prescription required? Yes, for some forms

Natural Sources

Almonds	Leafy, green	Snails
Bluefish	vegetables	Soybeans
Carp	Mackerel	Sunflower
Cod	Molasses	seeds
Flounder	Nuts	Swordfish
Halibut	Ocean perch	Wheat germ
Herring	Shrimp	

Reasons to Use

- Aids bone growth.
- Aids function of nerves and muscles, including regulation of normal heart rhythm.
- Keeps metabolism steady.
- Conducts nerve impulses.
- Works as laxative in large doses.
- Acts as antacid in small doses.
- Strengthens tooth enamel.

Unproved Speculated Benefits

- Cures alcoholism.
- Cures kidney stones.
- Alleviates heart disease.
- Helps control bad breath and body odor.
- Helps contain lead poisoning.

Who Needs Additional Amounts?

- Anyone with inadequate caloric or dietary intake or increased nutritional requirements.
- Those who abuse alcohol or other drugs.
- People with a chronic wasting illness or who have recently undergone surgery.
- Those with recent severe burns or injuries.

Deficiency Symptoms

Following symptoms occur rarely:
- Muscle contractions
- Convulsions
- Confusion and delirium
- Irritability
- Nervousness
- Skin problems
- Hardening of soft tissues

Unproved Speculated Symptoms

- Sudden heart failure

Lab Tests to Detect Deficiency

- Serum magnesium

Dosage and Usage Information

Recommended Dietary Allowance (RDA):
Estimate of adequate daily intake by the Food and Nutrition Board of the National Research Council, 1980 (See Glossary).

Age	RDA
0–6 months	50mg
6–12 months	70mg
1–3 years	150mg
4–6 years	200mg
7–10 years	250mg
Males	
11–14 years	350mg
15–18 years	400mg
18+ years	350mg
Females	
11+ years	300mg
Pregnant	+150mg
Lactating	+150mg

What this mineral does:
- Activates essential enzymes.
- Affects metabolism of proteins and nucleic acids.
- Helps transport sodium and potassium across cell membranes.
- Influences calcium levels inside cells.

Available as:
- Oral suspension: Dilute in at least 1/2 glass of water or other liquid. Take with meals or 1 to 1-1/2 hours after meals unless otherwise directed by your doctor.

➤➤➤

Dihydroxyaluminum Aminoacetate/ Magnesia/Alumina

Warnings and Precautions

Don't take if you have:
- Kidney failure.
- Heart block (unless you have a pacemaker).
- An ileostomy.

Consult your doctor if you have:
- Chronic constipation, colitis, diarrhea.
- Symptoms of appendicitis.
- Stomach or intestinal bleeding.

Over age 55:
- Adverse reactions/side effects more likely.

Pregnancy:
- Risk to fetus outweighs benefits. Don't use.

Breast-feeding:
- Avoid magnesium in large quantities.
- If you must take temporarily, discontinue breast-feeding. Consult doctor for advice on maintaining milk supply.
- Don't take megadoses.

Effect on lab tests:
- Makes test for stomach-acid secretion inaccurate.
- May increase or decrease serum-phosphate concentrations.
- May decrease serum and urine pH.

Storage:
- Store in cool, dry place away from direct light, but don't freeze.
- Store safely out of reach of children.
- Don't store in bathroom medicine cabinet. Heat and moisture may change action of mineral.

Others:
- Chronic kidney disease frequently causes body to retain excess magnesium.
- Adverse reactions, side effects and interactions with medicines, vitamins or minerals occur *only rarely* when too much magnesium is taken for too long or if you have kidney disease.

Overdose/Toxicity

Signs and symptoms:
Severe nausea and vomiting, extremely low blood pressure, extreme muscle weakness, difficulty breathing, heartbeat irregularity.

What to do:
For symptoms of overdosage: Discontinue mineral, and consult doctor immediately. Also see *Adverse Reactions or Side Effects* section below.

For accidental overdosage (such as child taking entire bottle): Dial 911 (emergency), 0 for operator or your nearest Poison Control Center.

Possible Adverse Reactions or Side Effects

Reaction or effect	What to do
Abdominal pain	Seek emergency treatment.
Abnormal bleeding	Seek emergency treatment.
Gastric ulceration (burning pain in upper chest relieved by food or antacid)	Discontinue. Call doctor immediately.
Mild diarrhea	Discontinue. Call doctor when convenient.
Nausea, vomiting	Discontinue. Call doctor immediately.

Interactions with Medicine, Vitamins or Minerals

Interacts With	Combined Effect
Calcium	Interferes with zinc absorption.
Copper	Decreases absorption of copper. Large doses of zinc must be taken to produce this effect.
Cortisone drugs	May interfere with lab tests measuring zinc.
Diuretics	Increases zinc excretion. Requires taking greater amounts.
Iron	Decreases absorption of iron. Large doses of zinc must be taken to produce this effect.
Oral contraceptives	Lowers zinc blood levels.
Tetracycline	Decreases amount of tetracycline absorbed into bloodstream.
Vitamin A	Assists in absorption of vitamin A.

Interaction with Other Substances

Alcohol can increase zinc excretion in urine and impair body's ability to combine zinc into its correct enzyme combinations in liver.

Coffee and zinc should not be consumed together because they may decrease zinc absorption.

Ferrous Fumarate

Basic Information

Ferrous fumarate is 33% elemental iron.
Brand names, see page 477.
Available from natural sources? Yes
Available from synthetic sources? Yes
Prescription required? Yes

Natural Sources

Bread, enriched	Molasses, black-strap
Cashews	Mussels
Caviar	Pistachios
Cheddar cheese	Pumpkin seeds
Egg yolk	Seaweed
Garbanzo beans	Walnuts
(chickpeas)	Wheat germ
Lentils	Whole-grain products

Note: Even iron-rich foods are poorly absorbed by humans. Only about 10% of food iron is absorbed from food consumed by an individual with normal iron stores. However, an iron-deficient person may absorb 20 to 30%.

Reasons to Use

- Prevents and treats iron-deficiency anemia due to dietary iron deficiency or other causes.
- Stimulates bone-marrow production of hemoglobin, the red-blood-cell pigment that carries oxygen to body cells.
- Forms part of several enzymes and proteins in the body.

Unproved Speculated Benefits

- Controls alcoholism.
- Helps alleviate menstrual discomfort.
- Stimulates immunity.
- Boosts physical performance.
- Prevents learning disorders in children.

Who Needs Additional Amounts?

- Many women of child-bearing age are mildly iron-deficient even when they get all their nutritional requirements.
- Anyone with inadequate caloric or dietary intake or increased nutritional requirements.
- Older people (over 55 years).
- Pregnant or breast-feeding women.
- Women with heavy menstrual flow, long menstrual periods or short menstrual cycles.
- Those who abuse alcohol or other drugs.
- People with a chronic wasting illness, excess stress for long periods or who have recently undergone surgery.
- Athletes and workers who participate in vigorous physical activities.
- Those with a portion of the gastrointestinal tract surgically removed.
- People with recent severe burns or injuries.
- Anyone who has lost blood recently, such as from heavy menstrual periods or from an accident.
- Vegetarians.
- Infants from 2 to 24 months.

Deficiency Symptoms

- Listlessness
- Heart palpitations upon exertion
- Fatigue
- Irritability
- Pale appearance to skin
- Cracking of lips and tongue
- Difficulty swallowing
- General feeling of poor health

Unproved Speculated Symptoms

- None

Lab Tests to Detect Deficiency

- Red-blood-cell count
- Microscopic exam of red blood cells
- Serum iron
- Hemoglobin determinations

Dosage and Usage Information

Recommended Dietary Allowance (RDA):
Estimate of adequate daily intake by the Food and Nutrition Board of the National Research Council, 1980 (See Glossary). ⧕▶

Ferrous Fumarate

Age	RDA
0–6 months	10mg
6–12 months	15mg
1–3 years	15mg
4–6 years	10mg
7–10 years	10mg
Males	
11–18 years	18mg
19+ years	10mg
Females	
11–50 years	18mg
51+ years	10mg
Pregnant	+30–60mg
Lactating	+30–60mg

What this mineral does:
- Iron is an essential component of hemoglobin, myoglobin and a co-factor of several essential enzymes. Of the total iron in the body, 60 to 70% is stored in hemoglobin (the red part of red blood cells).
- Hemoglobin is also a component of myoglobin, an iron-protein complex in muscles. This complex helps muscles get extra energy when they work hard.

Miscellaneous information:
- Iron-deficiency anemia in older men is usually considered to be due to slow loss of blood from a malignancy in the gastrointestinal tract until proved otherwise.
- Iron content of foods, especially acidic foods, can be dramatically increased by preparation in iron cookware.
- May require 3 weeks of treatment before you receive maximum benefit.
- Works best with vitamin C (ascorbic acid).

Available as:
- Tablets and capsules: Swallow whole with full glass of liquid. Don't chew or crush. Take with or immediately after food to decrease stomach irritation.
- Oral solution: Dilute in at least 1/2 glass water or other liquid. Take with meals or 1 to 1-1/2 hours after meals unless otherwise directed by your doctor.
- Chewable tablets: Chew well before swallowing.
- Enteric-coated tablets: Swallow whole with full glass of liquid. Take with meals or 1 to 1-1/2 hours after meals unless otherwise directed by your doctor.

Warnings and Precautions

Don't take if you:
- Are allergic to any iron supplement.
- Have acute hepatitis.
- Have hemosiderosis or hemochromatosis (conditions involving excess iron in body).
- Have hemolytic anemia.
- Have had repeated blood transfusions.

Consult your doctor if you have:
- Plans to become pregnant while taking medication.
- Had stomach surgery.
- Had peptic-ulcer disease, enteritis, colitis.
- Had pancreatitis or hepatitis.
- Alcoholism.
- Kidney disease.
- Rheumatoid arthritis.
- Intestinal disease.

Over age 55:
- Deficiency more likely. Check frequently with doctor for anemia symptoms or slow blood loss in stool.

Pregnancy:
- Pregnancy increases need. Check with doctor. During first 3 months of pregnancy, take *only* if doctor prescribes it.
- Don't take megadoses.

Breast-feeding:
- Supplements probably not needed if you are healthy and eat a balanced diet.
- Baby may need supplementation. Ask your doctor.
- Don't take megadoses.

Effect on lab tests:
- May cause abnormal results in serum bilirubin, serum calcium, serum iron, special radioactive studies of bones using technetium (Tc 99m-labeled agents), stool studies for blood.

Storage:
- Store in cool, dry place away from direct light, but don't freeze.
- Store safely out of reach of children. Iron tablets look like candy, and children love them.
- Don't store in bathroom medicine cabinet. Heat and moisture may change action of mineral.

Others:
- Iron can accumulate to harmful levels (hemosiderosis) in patients with chronic kidney failure, Hodgkins disease, rheumatoid arthritis.
- Prolonged use in high doses can cause hemochromatosis (iron-storage disease), leading to bronze skin, diabetes, liver damage, impotence, heart problems.

⟫➤

MINERAL

Ferrous Fumarate, Continued

 Overdose/Toxicity

Signs and symptoms:
Early signs: Diarrhea with blood, severe nausea, abdominal pain, vomiting with blood.
Late signs: Weakness, collapse, pallor, blue lips, blue hands, blue fingernails, shallow breathing, convulsions, coma, weak, rapid heartbeat.

What to do:
For symptoms of overdosage: Discontinue mineral, and consult doctor. Also see *Adverse Reactions or Side Effects* section below.
For accidental overdosage (such as child taking entire bottle): Dial 911 (emergency), 0 for operator or your nearest Poison Control Center.

 Adverse Reactions or Side Effects

Reaction or effect	What to do
Abdominal pain	Discontinue. Call doctor immediately.
Black or gray stools (always)	Nothing.
Blood in stools	Seek emergency treatment.
Chest pain	Seek emergency treatment.
Drowsiness	Discontinue. Call doctor when convenient.
Stained teeth (liquid forms)	Mix with water or juice to lessen effect. Brush teeth with baking soda or hydrogen peroxide to help remove stain.
Throat pain	Discontinue. Call doctor immediately.

 Interaction with Medicine, Vitamins or Minerals

Interacts with	Combined effect
Allopurinol	May cause excess iron storage in liver.
Antacids	Causes poor iron absorption.
Calcium	Combination necessary for efficient calcium absorption.
Cholestyramine	Decreases iron effect.
Copper	Assists in copper absorption.
Iron supplements (other)	May cause excess iron storage in liver.
Pancreatin	Decreases iron absorption.
Penicillamine	Decreases penicillamine effect.
Sulfasalazine	Decreases iron effect.
Tetracyclines	Decreases tetracycline effect. Take iron 3 hours before or 2 hours after taking tetracycline.
Vitamin C	Increases iron effect. Necessary for red-blood-cell and hemoglobin formation.
Vitamin E	Decreases iron absorption.
Zinc (large doses)	Decreases iron absorption.

 Interaction with Other Substances

Alcohol increases iron utilization. May cause organ damage. Avoid or use in moderation.

Beverages
• Milk decreases iron absorption.
• Tea decreases iron absorption.
• Coffee decreases iron absorption.

Basic Information

Ferrous gluconate is 11.6% elemental iron.
Brand names, see page 477.
Available from natural sources? Yes
Available from synthetic sources? Yes
Prescription required? Yes

Natural Sources

Bread, enriched	Molasses, black-strap
Cashews	Mussels
Caviar	Pistachios
Cheddar cheese	Pumpkin seeds
Egg yolk	Seaweed
Garbanzo beans	Walnuts
(chickpeas)	Wheat germ
Lentils	Whole-grain products

Note: Even iron-rich foods are poorly absorbed by humans. Only about 10% of food iron is absorbed from food consumed by an individual with normal iron stores. However, an iron-deficient person may absorb 20 to 30%.

Reasons to Use

- Prevents and treats iron-deficiency anemia due to dietary iron deficiency or other causes.
- Stimulates bone-marrow production of hemoglobin, the red-blood-cell pigment that carries oxygen to body cells.
- Forms part of several enzymes and proteins in the body.

Unproved Speculated Benefits

- Controls alcoholism.
- Helps alleviate menstrual discomfort.
- Stimulates immunity.
- Boosts physical performance.
- Prevents learning disorders in children.

Who Needs Additional Amounts?

- Many women of child-bearing age are mildly iron-deficient even when they get all their nutritional requirements.
- Anyone with inadequate caloric or dietary intake or increased nutritional requirements.
- Older people (over 55 years).
- Pregnant or breast-feeding women.
- Women with heavy menstrual flow, long menstrual periods or short menstrual cycles.
- Those who abuse alcohol or other drugs.
- People with a chronic wasting illness, excess stress for long periods or who have recently undergone surgery.
- Athletes and workers who participate in vigorous physical activities.
- Those with a portion of the gastrointestinal tract surgically removed.
- People with recent severe burns or injuries.
- Anyone who has lost blood recently, such as from heavy menstrual periods or from an accident.
- Vegetarians.
- Infants from 2 to 24 months.

Deficiency Symptoms

- Listlessness
- Heart palpitations upon exertion
- Fatigue
- Irritability
- Pale appearance to skin
- Cracking of lips and tongue
- Difficulty swallowing
- General feeling of poor health

Unproved Speculated Symptoms

- None

Lab Tests to Detect Deficiency

- Red-blood-cell count
- Microscopic exam of red blood cells
- Serum iron
- Hemoglobin determinations

⟫▶

Ferrous Gluconate, Continued

Dosage and Usage Information

Recommended Dietary Allowance (RDA):
Estimate of adequate daily intake by the Food and Nutrition Board of the National Research Council, 1980 (See Glossary).

Age	RDA
0–6 months	10mg
6–12 months	15mg
1–3 years	15mg
4–6 years	10mg
7–10 years	10mg
Males	
11–18 years	18mg
19+ years	10mg
Females	
11–50 years	18mg
51+ years	10mg
Pregnant	+30–60mg
Lactating	+30–60mg

What this mineral does:
• Iron is an essential component of hemoglobin, myoglobin and a co-factor of several essential enzymes. Of the total iron in the body, 60 to 70% is stored in hemoglobin (the red part of red blood cells).
• Hemoglobin is also a component of myoglobin, an iron-protein complex in muscles. This complex helps muscles get extra energy when they work hard.

Miscellaneous information:
• Iron-deficiency anemia in older men is usually considered to be due to slow loss of blood from a malignancy in the gastrointestinal tract until proved otherwise.
• Iron content of foods, especially acidic foods, can be dramatically increased by preparation in iron cookware.
• May require 3 weeks of treatment before you receive maximum benefit.
• Works best with vitamin C (ascorbic acid).

Available as:
• Capsules: Swallow whole with full glass of liquid. Don't chew or crush. Take with or immediately after food to decrease stomach irritation.
• Oral solution: Dilute in at least 1/2 glass water or other liquid. Take with meals or 1 to 1-1/2 hours after meals unless otherwise directed by your doctor.
• Tablets: Swallow whole with full glass of liquid. Don't chew or crush. Take with meals or 1 to 1-1/2 hours after meals unless otherwise directed by your doctor.

Warnings and Precautions

Don't take if you:
• Are allergic to any iron supplement.
• Have acute hepatitis.
• Have hemosiderosis or hemochromatosis (conditions involving excess iron in body).
• Have hemolytic anemia.
• Have had repeated blood transfusions.

Consult your doctor if you have:
• Plans to become pregnant while taking medication.
• Had stomach surgery.
• Had peptic-ulcer disease, enteritis, colitis.
• Had pancreatitis or hepatitis.
• Alcoholism.
• Kidney disease.
• Rheumatoid arthritis.
• Intestinal disease.

Over age 55:
• Deficiency more likely. Check frequently with doctor for anemia symptoms or slow blood loss in stool.

Pregnancy:
• Pregnancy increases need. Check with doctor. During first 3 months of pregnancy, take *only* if doctor prescribes it.
• Don't take megadoses.

Breast-feeding:
• Supplements probably not needed if you are healthy and eat a balanced diet.
• Baby may need supplementation. Ask your doctor.
• Don't take megadoses.

Effect on lab tests:
• May cause abnormal results in serum bilirubin, serum calcium, serum iron, special radioactive studies of bones using technetium (Tc 99m-labeled agents), stool studies for blood.

Storage:
• Store in cool, dry place away from direct light, but don't freeze.
• Store safely out of reach of children. Iron tablets look like candy, and children love them.
• Don't store in bathroom medicine cabinet. Heat and moisture may change action of mineral.

Others:
• Iron can accumulate to harmful levels (hemosiderosis) in patients with chronic kidney failure, Hodgkins disease, rheumatoid arthritis.
• Prolonged use in high doses can also cause hemochromatosis (iron-storage disease), leading to bronze skin, diabetes, liver damage, impotence, heart problems.

➤➤➤

Overdose/Toxicity

Signs and symptoms:
Early signs: Diarrhea with blood, severe nausea, abdominal pain, vomiting with blood.
Late signs: Weakness, collapse, pallor, blue lips, blue hands, blue fingernails, shallow breathing, convulsions, coma, weak, rapid heartbeat.

What to do:
For symptoms of overdosage: Discontinue mineral, and consult doctor. Also see *Adverse Reactions or Side Effects* section below.
For accidental overdosage (such as child taking entire bottle): Dial 911 (emergency), 0 for operator or your nearest Poison Control Center.

 ## Adverse Reactions or Side Effects

Reaction or effect	What to do
Abdominal pain	Discontinue. Call doctor immediately.
Black or gray stools (always)	Nothing.
Blood in stools	Seek emergency treatment.
Chest pain	Seek emergency treatment.
Drowsiness	Discontinue. Call doctor when convenient.
Stained teeth (liquid forms)	Mix with water or juice to lessen effect. Brush teeth with baking soda or hydrogen peroxide to help remove stain.
Throat pain	Discontinue. Call doctor immediately.

 ## Interaction with Medicine, Vitamins or Minerals

Interacts with	Combined effect
Allopurinol	May cause excess iron storage in liver.
Antacids	Causes poor iron absorption.
Calcium	Combination necessary for efficient calcium absorption.
Cholestyramine	Decreases iron effect.
Copper	Assists in copper absorption.
Iron supplements (other)	May cause excess iron storage in liver.
Pancreatin	Decreases iron absorption.
Penicillamine	Decreases penicillamine effect.
Sulfasalazine	Decreases iron effect.
Tetracyclines	Decreases tetracycline effect. Take iron 3 hours before or 2 hours after taking tetracycline.
Vitamin C	Increases iron effect. Necessary for red-blood-cell and hemoglobin formation.
Vitamin E	Decreases iron effect.
Zinc (large doses)	Decreases iron absorption.

 ## Interaction with Other Substances

Alcohol increases iron utilization. May cause organ damage. Avoid or use in moderation.

Beverages
• Milk decreases iron absorption.
• Tea decreases iron absorption.
• Coffee decreases iron absorption.

MINERAL

Ferrous Sulfate

Basic Information

Ferrous sulfate is 20% elemental iron.
Brand names, see page 477.
Available from natural sources? Yes
Available from synthetic sources? Yes
Prescription required? Yes

Natural Sources

Bread, enriched	Molasses, black-strap
Cashews	Mussels
Caviar	Pistachios
Cheddar cheese	Pumpkin seeds
Egg yolk	Seaweed
Garbanzo beans	Walnuts
(chickpeas)	Wheat germ
Lentils	Whole-grain products

Note: Even iron-rich foods are poorly absorbed by humans. Only about 10% of food iron is absorbed from food consumed by an individual with normal iron stores. However, an iron-deficient person may absorb 20-30%.

Reasons to Use

- Prevents and treats iron-deficiency anemia due to dietary iron deficiency or other causes.
- Stimulates bone-marrow production of hemoglobin, the red-blood-cell pigment that carries oxygen to body cells.
- Forms part of several enzymes and proteins in the body.

Unproved Speculated Benefits

- Controls alcoholism.
- Helps alleviate menstrual discomfort.
- Stimulates immunity.
- Boosts physical performance.
- Prevents learning disorders in children.

Who Needs Additional Amounts?

- Many women of child-bearing age are mildly iron-deficient even when they get all their nutritional requirements.
- Anyone with inadequate caloric or dietary intake or increased nutritional requirements.
- Older people (over 55 years).
- Pregnant or breast-feeding women.
- Women with heavy menstrual flow, long menstrual periods or short menstrual cycles.

- Those who abuse alcohol or other drugs.
- People with a chronic wasting illness, excess stress for long periods or who have recently undergone surgery.
- Athletes and workers who participate in vigorous physical activities.
- Those with a portion of the gastrointestinal tract surgically removed.
- People with recent severe burns or injuries.
- Anyone who has lost blood recently, such as from heavy menstrual periods or from an accident.
- Vegetarians.
- Infants from 2 to 24 months.

Deficiency Symptoms

- Listlessness
- Heart palpitations upon exertion
- Fatigue
- Irritability
- Pale appearance to skin
- Cracking of lips and tongue
- Difficulty swallowing
- General feeling of poor health

Unproved Speculated Symptoms

- None

Lab Tests to Detect Deficiency

- Red-blood-cell count
- Microscopic exam of red blood cells
- Serum iron
- Hemoglobin determinations

Dosage and Usage Information

Recommended Dietary Allowance (RDA):
Estimate of adequate daily intake by the Food and Nutrition Board of the National Research Council, 1980 (See Glossary). ⮞

Ferrous Sulfate

Age	RDA
0–6 months	10mg
6–12 months	15mg
1–3 years	15mg
4–6 years	10mg
7–10 years	10mg
Males	
11–18 years	18mg
19+ years	10mg
Females	
11–50 years	18mg
51+ years	10mg
Pregnant	+30–60mg
Lactating	+30–60mg

What this mineral does:
- Iron is an essential component of hemoglobin, myoglobin and a co-factor of several essential enzymes. Of the total iron in the body, 60 to 70% is stored in hemoglobin (the red part of red blood cells).
- Hemoglobin is also a component of myoglobin, an iron-protein complex in muscles. This complex helps muscles get extra energy when they work hard.

Miscellaneous information:
- Iron-deficiency anemia in older men is usually considered to be due to slow loss of blood from a malignancy in the gastrointestinal tract until proved otherwise.
- Iron content of foods, especially acidic foods, can be dramatically increased by preparation in iron cookware.
- May require 3 weeks of treatment before you receive maximum benefit.
- Works best with vitamin C (ascorbic acid).

Available as:
- Extended-release capsules or tablets: Swallow whole with full glass of liquid. Don't chew or crush. Take with or immediately after food to decrease stomach irritation.
- Oral solution: Dilute in at least 1/2 glass water or other liquid. Take with meals or 1 to 1-1/2 hours after meals unless otherwise directed by your doctor.
- Enteric-coated tablets: Swallow whole with full glass of liquid. Take with meals or 1 to 1-1/2 hours after meals unless otherwise directed by your doctor.
- Tablets: Swallow whole with full glass of liquid. Don't chew or crush. Take with meals or 1 to 1-1/2 hours after meals unless otherwise directed by your doctor.

 ## Warnings and Precautions

Don't take if you:
- Are allergic to any iron supplement.
- Have acute hepatitis.
- Have hemosiderosis or hemochromatosis (conditions involving excess iron in body).
- Have hemolytic anemia.
- Have had repeated blood transfusions.

Consult your doctor if you have:
- Plans to become pregnant while taking medication.
- Had stomach surgery.
- Had peptic-ulcer disease, enteritis, colitis.
- Had pancreatitis or hepatitis.
- Alcoholism.
- Kidney disease.
- Rheumatoid arthritis.
- Intestinal disease.

Over age 55:
- Deficiency more likely. Check frequently with doctor for anemia symptoms or slow blood loss in stool.

Pregnancy:
- Pregnancy increases need. Check with doctor. During first 3 months of pregnancy, take *only* if doctor prescribes it.
- Don't take megadoses.

Breast-feeding:
- Supplements probably not needed if you are healthy and eat a balanced diet.
- Baby may need supplementation. Ask your doctor.
- Don't take megadoses.

Effect on lab tests:
- May cause abnormal results in serum bilirubin, serum calcium, serum iron, special radioactive studies of bones using technetium (Tc 99m-labeled agents), stool studies for blood.

Storage:
- Store in cool, dry place away from direct light, but don't freeze.
- Store safely out of reach of children. Iron tablets look like candy, and children love them.
- Don't store in bathroom medicine cabinet. Heat and moisture may change action of mineral.

Others:
- Iron can accumulate to harmful levels (hemosiderosis) in patients with chronic kidney failure, Hodgkins disease, rheumatoid arthritis.
- Prolonged use in high doses can cause hemochromatosis (iron-storage disease), leading to bronze skin, diabetes, liver damage, impotence, heart problems.

⟫➤

MINERAL

Ferrous Sulfate, Continued

 Overdose/Toxicity

Signs and symptoms:
Early signs: Diarrhea with blood, severe nausea, abdominal pain, vomiting with blood.
Late signs: Weakness, collapse, pallor, blue lips, blue hands, blue fingernails, shallow breathing, convulsions, coma, weak, rapid heartbeat.

What to do:
For symptoms of overdosage: Discontinue mineral, and consult doctor. Also see *Adverse Reactions or Side Effects* section below.
For accidental overdosage (such as child taking entire bottle): Dial 911 (emergency), 0 for operator or your nearest Poison Control Center.

 Adverse Reactions or Side Effects

Reaction or effect	What to do
Abdominal pain	Discontinue. Call doctor immediately.
Black or gray stools (always)	Nothing.
Blood in stools	Seek emergency treatment.
Chest pain	Seek emergency treatment.
Drowsiness	Discontinue. Call doctor when convenient.
Stained teeth (liquid forms)	Mix with water or juice to lessen effect. Brush teeth with baking soda or hydrogen peroxide to help remove stain.
Throat pain	Discontinue. Call doctor immediately.

 Interaction with Medicine, Vitamins or Minerals

Interacts with	Combined effect
Allopurinol	May cause excess iron storage in liver.
Antacids	Causes poor iron absorption.
Calcium	Combination necessary to efficient calcium absorption.
Cholestyramine	Decreases iron effect.
Copper	Assists in copper absorption.
Iron supplements (other)	May cause excess iron storage in liver.
Pancreatin	Decreases iron absorption.
Penicillamine	Decreases penicillamine effect.
Sulfasalazine	Decreases iron effect.
Tetracyclines	Decreases tetracycline effect. Take iron 3 hours before or 2 hours after taking tetracycline.
Vitamin C	Increases iron effect. Necessary for red-blood-cell and hemoglobin formation.
Vitamin E	Decreases iron effect.
Zinc (large doses)	Decreases iron absorption.

 Interaction with Other Substances

Alcohol increases iron utilization. May cause organ damage. Avoid or use in moderation.

Beverages
• Milk decreases iron absorption.
• Tea decreases iron absorption.
• Coffee decreases iron absorption.

Fluoride

Basic Information

Fluoride is available commercially as sodium fluoride.
Brand names, see page 477.
Available from natural sources? Yes
Available from synthetic sources? Yes
Prescription required? Yes

Natural Sources

Apples	Kidneys
Calves' liver	Salmon, canned
Cod	Sardines, canned
Eggs	Tea

Note: The fluoride content of foods varies tremendously. It is relatively high where soils are rich and water is fluoridated and low otherwise.

Reasons to Use

- Prevents dental caries (cavities) in children when level of fluoride in water is inadequate.
- Treats osteoporosis with calcium and vitamin D, but use must be carefully monitored by a physician.

Unproved Speculated Benefits

- Prevents osteoporosis in older people.
- Prevents the most-common cause of hearing loss in the elderly by recalcifying ear's inner-bone structure.

Who Needs Additional Amounts?

- Anyone with inadequate caloric or dietary intake or increased nutritional requirements.
- People living in an area with low fluoride water content. Check with your doctor, dentist or local health department.

Deficiency Symptoms

- Significant increase in dental caries

Unproved Speculated Symptoms

- Softening of bones in post-menopausal women

Lab Tests to Detect Deficiency

- None available. Examinations of mouth for dental caries once or twice a year yields all necessary evidence.

Dosage and Usage Information

Recommended Dietary Allowance (RDA):
No RDA has been established. Estimated safe intake given below.

Age	Estimated Safe Intake
0–6 months	0.1–0.5mg/day
6–12 months	0.20–1.0mg/day
1–3 years	0.5–1.5mg/day
4–6 years	1.0–2.5mg/day
7–10 years	1.5–2.5mg/day
11+ years	1.5–4.0mg/day

What this mineral does:
- Contributes to solid bone and tooth formation by helping body retain calcium.
- Interferes with growth and development of bacteria that cause dental plaque.

Miscellaneous information:
- Taking fluoride does not remove need for good dental habits, including a good diet, brushing and flossing teeth and regular dental visits.
- If fluoride supplementation is needed in your area, continue until child is 16. Subsequent topical applications every year or two may be continued to prevent caries.
- Claims that persons residing in areas with fluoridated water supplies have a higher incidence of cancer have been refuted by the National Cancer Institute.

Available as:
- Tablets: Swallow whole with full glass of liquid. Don't chew or crush. Take with meals or 1 to 1-1/2 hours after meals unless otherwise directed by your doctor.
- Drops: Dilute in at least 1/2 glass of water or other liquid. Take with meals or 1 to 1-1/2 hours after meals unless otherwise directed by your doctor. Do not take with milk or dairy products.
- Rinses: Follow directions, and use just before bedtime, after proper brushing and flossing.
- Gels: Follow directions, and use just before bedtime, after proper brushing and flossing.
- Paste: Follow directions, and use just before bedtime, after proper brushing and flossing.

Fluoride, Continued

Warnings and Precautions

Don't take if:
- Sodium intake is restricted or fluoride intake from drinking water exceeds 0.7 parts fluoride/million. Too much fluoride stains teeth permanently.
- You have underactive thyroid function.

Consult your doctor if you have:
- Osteoporosis.

Over age 55:
- No problems expected.

Pregnancy:
- Reports do not agree regarding benefit and risk to unborn child. Follow doctor's instructions.
- Don't take megadoses.

Breast-feeding:
- No problems expected.
- Don't take megadoses.

Effect on lab tests:
- Serum acid phosphatase, serum calcium and protein-bound iodine may be falsely decreased.
- Serum aspartase aminotranfererase (SGOT) may be falsely increased. (See Glossary)

Storage:
- Store in cool, dry place away from direct light, but don't freeze. Keep in original plastic container. Fluoride decomposes glass.
- Store safely out of reach of children.
- Don't store in bathroom medicine cabinet. Heat and moisture may change action of mineral.

Others:
- High dosage needed to treat osteoporosis leads to likelihood of toxic effects, including increased number of bone fractures.

Overdose/Toxicity

Signs and symptoms:
- Stomach cramps or pain, faintness, vomiting (possibly bloody), diarrhea, black stools, shallow breathing, tremors, increased saliva, unusual excitement.

What to do:
For symptoms of overdosage: Discontinue mineral, and consult doctor. Also see *Adverse Reactions or Side Effects* section below.
For accidental overdosage (such as child taking entire bottle) Dial 911 (emergency), 0 for operator or your nearest Poison Control Center.

Adverse Reactions or Side Effects

Reaction or effect	What to do
Excessive amounts of fluoride can cause:	
Appetite loss	Discontinue. Call doctor when convenient.
Constipation	Discontinue. Call doctor when convenient.
Decreased calcium in body characterized by bone pain, leg cramps	Continue. Call doctor when convenient.
Mottling of teeth with brown, black or white discoloration	Continue. Call doctor when convenient.
Nausea	Discontinue. Call doctor when convenient.
Pain and aching in bones	Discontinue. Call doctor immediately.
Skin rash	Discontinue. Call doctor immediately.
Sores in mouth	Discontinue. Call doctor immediately.
Stiffness	Discontinue. Call doctor immediately.
Weight loss	Discontinue. Call doctor when convenient.

Interaction with Medicine, Vitamins or Minerals

Interacts with	Combined effect
Aluminum hydroxide	Decreases absorption of fluoride.
Calcium supplements	Decreases absorption of fluoride.

Interaction with Other Substances

Beverages
- Milk decreases absorption of fluoride. Take dose 2 hours before or after milk.

Iodine

Basic Information
Available from natural sources? Yes
Available from synthetic sources? No
Prescription required? Yes, for strengths
  over 130mg

Natural Sources

Cod	Salmon, canned
Cod-liver oil	Salt, table (iodized)
Haddock	and sea
Herring	Seaweed
Lobster	Shrimp
Oysters	Sunflower seeds

Reasons to Use

- Promotes normal function of thyroid gland.
- Promotes normal cell function.
- Shrinks thyroid prior to thyroid surgery.
- Tests thyroid function before and after administration of a radioactive form of iodine.
- Keeps skin, hair, nails healthy.
- Protects thyroid gland after accidental exposure to radiation.
- Prevents goiter.

Unproved Speculated Benefits

- Cures anemia.
- Treats angina pectoris.
- Treats arteriosclerosis.
- Treats arthritis.
- Treats erythema nodosum.
- Restores vigor.
- Solves hair problems.
- Treats sporotrichosis infection of skin.

Who Needs Additional Amounts?

- Anyone with inadequate caloric or nutritional dietary intake or increased nutritional requirements.
- Anyone who lives in a region where the soil is deficient in iodine. Deficiency is usually treated by using iodized table salt.
- People who eat large amounts of food that can cause thyroid goiter, such as spinach, lettuce, turnips, beets, rutabagas, kale.

Deficiency Symptoms

Childhood deficiencies:
- Depressed growth
- Delayed sexual development
- Mental retardation
- Deafness

Adult deficiencies:
- Goiter

Symptoms of low-thyroid-hormone level (children and adults):
- Listlessness
- Sluggish behavior

Unproved Speculated Symptoms

- Baldness
- Tiredness
- Chest pain

Lab Tests to Detect Deficiency

- Tests may indicate lower-than-normal thyroid function, implying a deficiency of iodine in some cases

Dosage and Usage Information

Recommended Dietary Allowance (RDA):
Estimate of adequate daily intake by the Food and Nutrition Board of the National Research Council, 1980 (See Glossary).

Age	RDA
0–6 months	40mcg
6–12 months	50mcg
1–3 years	70mcg
4–6 years	90mcg
7–10 years	120mcg
11+ years	150mcg
Pregnant	+25mcg
Lactating	+50mcg

What this mineral does:
- Iodine is an integral part of the thyroid hormones tetraiodothyronine (thyroxin) and triiodothyronine.

Miscellaneous information:
- Iodated salt and use of iodophores as antiseptics by the dairy industry are the main source of iodine in most diets.
- It is safe to consume 100–300mcg/day.

⤷➞

Iodine, Continued

Available as:
- Tablets: Swallow whole with full glass of liquid. Don't chew or crush. Take with meals or 1 to 1-1/2 hours after meals unless otherwise directed by your doctor.
- Oral solution: Dilute in at least 1/2 glass water or other liquid. Take with meals or 1 to 1-1/2 hours after meals unless otherwise directed by your doctor.
- Enteric-coated tablets are *not recommended*. They may cause obstruction, bleeding, perforation of small bowel.

Warnings and Precautions

Don't take if you:
- Have elevated serum potassium (determined by laboratory study).
- Have myotonia congenita.

Consult your doctor if you have:
- Hyperthyroidism.
- Kidney disease.
- Taken or are taking amiloride, antithyroid medications, lithium, spironolactone, triamterene.

Over age 55:
- No special considerations.

Pregnancy:
- If too much iodine is consumed during pregnancy, the infant may have thyroid enlargement, hypothyroidism or cretinism (dwarfism and mental deficiency).

Breast-feeding:
- Avoid supplements while nursing.
- Iodine in milk can cause skin rash and suppression of normal thyroid function in infant.
- Don't take megadoses.

Effect on lab tests:
- May cause false elevation in all thyroid-function studies.
- Interferes with test for naturally occurring steroids in urine.

Storage:
- Store in cool, dry place away from direct light, but don't freeze.
- Store safely out of reach of children.
- Don't store in bathroom medicine cabinet. Heat and moisture may change action of mineral.

Overdose/Toxicity

Signs and symptoms:
Irregular heartbeat, confusion, swollen neck or throat, bloody or black, tarry stools.

What to do:
For symptoms of overdosage: Discontinue mineral, and consult doctor. Also see *Adverse Reactions or Side Effects* section below.
For accidental overdosage (such as child taking entire bottle): Dial 911 (emergency), 0 for operator or your nearest Poison Control Center.

Adverse Reactions or Side Effects

Reaction or effect	What to do
Abdominal pain	Discontinue. Call doctor immediately.
Burning in mouth or throat	Discontinue. Call doctor immediately.
Diarrhea	Discontinue. Call doctor immediately.
Fever	Discontinue. Call doctor immediately.
Headache	Discontinue. Call doctor immediately.
Heavy legs	Discontinue. Call doctor when convenient.
Increased salivation	Discontinue. Call doctor immediately.
Metallic taste	Discontinue. Call doctor when convenient.
Nausea	Continue. Tell doctor at next visit.
Numbness, tingling or pain in hands or feet	Discontinue. Call doctor immediately.
Swelling of salivary gland	Seek emergency treatment.
Skin rash	Discontinue. Call doctor immediately.
Sore teeth or gums	Discontinue. Call doctor immediately.
Tiredness or weakness	Discontinue. Call doctor immediately.

 Interaction with Medicine, Vitamins or Minerals

Interacts with	Combined effect
Lithium carbonate for manic-depressive illness	Produces abnormally low thyroid activity. People taking lithium carbonate should avoid iodine, which suppresses the thyroid gland.

 Interaction with Other Substances

None known

MINERAL

Iron Dextran

Basic Information

Iron dextran contains 50mg elemental iron per milliliter. It is a special form of iron designed to be used as an injection deep into muscle, usually in the buttocks.
Available from natural sources? Yes
Available from synthetic sources? Yes
Prescription required? Yes

Natural Sources

Bread, enriched
Cashews
Caviar
Cheese, cheddar
Egg yolk
Garbanzo beans
  (chickpeas)
Lentils

Molasses, black-strap
Mussels
Pistachios
Pumpkin seeds
Seaweed
Walnuts
Wheat germ
Whole-grain products

Note: Even iron-rich foods are poorly absorbed by humans. Only about 10% of food iron is absorbed from food consumed by an individual with normal iron stores. However, an iron-deficient person, may absorb 20 to 30%.

Reasons to Use

- Prevents and treats iron-deficiency anemia due to dietary iron deficiency or other causes. Iron dextran is particularly useful for people who develop severe gastrointestinal symptoms when they take iron orally.
- Stimulates bone-marrow production of hemoglobin, the red-blood-cell pigment that carries oxygen to body cells.
- Forms part of several enzymes and proteins in the body.

Unproved Speculated Benefits

- Controls alcoholism.
- Helps alleviate menstrual discomfort.
- Stimulates immunity.
- Boosts physical performance.
- Prevents learning disorders in children.

Who Needs Additional Amounts?

- Many women of child-bearing age are mildly iron-deficient even when they get all their nutritional requirements.
- Anyone with inadequate caloric or dietary intake or increased nutritional requirements.

- Older people (over 55 years).
- Pregnant or breast-feeding women.
- Women with heavy menstrual flow, long menstrual periods or short menstrual cycles.
- Those who abuse alcohol or other drugs.
- People with a chronic wasting illness. excess stress for long periods or those who have recently undergone surgery.
- Athletes and workers who participate in vigorous physical activities.
- Those with a portion of the gastrointestinal tract surgically removed.
- People with recent severe burns or injuries.
- Anyone who has lost blood recently, such as from heavy menstrual periods or from an accident.
- Vegetarians.
- Infants from 2 to 24 months.

Deficiency Symptoms

- Listlessness
- Heart palpitations upon exertion
- Fatigue
- Irritability
- Paleness of skin
- Cracking of lips and tongue
- Difficulty swallowing
- General feeling of poor health

Unproved Speculated Symptoms

- None

Lab Tests to Detect Deficiency

- Red-blood-cell count
- Microscopic exam of red blood cells
- Serum iron
- Hemoglobin determinations

Dosage and Usage Information

Recommended Dietary Allowance (RDA):
Estimate of adequate daily intake by the Food and Nutrition Board of the National Research Council, 1980 (See Glossary). ⟫➤

Iron Dextran

Age	RDA
0–6 months	10mg
6–12 months	15mg
1–3 years	15mg
4–6 years	10mg
7–10 years	10mg
Males	
11–18 years	18mg
19+ years	10mg
Females	
11–50 years	18mg
51+ years	10mg
Pregnant	+30–60mg
Lactating	+30–60mg

What this mineral does:
- Iron is an essential component of hemoglobin, myoglobin and a co-factor of several essential enzymes. Of the total iron in the body, 60 to 70% is stored in hemoglobin (the red part of red blood cells).
- Hemoglobin is also a component of myoglobin, an iron-protein complex in muscles. This complex helps muscles get extra energy when they work hard.

Miscellaneous information:
- Iron-deficiency anemia in older men is usually considered to be due to slow loss of blood from a malignancy in the gastrointestinal tract until proved otherwise.
- Iron content of foods, especially acidic foods, can be dramatically increased by preparation in iron cookware.
- May require 3 weeks of treatment before you receive maximum benefit.
- Works best with vitamin C (ascorbic acid).

Available as:
- Injectable forms are administered by doctor or nurse.

 Warnings and Precautions

Don't take if you:
- Are allergic to any iron supplement.
- Have acute hepatitis.
- Have hemosiderosis or hemochromatosis (conditions involving excess iron in body).
- Have hemolytic anemia.
- Have had repeated blood transfusions.

Consult your doctor if you have:
- Plans to become pregnant while taking medication.
- Had stomach surgery.
- Had peptic-ulcer disease, enteritis, colitis.
- Had pancreatitis or hepatitis.
- Alcoholism.
- Kidney disease.
- Rheumatoid arthritis.
- Intestinal disease.

Over age 55:
- Deficiency more likely. Check frequently with doctor for anemia symptoms or slow blood loss in stool.

Pregnancy:
- Pregnancy increases need. Check with doctor. During first 3 months of pregnancy, take *only* if doctor prescribes it.
- Don't take megadoses.

Breast-feeding:
- Supplements probably not needed if you are healthy and eat a balanced diet.
- Baby may need supplementation. Ask your doctor.
- Don't take megadoses.

Effect on lab tests:
- May cause abnormal results in serum bilirubin, serum calcium, serum iron, special radioactive studies of bones using technetium (Tc 99m-labeled agents), stool studies for blood.

Storage:
- Store in cool, dry place away from direct light, but don't freeze.
- Store safely out of reach of children. Iron tablets look like candy, and children love them.
- Don't store in bathroom medicine cabinet. Heat and moisture may change action of mineral.

Others:
- Iron can accumulate to harmful levels (hemosiderosis) in patients with chronic kidney failure, Hodgkins disease, rheumatoid arthritis.
- Prolonged use in high doses can cause hemochromatosis (iron-storage disease), leading to bronze skin, diabetes, liver damage, impotence, heart problems.
- Intramuscular iron can leave unsightly black deposits in skin that mimic tattoo marks.

≫➤

Iron Dextran, Continued

Overdose/Toxicity

Signs and symptoms:
Early signs: Diarrhea with blood, severe nausea, abdominal pain, vomiting with blood.
Late signs: Weakness, collapse, pallor, blue lips, blue hands, blue fingernails, shallow breathing, convulsions, coma, weak, rapid heartbeat.

What to do:
For symptoms of overdosage: Discontinue mineral, and consult doctor. Also see *Adverse Reactions or Side Effects* section below.
For accidental overdosage (such as child taking entire bottle): Dial 911 (emergency), 0 for operator or your nearest Poison Control Center.

Adverse Reactions or Side Effects

Reaction or effect	What to do
Abdominal pain	Discontinue. Call doctor immediately.
Anaphylaxis (extremely rare)—symptoms include immediate severe itching, paleness, low blood pressure, loss of consciousness, coma	Yell for help. Don't leave victim. Begin CPR (cardiopulmonary resuscitation), mouth-to-mouth breathing and external cardiac massage. Have someone dial "0" (operator) or 911 (emergency). Don't stop CPR until help arrives.
Black or gray stools (always)	Nothing.
Blood in stools	Seek emergency treatment.
Chest pain	Seek emergency treatment.
Chills	Seek emergency treatment.
Drowsiness	Discontinue. Call doctor when convenient.
Hives	Seek emergency treatment.
Loss of consciousness	Seek emergency treatment.
Shortness of breath	Seek emergency treatment.
Skin rash	Seek emergency treatment.

Throat pain	Discontinue. Call doctor immediately.

Interaction with Medicine, Vitamins or Minerals

Interacts with	Combined effect
Allopurinol	May cause excess iron storage in liver.
Antacids	Causes poor iron absorption.
Calcium	Combination necessary for efficient calcium absorption.
Cholestyramine	Decreases iron effect.
Copper	Assists in copper absorption.
Iron supplements (other)	May cause excess iron storage in liver.
Pancreatin	Decreases iron absorption.
Penicillamine	Decreases penicillamine effect.
Sulfasalazine	Decreases iron effect.
Tetracyclines	Decreases tetracycline effect. Take iron 3 hours before or 2 hours after taking tetracycline.
Vitamin C	Increases iron effect. Necessary for red-blood-cell and hemoglobin formation.
Vitamin E	Decreases iron effect.
Zinc (large doses)	Decreases iron absorption.

Interaction with Other Substances

Alcohol increases iron utilization. May cause organ damage. Avoid or use in moderation.

Beverages
• Milk decreases iron absorption.
• Tea decreases iron absorption.
• Coffee decreases iron absorption.

Iron-Polysaccharide

Basic Information

Brand names, see page 477.
Available from natural sources? Yes
Available from synthetic sources? Yes
Prescription required? Yes

Natural Sources

Bread, enriched	Molasses, black-strap
Cashews	Mussels
Caviar	Pistachios
Cheddar cheese	Pumpkin seeds
Egg yolk	Seaweed
Garbanzo beans	Walnuts
(chickpeas)	Wheat germ
Lentils	Whole-grain products

Note: Even iron-rich foods are poorly absorbed by humans. Only about 10% of iron from food is absorbed from food consumed by an individual with normal iron stores. However, an iron-deficient person may absorb 20 to 30%.

Reasons to Use

- Prevents and treats iron-deficiency anemia due to dietary iron deficiency or other causes.
- Stimulates bone-marrow production of hemoglobin, the red-blood-cell pigment that carries oxygen to body cells.
- Forms part of several enzymes and proteins in the body.

Unproved Speculated Benefits

- Controls alcoholism.
- Helps alleviate menstrual discomfort.
- Stimulates immunity.
- Boosts physical performance.
- Prevents learning disorders in children.

Who Needs Additional Amounts?

- Many women of child-bearing age are mildly iron-deficient even when they get all their nutritional requirements.
- Anyone with inadequate caloric or nutritional dietary intake or increased nutritional requirements.
- Older people (over 55 years).
- Pregnant or breast-feeding women.
- Those who abuse alcohol or other drugs.
- People with a chronic wasting illness, excess stress for long periods or who have recently undergone surgery.

- Athletes and workers who participate in vigorous physical activities.
- Those with a portion of the gastrointestinal tract surgically removed.
- People with recent severe burns or injuries.
- Anyone who has lost blood recently, such as from heavy menstrual periods or from an accident.
- Vegetarians.
- Infants from 2 to 24 months.

Deficiency Symptoms

- Listlessness
- Heart palpitations upon exertion
- Fatigue
- Irritability
- Pale appearance to skin
- Cracking of lips and tongue
- Difficulty swallowing
- General feeling of poor health

Unproved Speculated Symptoms

- None

Lab Tests to Detect Deficiency

- Red-blood-cell count
- Microscopic exam of red blood cells
- Serum iron
- Hemoglobin determinations

≫➤

Iron-Polysaccharide, Continued

Dosage and Usage Information

Recommended Dietary Allowance (RDA):
Estimate of adequate daily intake by the Food and Nutrition Board of the National Research Council, 1980 (See Glossary).

Age	RDA
0–6 months	10mg
6–12 months	15mg
1–3 years	15mg
4–6 years	10mg
7–10 years	10mg
Males	
11–18 years	18mg
19+ years	10mg
Females	
11–50 years	18mg
51+ years	10mg
Pregnant	+30–60mg
Lactating	+30–60mg

What this mineral does:
• Iron is an essential component of hemoglobin, myoglobin and a co-factor of several essential enzymes. Of the total iron in the body, 60 to 70% is stored in hemoglobin (the red part of red blood cells).
• Hemoglobin is also a component of myoglobin, an iron-protein complex in muscles. This complex helps muscles get extra energy when they work hard.

Miscellaneous information:
• Iron-deficiency anemia in older men is usually considered to be due to slow loss of blood from a malignancy in the gastrointestinal tract until proved otherwise.
• Iron content of foods, especially acidic foods, can be dramatically increased by preparation in iron cookware.
• May require 3 weeks of treatment before you receive maximum benefit.
• Works best with vitamin C (ascorbic acid).

Available as:
• Capsules: Swallow whole with full glass of liquid. Don't chew or crush. Take with or immediately after food to decrease stomach irritation.
  Oral solution: Dilute in at least 1/2 glass water or other liquid. Take with meals or 1 to 1-1/2 hours after meals unless otherwise directed by your doctor.
• Tablets: Swallow whole with full glass of liquid. Don't chew or crush. Take with meals or 1 to 1-1/2 hours after meals unless otherwise directed by your doctor.

Warnings and Precautions

Don't take if you:
• Are allergic to any iron supplement.
• Have acute hepatitis.
• Have hemosiderosis or hemochromatosis (conditions involving excess iron in body).
• Have hemolytic anemia.
• Have had repeated blood transfusions.

Consult your doctor if you have:
• Plans to become pregnant while taking medication.
• Had stomach surgery.
• Had peptic-ulcer disease, enteritis or colitis.
• Had pancreatitis or hepatitis.
• Alcoholism.
• Kidney disease.
• Rheumatoid arthritis.
• Intestinal disease.

Over age 55:
• Deficiency more likely. Check frequently with doctor for anemia symptoms or slow blood loss in stool.

Pregnancy:
• Pregnancy increases need. Check with your doctor. During first 3 months of pregnancy, take *only* if your doctor prescribes it.
• Don't take megadoses.

Breast-feeding:
• Supplements probably not needed if you are healthy and eat a balanced diet.
• Baby may need supplement. Ask your doctor.
• Don't take megadoses.

Effect on lab tests:
• May cause abnormal results in serum bilirubin, serum calcium, serum iron, special radioactive studies of bones using technetium (Tc 99m-labeled agents), stool studies for blood.

Storage:
• Store in cool, dry place away from direct light, but don't freeze.
• Store safely out of reach of children. Iron tablets look like candy, and children love them.
• Don't store in bathroom medicine cabinet. Heat and moisture may change action of mineral.

Others:
• Iron can accumulate to harmful levels (hemosiderosis) in patients with chronic kidney failure, Hodgkins disease, rheumatoid arthritis.
• Prolonged use in high doses can cause hemochromatosis (iron-storage disease), leading to bronze skin, diabetes, liver damage, impotence, heart problems.

➤➤

Overdose/Toxicity

Signs and symptoms:
Early signs: Diarrhea with blood, severe nausea, abdominal pain, vomiting with blood.
Late signs: Weakness, collapse, pallor, blue lips, blue hands, blue fingernails, shallow breathing, convulsions, coma, weak, rapid heartbeat.

What to do:
For symptoms of overdosage: Discontinue mineral, and consult doctor. Also see *Adverse Reactions or Side Effects* section below.
For accidental overdosage (such as child taking entire bottle): Dial 911 (emergency), 0 for operator or your nearest Poison Control Center.

Adverse Reactions or Side Effects

Reaction or effect	What to do
Abdominal pain	Discontinue. Call doctor immediately.
Black stools or gray stools (always)	Nothing.
Blood in stools	Seek emergency treatment.
Chest pain	Seek emergency treatment.
Drowsiness	Discontinue. Call doctor when convenient.
Stained teeth (liquid forms)	Mix with water or juice to lessen effect. Brush teeth with baking soda or hydrogen peroxide to help remove stain.
Throat pain	Discontinue. Call doctor immediately.

Interaction with Medicine, Vitamins or Minerals

Interacts with	Combined effect
Allopurinol	Possible excess iron storage in liver.
Antacids	Poor iron absorption.
Calcium	Combination necessary for efficient calcium absorption.
Cholestyramine	Decreases iron effect.
Copper	Assists in copper absorption.
Iron supplements (other)	Possible excess iron storage in liver.
Pancreatin	Decreases iron absorption.
Penicillamine	Decreases penicillamine effect.
Sulfasalazine	Decreases iron effect.
Tetracyclines	Decreases tetracycline effect. Take iron 3 hours before or 2 hours after taking tetracycline.
Vitamin C	Increases iron effect. Contribution necessary for red-blood-cell and hemoglobin formation.
Vitamin E	Decreases iron absorption.
Zinc (large doses)	Decreases iron absorption.

Interaction with Other Substances

Alcohol increases iron utilization. May cause organ damage. Avoid or use in moderation.

Beverages
• Milk decreases iron absorption.
• Tea decreases iron absorption.
• Coffee decreases iron absorption.

MINERAL

Magnesium

Basic Information
Brand names, see page 477.
Available from natural sources? Yes
Available from synthetic sources? No
Prescription required? Yes, for some forms

Natural Sources

Almonds	Leafy, green	Snails
Bluefish	vegetables	Soybeans
Carp	Mackerel	Sunflower
Cod	Molasses	seeds
Flounder	Nuts	Swordfish
Halibut	Ocean perch	Wheat germ
Herring	Shrimp	

Reasons to Use

- Aids bone growth.
- Aids function of nerves and muscles, including regulation of normal heart rhythm.
- Keeps metabolism steady.
- Conducts nerve impulses.
- Works as laxative in large doses.
- Acts as antacid in small doses.
- Strengthens tooth enamel.

Unproved Speculated Benefits

- Cures alcoholism.
- Cures kidney stones.
- Alleviates heart disease.
- Helps control bad breath and body odor.
- Helps contain lead poisoning.

Who Needs Additional Amounts?

- Anyone with inadequate caloric or dietary intake or increased nutritional requirements.
- Those who abuse alcohol or other drugs.
- People with a chronic wasting illness or who have recently undergone surgery.
- Those with recent severe burns or injuries.

Deficiency Symptoms

Following symptoms occur rarely:
- Muscle contractions
- Convulsions
- Confusion and delirium
- Irritability
- Nervousness
- Skin problems
- Hardening of soft tissues

Unproved Speculated Symptoms

- Sudden heart failure

Lab Tests to Detect Deficiency

- Serum magnesium

Dosage and Usage Information

Recommended Dietary Allowance (RDA):
Estimate of adequate daily intake by the Food and Nutrition Board of the National Research Council, 1980 (See Glossary).

Age	RDA
0–6 months	50mg
6–12 months	70mg
1–3 years	150mg
4–6 years	200mg
7–10 years	250mg
Males	
11–14 years	350mg
15–18 years	400mg
18+ years	350mg
Females	
11+ years	300mg
Pregnant	+150mg
Lactating	+150mg

What this mineral does:
- Activates essential enzymes.
- Affects metabolism of proteins and nucleic acids.
- Helps transport sodium and potassium across cell membranes.
- Influences calcium levels inside cells.

Available as:
- A constituent of many multivitamin/mineral preparations.

》▶

Warnings and Precautions

Don't take if you have:
• Kidney failure.
• Heart block (unless you have a pacemaker).
• An ileostomy.

Consult your doctor if you have:
• Chronic constipation, colitis, diarrhea.
• Symptoms of appendicitis.
• Stomach or intestinal bleeding.

Over age 55:
• Adverse reactions/side effects more likely.

Pregnancy:
• Risk to fetus outweighs benefits. Don't use.

Breast-feeding:
• Avoid magnesium in large quantities.
• If you must take temporarily, discontinue breast-feeding. Consult doctor for advice on maintaining milk supply.
• Don't take megadoses.

Effect on lab tests:
• Makes test for stomach-acid secretion inaccurate.
• May increase or decrease serum-phosphate concentrations.
• May decrease serum and urine pH.

Storage:
• Store in cool, dry place away from direct light, but don't freeze.
• Store safely out of reach of children.
• Don't store in bathroom medicine cabinet. Heat and moisture may change action of mineral.

Others:
• Chronic kidney disease frequently causes body to retain excess magnesium.
• Adverse reactions, side effects and interactions with medicines, vitamins or minerals occur *only rarely* when too much magnesium is taken for too long or if you have kidney disease.

 ## Overdose/Toxicity

Signs and symptoms:
Severe nausea and vomiting, extremely low blood pressure, extreme muscle weakness, difficulty breathing, heartbeat irregularity.

What to do:
For symptoms of overdosage: Discontinue mineral, and consult doctor immediately. Also see *Adverse Reactions or Side Effects* section below.
For accidental overdosage (such as child taking entire bottle): Dial 911 (emergency), 0 for operator or your nearest Poison Control Center.

 ## Adverse Reactions or Side Effects

Reaction or effect	What to do
Abdominal pain	Discontinue. Call doctor immediately.
Appetite loss	Discontinue. Call doctor when convenient.
Diarrhea	Discontinue. Call doctor immediately.
Irregular heartbeat	Seek emergency treatment.
Mood changes or mental changes	Discontinue. Call doctor when convenient.
Nausea	Discontinue. Call doctor immediately.
Tiredness or weakness	Discontinue. Call doctor when convenient.
Urination discomfort	Discontinue. Call doctor when convenient.
Vomiting	Discontinue. Call doctor immediately.

 ## Interaction with Medicine, Vitamins or Minerals

Interacts with	Combined effect
Cellulose sodium phosphate	Decreases magnesium effect. Take 1 or more hours apart.
Fat-soluble vitamins (A, E, K)	Decreases absorption of mineral.
Ketoconazole	Reduces absorption of ketoconazole. Take 2 hours apart.
Mecamylamine	May slow urinary excretion of mecamylamine. Avoid combination.
Tetracycline	Decreases absorption of tetracycline.
Vitamin D	May raise magnesium level too high.

 ## Interaction with Other Substances

None known

Manganese

Basic Information

Brand names, see page 477.
Available from natural sources? Yes
Available from synthetic sources? Yes
Prescription required? Yes

Natural Sources

Avocados	Ginger
Barley	Hazelnuts (filberts)
Beans, dried	Oatmeal
Blackberries	Peanuts
Bran	Peas
Buckwheat	Pecans
Chestnuts	Seaweed
Cloves	Spinach
Coffee	

Reasons to Use

- Promotes normal growth and development.
- Promotes cell function.
- Helps many body enzymes generate energy. Without manganese they could not function.
- Used as a supplement for those receiving long-term nutrition intravenously or through a naso-gastric tube.

Unproved Speculated Benefits

- Alleviates asthma.
- Helps overcome infertility.
- Alleviates diabetes.
- Helps relieve fatigue.
- Acts as an anti-aging substance.
- Helps treat schizophrenia.
- Provides part of molecules necessary for reproduction.

Who Needs Additional Amounts?

- Anyone with inadequate caloric or nutritional dietary intake or increased nutritional requirements.
- Extremely ill people who must be fed intravenously or by naso-gastric tube.

Deficiency Symptoms

- Abnormal growth and development of children
- No proven symptoms caused by manganese deficiency in adults

Unproved Speculated Symptoms

- Changes in beard and hair growth—usually a slowing of growth
- Occasional nausea and vomiting
- Hypocholesterolemia
- Weight loss

Lab Tests to Detect Deficiency

- Serum manganese

Dosage and Usage Information

Recommended Dietary Allowance (RDA):
No RDA has been established. Estimated safe intake given below.

Age	Estimated Safe Intake
0–6 months	0.5–0.7mg/day
6–12 months	0.7–1.0mg/day
1–3 years	1.0–1.5mg/day
4–6 years	1.5–2.0mg/day
7–10 years	2.0–3.0mg/day
11+ years	2.5–5.0mg/day

What this mineral does:
- Manganese is concentrated in cells of pituitary gland, liver, pancreas, kidney and bone. It influences syntheses of muropolysaccharides, stimulates production of cholesterol by the liver and is a co-factor in many enzymes.

Miscellaneous information:
- Manganese is abundant in many foods.
- Miners and workers in some industries are at risk of toxicity from inhaling manganese. Chronic inhalation can lead to symptoms of Parkinson's disease, with the following signs and symptoms:
  Tremors, especially when not moving
  General muscle stiffness and soreness
  Awkward or shuffling walk
  Stooped posture
  Loss of facial expression
  Voice changes—voice becomes weak and high pitched

»→

Difficulty swallowing
Intellectual ability unchanged until advanced stages, when it deteriorates slowly.
* *Manganese* and *magnesium* are *not* related to each other!

Available as:
* Capsules: Swallow whole with full glass of liquid. Don't chew or crush. Take with or immediately after food to decrease stomach irritation.
* A constituent of many multivitamin/mineral preparations.

Warnings and Precautions

Don't take if you:
* Are healthy and eat regular, balanced meals.

Consult your doctor if you have:
* Liver disease.

Over age 55:
* No special problems expected.

Pregnancy:
* Don't take supplements with manganese unless prescribed by your doctor.
* Don't take megadoses.

Breast-feeding:
* Don't take supplements with manganese unless prescribed by your doctor.
* Don't take megadoses.

Effect on lab tests:
* Excess manganese can reduce serum iron.

Storage:
* Store in cool, dry place away from direct light, but don't freeze.
* Store safely out of reach of children.
* Don't store in bathroom medicine cabinet. Heat and moisture may change action of mineral.

Others:
* Check with your industrial health office if you are a miner or industrial worker to make sure your work environment does not contain toxic amounts of manganese.

Overdose/Toxicity

Signs and symptoms:
Delusions, hallucinations, insomnia, depression, impotence.

What to do:
For symptoms of overdosage: Discontinue mineral, and consult doctor. Also see *Adverse Reactions or Side Effects* section below.
For accidental overdosage (such as child taking entire bottle): Dial 911 (emergency), 0 for operator or your nearest Poison Control Center.

Adverse Reactions or Side Effects

Reaction or effect	What to do
Appetite loss	Discontinue. Call doctor when convenient.
Breathing problems	Seek emergency treatment.
Headaches	Discontinue. Call doctor when convenient.
Impotence	Discontinue. Call doctor when convenient.
Leg cramps	Discontinue. Call doctor immediately.
Unusual tiredness	Discontinue. Call doctor when convenient.

Interaction with Medicine, Vitamins or Minerals

Interacts with	Combined effect
Calcium (from food or supplements)	May decrease manganese absorption when taken in large doses.
Iron (from food or supplements)	Excess manganese interferes with iron absorption and can lead to iron-deficiency anemia.
Magnesium (from food or supplements)	May decrease manganese absorption when taken in large doses.
Oral contraceptives	Decreases manganese in blood.
Phosphate (from food or supplements)	When taken in large doses, may decrease manganese absorption.

Interaction with Other Substances

None known

Molybdenum

Basic Information

Available from natural sources? Yes
Available from synthetic sources? No
Prescription required? Yes

Natural Sources

Beans
Cereal grains
Dark-green, leafy vegetables
Organ meats (liver, kidney, sweetbreads)
Peas and other legumes
Note: Dietary concentration of molybdenum may vary according to status of soil in which grains and vegetables are raised.

Reasons to Use

- Promotes normal growth and development.
- Promotes normal cell function.
- Is a component of xanthine oxidase, an enzyme involved in converting nucleic acid to uric acid, a waste product eliminated in the urine.

Unproved Speculated Benefits

- Protects against cancer.
- Protects teeth.
- Prevents anemia by mobilizing iron.

Who Needs Additional Amounts?

- Anyone with inadequate caloric or nutritional dietary intake or increased nutritional requirements.
- People with recent severe burns or injuries.
- Extremely ill people who must be fed intravenously or by naso-gastric tube.

Deficiency Symptoms

- None

Unproved Speculated Symptoms

- Rapid heartbeat
- Rapid breathing
- Night blindness
- Irritability

Lab Tests to Detect Deficiency

- None available, except for experimental purposes.

Dosage and Usage Information

Recommended Dietary Allowance (RDA):
No RDA has been established. Estimated safe intake given below.

Age	Estimated Safe Intake
0–6 months	0.03–0.06mg/day
6–12 months	0.04–0.08mg/day
1–3 years	0.05–0.10mg/day
4–6 years	0.06–0.15mg/day
7–10 years	0.10–0.30mg/day
11+ years	0.15–0.50mg/day

What this mineral does:
- Becomes a part of bones, liver, kidney.
- Forms part of the enzyme system of xanthine oxidase.

Miscellaneous information:
- Balanced diet provides all the molybdenum that is necessary in a healthy child or adult.

Available as:
- Capsules: Swallow whole with full glass of liquid. Don't chew or crush. Take with meals or 1 to 1-1/2 hours after meals unless otherwise directed by your doctor.
- A constituent of many multivitamin/mineral preparations.

»→

Molybdenum

Warnings and Precautions

Don't take if you:
• No absolute contraindications to 0.15 to 0.5mg/day. Don't take higher doses without doctor's prescription.

Consult your doctor if you have:
• High levels of uric acid.
• Gout.

Over age 55:
• No problems expected.

Pregnancy:
• Don't take.

Breast-feeding:
• Don't take.

Effect on lab tests:
• Excess molybdenum causes serum copper to drop.

Storage:
• Store in cool, dry place away from direct light, but don't freeze.
• Store safely out of reach of children.
• Don't store in bathroom medicine cabinet. Heat and moisture may change action of mineral.

Overdose/Toxicity

Signs and symptoms:
Gout can be produced by massive intake (10 to 15mg/daily). Moderate excess (up to 0.54mg/day) can cause excess loss of copper in urine.

Possible Consequences of Overdose:
Daily intake of 10 to 15mg of molybdenum has been associated with a gout-like syndrome. A moderate excess of 0.54mg/daily may be associated with significant urinary loss of copper.

What to do:
For symptoms of overdosage: Discontinue mineral, and consult doctor.
For accidental overdosage (such as child taking entire bottle): Dial 911 (emergency), 0 for operator or your nearest Poison Control Center.

Adverse Reactions or Side Effects

None expected

Interaction with Medicine, Vitamins or Minerals

Interacts with	Combined effect
Copper	Maintains appropriate ratio of molybdenum and copper in body. With excess molybdenum, copper level drops. With excess copper, molybdenum level drops.
Sulfur	Increased sulfur intake causes decline in molybdenum concentration.

Interaction with Other Substances

None known

MINERAL

149

Potassium Acetate/Bicarbonate/Citrate

Basic Information

This combination is also called trikates.
Brand names, see page 477.
Available from natural sources? Yes
Available from synthetic sources? Yes
Prescription required? Some yes; others no

Natural Sources

Avocados
Bananas
Chard
Citrus fruits
Juices
  grapefruit, tomato,
  orange
Lentils, dried
Milk
Molasses

Nuts
  almonds, Brazil,
  cashews, peanuts,
  pecans, walnuts
Parsnips
Peaches, dried
Potatoes
Raisins
Sardines, canned
Spinach, fresh
Whole-grain cereals

Reasons to Use

- Promotes regular heartbeat.
- Promotes normal muscle contraction.
- Regulates transfer of nutrients to cells.
- Maintains water balance in body tissues and cells.
- Preserves or restores normal function of nerve cells, heart cells, skeletal-muscle cells, kidneys, stomach-juice secretion.
- Treats potassium deficiency from illness or taking diuretics (water pills), cortisone drugs or digitalis preparations.

Unproved Speculated Benefits

- Cures alcoholism.
- Cures acne.
- Cures allergies.
- Cures heart disease.
- Helps heal burns.
- Prevents high blood pressure.

Who Needs Additional Amounts?

- People who take diuretics, cortisone drugs or digitalis preparations.
- Anyone with inadequate caloric or nutritional dietary intake or increased nutritional requirements.
- Older people (over 55 years).

- Pregnant or breast-feeding women.
- Women taking oral contraceptives.
- People who abuse alcohol or other drugs.
- Tobacco smokers.
- People with a chronic wasting illness, excess stress for long periods or who have recently undergone surgery.
- Athletes and workers who participate in vigorous physical activities, especially when endurance is an important aspect of the activity.
- Those with part of the gastrointestinal tract surgically removed.
- People with malabsorption illnesses (See Glossary).
- Those with recent severe burns or injuries.
- Vegetarians.

Deficiency Symptoms

- Hypokalemia
- Weakness, paralysis
- Low blood pressure
- Life-threatening, irregular or rapid heartbeat that can lead to cardiac arrest and death

Unproved Speculated Symptoms

- Acne
- Allergies
- High blood pressure

Lab Tests to Detect Deficiency

- Serum-potassium determinations
- Serum creatinine
- Electrocardiograms
- Serum-pH determinations

Dosage and Usage Information

Recommended Dietary Allowance (RDA):
No RDA has been established. Nutritionists recommend a *decrease* in sodium (table salt) intake and an *increase* in foods high in potassium for a total daily intake of 40 to 150 milliequivalents per day.

⟫➤

Potassium Acetate/Bicarbonate/Citrate

What this mineral does:
- Potassium is the predominant positive electrolyte in body cells. An enzyme (adenosinetriphosphatase) controls flow of potassium and sodium into and out of cells to maintain normal function of heart, brain, skeletal muscles, normal kidney function, acid-base balance.

Miscellaneous Information:
- Normal potassium content is reduced when foods are canned or frozen.
- Avoid peeling food.
- Avoid cooking food in large amounts of water.
- Keep meat drippings and use as gravies.

Available as:
- Oral solution: Dilute in at least 1/2 glass water or other liquid. Take with meals or 1 to 1-1/2 hours after meals unless otherwise directed by your doctor.
- Potassium is not recommended for children.
- Some forms are available by generic name.

Warnings and Precautions

Don't take if you:
- Take potassium-sparing diuretics, such as spironolactone, triamterene or amiloride.
- Are allergic to any potassium supplement.
- Have kidney disease.

Consult your doctor if you have:
- Addison's disease.
- Heart disease.
- Intestinal blockage.
- A stomach ulcer.
- To use diuretics.
- To use heart medicine.
- To use laxatives or if you have chronic diarrhea.
- To use salt substitutes or low-salt milk.

Over age 55:
- Observe dose schedule strictly. Potassium balance is critical. Deviation above or below normal can have serious results.

Pregnancy:
- No problems expected, except with megadoses.

Breast-feeding:
- Studies inconclusive on harm to infant. Consult doctor about supplements.
- Don't take megadoses.

Effect on lab tests:
- ECG and kidney function studies can be affected by too much or too little potassium.
- None expected on blood studies, except serum-potassium levels.

Storage:
- Store in cool, dry area away from direct light, but don't freeze.
- Store safely out of reach of children.
- Don't store in bathroom medicine cabinet. Heat and moisture may change action of the mineral

Others:
- Take with meals or with food.

Overdose/Toxicity

Signs and symptoms:
Irregular or fast heartbeat, paralysis of arms and legs, blood-pressure drop, convulsions, coma, cardiac arrest.

What to do:
For symptoms of overdosage: Discontinue mineral, and consult doctor. Also see *Adverse Reactions or Side Effects* section below.
For accidental overdosage (such as child taking entire bottle): Dial 911 (emergency), 0 for operator or your nearest Poison Control Center. If person's heart has stopped beating, render CPR until trained help arrives.

Potassium Acetate/Bicarbonate/Citrate, Continued

 Adverse Reactions or Side Effects

Reaction or effect	What to do
Black, tarry stool	Seek emergency treatment.
Bloody stool	Seek emergency treatment.
Breathing difficulty	Seek emergency treatment.
Confusion	Discontinue. Call doctor immediately.
Diarrhea	Discontinue. Call doctor immediately.
Extreme fatigue	Discontinue. Call doctor when convenient.
Heaviness in legs	Discontinue. Call doctor when convenient.
Irregular heartbeat	Seek emergency treatment.
Nausea	Discontinue. Call doctor when convenient.
Numbness in hands or feet	Discontinue. Call doctor when convenient.
Stomach discomfort	Discontinue. Call doctor when convenient.
Tingling in hands and feet	Discontinue. Call doctor when convenient.
Vomiting	Discontinue. Call doctor immediately.
Weakness	Discontinue. Call doctor immediately.

 Interaction with Medicine, Vitamins or Minerals

Interacts with	Combined effect
Amiloride	Causes dangerous rise in blood potassium.
Atropine	Increases possibility of intestinal ulcers, which may occur with oral potassium tablets.
Belladonna	Increases possibility of intestinal ulcers, which may occur with oral potassium.
Calcium	Increases possibility of heartbeat irregularities.
Captopril	Increases chance of excessive amounts of potassium.
Cortisone	Decreases effect of potassium.
Digitalis preparations	May cause irregular heartbeat.
Enalapril	Increases chance of excessive amounts of potassium.
Laxatives	May decrease potassium effect.
Spironolactone	Increases blood potassium.
Triamterene	Increases blood potassium.
Vitamin B-12	Extended-release tablets may decrease vitamin B-12 absorption and increase vitamin B-12 requirements.

 Interaction with Other Substances

Tobacco decreases absorption. Smokers may require supplemental potassium.

Alcohol intensifies gastrointestinal symptoms.

Cocaine may cause irregular heartbeat.

Marijuana may cause irregular heartbeat.

Beverages
- Salty drinks, such as tomato juice and commercial thirst quenchers, cause increased fluid retention.
- Coffee decreases potassium absorption and intensifies gastrointestinal symptoms.
- Low-salt milk increases fluid retention.

Foods
- Salty foods increase fluid retention.
- Sugar decreases potassium absorption.

Potassium Bicarbonate

Basic Information

Brand names, see page 477.
Available from natural sources? Yes
Available from synthetic sources? Yes
Prescription required? Some yes; others no

 Natural Sources

Avocados
Bananas
Chard
Citrus fruit
Juices
  grapefruit, tomato,
  orange
Lentils, dried
Milk
Molasses

Nuts
  almonds, Brazil,
  cashews, peanuts,
  pecans, walnuts
Parsnips
Peaches, dried
Potatoes
Raisins
Sardines, canned
Spinach, fresh
Whole-grain cereals

 Reasons to Use

- Promotes regular heartbeat.
- Promotes normal muscle contraction.
- Regulates transfer of nutrients to cells.
- Maintains water balance in body tissues and cells.
- Preserves or restores normal function of nerve cells, heart cells, skeletal-muscle cells, kidneys, stomach-juice secretion.
- Treats potassium deficiency from illness or taking diuretics (water pills), cortisone drugs or digitalis preparations.

 Unproved Speculated Benefits

- Cures alcoholism.
- Cures acne.
- Cures allergies.
- Cures heart disease.
- Helps heal burns.
- Prevents high blood pressure.

 Who Needs Additional Amounts?

- People who take diuretics, cortisone drugs or digitalis preparations.
- Anyone with inadequate caloric or nutritional dietary intake or increased nutritional requirements.
- Older people (over 55 years).
- Pregnant or breast-feeding women.
- Women taking oral contraceptives.

- People who abuse alcohol or other drugs.
- Tobacco smokers.
- People with a chronic wasting illness, excess stress for long periods or who have recently undergone surgery.
- Athletes and workers who participate in vigorous physical activities, especially when endurance is an important aspect of the activity.
- Those with part of the gastrointestinal tract surgically removed.
- People with malabsorption illnesses (See Glossary).
- Those with recent severe burns or injuries.
- Vegetarians.

 Deficiency Symptoms

- Hypokalemia
- Weakness, paralysis
- Low blood pressure
- Life-threatening, irregular or rapid heartbeat that can lead to cardiac arrest and death

 Unproved Speculated Symptoms

- Acne
- Allergies
- High blood pressure

 Lab Tests to Detect Deficiency

- Serum-potassium determinations
- Serum creatinine
- Electrocardiograms
- Serum-pH determinations

 Dosage and Usage Information

Recommended Dietary Allowance (RDA):
No RDA has been established. Nutritionists recommend a *decrease* in sodium (table salt) intake and an *increase* in foods high in potassium for a total daily intake of 40 to 150 milliequivalents per day.

⟫➤

Potassium Bicarbonate, Continued

What this mineral does:
- Potassium is the predominant positive electrolyte in body cells. An enzyme (adenosinetriphosphatase) controls flow of potassium and sodium into and out of cells to maintain normal function of heart, brain, skeletal muscles, normal kidney function, acid-base balance.

Miscellaneous information:
- Normal potassium content is reduced when foods are canned or frozen.
- Avoid peeling food.
- Avoid cooking food in large amounts of water.
- Keep meat drippings and use as gravies.

Available as:
- Effervescent tablets: Dilute in at least 1/2 glass water or other liquid. Take with meals or 1 to 1-1/2 hours after meals unless otherwise directed by your doctor.
- Potassium is not recommended for children.
- Some forms available by generic name.

Warnings and Precautions

Don't take if you:
- Take potassium-sparing diuretics, such as spironolactone, triamterene or amiloride.
- Are allergic to any potassium supplement.
- Have kidney disease.

Consult your doctor if you have:
- Addison's disease.
- Heart disease.
- Intestinal blockage.
- A stomach ulcer.
- To use diuretics.
- To use heart medicine.
- To use laxatives or if you have chronic diarrhea.
- To use salt substitutes or low-salt milk.

Over age 55:
- Observe dose schedule strictly. Potassium balance is critical. Deviation above or below normal can have serious results.

Pregnancy:
- No problems expected, except with megadoses.

Breast-feeding:
- Studies inconclusive on harm to infant. Consult doctor about supplement.
- Don't take megadoses.

Effect on lab tests:
- ECG and kidney function studies can be affected by too much or too little potassium.
- None expected on blood studies, except serum-potassium levels.

Storage:
- Store in cool, dry area away from direct light, but don't freeze.
- Store safely out of reach of children.
- Don't store in bathroom medicine cabinet. Heat and moisture may change action of the mineral.

Others:
- Take with meals or with food.

Overdose/Toxicity

Signs and symptoms:
Irregular or fast heartbeat, paralysis of arms and legs, blood-pressure drop, convulsions, coma, cardiac arrest.

What to do:
For symptoms of overdosage: Discontinue mineral, and consult doctor. Also see *Adverse Reactions or Side Effects* section below.

For accidental overdosage (such as child taking entire bottle): Dial 911 (emergency), 0 for operator or your nearest Poison Control Center. If person's heart has stopped beating, render CPR until trained help arrives.

⟫▸

Potassium Bicarbonate, Continued

 Adverse Reactions or Side Effects

Reaction or effect	What to do
Black, tarry stool	Seek emergency treatment.
Bloody stool	Seek emergency treatment.
Breathing difficulty	Seek emergency treatment.
Confusion	Discontinue. Call doctor immediately.
Diarrhea	Discontinue. Call doctor immediately.
Extreme fatigue	Discontinue. Call doctor when convenient.
Heaviness in legs	Discontinue. Call doctor when convenient.
Irregular heartbeat	Seek emergency treatment.
Nausea	Discontinue. Call doctor when convenient.
Numbness in hands or feet	Discontinue. Call doctor when convenient.
Stomach discomfort	Discontinue. Call doctor when convenient.
Tingling in hands and feet	Discontinue. Call doctor when convenient.
Vomiting	Discontinue. Call doctor immediately.
Weakness	Discontinue. Call doctor immediately.

 Interaction with Medicine, Vitamins or Minerals

Interacts with	Combined effect
Amiloride	Causes dangerous rise in blood potassium.
Atropine	Increases possibility of intestinal ulcers, which may occur with oral potassium.
Belladonna	Increases possibility of intestinal ulcers, which may occur with oral potassium.
Calcium	Increases possibility of heartbeat irregularities.
Captopril	Increases chance of excessive amounts of potassium.
Cortisone	Decreases effect of potassium.
Digitalis preparations	May cause irregular heartbeat.
Enalapril	Increases chance of excessive amounts of potassium.
Laxatives	May decrease potassium effect.
Spironolactone	Increases blood potassium.
Triamterene	Increases blood potassium.
Vitamin B-12	Extended-release tablets may decrease vitamin B-12 absorption and increase vitamin B-12 requirements.

 Interaction with Other Substances

Tobacco decreases absorption. Smokers may require supplemental potassium.

Alcohol intensifies gastrointestinal symptoms.

Cocaine may cause irregular heartbeat.

Marijuana may cause irregular heartbeat.

Beverages
• Salty drinks, such as tomato juice and commercial thirst quenchers, cause increased fluid retention.
• Coffee decreases potassium absorption and intensifies gastrointestinal symptoms.
• Low-salt milk increases fluid retention.

Foods
• Salty foods increase fluid retention.
• Sugar decreases potassium absorption.

MINERAL

Potassium Bicarbonate/Chloride

Basic Information

Brand names, see page 477.
Available from natural sources? Yes
Available from synthetic sources? Yes
Prescription required? Some yes; others no

 Natural Sources

Avocados
Bananas
Chard
Citrus fruit
Juices
  grapefruit, tomato,
  orange
Lentils, dried
Milk
Molasses

Nuts
  almonds, Brazil,
  cashews, peanuts,
  pecans, walnuts
Parsnips
Peaches, dried
Potatoes
Raisins
Sardines, canned
Spinach, fresh
Whole-grain cereals

 Reasons to Use

- Promotes regular heartbeat.
- Promotes normal muscle contraction.
- Regulates transfer of nutrients to cells.
- Maintains water balance in body tissues and cells.
- Preserves or restores normal function of nerve cells, heart cells, skeletal-muscle cells, kidneys, stomach-juice secretion.
- Treats potassium deficiency from illness or taking diuretics (water pills), cortisone drugs or digitalis preparations.

 Unproved Speculated Benefits

- Cures alcoholism.
- Cures acne.
- Cures allergies.
- Cures heart disease.
- Helps heal burns.
- Prevents high blood pressure.

 Who Needs Additional Amounts?

- People who take diuretics, cortisone drugs or digitalis preparations.
- Anyone with inadequate caloric or nutritional dietary intake or increased nutritional requirements.
- Older people (over 55 years).
- Pregnant or breast-feeding women.
- Women taking oral contraceptives.

- People who abuse alcohol or other drugs.
- Tobacco smokers.
- People with a chronic wasting illness, excess stress for long periods or who have recently undergone surgery.
- Athletes and workers who participate in vigorous physical activities, especially when endurance is an important aspect of the activity.
- Those with part of the gastrointestinal tract surgically removed.
- People with malabsorption illnesses (See Glossary).
- Those with recent severe burns or injuries.
- Vegetarians.

 Deficiency Symptoms

- Hypokalemia
- Weakness, paralysis
- Low blood pressure
- Life-threatening, irregular or rapid heartbeat that can lead to cardiac arrest and death

 Unproved Speculated Symptoms

- Acne
- Allergies
- High blood pressure

 Lab Tests to Detect Deficiency

- Serum-potassium determinations
- Serum creatinine
- Electrocardiograms
- Serum-pH determinations

 Dosage and Usage Information

Recommended Dietary Allowance (RDA):
No RDA has been established. Nutritionists recommend a *decrease* in sodium (table salt) intake and an *increase* in foods high in potassium for a total daily intake of 40 to 150 milliequivalents per day.

————————————— »→

Potassium Bicarbonate/Chloride

What this mineral does:
• Potassium is the predominant positive electrolyte in body cells. An enzyme (adenosinetriphosphatase) controls flow of potassium and sodium into and out of cells to maintain normal function of heart, brain, skeletal muscles, normal kidney function, acid-base balance.

Miscellaneous Information:
• Normal potassium content is reduced when foods are canned or frozen.
• Avoid peeling food.
• Avoid cooking food in large amounts of water.
• Keep meat drippings and use as gravies.

Available as:
• Tablets: Swallow whole with full glass of liquid. Don't chew or crush. Take with or immediately after food to decrease stomach irritation.
• Powder for effervescent oral solution: Dissolve powder in cold water or juice. Take with meals or 1 to 1-1/2 hours after meals unless otherwise directed by your doctor.
• Potassium is not recommended for children.
• Not available by generic name.

Warnings and Precautions

Don't take if you:
• Take potassium-sparing diuretics, such as spironolactone, triamterene or amiloride.
• Are allergic to any potassium supplement.
• Have kidney disease.

Consult your doctor if you have:
• Addison's disease.
• Heart disease.
• Intestinal blockage.
• A stomach ulcer.
• To use diuretics.
• To use heart medicine.
• To use laxatives or if you have chronic diarrhea.
• To use salt substitutes or low-salt milk.

Over age 55:
• Observe dose schedule strictly. Potassium balance is critical. Deviation above or below normal can have serious results.

Pregnancy:
• No problems expected, except with megadoses

Breast-feeding:
• Studies inconclusive on harm to infant. Consult doctor about supplement.
• Don't take megadoses.

Effect on lab tests:
• ECG and kidney function studies can be affected by too much or too little potassium.
• None expected on blood studies, except serum-potassium levels.

Storage:
• Store in cool, dry area away from direct light, but don't freeze.
• Store safely out of reach of children.
• Don't store in bathroom medicine cabinet. Heat and moisture may change action of the mineral.

Others:
• Take with meals or with food.

Overdose

Signs and symptoms:
Irregular or fast heartbeat, paralysis of arms and legs, blood-pressure drop, convulsions, coma, cardiac arrest.

What to do:
For symptoms of overdosage: Discontinue mineral, and consult doctor. Also see *Adverse Reactions or Side Effects* section below.
For accidental overdosage (such as child taking entire bottle): Dial 911 (emergency), 0 for operator or your nearest Poison Control Center. If person's heart has stopped beating, render CPR until trained help arrives.

»→

MINERAL

Potassium Bicarbonate/Chloride, Continued

 Adverse Reactions or Side Effects

Reaction or effect	What to do
Black, tarry stool	Seek emergency treatment.
Bloody stool	Seek emergency treatment.
Breathing difficulty	Seek emergency treatment.
Confusion	Discontinue. Call doctor immediately.
Diarrhea	Discontinue. Call doctor immediately.
Extreme fatigue	Discontinue. Call doctor when convenient.
Heaviness in legs	Discontinue. Call doctor when convenient.
Irregular heartbeat	Seek emergency treatment.
Nausea	Discontinue. Call doctor when convenient.
Numbness in hands or feet	Discontinue. Call doctor when convenient.
Stomach discomfort	Discontinue. Call doctor when convenient.
Tingling in hands and feet	Discontinue. Call doctor when convenient.
Vomiting	Discontinue. Call doctor immediately.
Weakness	Discontinue. Call doctor immediately.

 Interaction with Medicine, Vitamins or Minerals

Interacts with	Combined effect
Amiloride	Causes dangerous rise in blood potassium.
Atropine	Increases possibility of intestinal ulcers, which may occur with oral potassium tablets.
Belladonna	Increases possibility of intestinal ulcers, which may occur with oral potassium.
Calcium	Increases possibility of heartbeat irregularities.
Captopril	Increases chance of excessive amounts of potassium.
Cortisone	Decreases effect of potassium.
Digitalis preparations	May cause irregular heartbeat.
Enalapril	Increases chance of excessive amounts of potassium.
Laxatives	May decrease potassium effect.
Spironolactone	Increases blood potassium.
Triamterene	Increases blood potassium.
Vitamin B-12	Extended-release tablets may decrease vitamin B-12 absorption and increase vitamin B-12 requirements.

 Interaction with Other Substances

Tobacco decreases absorption. Smokers may require supplemental potassium.

Alcohol intensifies gastrointestinal symptoms.

Cocaine may cause irregular heartbeat.

Marijuana may cause irregular heartbeat.

Beverages
- Salty drinks, such as tomato juice and commercial thirst quenchers, cause increased fluid retention.
- Coffee decreases potassium absorption and intensifies gastrointestinal symptoms.
- Low-salt milk increases fluid retention.

Foods
- Salty foods increase fluid retention.
- Sugar decreases potassium absorption.

Potassium Bicarbonate/Citrate

Basic Information

Brand names, see page 477.
Available from natural sources? Yes
Available from synthetic sources? Yes
Prescription required? Some yes; others no

Natural Sources

Avocados
Bananas
Chard
Citrus fruit
Juices
  grapefruit, tomato,
  orange
Lentils, dried
Milk
Molasses

Nuts
  almonds, Brazil,
  cashews, peanuts,
  pecans, walnuts
Parsnips
Peaches, dried
Potatoes
Raisins
Sardines, canned
Spinach, fresh
Whole-grain cereals

Reasons to Use

- Promotes regular heartbeat.
- Promotes normal muscle contraction.
- Regulates transfer of nutrients to cells.
- Maintains water balance in body tissues and cells.
- Preserves or restores normal function of nerve cells, heart cells, skeletal-muscle cells, kidneys, stomach-juice secretion.
- Treats potassium deficiency from illness or taking diuretics (water pills), cortisone drugs or digitalis preparations.

Unproved Speculated Benefits

- Cures alcoholism.
- Cures acne.
- Cures allergies.
- Cures heart disease.
- Helps heal burns.
- Prevents high blood pressure.

Who Needs Additional Amounts?

- People who take diuretics, cortisone drugs or digitalis preparations.
- Anyone with inadequate caloric or nutritional dietary intake or increased nutritional requirements.
- Older people (over 55 years).
- Pregnant or breast-feeding women.
- Women taking oral contraceptives.
- People who abuse alcohol or other drugs.
- Tobacco smokers.
- People with a chronic wasting illness, excess stress for long periods or who have recently undergone surgery.
- Athletes and workers who participate in vigorous physical activities, especially when endurance is an important aspect of the activity.
- Those with part of the gastrointestinal tract surgically removed.
- People with malabsorption illnesses (See Glossary).
- Those with recent severe burns or injuries.
- Vegetarians.

Deficiency Symptoms

- Hypokalemia
- Weakness, paralysis
- Low blood pressure
- Life-threatening, irregular or rapid heartbeat that can lead to cardiac arrest and death

Unproved Speculated Symptoms

- Acne
- Allergies
- High blood pressure

Lab Tests to Detect Deficiency

- Serum-potassium determinations
- Serum creatinine
- Electrocardiograms
- Serum-pH determinations

Dosage and Usage Information

Recommended Dietary Allowance (RDA):
No RDA has been established. Nutritionists recommend a *decrease* in sodium (table salt) intake and an *increase* in foods high in potassium for a total daily intake of 40 to 150 milliequivalents per day.

➤➤➤

Potassium Bicarbonate/Citrate, Continued

What this mineral does:
- Potassium is the predominant positive electrolyte in body cells. An enzyme (adenosinetriphosphatase) controls flow of potassium and sodium into and out of cells to maintain normal function of heart, brain, skeletal muscles, normal kidney function, acid-base balance.

Miscellaneous Information:
- Normal potassium content is reduced when foods are canned or frozen.
- Avoid peeling food.
- Avoid cooking food in large amounts of water.
- Keep meat drippings and use as gravies.

Available as:
- Effervescent tablets: Dissolve in cold water to make a palatable, bubbly solution. Take with or immediately after food to decrease stomach irritation.
- Potassium is not recommended for children.
- Not available by generic name.

Warnings and Precautions

Don't take if you:
- Take potassium-sparing diuretics, such as spironolactone, triamterene or amiloride.
- Are allergic to any potassium supplement.
- Have kidney disease.

Consult your doctor if you have:
- Addison's disease.
- Heart disease.
- Intestinal blockage.
- A stomach ulcer.
- To use diuretics.
- To use heart medicine.
- To use laxatives or if you have chronic diarrhea.
- To use salt substitutes or low-salt milk.

Over age 55:
- Observe dose schedule strictly. Potassium balance is critical. Deviation above or below normal can have serious results.

Pregnancy:
- No problems expected, except with megadoses.

Breast-feeding:
- Studies inconclusive on harm to infant. Consult doctor about supplement.
- Don't take megadoses.

Effect on lab tests:
- ECG and kidney function studies can be affected by too much or too little potassium.
- None expected on blood studies, except serum-potassium levels.

Storage:
- Store in cool, dry area away from direct light, but don't freeze.
- Store safely out of reach of children.
- Don't store in bathroom medicine cabinet. Heat and moisture may change action of the mineral.

Others:
- Take with meals or with food.

Overdose/Toxicity

Signs and symptoms:
Irregular or fast heartbeat, paralysis of arms and legs, blood-pressure drop, convulsions, coma, cardiac arrest.

What to do:
For symptoms of overdosage: Discontinue mineral, and consult doctor. Also see *Adverse Reactions or Side Effects* section below.

For accidental overdosage (such as child taking entire bottle): Dial 911 (emergency), 0 for operator or your nearest Poison Control Center. If person's heart has stopped beating, render CPR until trained help arrives.

⟫▶

Potassium Bicarbonate/Citrate, Continued

 Adverse Reactions or Side Effects

Reaction or effect	What to do
Black, tarry stool	Seek emergency treatment.
Bloody stool	Seek emergency treatment.
Breathing difficulty	Seek emergency treatment.
Confusion	Discontinue. Call doctor immediately.
Diarrhea	Discontinue. Call doctor immediately.
Extreme fatigue	Discontinue. Call doctor when convenient.
Heaviness in legs	Discontinue. Call doctor when convenient.
Irregular heartbeat	Seek emergency treatment.
Nausea	Discontinue. Call doctor when convenient.
Numbness in hands or feet	Discontinue. Call doctor when convenient.
Stomach discomfort	Discontinue. Call doctor when convenient.
Tingling in hands and feet	Discontinue. Call doctor when convenient.
Vomiting	Discontinue. Call doctor immediately.
Weakness	Discontinue. Call doctor immediately.

 Interaction with Medicine, Vitamins or Minerals

Interacts with	Combined effect
Amiloride	Causes dangerous rise in blood potassium.
Atropine	Increases possibility of intestinal ulcers, which may occur with oral potassium tablets.
Belladonna	Increases possibility of intestinal ulcers, which may occur with oral potassium.
Calcium	Increases possibility of heartbeat irregularities.
Captopril	Increases chance of excessive amounts of potassium.
Cortisone	Decreases effect of potassium.
Digitalis preparations	May cause irregular heartbeat.
Enalapril	Increases chance of excessive amounts of potassium.
Laxatives	May decrease potassium effect.
Spironolactone	Increases blood potassium.
Triamterene	Increases blood potassium.
Vitamin B-12	Extended-release tablets may decrease vitamin B-12 absorption and increase vitamin B-12 requirements.

 Interaction with Other Substances

Tobacco decreases absorption. Smokers may require supplemental potassium.

Alcohol intensifies gastrointestinal symptoms.

Cocaine may cause irregular heartbeat.

Marijuana may cause irregular heartbeat.

Beverages
- Salty drinks, such as tomato juice and commercial thirst quenchers, cause increased fluid retention.
- Coffee decreases potassium absorption and intensifies gastrointestinal symptoms.
- Low-salt milk increases fluid retention.

Foods
- Salty foods increase fluid retention.
- Sugar decreases potassium absorption.

MINERAL

Potassium Chloride

Basic Information

Brand names, see page 477.
Available from natural sources? Yes
Available from synthetic sources? Yes
Prescription required? Some yes; others no

Natural Sources

Avocados	Nuts
Bananas	almonds, Brazil,
Chard	cashews, peanuts,
Citrus fruit	pecans, walnuts
Juices	Parsnips
grapefruit, tomato,	Peaches, dried
orange	Potatoes
Lentils, dried	Raisins
Milk	Sardines, canned
Molasses	Spinach, fresh
	Whole-grain cereals

Reasons to Use

- Promotes regular heartbeat.
- Promotes normal muscle contraction.
- Regulates transfer of nutrients to cells.
- Maintains water balance in body tissues and cells.
- Preserves or restores normal function of nerve cells, heart cells, skeletal-muscle cells, kidneys, stomach-juice secretion.
- Treats potassium deficiency from illness or taking diuretics (water pills), cortisone drugs or digitalis preparations.

Unproved Speculated Benefits

- Cures alcoholism.
- Cures acne.
- Cures allergies.
- Cures heart disease.
- Helps heal burns.
- Prevents high blood pressure.

Who Needs Additional Amounts?

- People who take diuretics, cortisone drugs or digitalis preparations.
- Anyone with inadequate caloric or nutritional dietary intake or increased nutritional requirements.
- Older people (over 55 years).
- Pregnant or breast-feeding women.

- Women taking oral contraceptives.
- People who abuse alcohol or other drugs.
- Tobacco smokers.
- People with a chronic wasting illness, excess stress for long periods or who have recently undergone surgery.
- Athletes and workers who participate in vigorous physical activities, especially when endurance is an important aspect of the activity.
- Those with part of the gastrointestinal tract surgically removed.
- People with malabsorption illnesses (See Glossary).
- People with recent severe burns or injuries.
- Vegetarians.

Deficiency Symptoms

- Hypokalemia
- Weakness, paralysis
- Low blood pressure
- Life-threatening, irregular or rapid heartbeat that can lead to cardiac arrest and death

Unproved Speculated Symptoms

- Acne
- Allergies
- High blood pressure

Lab Tests to Detect Deficiency

- Serum-potassium determinations
- Serum creatinine
- Electrocardiograms
- Serum-pH determinations

Dosage and Usage Information

Recommended Dietary Allowance (RDA):
No RDA has been established. Nutritionists recommend a *decrease* in sodium (table salt) intake and an *increase* in foods high in potassium for a total daily intake of 40 to 150 milliequivalents per day.

》▶

Potassium Chloride

What this mineral does:
• Potassium is the predominant positive electrolyte in body cells. An enzyme (adenosinetriphosphatase) controls flow of potassium and sodium into and out of cells to maintain normal function of heart, brain, skeletal muscles, normal kidney function, acid-base balance.

Miscellaneous Information:
• Normal potassium content is reduced when foods are canned or frozen.
• Avoid peeling food.
• Avoid cooking food in large amounts of water.
• Keep meat drippings and use as gravies.

Available as:
• Extended-release tablets or capsules: Swallow whole with full glass of liquid. Don't chew or crush. Take with meals or 1 to 1-1/2 hours after meals unless otherwise directed by your doctor.
• Oral solution: Dilute in 1/2 glass water or other liquid. Take with meals or 1 to 1-1/2 hours after meals unless otherwise directed by your doctor.
• Potassium chloride for oral solution: Dissolve powder in cold water or juice. Take with meals or 1 to 1-1/2 hours after meals unless otherwise directed by your doctor.
• Enteric-coated tablets are no longer recommended.
• Potassium is not recommended for children.
• Some forms available by generic name.

Warnings and Precautions

Don't take if you:
• Take potassium-sparing diuretics, such as spironolactone, triamterene or amiloride.
• Are allergic to any potassium supplement.
• Have kidney disease.

Consult your doctor if you have:
• Addison's disease.
• Heart disease.
• Intestinal blockage.
• A stomach ulcer.
• To use diuretics.
• To use heart medicine.
• To use laxatives or if you have chronic diarrhea.
• To use salt substitutes or low-salt milk.

Over age 55:
• Observe dose schedule strictly. Potassium balance is critical. Deviation above or below normal can have serious results.

Pregnancy:
• No problems expected, except with megadoses.

Breast-feeding:
• Studies inconclusive on harm to infant. Consult doctor about supplement.
• Don't take megadoses.

Effect on lab tests:
• ECG and kidney function studies can be affected by too much or too little potassium.
• None expected on blood studies, except serum-potassium levels.

Storage:
• Store in cool, dry area away from direct light, but don't freeze.
• Store safely out of reach of children.
• Don't store in bathroom medicine cabinet. Heat and moisture may change action of the mineral.

Others:
• Take with meals or with food.

Overdose/Toxicity

Signs and symptoms:
Irregular or fast heartbeat, paralysis of arms and legs, blood-pressure drop, convulsions, coma, cardiac arrest.

What to do:
For symptoms of overdosage: Discontinue mineral, and consult doctor. Also see *Adverse Reactions or Side Effects* section below.

For accidental overdosage (such as child taking entire bottle): Dial 911 (emergency), 0 for operator or your nearest Poison Control Center. If person's heart has stopped beating, render CPR until trained help arrives.

》》→

Potassium Chloride, Continued

 ## Adverse Reactions or Side Effects

Reaction or effect	What to do
Black, tarry stool	Seek emergency treatment.
Bloody stool	Seek emergency treatment.
Breathing difficulty	Seek emergency treatment.
Confusion	Discontinue. Call doctor immediately.
Diarrhea	Discontinue. Call doctor immediately.
Extreme fatigue	Discontinue. Call doctor when convenient.
Heaviness in legs	Discontinue. Call doctor when convenient.
Irregular heartbeat	Seek emergency treatment.
Nausea	Discontinue. Call doctor when convenient.
Numbness in hands or feet	Discontinue. Call doctor when convenient.
Stomach discomfort	Discontinue. Call doctor when convenient.
Tingling in hands and feet	Discontinue. Call doctor when convenient.
Vomiting	Discontinue. Call doctor immediately.
Weakness	Discontinue. Call doctor immediately.

 ## Interaction with Medicine, Vitamins or Minerals

Interacts with	Combined effect
Amiloride	Causes dangerous rise in blood potassium.
Atropine	Increases possibility of intestinal ulcers, which may occur with oral potassium tablets.
Belladonna	Increases possibility of intestinal ulcers, which may occur with oral potassium.
Calcium	Increases possibility of heartbeat irregularities.
Captopril	Increases chance of excessive amounts of potassium.
Cortisone	Decreases effect of potassium.
Digitalis preparations	May cause irregular heartbeat.
Enalapril	Increases chance of excessive amounts of potassium.
Laxatives	May decrease potassium effect.
Spironolactone	Increases blood potassium.
Triamterene	Increases blood potassium.
Vitamin B-12	Extended-release tablets may decrease vitamin B-12 absorption and increase vitamin B-12 requirements.

 ## Interaction with Other Substances

Tobacco decreases absorption. Smokers may require supplemental potassium.

Alcohol intensifies gastrointestinal symptoms.

Cocaine may cause irregular heartbeat.

Marijuana may cause irregular heartbeat.

Beverages
- Salty drinks, such as tomato juice and commercial thirst quenchers, cause increased fluid retention.
- Coffee decreases potassium absorption and intensifies gastrointestinal symptoms.
- Low-salt milk increases fluid retention.

Foods
- Don't take dairy products within 2 hours of taking potassium chloride or potassium iodide.
- Salty foods increase fluid retention.
- Sugar decreases potassium absorption.

Potassium Chloride/Bicarbonate/Citrate

Basic Information

Brand names, see page 477.
Available from natural sources? Yes
Available from synthetic sources? Yes
Prescription required? Some yes; others no

Natural Sources

Avocados	Nuts
Bananas	almonds, Brazil,
Chard	cashews, peanuts,
Citrus fruit	pecans, walnuts
Juices	Parsnips
grapefruit, tomato,	Peaches, dried
orange	Potatoes
Lentils, dried	Raisins
Milk	Sardines, canned
Molasses	Spinach, fresh
	Whole-grain cereals

Reasons to Use

- Promotes regular heartbeat.
- Promotes normal muscle contraction.
- Regulates transfer of nutrients to cells.
- Maintains water balance in body tissues and cells.
- Preserves or restores normal function of nerve cells, heart cells, skeletal-muscle cells, kidneys, stomach-juice secretion.
- Treats potassium deficiency from illness or taking diuretics (water pills), cortisone drugs or digitalis preparations.

Unproved Speculated Benefits

- Cures alcoholism.
- Cures acne.
- Cures allergies.
- Cures heart disease.
- Helps heal burns.
- Prevents high blood pressure.

Who Needs Additional Amounts?

- People who take diuretics, cortisone drugs or digitalis preparations.
- Anyone with inadequate caloric or nutritional dietary intake or increased nutritional requirements.
- Older people (over 55 years).
- Pregnant or breast-feeding women.

- Women taking oral contraceptives.
- People who abuse alcohol or other drugs.
- Tobacco smokers.
- People with a chronic wasting illness, excess stress for long periods or who have recently undergone surgery.
- Athletes and workers who participate in vigorous physical activities.
- Those with part of the gastrointestinal tract surgically removed.
- People with malabsorption illnesses (See Glossary).
- People with recent severe burns or injuries.
- Vegetarians.

Deficiency Symptoms

- Hypokalemia
- Weakness, paralysis
- Low blood pressure
- Life-threatening, irregular or rapid heartbeat that can lead to cardiac arrest and death

Unproved Speculated Symptoms

- Acne
- Allergies
- High blood pressure

Lab Tests to Detect Deficiency

- Serum-potassium determinations
- Serum creatinine
- Electrocardiograms
- Serum-pH determinations

»→

MINERAL

Potassium Chloride/Bicarbonate/Citrate, Continued

Dosage and Usage Information

Recommended Dietary Allowance (RDA):
No RDA has been established. Nutritionists recommend a *decrease* in sodium (table salt) intake and an *increase* in foods high in potassium for a total daily intake of 40 to 150 milliequivalents per day.

What this mineral does:
• Potassium is the predominant positive electrolyte in body cells. An enzyme (adenosinetriphosphatase) controls flow of potassium and sodium into and out of cells to maintain normal function of heart, brain, skeletal muscles, normal kidney function, acid-base balance.

Miscellaneous Information:
• Normal potassium content is reduced when foods are canned or frozen.
• Avoid peeling food.
• Avoid cooking food in large amounts of water.
• Keep meat drippings and use as gravies.

Available as:
• Effervescent tablets: Dissolve in cold water or other liquid. Take with meals or 1 to 1-1/2 hours after meals unless otherwise directed by your doctor.
• Potassium is not recommended for children.
• Not available by generic name.

Warnings and Precautions

Don't take if you:
• Take potassium-sparing diuretics, such as spironolactone, triamterene or amiloride.
• Are allergic to any potassium supplement.
• Have kidney disease.

Consult your doctor if you have:
• Addison's disease.
• Heart disease.
• Intestinal blockage.
• A stomach ulcer.
• To use diuretics.
• To use heart medicine.
• To use laxatives or if you have chronic diarrhea.
• To use salt substitutes or low-salt milk.

Over age 55:
• Observe dose schedule strictly. Potassium balance is critical. Deviation above or below normal can have serious results.

Pregnancy:
• No problems expected, except with megadoses.

Breast-feeding:
• Studies inconclusive on harm to infant. Consult doctor about supplement.
• Don't take megadoses.

Effect on lab tests:
• ECG and kidney function studies can be affected by too much or too little potassium.
• None expected on blood studies, except serum-potassium levels.

Storage:
• Store in cool, dry area away from direct light, but don't freeze.
• Store safely out of reach of children.
• Don't store in bathroom medicine cabinet. Heat and moisture may change action of the mineral.

Others:
• Take with meals or with food.

Overdose/Toxicity

Signs and symptoms:
Irregular or fast heartbeat, paralysis of arms and legs, blood-pressure drop, convulsions, coma, cardiac arrest.

What to do:
For symptoms of overdosage: Discontinue mineral, and consult doctor. Also see *Adverse Reactions or Side Effects* section below.
For accidental overdosage (such as child taking entire bottle): Dial 911 (emergency), 0 for operator or your nearest Poison Control Center. If person's heart has stopped beating, render CPR until trained help arrives.

Adverse Reactions or Side Effects

Reaction or effect	What to do
Black, tarry stool	Seek emergency treatment.
Bloody stool	Seek emergency treatment.
Breathing difficulty	Seek emergency treatment.
Confusion	Discontinue. Call doctor immediately.
Diarrhea	Discontinue. Call doctor immediately.
Extreme fatigue	Discontinue. Call doctor when convenient.
Heaviness in legs	Discontinue. Call doctor when convenient.
Irregular heartbeat	Seek emergency treatment.
Nausea	Discontinue. Call doctor when convenient.
Numbness in hands or feet	Discontinue. Call doctor when convenient.
Stomach discomfort	Discontinue. Call doctor when convenient.
Tingling in hands and feet	Discontinue. Call doctor when convenient.
Vomiting	Discontinue. Call doctor immediately.
Weakness	Discontinue. Call doctor immediately.

 ## Interaction with Medicine, Vitamins or Minerals

Interacts with	Combined effect
Amiloride	Causes dangerous rise in blood potassium.
Atropine	Increases possibility of intestinal ulcers, which may occur with oral potassium tablets.
Belladonna	Increases possibility of intestinal ulcers, which may occur with oral potassium.
Calcium	Increases possibility of heartbeat irregularities.
Captopril	Increases chance of excessive amounts of potassium.
Cortisone	Decreases effect of potassium.
Digitalis preparations	May cause irregular heartbeat.
Enalapril	Increases chance of excessive amounts of potassium.
Laxatives	May decrease potassium effect.
Spironolactone	Increases blood potassium.
Triamterene	Increases blood potassium.
Vitamin B-12	Extended-release tablets may decrease vitamin B-12 absorption and increase vitamin B-12 requirements.

 ## Interaction with Other Substances

Tobacco decreases absorption. Smokers may require supplemental potassium.

Alcohol intensifies gastrointestinal symptoms.

Cocaine may cause irregular heartbeat.

Marijuana may cause irregular heartbeat.

Beverages
- Salty drinks, such as tomato juice and commercial thirst quenchers, cause increased fluid retention.
- Coffee decreases potassium absorption and intensifies gastrointestinal symptoms.
- Low-salt milk increases fluid retention.

Foods
- Don't take dairy products within 2 hours of taking potassium chloride.
- Salty foods increase fluid retention.
- Sugar decreases potassium absorption.

MINERAL

Potassium Gluconate

Basic Information

Brand names, see page 477.
Available from natural sources? Yes
Available from synthetic sources? Yes
Prescription required? Some yes; others no

 Natural Sources

Avocados	Nuts
Bananas	almonds, Brazil,
Chard	cashews, peanuts,
Citrus fruit	pecans, walnuts
Juices	Parsnips
grapefruit, tomato,	Peaches, dried
orange	Potatoes
Lentils, dried	Raisins
Milk	Sardines, canned
Molasses	Spinach, fresh
	Whole-grain cereals

 Reasons to Use

- Promotes regular heartbeat.
- Promotes normal muscle contraction.
- Regulates transfer of nutrients to cells.
- Maintains water balance in body tissues and cells.
- Preserves or restores normal function of nerve cells, heart cells, skeletal-muscle cells, kidneys, stomach-juice secretion.
- Treats potassium deficiency from illness or taking diuretics (water pills), cortisone drugs or digitalis preparations.

 Unproved Speculated Benefits

- Cures alcoholism.
- Cures acne.
- Cures allergies.
- Cures heart disease.
- Helps heal burns.
- Prevents high blood pressure.

 Who Needs Additional Amounts?

- People who take diuretics, cortisone drugs or digitalis preparations.
- Anyone with inadequate caloric or nutritional dietary intake or increased nutritional requirements.
- Older people (over 55 years).
- Pregnant or breast-feeding women.
- Women taking oral contraceptives.
- People who abuse alcohol or other drugs.
- Tobacco smokers.
- People with a chronic wasting illness, excess stress for long periods or who have recently undergone surgery.
- Athletes and workers who participate in vigorous physical activities, especially when endurance is an important aspect of the activity.
- Those with part of the gastrointestinal tract surgically removed.
- People with malabsorption illnesses (See Glossary).
- People with recent severe burns or injuries.
- Vegetarians.

 Deficiency Symptoms

- Hypokalemia
- Weakness, paralysis
- Low blood pressure
- Life-threatening, irregular or rapid heartbeat that can lead to cardiac arrest and death

 Unproved Speculated Symptoms

- Acne
- Allergies
- High blood pressure

 Lab Tests to Detect Deficiency

- Serum-potassium determinations
- Serum creatinine
- Electrocardiograms
- Serum-pH determinations

 Dosage and Usage Information

Recommended Dietary Allowance (RDA):
No RDA has been established. Nutritionists recommend a *decrease* in sodium (table salt) intake and an *increase* in foods high in potassium for a total daily intake of 40 to 150 milliequivalents per day.

What this mineral does:
- Potassium is the predominant positive electrolyte in body cells. An enzyme (adenosinetriphosphatase) controls flow of potassium and sodium into and out of cells ▸▸▸

to maintain normal function of heart, brain, skeletal muscles, normal kidney function, acid-base balance.

Miscellaneous Information:
- Normal potassium content is reduced when foods are canned or frozen.
- Avoid peeling food.
- Avoid cooking food in large amounts of water.
- Keep meat drippings and use as gravies.

Available as:
- Extended-release capsules or tablets: Swallow whole with full glass of liquid. Don't chew or crush. Take with meals or 1 to 1-1/2 hours after meals unless otherwise directed by your doctor.
- Elixir for oral solution: Dilute in at least 1/2 glass water or other liquid. Take with meals or 1 to 1-1/2 hours after meals unless otherwise directed by your doctor.
- Potassium is not recommended for children.
- Some forms available by generic name.

Warnings and Precautions

Don't take if you:
- Take potassium-sparing diuretics, such as spironolactone, triamterene or amiloride.
- Are allergic to any potassium supplement.
- Have kidney disease.

Consult your doctor if you have:
- Addison's disease.
- Heart disease.
- Intestinal blockage.
- A stomach ulcer.
- To use diuretics.
- To use heart medicine.
- To use laxatives or if you have chronic diarrhea.
- To use salt substitutes or low-salt milk.

Over age 55:
- Observe dose schedule strictly. Potassium balance is critical. Deviation above or below normal can have serious results.

Pregnancy:
- No problems expected, except with megadoses.

Breast-feeding:
- Studies inconclusive on harm to infant. Consult doctor about supplement.
- Don't take megadoses.

Effect on lab tests:
- ECG and kidney function studies can be affected by too much or too little potassium.
- None expected on blood studies, except serum-potassium levels.

Storage:
- Store in cool, dry area away from direct light, but don't freeze.
- Store safely out of reach of children.
- Don't store in bathroom medicine cabinet. Heat and moisture may change action of the mineral.

Others:
- Take with meals or with food.

Overdose/Toxicity

Signs and symptoms:
Irregular or fast heartbeat, paralysis of arms and legs, blood-pressure drop, convulsions, coma, cardiac arrest.

What to do:
For symptoms of overdosage: Discontinue mineral, and consult doctor. Also see *Adverse Reactions or Side Effects* section below.
For accidental overdosage (such as child taking entire bottle): Dial 911 (emergency), 0 for operator or your nearest Poison Control Center. If person's heart has stopped beating, render CPR until trained help arrives.

————————————————— ≫➤

Potassium Gluconate, Continued

 Adverse Reactions or Side Effects

Reaction or effect	What to do
Black, tarry stool	Seek emergency treatment.
Bloody stool	Seek emergency treatment.
Breathing difficulty	Seek emergency treatment.
Confusion	Discontinue. Call doctor immediately.
Diarrhea	Discontinue. Call doctor immediately.
Extreme fatigue	Discontinue. Call doctor when convenient.
Heaviness in legs	Discontinue. Call doctor when convenient.
Irregular heartbeat	Seek emergency treatment.
Nausea	Discontinue. Call doctor when convenient.
Numbness in hands or feet	Discontinue. Call doctor when convenient.
Stomach discomfort	Discontinue. Call doctor when convenient.
Tingling in hands and feet	Discontinue. Call doctor when convenient.
Vomiting	Discontinue. Call doctor immediately.
Weakness	Discontinue. Call doctor immediately.

 Interaction with Medicine, Vitamins or Minerals

Interacts with	Combined effect
Amiloride	Causes dangerous rise in blood potassium.
Atropine	Increases possibility of intestinal ulcers, which may occur with oral potassium tablets.
Belladonna	Increases possibility of intestinal ulcers, which may occur with oral potassium.
Calcium	Increases possibility of heartbeat irregularities.
Captopril	Increases chance of excessive amounts of potassium.
Cortisone	Decreases effect of potassium.
Digitalis preparations	May cause irregular heartbeat.
Enalapril	Increases chance of excessive amounts of potassium.
Laxatives	May decrease potassium effect.
Spironolactone	Increases blood potassium.
Triamterene	Increases blood potassium.
Vitamin B-12	Extended-release tablets may decrease vitamin B-12 absorption and increase vitamin B-12 requirements.

 Interaction with Other Substances

Tobacco decreases absorption. Smokers may require supplemental potassium.

Alcohol intensifies gastrointestinal symptoms.

Cocaine may cause irregular heartbeat.

Marijuana may cause irregular heartbeat.

Beverages
- Salty drinks, such as tomato juice and commercial thirst quenchers, cause increased fluid retention.
- Coffee decreases potassium absorption and intensifies gastrointestinal symptoms.
- Low-salt milk increases fluid retention.

Foods
- Salty foods increase fluid retention.
- Sugar decreases potassium absorption.

Potassium Gluconate/Chloride

Basic Information

Brand names, see page 477.
Available from natural sources? Yes
Available from synthetic sources? Yes
Prescription required? Some yes; others no

Natural Sources

Avocados
Bananas
Chard
Citrus fruit
Juices
  grapefruit, tomato,
  orange
Lentils, dried
Milk
Molasses

Nuts
  almonds, Brazil,
  cashews, peanuts,
  pecans, walnuts
Parsnips
Peaches, dried
Potatoes
Raisins
Sardines, canned
Spinach, fresh
Whole-grain cereals

Reasons to Use

- Promotes regular heartbeat.
- Promotes normal muscle contraction.
- Regulates transfer of nutrients to cells.
- Maintains water balance in body tissues and cells.
- Preserves or restores normal function of nerve cells, heart cells, skeletal-muscle cells, kidneys, stomach-juice secretion.
- Treats potassium deficiency from illness or taking diuretics (water pills), cortisone drugs or digitalis preparations.

Unproved Speculated Benefits

- Cures alcoholism.
- Cures acne.
- Cures allergies.
- Cures heart disease.
- Helps heal burns.
- Prevents high blood pressure.

Who Needs Additional Amounts?

- People who take diuretics, cortisone drugs or digitalis preparations.
- Anyone with inadequate caloric or nutritional dietary intake or increased nutritional requirements.
- Older people (over 55 years).
- Pregnant or breast-feeding women.
- Women taking oral contraceptives.

- People who abuse alcohol or other drugs.
- Tobacco smokers.
- People with a chronic wasting illness, excess stress for long periods or who have recently undergone surgery.
- Athletes and workers who participate in vigorous physical activities, especially when endurance is an important aspect of the activity.
- Those with part of the gastrointestinal tract surgically removed.
- People with malabsorption illnesses (See Glossary).
- People with recent severe burns or injuries.
- Vegetarians.

Deficiency Symptoms

- Hypokalemia
- Weakness, paralysis
- Low blood pressure
- Life-threatening, irregular or rapid heartbeat that can lead to cardiac arrest and death

Unproved Speculated Symptoms

- Acne
- Allergies
- High blood pressure

Lab Tests to Detect Deficiency

- Serum-potassium determinations
- Serum creatinine
- Electrocardiograms
- Serum-pH determinations

Potassium Gluconate/Chloride, Continued

Dosage and Usage Information

Recommended Dietary Allowance (RDA):
No RDA has been established. Nutritionists recommend a *decrease* in sodium (table salt) intake and an *increase* in foods high in potassium for a total daily intake of 40 to 150 milliequivalents per day.

What this mineral does:
• Potassium is the predominant positive electrolyte in body cells. An enzyme (adenosinetriphosphatase) controls flow of potassium and sodium into and out of cells to maintain normal function of heart, brain, skeletal muscles, normal kidney function, acid-base balance.

Miscellaneous Information:
• Normal potassium content is reduced when foods are canned or frozen.
• Avoid peeling food.
• Avoid cooking food in large amounts of water.
• Keep meat drippings and use as gravies.

Available as:
• Oral solution: Dilute in at least 1/2 glass water or other liquid. Take with meals or 1 to 1-1/2 hours after meals unless otherwise directed by your doctor.
• Powder for oral solution: Dissolve powder in cold water or juice. Take with meals or 1 to 1-1/2 hours after meals unless othewise directed by your doctor.
• Special instructions for children.
• Not available by generic name.

Warnings and Precautions

Don't take if you:
• Take potassium-sparing diuretics, such as spironolactone, triamterene or amiloride.
• Are allergic to any potassium supplement.
• Have kidney disease.

Consult your doctor if you have:
• Addison's disease.
• Heart disease.
• Intestinal blockage.
• A stomach ulcer.
• To use diuretics.
• To use heart medicine.
• To use laxatives or if you have chronic diarrhea.
• To use salt substitutes or low-salt milk.

Over age 55:
• Observe dose schedule strictly. Potassium balance is critical. Deviation above or below normal can have serious results.

Pregnancy:
• No problems expected, except with megadoses.

Breast-feeding:
• Studies inconclusive on harm to infant. Consult doctor about supplement.
• Don't take megadoses.

Effect on lab tests
• ECG and kidney function studies can be affected by too much or too little potassium.
• None expected on blood studies, except serum-potassium levels.

Storage:
• Store in cool, dry area away from direct light, but don't freeze.
• Store safely out of reach of children.
• Don't store in bathroom medicine cabinet. Heat and moisture may change action of the mineral.

Others:
• Take with meals or with food.

Overdose/Toxicity

Signs and symptoms:
Irregular or fast heartbeat, paralysis of arms and legs, blood-pressure drop, convulsions, coma, cardiac arrest.

What to do:
For symptoms of overdosage: Discontinue mineral, and consult doctor. Also see *Adverse Reactions or Side Effects* section below.
For accidental overdosage (such as child taking entire bottle): Dial 911 (emergency), 0 for operator or your nearest Poison Control Center. If person's heart has stopped beating, render CPR until trained help arrives.

⟫➤

Adverse Reactions or Side Effects

Reaction or effect	What to do
Black, tarry stool	Seek emergency treatment.
Bloody stool	Seek emergency treatment.
Breathing difficulty	Seek emergency treatment.
Confusion	Discontinue. Call doctor immediately.
Diarrhea	Discontinue. Call doctor immediately.
Extreme fatigue	Discontinue. Call doctor when convenient.
Heaviness in legs	Discontinue. Call doctor when convenient.
Irregular heartbeat	Seek emergency treatment.
Nausea	Discontinue. Call doctor when convenient.
Numbness in hands or feet	Discontinue. Call doctor when convenient.
Stomach discomfort	Discontinue. Call doctor when convenient.
Tingling in hands and feet	Discontinue. Call doctor when convenient.
Vomiting	Discontinue. Call doctor immediately.
Weakness	Discontinue. Call doctor immediately.

Interaction with Medicine, Vitamins or Minerals

Interacts with	Combined effect
Amiloride	Causes dangerous rise in blood potassium.
Atropine	Increases possibility of intestinal ulcers, which may occur with oral potassium tablets.
Belladonna	Increases possibility of intestinal ulcers, which may occur with oral potassium.
Calcium	Increases possibility of heartbeat irregularities.
Captopril	Increases chance of excessive amounts of potassium.
Cortisone	Decreases effect of potassium.
Digitalis preparations	May cause irregular heartbeat.
Enalapril	Increases chance of excessive amounts of potassium.
Laxatives	May decrease potassium effect.
Spironolactone	Increases blood potassium.
Triamterene	Increases blood potassium.
Vitamin B-12	Extended-release tablets may decrease vitamin B-12 absorption and increase vitamin B-12 requirements.

Interaction with Other Substances

Tobacco decreases absorption. Smokers may require supplemental potassium.

Alcohol intensifies gastrointestinal symptoms.

Cocaine may cause irregular heartbeat.

Marijuana may cause irregular heartbeat.

Beverages
- Salty drinks, such as tomato juice and commercial thirst quenchers, cause increased fluid retention.
- Coffee decreases potassium absorption and intensifies gastrointestinal symptoms.
- Low-salt milk increases fluid retention.

Foods
- Don't take dairy products within 2 hours of taking potassium chloride.
- Salty foods increase fluid retention.
- Sugar decreases potassium absorption.

MINERAL

Potassium Gluconate/Citrate

Basic Information

Brand names, see page 477.
Available from natural sources? Yes
Available from synthetic sources? Yes
Prescription required? Some yes; others no

 Natural Sources

Avocados	Nuts
Bananas	almonds, Brazil,
Chard	cashews, peanuts,
Citrus fruit	pecans, walnuts
Juices	Parsnips
grapefruit, tomato,	Peaches, dried
orange	Potatoes
Lentils, dried	Raisins
Milk	Sardines, canned
Molasses	Spinach, fresh
	Whole-grain cereals

 Reasons to Use

- Promotes regular heartbeat.
- Promotes normal muscle contraction.
- Regulates transfer of nutrients to cells.
- Maintains water balance in body tissues and cells.
- Preserves or restores normal function of nerve cells, heart cells, skeletal-muscle cells, kidneys, stomach-juice secretion.
- Treats potassium deficiency from illness or taking diuretics (water pills), cortisone drugs or digitalis preparations.

 Unproved Speculated Benefits

- Cures alcoholism.
- Cures acne.
- Cures allergies.
- Cures heart disease.
- Helps heal burns.
- Prevents high blood pressure.

 Who Needs Additional Amounts?

- People who take diuretics, cortisone drugs or digitalis preparations.
- Anyone with inadequate caloric or nutritional dietary intake or increased nutritional requirements.
- Older people (over 55 years).
- Pregnant or breast-feeding women.
- Women taking oral contraceptives.
- People who abuse alcohol or other drugs.
- Tobacco smokers.
- People with a chronic wasting illness, excess stress for long periods or who have recently undergone surgery.
- Athletes and workers who participate in vigorous physical activities, especially when endurance is an important aspect of the activity.
- Those with part of the gastrointestinal tract surgically removed.
- People with malabsorption illnesses (See Glossary).
- People with recent severe burns or injuries.
- Vegetarians.

 Deficiency Symptoms

- Hypokalemia
- Weakness, paralysis
- Low blood pressure
- Life-threatening, irregular or rapid heartbeat that can lead to cardiac arrest and death

 Unproved Speculated Symptoms

- Acne
- Allergies
- High blood pressure

 Lab Tests to Detect Deficiency

- Serum-potassium determinations
- Serum creatinine
- Electrocardiograms
- Serum-pH determinations

 Dosage and Usage Information

Recommended Dietary Allowance (RDA):
No RDA has been established. Nutritionists recommend a *decrease* in sodium (table salt) intake and an *increase* in foods high in potassium for a total daily intake of 40 to 150 milliequivalents per day.

What this mineral does:
- Potassium is the predominant positive electrolyte in body cells. An enzyme (adenosinetriphosphatase) controls flow of potassium and sodium into and out of cells to maintain normal function of heart, brain,

➠➠

skeletal muscles, normal kidney function, acid-base balance.

Miscellaneous Information:
- Normal potassium content is reduced when foods are canned or frozen.
- Avoid peeling food.
- Avoid cooking food in large amounts of water.
- Keep meat drippings and use as gravies.

Available as:
- Oral solution: Dilute in at least 1/2 glass water or other liquid. Take with meals or 1 to 1-1/2 hours after meals unless otherwise directed by your doctor.
- Special instructions for children.
- Not available by generic name.

Warnings and Precautions

Don't take if you:
- Take potassium-sparing diuretics, such as spironolactone, triamterene or amiloride.
- Are allergic to any potassium supplement.
- Have kidney disease.

Consult your doctor if you have:
- Addison's disease.
- Heart disease.
- Intestinal blockage.
- A stomach ulcer.
- To use diuretics.
- To use heart medicine.
- To use laxatives or if you have chronic diarrhea.
- To use salt substitutes or low-salt milk.

Over age 55:
- Observe dose schedule strictly. Potassium balance is critical. Deviation above or below normal can have serious results.

Pregnancy:
- No problems expected, except with megadoses.

Breast-feeding:
- Studies inconclusive on harm to infant. Consult doctor about supplement.
- Don't take megadoses.

Effect on lab tests:
- ECG and kidney function studies can be affected by too much or too little potassium.
- None expected on blood studies, except serum-potassium levels.

Storage:
- Store in cool, dry area away from direct light, but don't freeze.
- Store safely out of reach of children.
- Don't store in bathroom medicine cabinet. Heat and moisture may change action of the mineral.

Others:
- Take with meals or with food.

Overdose/Toxicity

Signs and symptoms:
Irregular or fast heartbeat, paralysis of arms and legs, blood-pressure drop, convulsions, coma, cardiac arrest.

What to do:
For symptoms of overdosage: Discontinue mineral, and consult doctor. Also see *Adverse Reactions or Side Effects* section below.
For accidental overdosage (such as child taking entire bottle): Dial 911 (emergency), 0 for operator or your nearest Poison Control Center. If person's heart has stopped beating, render CPR until trained help arrives.

MINERAL

Potassium Gluconate/Citrate, Continued

Adverse Reactions or Side Effects

Reaction or effect	What to do
Black, tarry stool	Seek emergency treatment.
Bloody stool	Seek emergency treatment.
Breathing difficulty	Seek emergency treatment.
Confusion	Discontinue. Call doctor immediately.
Diarrhea	Discontinue. Call doctor immediately.
Extreme fatigue	Discontinue. Call doctor when convenient.
Heaviness in legs	Discontinue. Call doctor when convenient.
Irregular heartbeat	Seek emergency treatment.
Nausea	Discontinue. Call doctor when convenient.
Numbness in hands or feet	Discontinue. Call doctor when convenient.
Stomach discomfort	Discontinue. Call doctor when convenient.
Tingling in hands and feet	Discontinue. Call doctor when convenient.
Vomiting	Discontinue. Call doctor immediately.
Weakness	Discontinue. Call doctor immediately.

Interaction with Medicine, Vitamins or Minerals

Interacts with	Combined effect
Amiloride	Causes dangerous rise in blood potassium.
Atropine	Increases possibility of intestinal ulcers, which may occur with oral potassium tablets.
Belladonna	Increases possibility of intestinal ulcers, which may occur with oral potassium.
Calcium	Increases possibility of heartbeat irregularities.
Captopril	Increases chance of excessive amounts of potassium.
Cortisone	Decreases effect of potassium.
Digitalis preparations	May cause irregular heartbeat.
Enalapril	Increases chance of excessive amounts of potassium.
Laxatives	May decrease potassium effect.
Spironolactone	Increases blood potassium.
Triamterene	Increases blood potassium.
Vitamin B-12	Extended-release tablets may decrease vitamin B-12 absorption and increase vitamin B-12 requirements.

Interaction with Other Substances

Tobacco decreases absorption. Smokers may require supplemental potassium.

Alcohol intensifies gastrointestinal symptoms.

Cocaine may cause irregular heartbeat.

Marijuana may cause irregular heartbeat.

Beverages
• Salty drinks, such as tomato juice and commercial thirst quenchers, cause increased fluid retention.
• Coffee decreases potassium absorption and intensifies gastrointestinal symptoms.
• Low-salt milk increases fluid retention.

Foods
• Salty foods increase fluid retention.
• Sugar decreases potassium absorption.

Potassium Gluconate/Citrate/Ammonium Chloride

Basic Information

Brand names, see page 477.
Available from natural sources? Yes
Available from synthetic sources? Yes
Prescription required? Some yes; others no

Natural Sources

Avocados
Bananas
Chard
Citrus fruit
Juices
  grapefruit, tomato,
  orange
Lentils, dried
Milk
Molasses

Nuts
  almonds, Brazil,
  cashews, peanuts,
  pecans, walnuts
Parsnips
Peaches, dried
Potatoes
Raisins
Sardines, canned
Spinach, fresh
Whole-grain cereals

Reasons to Use

- Promotes regular heartbeat.
- Promotes normal muscle contraction.
- Regulates transfer of nutrients to cells.
- Maintains water balance in body tissues and cells.
- Preserves or restores normal function of nerve cells, heart cells, skeletal-muscle cells, kidneys, stomach-juice secretion.
- Treats potassium deficiency from illness or taking diuretics (water pills), cortisone drugs or digitalis preparations.

Unproved Speculated Benefits

- Cures alcoholism.
- Cures acne.
- Cures allergies.
- Cures heart disease.
- Helps heal burns.
- Prevents high blood pressure.

Who Needs Additional Amounts?

- People who take diuretics, cortisone drugs or digitalis preparations.
- Anyone with inadequate caloric or nutritional dietary intake or increased nutritional requirements.
- Older people (over 55 years).
- Pregnant or breast-feeding women.
- Women taking oral contraceptives.
- People who abuse alcohol or other drugs.
- Tobacco smokers.
- People with a chronic wasting illness, excess stress for long periods or who have recently undergone surgery.
- Athletes and workers who participate in vigorous physical activities, especially when endurance is an important aspect of the activity.
- Those with part of the gastrointestinal tract surgically removed.
- People with malabsorption illnesses (See Glossary).
- People with recent severe burns or injuries.
- Vegetarians.

Deficiency Symptoms

- Hypokalemia
- Weakness, paralysis
- Low blood pressure
- Life-threatening, irregular or rapid heartbeat that can lead to cardiac arrest and death

Unproved Speculated Symptoms

- Acne
- Allergies
- High blood pressure

Lab Tests to Detect Deficiency

- Serum-potassium determinations
- Serum creatinine
- Electrocardiograms
- Serum-pH determinations

》▶

Potassium Gluconate/Citrate/Ammonium Chloride,
Continued

 ## Dosage and Usage Information

Recommended Dietary Allowance (RDA):
No RDA has been established. Nutritionists recommend a *decrease* in sodium (table salt) intake and an *increase* in foods high in potassium for a total daily intake of 40 to 150 milliequivalents per day.

What this mineral does:
• Potassium is the predominant positive electrolyte in body cells. An enzyme (adenosinetriphosphatase) controls flow of potassium and sodium into and out of cells to maintain normal function of heart, brain, skeletal muscles, normal kidney function, acid-base balance.

Miscellaneous Information:
• Normal potassium content is reduced when foods are canned or frozen.
• Avoid peeling food.
• Avoid cooking food in large amounts of water.
• Keep meat drippings and use as gravies.

Available as:
• Oral solution: Dilute in at least 1/2 glass water or other liquid. Take with meals or 1 to 1-1/2 hours after meals unless otherwise directed by your doctor.
• Special instructions for children.
• Not available by generic name.

 ## Warnings and Precautions

Don't take if you:
• Take potassium-sparing diuretics, such as spironolactone, triamterene or amiloride.
• Are allergic to any potassium supplement.
• Have kidney disease.

Consult your doctor if you have:
• Addison's disease.
• Heart disease.
• Intestinal blockage.
• A stomach ulcer.
• To use diuretics.
• To use heart medicine.
• To use laxatives or if you have chronic diarrhea.
• To use salt substitutes or low-salt milk.

Over age 55:
• Observe dose schedule strictly. Potassium balance is critical. Deviation above or below normal can have serious results.

Pregnancy:
• No problems expected, except with megadoses.

Breast-feeding:
• Studies inconclusive on harm to infant. Consult doctor about supplement.
• Don't take megadoses.

Effect on lab tests:
• ECG and kidney function studies can be affected by too much or too little potassium.
• None expected on blood studies, except serum-potassium levels.

Storage:
• Store in cool, dry area away from direct light, but don't freeze.
• Store safely out of reach of children.
• Don't store in bathroom medicine cabinet. Heat and moisture may change action of the mineral.

Others:
• Take with meals or with food.

 ## Overdose/Toxicity

Signs and symptoms:
Irregular or fast heartbeat, paralysis of arms and legs, blood-pressure drop, convulsions, coma, cardiac arrest.

What to do:
For symptoms of overdosage: Discontinue mineral, and consult doctor. Also see *Adverse Reactions or Side Effects* section below.
For accidental overdosage (such as child taking entire bottle): Dial 911 (emergency), 0 for operator or your nearest Poison Control Center. If person's heart has stopped beating, render CPR until trained help arrives.

Adverse Reactions or Side Effects

Reaction or effect	What to do
Black, tarry stool	Seek emergency treatment.
Bloody stool	Seek emergency treatment.
Breathing difficulty	Seek emergency treatment.
Confusion	Discontinue. Call doctor immediately.
Diarrhea	Discontinue. Call doctor immediately.
Extreme fatigue	Discontinue. Call doctor when convenient.
Heaviness in legs	Discontinue. Call doctor when convenient.
Irregular heartbeat	Seek emergency treatment.
Nausea	Discontinue. Call doctor when convenient.
Numbness in hands or feet	Discontinue. Call doctor when convenient.
Stomach discomfort	Discontinue. Call doctor when convenient.
Tingling in hands and feet	Discontinue. Call doctor when convenient.
Vomiting	Discontinue. Call doctor immediately.
Weakness	Discontinue. Call doctor immediately.

 Interaction with Medicine, Vitamins or Minerals

Interacts with	Combined effect
Amiloride	Causes dangerous rise in blood potassium.
Atropine	Increases possibility of intestinal ulcers, which may occur with oral potassium tablets.
Belladonna	Increases possibility of intestinal ulcers, which may occur with oral potassium.
Calcium	Increases possibility of heartbeat irregularities.
Captopril	Increases chance of excessive amounts of potassium.
Cortisone	Decreases effect of potassium.
Digitalis preparations	May cause irregular heartbeat.
Enalapril	Increases chance of excessive amounts of potassium.
Laxatives	May decrease potassium effect.
Spironolactone	Increases blood potassium.
Triamterene	Increases blood potassium.
Vitamin B-12	Extended-release tablets may decrease vitamin B-12 absorption and increase vitamin B-12 requirements.

 Interaction with Other Substances

Tobacco decreases absorption. Smokers may require supplemental potassium.

Alcohol intensifies gastrointestinal symptoms.

Cocaine may cause irregular heartbeat.

Marijuana may cause irregular heartbeat.

Beverages
- Salty drinks, such as tomato juice and commercial thirst quenchers, cause increased fluid retention.
- Coffee decreases potassium absorption and intensifies gastrointestinal symptoms.
- Low-salt milk increases fluid retention.

Foods
- Salty foods increase fluid retention.
- Sugar decreases potassium absorption.

MINERAL

Potassium Phosphate

Basic Information

Potassium phosphate is a phosphate supplement. It does not function as a potassium supplement.
Brand names, see page 477.
Available from natural sources? Yes
Available from synthetic sources? No
Prescription required? Yes, for medical purposes

Natural Sources

Almonds	Peanuts
Beans, dried	Peas
Calves' liver	Poultry
Cheese, cheddar	Pumpkin seeds
Cheese, pasteurized	Red meat
process	Sardines, canned
Eggs	Scallops
Fish	Soybeans
Milk	Sunflower seeds
Milk products	Tuna
	Whole-grain products

Reasons to Use

• Builds strong bones and teeth (with calcium).
• Promotes energy metabolism.
• Promotes growth, maintenance and repair of all body tissues.
• Buffers body fluids for acid-base balance.
• Acidifies urine and reduces possibility of kidney stones.

Unproved Speculated Benefits

• Reduces effects of stress.
• Accelerates growth in children.
• Helps reduce pain of arthritis.

Who Needs Additional Amounts?

• Anyone suffering prolonged vomiting.
• Those with inadequate caloric or dietary intake or increased nutritional requirements.
• Those who take excessive amounts of antacid.
• Older people (over 55 years).
• Those who abuse alcohol or other drugs. Alcoholics probably need phosphate supplementation.
• People with a chronic wasting illness, excess stress for long periods or who have recently undergone surgery.

• Those with liver disease.
• People with hyperparathyroidism.

Deficiency Symptoms

• Bone pain
• Loss of appetite
• Weakness
• Easily broken bones

Unproved Speculated Symptoms

• Rickets

Lab Tests to Detect Deficiency

• Serum phosphorous

Dosage and Usage Information

Recommended Dietary Allowance (RDA):
Estimate of adequate daily intake by the Food and Nutrition Board of the National Research Council, 1980 (See Glossary).

Age	RDA
0–6 months	240mg
6–12 months	360mg
1–10 years	800mg
11–17 years	1200mg
18+ years	800mg
Pregnant	+400mg
Lactating	+400mg

What this mineral does:
• Necessary for utilization of many B-complex vitamins.
• An important constituent of all fats, proteins, carbohydrates and many enzymes.

Available as:
• Tablets: Swallow whole with full glass of liquid. Don't chew or crush. Take with meals or 1 to 1-1/2 hours after meals unless otherwise directed by your doctor.
• Capsules for oral solution: Empty contents into at least 1/2 glass water or other liquid. Don't swallow filled capsule. Take with meals or 1 to 1-1/2 hours after meals unless otherwise directed by your doctor.
• Oral solution: Dilute in at least 1/2 glass water or other liquid. Take with meals or 1 to 1-1/2 hours after meals unless otherwise directed by your doctor.

»→

• A constituent of many multivitamin/mineral preparations.

Warnings and Precautions

Don't take if you:
• Have severe kidney disease.
• Have kidney stones and analysis has shown their composition to be magnesium ammonium phosphate.

Consult your doctor if you have:
• Hypoparathyroidism.
• Osteomalacia.
• Acute pancreatitis.
• Chronic kidney disease.
• Rickets.
• Adrenal insufficiency (Addison's disease).
• Dehydration.
• Severe burns.
• Heart disease.

Over age 55:
• No special problems expected.

Pregnancy:
• Take under doctor's supervision only.
• Don't take megadoses.

Breast-feeding:
• Take under doctor's supervision only.
• Don't take megadoses.

Effect on lab tests:
• May show false decrease in bone uptake in technetium-labeled diagnostic-imaging tests.

Storage:
• Store in cool, dry place away from direct light, but don't freeze.
• Store safely out of reach of children.
• Don't store in bathroom medicine cabinet. Heat and moisture may change action of mineral.

Overdose/Toxicity

Signs and symptoms:
Seizures, heartbeat irregularities, shortness of breath.

What to do:
For symptoms of overdosage: Discontinue mineral, and seek emergency treatment. Also see *Adverse Reactions or Side Effects* section below.
For accidental overdosage (such as child taking entire bottle): Dial 911 (emergency), 0 for operator or your nearest Poison Control Center.

Adverse Reactions or Side Effects

Reaction or effect	What to do
Abdominal pain	Discontinue. Call doctor immediately.
Bone or joint pain	Discontinue. Call doctor immediately.
Confusion	Discontinue. Call doctor immediately.
Decreased volume of urine in one day	Seek emergency treatment.
Diarrhea	Discontinue. Call doctor immediately.
Easy fatigue	Discontinue. Call doctor when convenient.
Edema of feet or legs	Discontinue. Call doctor when convenient.
Headaches	Discontinue. Call doctor immediately.
Muscle cramps	Discontinue. Call doctor when convenient.
Numbness or tingling in hands or feet	Discontinue. Call doctor when convenient.
Unusual thirst	Discontinue. Call doctor when convenient.

≫➤

MINERAL

Potassium Phosphate, Continued

 Interaction with Medicine, Vitamins or Minerals

Interacts with	Combined effect
Anabolic steroids	Increases risk of edema.
Antacids with aluminum or magnesium	May prevent absorption of phosphates.
Calcium-containing supplements and antacids	Increases risk of depositing calcium in soft tissues. Decreases phosphate absorption.
Captopril	Increases risk of too much potassium (hyperkalemia).
Cortisone drugs or ACTH	Increases serum sodium.
Digitalis preparations	Increases risk of too much potassium (hyperkalemia).
Diuretics, potassium-conserving (amiloride, spironelactene, triamterene)	Increases risk of too much potassium (hyperkalemia).
Emalapril	Increases risk of too much potassium (hyperkalemia).
Salicylates	May increase plasma concentration of salicylates.
Testosterone	Increases risk of edema.
Vitamin D	Phosphate absorption enhanced, but may increase chance of too much phosphorous in blood and body cells.

 Interaction with Other Substances

Beverages
- Alcoholic beverages decrease available phosphorous for vital body functions.
- Overconsumption of soft drinks may adversely affect absorption of phosphorous and calcium.

Foods
- Overconsumption of rhubarb, spinach and bran may decrease absorption of potassium phosphates.
- Overconsumption of meats and convenience foods may adversely affect absorption of phosphorous and calcium.

Basic Information

Potassium/sodium phosphate is a phosphate supplement. It does not function as a potassium supplement.
Brand names, see page 477.
Available from natural sources? Yes
Available from synthetic sources? No
Prescription required? Yes for medical purposes

Natural Sources

Almonds	Peanuts
Beans, dried	Peas
Calves' liver	Poultry
Cheese, cheddar	Pumpkin seeds
Cheese, pasteurized	Red meat
process	Sardines, canned
Eggs	Scallops
Fish	Soybeans
Milk	Sunflower seeds
Milk products	Tuna
	Whole-grain products

Reasons to Use

- Builds strong bones and teeth (with calcium).
- Promotes energy metabolism.
- Promotes growth, maintenance and repair of all body tissues.
- Buffers body fluids for acid-base balance.
- Acidifies urine and reduces possibility of kidney stones.

Unproved Speculated Benefits

- Reduces effects of stress.
- Accelerates growth in children.
- Helps reduce pain of arthritis.

Who Needs Additional Amounts?

- People suffering prolonged vomiting.
- Anyone with inadequate caloric or nutritional dietary intake or increased nutritional requirements.
- Those who take excessive amounts of antacids.
- Older people (over 55 years).
- Those who abuse alcohol or other drugs. Alcoholics most probably need phosphate supplementation.

- People with a chronic wasting illness, excess stress for long periods or who have recently undergone surgery.
- Those with liver disease.
- People with hyperparathyroidism.

Deficiency Symptoms

Proven symptoms:
- Bone pain
- Loss of appetite
- Weakness
- Easily broken bones

Unproved Speculated Symptoms

- Rickets

Lab Tests to Detect Deficiency

- Serum phosphorous

Dosage and Usage Information

Recommended Dietary Allowance (RDA):
Estimate of adequate daily intake by the Food and Nutrition Board of the National Research Council, 1980 (See Glossary).

Age	RDA
0–6 months	240mg
6–12 months	360mg
1–10 years	800mg
11–17 years	1200mg
18+ years	800mg
Pregnant	+400mg
Lactating	+400mg

What this mineral does:
- Necessary for utilization of many B-complex vitamins.
- An important constituent of all fats, proteins, carbohydrates and many enzymes.

Available as:
- Tablets: Swallow whole with full glass of liquid. Don't chew or crush. Take with meals or 1 to 1-1/2 hours after meals unless otherwise directed by your doctor.
- Capsules for oral solution: Empty contents into at least 1/2 glass water or other liquid. Don't swallow filled capsule. Take with meals or 1 to 1-1/2 hours after meals unless otherwise directed by your doctor.

≫➤

Potassium/Sodium Phosphate, Continued

- Oral solution: Dilute in at least 1/2 glass water or other liquid. Take with meals or 1 to 1-1/2 hours after meals unless otherwise directed by your doctor.
- A constituent of many multivitamin/mineral preparations.

Warnings and Precautions

Don't take if you have:
- Severe kidney disease.
- Kidney stones and analysis has shown their composition to be magnesium ammonium phosphate.

Consult your doctor if you have:
- Hypoparathyroidism.
- Osteomalacia.
- Acute pancreatitis.
- Chronic kidney disease.
- Rickets.
- Adrenal insufficiency (Addison's disease).
- Dehydration.
- Severe burns.
- Heart disease.
- Congestive heart failure.
- Liver cirrhosis.
- Edema.
- Increased sodium in blood.
- High blood pressure.
- Toxemia of pregnancy.

Over age 55:
- No special problems expected.

Pregnancy:
- Take under doctor's supervision only.
- Don't take megadoses.

Breast-feeding:
- Take under doctor's supervision only.
- Don't take megadoses.

Effect on lab tests:
- May show false decrease in bone uptake in technetium-labeled diagnostic-imaging tests.

Storage:
- Store in cool, dry place away from direct light, but don't freeze.
- Store safely out of reach of children.
- Don't store in bathroom medicine cabinet. Heat and moisture may change action of mineral.

Overdose/Toxicity

Signs and symptoms:
Seizures, heartbeat irregularities, shortness of breath.

What to do:
For symptoms of overdosage: Discontinue mineral, and seek emergency treatment. Also see *Adverse Reactions or Side Effects* section below.
For accidental overdosage (such as child taking entire bottle): Dial 911 (emergency), 0 for operator or your nearest Poison Control Center.

Adverse Reactions or Side Effects

Reaction or effect	What to do
Abdominal pain	Discontinue. Call doctor immediately.
Bone or joint pain	Discontinue. Call doctor immediately.
Confusion	Discontinue. Call doctor immediately.
Decreased volume of urine in one day	Seek emergency treatment.
Diarrhea	Discontinue. Call doctor immediately.
Easy fatigue	Discontinue. Call doctor when convenient.
Edema of feet or legs	Discontinue. Call doctor when convenient.
Headaches	Discontinue. Call doctor immediately.
Muscle cramps	Discontinue. Call doctor when convenient.
Numbness or tingling in hands or feet	Discontinue. Call doctor when convenient.
Unusual thirst	Discontinue. Call doctor when convenient.

≫➤

Interaction with Medicine, Vitamins or Minerals

Interacts with	Combined effect
Anabolic steroids	Increases risk of edema.
Antacids with aluminum or magnesium	May prevent absorption of phosphates.
Calcium-containing supplements and antacids	Increases risk of depositing calcium in soft tissues. Decreases phosphate absorption.
Captopril	Increases risk of too much potassium (hyperkalemia).
Cortisone drugs or ACTH	Increases serum sodium.
Digitalis preparations	Increases risk of too much potassium (hyperkalemia).
Diuretics, potassium-conserving (amiloride, spironelactene, triamterene)	Increases risk of too much potassium (hyperkalemia).
Emalapril	Increases risk of too much potassium (hyperkalemia).
Salicylates	May increase plasma concentration of salicylates.
Testosterone	Increases risk of edema.
Vitamin D	Phosphate absorption enhanced, but may increase chance of too much phosphorous in blood and body cells.

 Interaction with Other Substances

Beverages:
- Alcoholic beverages decrease available phosphorous for vital body functions.
- Overconsumption of soft drinks may adversely affect absorption of phosphorous and calcium.

Foods
- Overconsumption of meats and convenience foods may adversely affect absorption of phosphorous and calcium.
- Overconsumption of rhubarb, spinach and bran may decrease absorption of potassium phosphates.

MINERAL

Selenium

Basic Information

Brand names, see page 477.
Available from natural sources? Yes
Available from synthetic sources? No
Prescription required? No

Natural Sources

Bran	Liver
Broccoli	Milk
Cabbage	Mushrooms
Celery	Onions
Chicken	Seafood
Cucumbers	Tuna
Egg yolk	Wheat germ
Garlic	Whole-grain products
Kidney	

Note: The selenium content of food varies greatly because of the wide variability of this element in the soil. Accurate levels in food are not available.

Reasons to Use

- Complements vitamin E to act as an efficient anti-oxidant.
- Promotes normal growth and development.
- Functions as anti-oxidant itself.

Unproved Speculated Benefits

- Stimulates immune system.
- Cures cancer.
- Cures arthritis.
- Protects against all hypothesized aging mechanisms.
- Protects against cardiovascular disease, strokes and heart attacks.
- Decreases platelet clumping in bloodstream, and prevents clots at site of blood-vessel damage in heart and brain.
- Increases elasticity and youthfulness of skin.
- Helps control dandruff (selenium sulfide) when applied to scalp. Used this way it exerts anti-fungal and anti-bacterial effects.
- Acts as an aphrodisiac.
- Increases fertility.
- Removes age spots when rubbed on skin.
- Protects against damage caused by tobacco smoking.

Who Needs Additional Amounts?

- Anyone with inadequate caloric or nutritional dietary intake or increased nutritional requirements.
- People who live in areas where soil is selenium-deficient, such as China, New Zealand and central and eastern United States. In the United States check with your local county agricultural agent.

Deficiency Symptoms

- Selenium deficiency in the soil and water has resulted in cardiomyopathy and myocardial deaths in humans

Unproved Speculated Symptoms

- Keshan's disease, a fatal heart disease found in children living in certain sections of China
- Cataracts
- Muscular dystrophy
- Retarded growth
- Liver problems
- Infertility
- Some forms of cancer

Lab Tests to Detect Deficiency

- 24-hour urine collection

Dosage and Usage Information

Recommended Dietary Allowance (RDA):
No RDA has been established. Estimated safe intake per day given below:

Age	Estimated Safe Intake
0–6 months	0.01–0.04mg
6–12 months	0.02–0.06mg
1–3 years	0.02–0.08mg
4–6 years	0.03–0.12mg
7–10 years	0.05–0.20mg
11+ years	0.05–0.20mg

What this mineral does:
- Selenium helps defend against damage from oxidation.

Miscellaneous information:
- Should be part of a well-balanced vitamin-mineral regimen.

⫸➡

- Protection from human degenerative disorders has yet to be proved.
- Experimental studies are trying to prove selenium plays a big part as an "anti-oxidant nutrient" to help protect against damaging "free radicals."
- Organic forms (from foods or brewer's yeast) are less toxic than inorganic sodium selenite.
- No one can be sure of correct amount to be ingested each day. People who eat a balanced diet of food grown in the western United States probably get enough from food.

Available as:
- Tablets or capsules: Swallow whole with full glass of liquid. Don't chew or crush. Take with meals or 1 to 1-1/2 hours after meals unless otherwise directed by your doctor.
- A constituent of many multivitamin/mineral preparations.

Warnings and Precautions

Don't take if you:
- Plan to use it on scalp or skin for seborrheic dermatitis or dandruff if you have any inflammation or oozing.

Consult your doctor if you have:
- Plans to take more than the dose recommended by the manufacturer.

Over age 55:
- No problems expected with usual doses.

Pregnancy:
- No problems expected with usual doses.
- Don't take megadoses.

Breast-feeding:
- No problems expected with usual doses.
- Don't take megadoses.

Effect on lab tests:
- May decrease serum vitamin C.

Storage:
- Store in cool, dry place away from direct light, but don't freeze.
- Store safely out of reach of children.
- Don't store in bathroom medicine cabinet. Heat and moisture may change action of mineral.

Others:
- When used on hair, rinse hair carefully to prevent discoloration.
- Workers at industrial sites that manufacture glass, pesticides, rubber, semi-conductors, copper and film are at increased risk of developing toxic symptoms from inhalation, absorption through the skin and ingestion. These may include bronchial pneumonia,

asthma, precipitous drop in blood pressure, red eyes, garlic odor on breath and in urine, headaches, metallic taste, nose and throat irritation, difficulty breathing, vomiting, weakness.

Overdose/Toxicity

Signs and symptoms:
Unlikely to develop if organic selenium is not consumed at a rate greater than dose recommended by the manufacturer.

Possible Consequences of Overdose:
- Individuals in industrial settings have been reported to suffer toxic symptoms of selenium overdoses, including liver disease and cardiomyopathy. Children raised in selenium-rich areas show a higher incidence of decayed, missing and filled teeth.
- Selenium is toxic in megadoses and may cause alopecia, loss of nails, fatigue, nausea, vomiting, sour-milk breath.

What to do:
For symptoms of overdosage: Discontinue mineral, and consult doctor. Also see *Adverse Reactions or Side Effects* section below.
For accidental overdosage (such as child taking entire bottle): Dial 911 (emergency), 0 for operator or your nearest Poison Control Center.

⟩⟩▶

Selenium, Continued

Adverse Reactions or Side Effects

Reaction or effect	What to do
Dizziness and nausea, without other apparent cause	Discontinue. Call doctor immediately.
Fragile or black fingernails	Discontinue. Call doctor when convenient.
Persistent garlic odor on breath and skin	Discontinue. Call doctor when convenient.
Unusual dryness when used on scalp or skin	Discontinue. Call doctor when convenient.
Unusual hair loss or discoloration of hair	Discontinue. Call doctor when convenient.

Interaction with Medicine, Vitamins or Minerals

Interacts with	Combined effect
Vitamin C	May decrease selenium absorption if taken with an inorganic form of selenium.
Vitamin E	Prevents oxidation that might cause breakdown of body chemicals.

Interaction with Other Substances

None known

Basic Information

Available from natural sources? Yes
Available from synthetic sources? No
Prescription required? No

Natural Sources

Bacon	Ham
Beef, dried and fresh	Margarine
Bread	Milk
Butter	Sardines, canned
Clams	Table salt (chief source
Green beans	of sodium)
	Tomatoes, canned

Note: In most commercially canned vegetables, frozen foods and processed foods, salt is added to improve taste. "Highly processed" foods (also high in sodium) include soups, bouillon, pickles, potato chips, snack foods, ham.

Reasons to Use

- Helps regulate water balance in body.
- Plays a crucial role in maintaining blood pressure.
- Aids muscle contraction and nerve transmission.
- Regulates body's acid-base balance.

Unproved Speculated Benefits

- Lowers fevers
- Prevents heatstroke

Who Needs Additional Amounts?

- People with a chronic wasting illness, excess stress for long periods or who have recently undergone surgery.
- Anyone who suffers prolonged loss of body fluids from vomiting or diarrhea.
- Those with Addison's disease.
- People suffering congestive heart failure who take diuretics.
- Those who drink water excessively for prolonged periods. (This is usually a psychiatric condition.)
- People who suffer some types of cancers of the adrenal glands.
- Anyone who suffers infections with high fever.

- Those who have excessive sweating (rare cause).
- People who use diuretics.
- Anyone who cannot eat or drink, such as those with stroke or gastrointestinal upset.
- Those with chronic kidney disease.

Deficiency Symptoms

- Muscle and stomach cramps
- Nausea
- Fatigue
- Mental apathy
- Muscle twitching and cramping (usually in legs)
- Appetite loss

Unproved Speculated Symptoms

- Neuralgia

Lab Tests to Detect Deficiency

- Serum sodium

Dosage and Usage Information

Recommended Dietary Allowance (RDA):
No RDA has been established. Estimated safe intake per day given below.

Age	Estimated Safe Intake
0–6 months	0.115–0.35g
6–12 months	0.25–0.75g
1–3 years	0.325–0.975g
4–6 years	0.45–1.35g
7–10 years	0.60–1.80g
11–17 years	0.90–2.270g
18+ years	1.10–3.30g

What this mineral does:
- As an electrolyte, sodium is present in all body cells. Its most important function is to regulate the balance of water inside and outside cells.

Miscellaneous information:
- We consume most of our sodium as sodium chloride—ordinary table salt.
- The most common problem with sodium in a healthy person is "too-much," rather than "too-little." A typical diet contains 3,000 to 12,000mg of sodium a day. For normal function, we only need 3,000mg.

≫➤

Sodium, Continued

- Excessive amounts of sodium can be a major factor in development of high blood pressure. Decreasing sodium intake helps control high blood pressure.

Available as:
- Sodium-chloride tablets, but these may cause stomach distress and an overload on the kidneys.

 Warnings and Precautions

Don't take if you have:
- Congestive heart failure.
- Hepatic cirrhosis.
- Hypertension.
- Edema from any cause.
- A family history of high blood pressure.

Consult your doctor if you have:
- Any heart or blood-vessel disease.
- Bleeding problems.
- Epilepsy.
- Kidney disease.

Over age 55:
- No special problems expected if healthy.

Pregnancy:
- Dietary restriction of sodium in healthy women during pregnancy is not recommended.
- Don't take megadoses.

Breast-feeding:
- Dietary restriction of sodium in healthy women during lactation is not recommended.
- Don't take megadoses.

Effect on lab tests:
- None expected.

Storage:
- Store in cool, dry place away from direct light, but don't freeze.
- Store safely out of reach of children.
- Don't store in bathroom medicine cabinet. Heat and moisture may change action of mineral.

Others:
- Too little sodium occurs almost entirely in people desperately ill with dehydration or those recovering from recent surgery or after excessive sweating from heavy physical activity in a hot environment.
- Proper replacement of sodium deficiencies requires care by your doctor and frequent laboratory studies.

 Overdose/Toxicity

Signs and symptoms:
- Tissue swelling (edema), stupor, coma.

What to do:
For symptoms of overdosage: Discontinue mineral, and consult doctor. Also see *Adverse Reactions or Side Effects* section below.
For accidental overdosage (such as child taking entire bottle): Dial 911 (emergency), 0 for operator or your nearest Poison Control Center.

 Adverse Reactions or Side Effects

Reaction or effect	What to do
With excessive amounts of sodium:	
Anxiety	Discontinue. Call doctor immediately.
Confusion	Discontinue. Call doctor immediately.
Edema	Discontinue. Call doctor immediately.
Nausea	Discontinue. Call doctor immediately.
Restlessness	Discontinue. Call doctor immediately.
Vomiting	Seek emergency treatment.
Weakness	Discontinue. Call doctor immediately.

 Interaction with Medicine, Vitamins or Minerals

None expected

 Interaction with Other Substances

None known

Sulfur

Basic Information

Available from natural sources? Yes
Available from synthetic sources? No
Prescription required? No

Natural Sources

Cabbage	Fish
Clams	Lean beef
Beans, dried	Milk
Eggs	Wheat germ

Reasons to Use

- Plays a role in oxidation-reduction reactions.
- Aids bile secretion in liver.

Unproved Speculated Benefits

- Extends life span.
- Protects against toxic substances.

Who Needs Additional Amounts?

- Supplements probably not needed. No recorded deficiency states.

Deficiency Symptoms

- None

Unproved Speculated Symptoms

- None

Lab Tests to Detect Deficiency

- None available, except for experimental uses.

Dosage and Usage Information

Recommended Dietary Allowance (RDA):
No RDA has been established.

What this mineral does:
- Sulfur is part of the chemical structure of cysteine, methionine, taurine, glutathione.

Available as:
- A constituent of many multivitamin/mineral preparations.

Warnings and Precautions

Don't take if you:
- No known contraindications.

Consult your doctor if you have:
- No known contraindications.

Over age 55:
- No known contraindications.

Pregnancy:
- No known contraindications.
- Don't take megadoses.

Breast-feeding:
- No known contraindications.
- Don't take megadoses.

Effect on lab tests:
- None expected.

Storage:
- Store in cool, dry place away from direct light, but don't freeze.
- Store safely out of reach of children.
- Don't store in bathroom medicine cabinet. Heat and moisture may change action of mineral.

Overdose/Toxicity

Signs and symptoms:
Unlikely to threaten life or cause significant symptoms.

What to do:
For symptoms of overdosage: Discontinue mineral, and consult doctor.
For accidental overdosage (such as child taking entire bottle): Dial 911 (emergency), 0 for operator or your nearest Poison Control Center.

Adverse Reactions or Side Effects

None known

Sulfur, Continued

 Interaction with Medicine, Vitamins or Minerals

None known

 Interaction with Other Substances

Tobacco decreases absorption. Smokers may require supplemental sulfur.

Basic Information

Available from natural sources? Yes
Available from synthetic sources? No
Prescription required? No

Natural Sources

Fish

Reasons to Use

• Plays role in metabolism of bones and teeth.

Unproved Speculated Benefits

• Aids in preventing heart attacks.

Who Needs Additional Amounts?

• Supplements probably not needed. No recorded deficiency states.

Deficiency Symptoms

• A vanadium-deficient diet fed to laboratory animals resulted in impaired reproductive ability and increased infant mortality.

Unproved Speculated Symptoms

• None

Lab Tests to Detect Deficiency

• None available, except for experimental purposes.

Dosage and Usage Information

Recommended Dietary Allowance (RDA):
No RDA has been established. Estimated requirements for adults are 0.1 to 0.3mg/day. Dietary intake of vanadium averages 4mg/day.

What this mineral does:
• Unknown in humans, but believed to be essential.

Miscellaneous information:
• Even the most nutritionally inadequate diet contains sufficient quantities to prevent deficiency.

Available as:
• Capsules: Swallow whole with full glass of liquid. Don't chew or crush. Take with meals or 1 to 1-1/2 hours after meals unless otherwise directed by your doctor.
• A constituent of many multivitamin/mineral preparations.

Warnings and Precautions

Don't take if you:
• No known contraindications.

Consult your doctor if you have:
• No known contraindications.

Over age 55:
• No known contraindications.

Pregnancy:
• No known contraindications.
• Don't take megadoses.

Breast-feeding:
• No known contraindications.
• Don't take megadoses.

Effect on lab tests:
• None expected.

Storage:
• Store in cool, dry place away from direct light, but don't freeze.
• Store safely out of reach of children.
• Don't store in bathroom medicine cabinet. Heat and moisture may change action of mineral.

Overdose/Toxicity

Signs and symptoms:
Unlikely to threaten life or cause significant symptoms.

What to do:
For symptoms of overdosage: Discontinue mineral, and consult doctor.
For accidental overdosage (such as child taking entire bottle): Dial 911 (emergency), 0 for operator or your nearest Poison Control Center.

Vanadium, Continued

 Adverse Reactions or Side Effects

None expected

 Interaction with Medicine, Vitamins or Minerals

Interacts with	Combined effect
Chromium	Chromium and vanadium may interfere with each other.

 Interaction with Other Substances

Tobacco decreases absorption. Smokers may require supplemental vanadium.

Basic Information

Brand names, see page 477.
Available from natural sources? Yes
Available from synthetic sources? No
Prescription required? No

Natural Sources

Beef, lean	Pork
Chicken heart	Sesame seeds
Egg yolk	Soybeans
Fish	Sunflower seeds
Herring	Turkey
Lamb	Wheat bran
Maple syrup	Wheat germ
Milk	Whole-grain products
Molasses, black-strap	Yeast
Oysters	

Reasons to Use

- Functions as anti-oxidant.
- Maintains normal taste and smell.
- Promotes normal growth and development.
- Aids wound healing.
- Promotes normal fetal growth.
- Helps synthesize DNA and RNA.
- Promotes cell division, cell repair, cell growth.
- Maintains normal level of vitamin A in blood.

Unproved Speculated Benefits

- Relieves angina.
- Relieves cirrhosis of liver.
- Boosts immunity.
- Prevents cancer.
- Increases male potency and sex drive.
- Enhances other treatments for diabetes mellitus.
- Treats acne.
- Treats arthritis.
- Retards aging.

Who Needs Additional Amounts?

- Anyone with inadequate caloric or nutritional dietary intake or increased nutritional requirements, such as vegetarians.
- Preschool children.
- Older people (over 55 years).
- Pregnant or breast-feeding women.
- Those who abuse alcohol or other drugs.
- People with a chronic wasting illness, excess stress for long periods or those who have recently undergone surgery.
- Those with a portion of the gastrointestinal tract surgically removed.
- People with recent severe burns or injuries.
- Anyone taking diuretics (water pills) for any reason, such as high blood pressure, congestive heart failure, liver disease.
- Women taking oral contraceptives.
- Those who live in areas where soil is deficient in zinc.

Deficiency Symptoms

Moderate deficiency:
- Loss of taste and smell
- Suboptimal growth in children
- Alopecia
- Rashes
- Multiple skin lesions
- Glossitis (See Glossary)
- Stomatitis (See Glossary)
- Blepharitis (See Glossary)
- Paronychia (See Glossary)
- Sterility
- Low sperm count
- Delayed wound healing

Serious deficiency:
- Delayed bone maturation
- Enlarged spleen or liver
- Decreased size of testicles
- Testicular function less than normal
- Decreased growth or dwarfism

Unproved Speculated Symptoms

- Infertility
- Symptoms of immunodeficient diseases, such as recurrent infections, fatigue, diarrhea, unexplained weight loss, unexplained fever, swollen lymph glands

Lab Tests to Detect Deficiency

- Serum zinc (by atomic absorption spectroscopy)

 ⟫→

Zinc, Continued

Dosage and Usage Information

Recommended Dietary Allowance (RDA):
Estimate of adequate daily intake by the Food and Nutrition Board of the National Research Council, 1980 (See Glossary).

Age	RDA
0–6 months	3mg
6–12 months	5mg
1–10 years	10mg
11+ years	15mg
Pregnant	+5mg
Lactating	+10mg

What this mineral does:
• Zinc is a part of the molecular structure of 80 or more known enzymes. These particular enzymes work with red blood cells to move carbon dioxide from tissues to lungs.

Miscellaneous information:
• Zinc toxicity from inhalation is rare but can occur in the following industries and occupations—alloy manufacturing, brass foundry, bronze foundry, electric-fuse manufacturing, gas welding, electroplating, galvanizing, paint manufacturing, metal cutting, metal spraying, rubber manufacturing, roof manufacturing, zinc manufacturing.
• If you take zinc supplements, take with food to decrease gastric irritation.

Available as:
• Tablets: Swallow whole with full glass of liquid. Don't chew or crush. Take with meals or 1 to 1-1/2 hours after meals unless otherwise directed by your doctor.
• A constituent of many multivitamin/mineral preparations.

Warnings and Precautions

Don't take if you have:
• Stomach or duodenal ulcers.

Consult your doctor if you have:
• Plans to take more than the manufacturer's recommended dose.
• To take any calcium supplement or tetracycline drugs. Zinc may interfere with absorption of these medicines.

Over age 55:
• Deficiency more likely.

Pregnancy:
• Many diets are marginally low in zinc and may not supply the zinc estimated to be required during pregnancy. Ask your doctor about supplementation.
• *Overconsumption* is dangerous and can lead to premature labor or stillbirth.
• Don't take megadoses.

Breast-feeding:
• Some diets are marginally low in zinc and may not supply the zinc estimated to be required while breast-feeding. Ask your doctor about supplementation.
• Don't take megadoses.

Effect on lab tests:
• Decreases high-density lipoprotein levels in young males. High-density lipoproteins decrease risk of coronary-artery disease.
• High doses decrease copper in blood.

Storage:
• Store in cool, dry place away from direct light, but don't freeze.
• Store safely out of reach of children.
• Don't store in bathroom medicine cabinet. Heat and moisture may change action of mineral.

Overdose

Signs and symptoms:
Toxicity at RDA doses highly unlikely. Toxic symptoms are extremes of the *Adverse Reactions or Side Effects* listed below. Overdose produces drowsiness, lethargy, lightheadedness, difficulty writing, staggering gait, restlessness, excessive vomiting leading to dehydration.

What to do:
For symptoms of overdosage: Discontinue mineral, and consult doctor. Also see *Adverse Reactions or Side Effects* section below.
For accidental overdosage (such as child taking entire bottle): Dial 911 (emergency), 0 for operator or your nearest Poison Control Center.

⟫▶

Adverse Reactions or Side Effects

Reaction or effect	What to do
Abdominal pain	Seek emergency treatment.
Abnormal bleeding	Seek emergency treatment.
Gastric ulceration (burning pain in upper chest relieved by food or antacid)	Discontinue. Call doctor immediately.
Mild diarrhea	Discontinue. Call doctor when convenient.
Nausea	Discontinue. Call doctor immediately.
Vomiting	Discontinue. Call doctor immediately.

 ## Interaction with Medicine, Vitamins or Minerals

Interacts with	Combined effect
Calcium	Interferes with calcium absorption.
Copper	Decreases absorption of copper. Large doses of zinc must be taken to produce this effect.
Cortisone drugs	May interfere with lab tests measuring zinc.
Diuretics	Increases zinc excretion. Requires taking greater amounts.
Iron	Decreases absorption of iron. Large doses of zinc must be taken to produce this effect.
Oral contraceptives	Lowers zinc blood levels.
Tetracycline	Decreases amount of tetracycline absorbed into bloodstream. Zinc and tetracycline should *not* be mixed. Take at least 2 hours apart.
Vitamin A	Assists in absorption of vitamin A.

 ## Interaction with Other Substances

Alcohol, even in moderate amounts, can increase the excretion of zinc in urine and can impair body's ability to combine zinc into its proper enzyme combinations in the liver.

Beverages
• Coffee should not be consumed at the same time as zinc because it may decrease absorption of zinc.

Amino Acids and Nucleic Acids

Amino acids are the 20 essential amino-acid molecules necessary for the body to synthesize proteins. These chemical molecules participate in building of all living structures.

Recent medical experiments suggest certain amino acids play a vital role in the central nervous system at transmission sites between nerve cells. They are called *neurotransmitters*. In addition, reports suggest some amino acids may protect against cancer and stimulate the immune system. This remains to be proved.

Amino acids listed in this book are available as supplements. Their usefulness as components of our bodies is unquestioned. Their usefulness as supplements remains to be proved.

Nucleic acids are large molecules encoded in the genes and are part of each living cell. They determine what kind of life form a cell will be, such as human, plant or animal. The use of nucleic acids is based on the unproved theory that an extra amount gives added life to "worn-out" cells and tissues.

Nucleic acids taken orally do no good because they are changed or destroyed in the intestinal tract before they can be absorbed. Injecting cells from young animals into our bodies is a dangerous practice. An injection of animal protein into humans may cause anaphylaxis—a serious allergic reaction causing immediate itching, severe drop in blood pressure, loss of consciousness and sometimes death.

Adenosine

Basic Information

Adenosine is a nucleic acid.
Available from natural sources? Yes
Available from synthetic sources? No
Prescription required? No

 Natural Sources

All foods

 Reasons to Use

- Functions as essential part of every living cell, but supplemental products taken orally are useless.

 Unproved Speculated Benefits

Note: These claims are from one researcher, the late Dr. Benjamin Frank. The scientific community does not accept his results because they have never been proved in other studies.
Injectable form:
- Treats congestive heart failure.
- Relieves angina (See Glossary).
- Increases vigor.
- Permits greater exercise endurance.
- Increases life span.
- Improves liver function.
- Enhances memory.
- Relieves problems caused by emphysema.
- Improves skin quality.

 Who Needs Additional Amounts?

- No one

 Deficiency Symptoms

- None   .

 Unproved Speculated Symptoms

- Aging
- All forms of degenerative diseases

 Lab Tests to Detect Deficiency

- None available, except for experimental purposes.

 Dosage and Usage Information

Recommended Dietary Allowance (RDA):
No RDA has been established. Oral supplements are destroyed in the intestine and do not get absorbed. Therefore they can exert *no* influence.

What this nucleic acid does:
- Nucleic acids form the substance of DNA (desoxyribonucleic acid) and RNA (ribonucleic acid). "Messages" are transferred because of the actions of the purine and pyrimidine bases of DNA and RNA, which include adenine, guanine, cytosine and thymine. These messages form encoded genetic instructions that guide the development of all living cells.
- Supplements have been advertised for oral use, topical use (as in cosmetics) and injectable use. "Cellular therapy," once popular in Europe, is the practice of injecting preparations of cells from young animals into humans with the false promise of replacing "worn-out" tissues in aging human bodies. These treatments are extremely expensive, dangerous and have been completely discredited.
- No positive effects can be possible with oral forms.

Available as:
- Tablets and capsules: Swallow whole with full glass of liquid. Don't chew or crush. Take with meals or 1 to 1-1/2 hours after meals unless otherwise directed by your doctor. These are normally found in health-food stores.
- Injectable forms are administered by doctor or nurse.

 Warnings and Precautions

Don't take if you:
- Have any medical problem listed under *Unproved Speculated Benefits.* There may be safer, more-effective treatments.

Consult your doctor if you have:
- Any medical problems listed under *Unproved Speculated Benefits.* There may be safer, more-effective treatments.

»▸

Over age 55:
• There may be safer, more-effective treatments.

Pregnancy:
• There may be safer, more-effective treatments.
• Don't take megadoses.

Breast-feeding:
• There may be safer, more-effective treatments.
• Don't take megadoses.

Effect on lab tests:
• None known.

Storage:
• Store in cool, dry place away from direct light, but don't freeze.
• Store safely out of reach of children.
• Don't store in bathroom medicine cabinet. Heat and moisture may change action of nucleic acid.

Others:
• Injectable forms can cause serious reactions, including anaphylaxis, serum sickness, transfer of disease.

Overdose/Toxicity

Signs and symptoms:
None for oral forms. For injectable forms, see *Adverse Reactions or Side Effects* section below.

What to do:
For symptoms of overdosage: Discontinue nucleic acid, and consult doctor. Also see *Adverse Reactions or Side Effects* section below.
For accidental overdosage (such as child taking entire bottle): Dial 911 (emergency), 0 for operator or your nearest Poison Control Center.

Adverse Reactions or Side Effects

Reaction or effect	What to do
For injectable forms:	
Anaphylaxis— symptoms include immediate severe itching, paleness, low blood pressure, fainting, coma	Yell for help. Don't leave victim. Begin CPR (cardio-pulmonary resuscitation), mouth-to-mouth breathing and external cardiac massage. Have someone call "0" (Operator) or 911 (Emergency). Don't stop CPR until help arrives.
Serum sickness, characterized by fever, edema of face and ankles, decreased urine output, skin rash	Seek emergency treatment.
Many viral illnesses, such as hepatitis or AIDS, if material is derived from human tissue or injected with contaminated needle	Discontinue. Call doctor immediately.

Interaction with Medicine, Vitamins or Minerals

None known

Interaction with Other Substances

None known

Arginine

Basic Information

Arginine is an amino acid.
Available from natural sources? Yes
Available from synthetic sources? Yes
Prescription required? No

Natural Sources

Brown rice	Raisins
Carob	Raw cereals
Chocolate	Sesame seeds
Nuts	Sunflower seeds
Oatmeal	Whole-wheat products
Popcorn	

Reasons to Use

- Functions as building block of all proteins.
- Stimulates human-growth hormone.

Unproved Speculated Benefits

- Increases metabolism in fat cells to decrease obesity.
- Builds muscle.
- Speeds wound healing.
- Stimulates immune system.
- Inhibits cancer.
- Increases sperm count in males.

Who Needs Additional Amounts?

- Single amino-acid deficiencies are unknown except in people on crash diets consisting of only a few foods.
- Amino-acid deficiencies appear more commonly as a result of total protein deficiency, which is rare in the United States and Canada.
- Anyone with inadequate caloric or nutritional dietary intake or increased nutritional requirements.
- Those with inadequate protein dietary intake.
- Children, pregnant or lactating women who are vegan vegetarians.
- People with recent severe burns or injuries.
- Premature infants.

Deficiency Symptoms

- None expected

Unproved Speculated Symptoms

- Male infertility

Lab Tests to Detect Deficiency

- None available, except for experimental purposes.

Dosage and Usage Information

Recommended Dietary Allowance (RDA):
No RDA has been established.

What this amino acid does:
- Provides part of all proteins.

Miscellaneous information:
- Arginine has been reported to increase the activity of some herpes viruses and inhibit others.
- If you take arginine as a supplement, take it on an empty stomach before retiring at night.
- Poorly nourished people have a greater chance of adverse side effects from taking amino-acid supplements, including an amino-acid imbalance.
- The poorer the diet, the greater the chance of an amino-acid supplement creating a harmful combination.

Available as:
- Tablets or capsules: Swallow whole with full glass of liquid. Don't chew or crush. Take with meals or 1 to 1-1/2 hours after meals unless otherwise directed by your doctor.
- Powder for oral solution: Dissolve powder in cold water or juice. Take with meals or 1 to 1-1/2 hours after meals unless otherwise directed by your doctor.

Warnings and Precautions

Don't take if you:
- Are a child or adolescent not fully grown.
- Are allergic to any food protein, such as eggs, milk, wheat.
- Are at risk of poor nutrition for any reason.

Arginine

Consult your doctor if you have:
- Any bone disease.
- Herpes infection (genital or oral).

Over age 55:
- Don't take amino-acid supplements if you are healthy.

Pregnancy:
- Don't take amino-acid supplements if you are healthy and eat an adequate diet.
- Don't take megadoses.

Breast-feeding:
- Don't take amino-acid supplements if you are healthy.
- Don't take megadoses.

Effect on lab tests:
- None known.

Storage:
- Store in cool, dry place away from direct light, but don't freeze.
- Store safely out of reach of children.
- Don't store in bathroom medicine cabinet. Heat and moisture may change action of amino acid.

Others:
- Children and adolescents should *not* take any arginine supplement. It may cause bone deformities.

Overdose/Toxicity

Signs and symptoms:
Unlikely to threaten life or cause significant symptoms.

What to do:
For symptoms of overdosage: Discontinue amino acid, and consult doctor. Also see *Adverse Reactions or Side Effects* section below.
For accidental overdosage (such as child taking entire bottle): Dial 911 (emergency), 0 for operator or your nearest Poison Control Center.

Adverse Reactions or Side Effects

Reaction or effect	What to do
Diarrhea (from large doses)	Decrease dose or discontinue.
Nausea (from large doses)	Decrease dose or discontinue.

Interaction with Medicine, Vitamins or Minerals

None known

Interaction with Other Substances

None known

AMINO ACID

DNA & RNA

Basic Information
DNA and RNA are nucleic acids.
Available from natural sources? Yes
Available from synthetic sources? No
Prescription required? No

Natural Sources

All foods

Reasons to Use

- Functions as essential part of every living cell, but supplemental products taken orally are useless.

Unproved Speculated Benefits

Note: These claims are from one researcher, the late Dr. Benjamin Frank. The scientific community does not accept his results because they have never been proved in other studies.

Injectable form:
- Treats congestive heart failure.
- Relieves angina (See Glossary).
- Increases vigor.
- Permits greater exercise endurance.
- Increases life span.
- Improves liver function.
- Enhances memory.
- Relieves problems caused by emphysema.
- Improves skin quality.

Who Needs Additional Amounts?

- No one

Deficiency Symptoms

- None

Unproved Speculated Symptoms

- Aging
- All forms of degenerative diseases

Lab Tests to Detect Deficiency

- None available, except for experimental purposes.

Dosage and Usage Information

Recommended Dietary Allowance (RDA):
No RDA has been established. Oral supplements are destroyed in the intestine and do not get absorbed. Therefore they can exert *no* influence.

What this nucleic acid does:
- Nucleic acids form the substance of DNA (desoxyribonucleic acid) and RNA (ribonucleic acid). "Messages" are transferred because of the actions of the purine and pyrimidine bases of DNA and RNA, which include adenine, guanine, cytosine and thymine. These messages form encoded genetic instructions that guide the development of all living cells.
- Supplements have been advertised for oral use, topical use (as in cosmetics) and injectable use. "Cellular therapy," once popular in Europe, is the practice of injecting preparations of cells from young animals into humans with the false promise of replacing "worn-out" tissues in aging human bodies. These treatments are extremely expensive, dangerous and have been completely discredited.
- No positive effects can be possible with oral forms.

Available as:
- Tablets and capsules: Swallow whole with full glass of liquid. Don't chew or crush. Take with meals or 1 to 1-1/2 hours after meals unless otherwise directed by your doctor. These are normally found in health-food stores.
- Injectable forms are administered by doctor or nurse.

Warnings and Precautions

Don't take if you:
- Have any medical problem listed under *Unproved Speculated Benefits.* There may be safer, more-effective treatments.

Consult your doctor if you have:
- Any medical problem listed under *Unproved Speculated Benefits.* There may be safer, more-effective treatments.

➤

Over age 55:
• There may be safer, more-effective treatments.

Pregnancy:
• There may be safer, more-effective treatments.
• Don't take megadoses.

Breast-feeding:
• There may be safer, more-effective treatments.
• Don't take megadoses.

Effect on lab tests:
• None known.

Storage:
• Store in cool, dry place away from direct light, but don't freeze.
• Store safely out of reach of children.
• Don't store in bathroom medicine cabinet. Heat and moisture may change action of nucleic acid.

Others:
• Injectable forms can cause serious reactions, including anaphylaxis, serum sickness and transfer of disease.

Overdose/Toxicity

Signs and symptoms:
None for oral forms. For injectable forms, see *Adverse Reactions or Side Effects* section below.

What to do:
For symptoms of overdosage: Discontinue nucleic acid, and consult doctor. Also see *Adverse Reactions or Side Effects* section below.
For accidental overdosage (such as child taking entire bottle): Dial 911 (emergency), 0 for operator or your nearest Poison Control Center.

Adverse Reactions or Side Effects

Reaction or effect	What to do
For injectable forms:	
Anaphylaxis—symptoms include immediate severe itching, paleness, low blood pressure, fainting, coma	Yell for help. Don't leave victim. Begin CPR (cardio-pulmonary resuscitation), mouth-to-mouth breathing and external cardiac massage. Have someone call "0" (Operator) or 911 (Emergency). Don't stop CPR until help arrives.
Serum sickness, characterized by fever, edema of face and ankles, decreased urine output, skin rash	Seek emergency treatment.
Many viral illnesses, such as hepatitis or AIDS, if material is derived from human tissue or injected with contaminated needle	Discontinue. Call doctor immediately.

Interaction with Medicine, Vitamins or Minerals

None known

Interaction with Other Substances

None known

Inosine

Basic Information

Inosine is a nucleic acid.
Available from natural sources? Yes
Available from synthetic sources? No
Prescription required? No

Natural Sources

All foods

Reasons to Use

- Functions as essential part of every living cell, but supplemental products taken orally are useless.

Unproved Speculated Benefits

Note: These claims are from one researcher, the late Dr. Benjamin Frank. The scientific community does not accept his results because they have never been proved in other studies.
Injectable form:
- Treats congestive heart failure.
- Relieves angina (See Glossary).
- Increases vigor.
- Permits greater exercise endurance.
- Increases life span.
- Improves liver function.
- Enhances memory.
- Relieves problems caused by emphysema.
- Improves skin quality.

Who Needs Additional Amounts?

- No one

Deficiency Symptoms

- None

Unproved Speculated Symptoms

- Aging
- All forms of degenerative diseases

Lab Tests to Detect Deficiency

- None available, except for experimental purposes.

Dosage and Usage Information

Recommended Dietary Allowance (RDA):
No RDA has been established. Oral supplements are destroyed in the intestine and do not get absorbed. Therefore they can exert *no* influence.

What this nucleic acid does:
- Nucleic acids form the substance of DNA (desoxyribonucleic acid) and RNA (ribonucleic acid). "Messages" are transferred because of the actions of the purine and pyrimidine bases of DNA and RNA, which include adenine, guanine, cytosine and thymine. These messages form encoded genetic instructions that guide the development of all living cells.
- Supplements have been advertised for oral use, topical use (as in cosmetics) and injectable use. "Cellular therapy," once popular in Europe, is the practice of injecting preparations of cells from young animals into humans with the false promise of replacing "worn-out" tissues in aging human bodies. These treatments are extremely expensive, dangerous and have been completely discredited.
- No positive effects can be possible with oral forms.

Available as:
- Tablets and capsules: Swallow whole with full glass of liquid. Don't chew or crush. Take with meals or 1 to 1-1/2 hours after meals unless otherwise directed by your doctor. These are normally found in health-food stores.
- Injectable forms are administered by doctor or nurse.

Warnings and Precautions

Don't take if you:
- Have any medical problems listed under *Unproved Speculated Benefits.* There may be safer, more-effective treatments.

Consult your doctor if you have:
- Any medical problems listed under *Unproved Speculated Benefits.* There may be safer, more-effective treatments.

»»▶

Over age 55:
- There may be safer, more-effective treatments.

Pregnancy:
- There may be safer, more-effective treatments.
- Don't take megadoses.

Breast-feeding:
- There may be safer, more-effective treatments.
- Don't take megadoses.

Effect on lab tests:
- None known.

Storage:
- Store in cool, dry place away from direct light, but don't freeze.
- Store safely out of reach of children.
- Don't store in bathroom medicine cabinet. Heat and moisture may change action of nucleic acid.

Others:
- Injectable forms can cause serious reactions, including anaphylaxis, serum sickness and transfer of disease.

Overdose/Toxicity

Signs and symptoms:
None for oral forms. For injectable forms, see *Adverse Reactions or Side Effects* section below.

What to do:
For symptoms of overdosage: Discontinue nucleic acid, and consult doctor. Also see below.
For accidental overdosage (such as child taking entire bottle): Dial 911 (emergency), 0 for operator or your nearest Poison Control Center.

Adverse Reactions or Side Effects

Reaction or effect	What to do
For injectable forms:	
Anaphylaxis— symptoms include immediate severe itching, paleness, low blood pressure, fainting, coma	Yell for help. Don't leave victim. Begin CPR (cardio-pulmonary resuscitation), mouth-to-mouth breathing and external cardiac massage. Have someone call "0" (Operator) or 911 (Emergency). Don't stop CPR until help arrives.
Serum sickness, characterized by fever, edema of face and ankles, decreased urine output, skin rash	Seek emergency treatment.
Many viral illnesses, such as hepatitis or AIDS, if material is derived from human tissue or injected with contaminated needle	Discontinue. Call doctor immediately.

Interaction with Medicine, Vitamins or Minerals

None known

Interaction with Other Substances

None known

L-Cysteine

Basic Information
L-cysteine is an amino acid.
Available from natural sources? Yes
Available from synthetic sources? Yes
Prescription required? No

Natural Sources

Dairy products
Eggs
Meat
Some cereals

Reasons to Use

• Functions as building block of all proteins.
• Eliminates certain toxic chemicals rendering them harmless (anti-oxidant).
• One of the amino acids containing sulfur in a form believed to inactivate free radicals. If so, it protects and preserves cells.

Unproved Speculated Benefits

• Helps build muscle.
• Burns fat.
• Protects against toxins and pollutants, including some found in cigarette smoke and alcohol.
• Combats arthritis.
• May participate in some forms of DNA repair and theoretically extend life span.

Who Needs Additional Amounts?

• Single amino-acid deficiencies are unknown except in people on crash diets consisting of only a few foods.
• Amino-acid deficiencies appear more commonly as a result of total protein deficiency, which is rare in the United States and Canada.
• Anyone with inadequate caloric or nutritional dietary intake or increased nutritional requirements.
• Those with inadequate protein dietary intake.
• Children, pregnant or breast-feeding women who are vegan vegetarians.
• People with recent severe burns or injuries.
• Premature infants.

Deficiency Symptoms

In moderate deficiencies:
• Slowed growth in children
• Low levels of essential proteins in blood
In severe deficiencies:
• Apathy
• Depigmentation of hair
• Edema
• Lethargy
• Liver damage
• Loss of muscle and fat
• Skin lesions
• Weakness

Unproved Speculated Symptoms

• None

Lab Tests to Detect Deficiency

• None available, except for experimental purposes.

Dosage and Usage Information

Recommended Dietary Allowance (RDA):
No RDA has been established.

What this amino acid does:
• Provides part of all proteins.
• Functions in synthesis of glutathione, a substance that may neutralize environmental pollutants including tobacco.

Miscellaneous information:
• Poorly nourished people have a greater chance of adverse side effects from taking amino-acid supplements, including an amino-acid imbalance.
• The poorer the diet, the greater the chance of an amino-acid supplement creating a harmful combination.
• Take L-cysteine supplements with vitamin C. Take 2 to 3 times as much vitamin C as cysteine, milligram to milligram, as a precaution against kidney- and/or bladder-stone formation.

Available as:
• Capsules: Swallow whole with full glass of liquid. Don't chew or crush. Take with meals or 1 to 1-1/2 hours after meals unless otherwise directed by your doctor.

Warnings and Precautions

Don't take if you:
- Are allergic to any food protein, such as eggs, milk, wheat.
- Are at risk of poor nutrition for any reason.
- Have diabetes.
- Are self-prescribing without medical supervision.

Consult your doctor if you have:
- Diabetes mellitus.

Over age 55:
- Don't take amino-acid supplements if you are healthy.

Pregnancy:
- Don't take amino-acid supplements if you are healthy.

Breast-feeding:
- Don't take amino-acid supplements if you are healthy.

Effect on lab tests:
- None known.

Storage:
- Store in cool, dry place away from direct light, but don't freeze.
- Store safely out of reach of children.
- Don't store in bathroom medicine cabinet. Heat and moisture may change action of amino acid.

Overdose/Toxicity

Signs and symptoms:
Unlikely to threaten life or cause significant symptoms.

What to do:
For symptoms of overdosage: Discontinue amino acid, and consult doctor.
For accidental overdosage (such as child taking entire bottle): Dial 911 (emergency), 0 for operator or your nearest Poison Control Center.

Adverse Reactions or Side Effects

None expected

Interaction with Medicine, Vitamins or Minerals

Interacts with	Combined effect
Monosodium-glutamate	L-cysteine may increase toxicity of monosodium-glutamate in individuals who suffer from the "Chinese-restaurant syndrome." Causes headache, dizziness, disorientation, burning sensations.
Vitamin C	Taken with L-cysteine, vitamin C helps prevent L-cysteine from converting to *cystine*, which may cause bladder and/or kidney stones.

Interaction with Other Substances

None known

AMINO ACID

L-Lysine

Basic Information
L-lysine is an amino acid.
Available from natural sources? Yes
Available from synthetic sources? Yes
Prescription required? No

Natural Sources

Cheese	Potatoes
Eggs	Red meat
Fish	Soy products
Lima beans	Yeast
Milk	

Reasons to Use

- Functions as essential building block of all proteins.
- Promotes growth, tissue repair and production of antibodies, hormones, enzymes.

Unproved Speculated Benefits

- Protects against some sexually transmissible herpes viruses.

Who Needs Additional Amounts?

- Single amino-acid deficiencies are unknown except in people on crash diets consisting of only a few foods.
- Amino-acid deficiencies appear more commonly as a result of total protein deficiency, which is rare in the United States and Canada.
- Anyone with inadequate caloric or nutritional dietary intake or increased nutritional requirements.
- Those with inadequate protein dietary intake.
- Children, pregnant or breast-feeding women who are vegan vegetarians.
- People with recent severe burns or injuries.
- Premature infants.

 Deficiency Symptoms

In moderate deficiencies:
- Slowed growth in children
- Low levels of essential proteins in blood

In severe deficiencies:
- Apathy
- Depigmentation of hair
- Edema
- Lethargy
- Liver damage
- Loss of muscle and fat
- Skin lesions
- Weakness

Unproved Speculated Symptoms

- None

Lab Tests to Detect Deficiency

- None available, except for experimental purposes.

Dosage and Usage Information

Recommended Dietary Allowance (RDA):
No RDA has been established.

What this amino acid does:
- This is one of eight essential amino acids that the body does not manufacture. All biological amino acids participate in the synthesis of proteins in animal bodies.

Miscellaneous information:
- There is no scientific evidence supplements are needed or helpful.
- Poorly nourished people have a greater chance of adverse side effects from taking amino-acid supplements, including an amino-acid imbalance.
- The poorer the diet, the greater the chance of an amino-acid supplement creating a harmful combination.

Available as:
- Capsules: Swallow whole with full glass of liquid. Don't chew or crush. Take with meals or 1 to 1-1/2 hours after meals unless otherwise directed by your doctor.
- A constituent of many multivitamin/mineral preparations.

 Warnings and Precautions

Don't take if you:
- Are allergic to any food protein, such as eggs, milk, wheat.
- Are at risk of poor nutrition for any reason. ⟫▶

- Have diabetes.
- Are self-prescribing without medical supervision.

Consult your doctor if you have:
- Diabetes mellitus.

Over age 55:
- Don't take amino-acid supplements if you are healthy.

Pregnancy:
- Don't take amino-acid supplements if you are healthy.

Breast-feeding:
- Don't take amino-acid supplements if you are healthy.

Effect on lab tests:
- None known.

Storage:
- Store in cool, dry place away from direct light, but don't freeze.
- Store safely out of reach of children.
- Don't store in bathroom medicine cabinet. Heat and moisture may change action of amino acid.

 ## Overdose/Toxicity

Signs and symptoms:
Unlikely to threaten life or cause significant symptoms.

What to do:
For symptoms of overdosage: Discontinue amino acid, and consult doctor.
For accidental overdosage (such as child taking entire bottle): Dial 911 (emergency), 0 for operator or your nearest Poison Control Center.

 ## Adverse Reactions or Side Effects

None expected

 ## Interaction with Medicine, Vitamins or Minerals

None expected

 ## Interaction with Other Substances

None known

Methionine

Basic Information

Methionine is an amino acid.
Available from natural sources? Yes
Available from synthetic sources? Yes
Prescription required? No

Natural Sources

Eggs
Fish
Meat
Milk
Note: Not available from plant sources.

Reasons to Use

- Functions as building block of all proteins.
- Cysteine and taurine may rely on methionine for synthesis in the human body.

Unproved Speculated Benefits

- Helps eliminate fatty substances that might obstruct arteries, including those that supply the brain, heart, kidneys.

Who Needs Additional Amounts?

- Single amino-acid deficiencies are unknown except in people on crash diets consisting of only a few foods.
- Amino-acid deficiencies appear more commonly as a result of total protein deficiency, which is rare in the United States and Canada.
- Anyone with inadequate caloric or nutritional dietary intake or increased nutritional requirements.
- Those with inadequate protein dietary intake.
- Children, pregnant or breast-feeding women who are vegan vegetarians.
- People with recent severe burns or injuries.
- Premature infants.

Deficiency Symptoms

In moderate deficiencies:
- Slowed growth in children
- Low levels of essential proteins in blood
In severe deficiencies:
- Apathy
- Depigmentation of hair
- Edema
- Lethargy
- Liver damage
- Loss of muscle and fat
- Skin lesions
- Weakness

Unproved Speculated Symptoms

- None

Lab Tests to Detect Deficiency

- None available, except for experimental purposes.

Dosage and Usage Information

Recommended Dietary Allowance (RDA):
No RDA has been established.

What this amino acid does:
- Provides part of all proteins.

Miscellaneous information:
- This sulfur-containing amino acid (like choline and taurine) may help eliminate fatty substances that could cause occlusion of vital arteries.
- Poorly nourished people have a greater chance of adverse side effects from taking amino-acid supplements, including an amino-acid imbalance.
- The poorer the diet, the greater the chance of an amino-acid supplement creating a harmful combination.

Available as:
- Tablets: Swallow whole with full glass of liquid. Don't chew or crush. Take with meals or 1 to 1-1/2 hours after meals unless otherwise directed by your doctor.
- Capsules: Swallow whole with full glass of liquid. Don't chew or crush. Take with or immediately after food to decrease stomach irritation.

Warnings and Precautions

Don't take if you:
• Are allergic to any food protein, such as eggs, milk, wheat.
• Are at risk of poor nutrition for any reason.

Consult your doctor if you have:
• Self-prescribed methionine without medical supervision.

Over age 55:
• Don't take amino-acid supplements if you are healthy.

Pregnancy:
• Don't take amino-acid supplements if you are healthy.

Breast-feeding:
• Don't take amino-acid supplements if you are healthy.

Effect on lab tests:
• None known.

Storage:
• Store in cool, dry place away from direct light, but don't freeze.
• Store safely out of reach of children.
• Don't store in bathroom medicine cabinet. Heat and moisture may change action of amino acid.

Overdose/Toxicity

Signs and symptoms:
Unlikely to threaten life or cause significant symptoms.

What to do:
For symptoms of overdosage: Discontinue amino acid, and consult doctor.
For accidental overdosage (such as child taking entire bottle): Dial 911 (emergency), 0 for operator or your nearest Poison Control Center.

Adverse Reactions or Side Effects

None expected

Interaction with Medicine, Vitamins or Minerals

None known

Interaction with Other Substances

None known

Orotate

Basic Information
Orotate is an nucleic acid.
Available from natural sources? Yes
Available from synthetic sources? No
Prescription required? No

 Natural Sources

All foods

 Reasons to Use

• Functions as essential part of every living cell, but supplemental products taken orally are useless.

 Unproved Speculated Benefits

Note: These claims are from one researcher, the late Dr. Benjamin Frank. The scientific community does not accept his results because they have never been proved in other studies.
Injectable form:
• Treats congestive heart failure.
• Relieves angina (See Glossary).
• Increases vigor.
• Permits greater exercise endurance.
• Increases life span.
• Improves liver function.
• Enhances memory.
• Improves emphysema.
• Improves skin quality.

 Who Needs Additional Amounts?

• No one

 Deficiency Symptoms

• None

 Unproved Speculated Symptoms

• Aging
• All forms of degenerative diseases

 Lab Tests to Detect Deficiency

• None available, except for experimental purposes.

 Dosage and Usage Information

Recommended Dietary Allowance (RDA):
No RDA has been established. Oral supplements are destroyed in the intestine and do not get absorbed. Therefore they can exert *no* influence.

What this nucleic acid does:
• Nucleic acids form the substance of DNA (desoxyribonucleic acid) and RNA (ribonucleic acid). "Messages" are transferred because of the actions of the purine and pyrimidine bases of DNA and RNA, which include adenine, guanine, cytosine and thymine. These messages form encoded genetic instructions that guide the development of all living cells.
• Supplements have been advertised for oral use, topical use (as in cosmetics) and injectable use. "Cellular therapy," once popular in Europe, is the practice of injecting preparations of cells from young animals into humans with the false promise of replacing "worn-out" tissues in aging human bodies. These treatments are extremely expensive, dangerous and have been completely discredited.
• No positive effects can be possible with oral forms.

Available as:
• Tablets and capsules: Swallow whole with full glass of liquid. Don't chew or crush. Take with meals or 1 to 1-1/2 hours after meals unless otherwise directed by your doctor. These are normally found in health-food stores.
• Injectable forms are administered by doctor or nurse.

 Warnings and Precautions

Don't take if you:
• Have any medical problem listed under *Unproved Speculated Benefits.* There may be safer, more-effective treatments.

Consult your doctor if you have:
• Any medical problem listed under *Unproved Speculated Benefits.* There may be safer, more-effective treatments.

⟩⟩→

Over age 55:
• There may be safer, more-effective treatments.

Pregnancy:
• There may be safer, more-effective treatments.
• Don't take megadoses.

Breast-feeding:
• There may be safer, more-effective treatments.
• Don't take megadoses.

Effect on lab tests:
• None known.

Storage:
• Store in cool, dry place away from direct light, but don't freeze.
• Store safely out of reach of children.
• Don't store in bathroom medicine cabinet. Heat and moisture may change action of nucleic acid.

Others:
• Injectable forms can cause serious reactions, including anaphylaxis, serum sickness, transfer of disease.

 ## Overdose/Toxicity

Signs and symptoms:
None for oral forms. For injectable forms, see *Adverse Reactions or Side Effects* section below.

What to do:
For symptoms of overdosage: Discontinue nucleic acid, and consult doctor. Also see *Adverse Reactions or Side Effects* section below.
For accidental overdosage (such as child taking entire bottle): Dial 911 (emergency), 0 for operator or your nearest Poison Control Center.

 ## Adverse Reactions or Side Effects

Reaction or effect	What to do
For injectable forms:	
Anaphylaxis—symptoms include immediate severe itching, paleness, low blood pressure, fainting, coma	Yell for help. Don't leave victim. Begin CPR (cardio-pulmonary resuscitation), mouth-to-mouth breathing and external cardiac massage. Have someone call "0" (Operator) or 911 (Emergency). Don't stop CPR until help arrives.
Serum sickness, characterized by fever, edema of face and ankles, decreased urine output, skin rash	Seek emergency treatment.
Many viral illnesses, such as hepatitis or AIDS, if material is derived from human tissue or injected with contaminated needle	Discontinue. Call doctor immediately.

 ## Interaction with Medicine, Vitamins or Minerals

None known

 ## Interaction with Other Substances

None known

Phenylalanine

Basic Information

Phenylalanine is an amino acid.
Available from natural sources? Yes
Available from synthetic sources? Yes
Prescription required? No

Natural Sources

Almonds	Non-fat dried milk
Avocado	Peanuts
Bananas	Pickled herring
Cheese	Pumpkin seeds
Cottage cheese	Sesame seeds
Lima beans	

Reasons to Use

- Functions as building block of all proteins.
- Can induce significant short-term increases of blood levels of norepinephrine, dopamine and epinephrine. May be harmful at times and helpful at others. Don't take without medical supervision!

Unproved Speculated Benefits

- Treats mental depression.
- Improves memory.
- Diminishes pain.
- Increases mental alertness.
- Promotes sexual interest.
- Releases hormones that suppress appetite.
- Treats Parkinson's disease.

Who Needs Additional Amounts?

- Single amino-acid deficiencies are unknown except in people on crash diets consisting of only a few foods.
- Amino-acid deficiencies appear more commonly as a result of total protein deficiency, which is rare in the United States and Canada.
- Anyone with inadequate caloric or nutritional dietary intake or increased nutritional requirements.
- Those with inadequate protein dietary intake.
- Children, pregnant or breast-feeding women who are vegan vegetarians.
- People with recent severe burns or injuries.
- Premature infants.

Deficiency Symptoms

In moderate deficiencies:
- Slowed growth in children
- Low levels of essential proteins in blood

In severe deficiencies:
- Apathy
- Depigmentation of hair
- Edema
- Lethargy
- Liver damage
- Loss of muscle and fat
- Skin lesions
- Weakness

Unproved Speculated Symptoms

- Lack of sexual interest
- Impotence
- Poor memory
- Obesity

Lab Tests to Detect Deficiency

- None available, except for experimental purposes.

Dosage and Usage Information

Recommended Dietary Allowance (RDA):
No RDA has been established.

What this amino acid does:
- It is involved in production of dopamine and epinephrine, which affect transmission of impulses in the human brain and other parts of the nervous system.

Miscellaneous information:
- Supplements taken by healthy people will not make them healthier.
- Poorly nourished people have a greater chance of adverse side effects from taking amino-acid supplements, including an amino-acid imbalance.
- The poorer the diet, the greater the chance of an amino-acid supplement creating a harmful combination.

Available as:
- Tablets: Swallow whole with full glass of liquid. Don't chew or crush. Take with meals or 1 to 1-1/2 hours after meals unless otherwise directed by your doctor.

Warnings and Precautions

Don't take if you:
- Are allergic to any food protein, such as eggs, milk, wheat.
- Are at risk of poor nutrition for any reason.
- Suffer from migraine headaches.
- Have phenylketonuria (PKU).
- Have pigmented malignant melanoma, a deadly form of skin cancer.
- Take any monamine oxidase inhibitor as an anti-depressant, including pargyline, isocarboxazid, phenelzine, procarbazine, tranylcypromine.

Consult your doctor if you have:
- High blood pressure.
- Self-medicated with phenylalanine for any reason without medical supervision.

Over age 55:
- Don't take amino-acid supplements if you are healthy.

Pregnancy:
- Don't take amino-acid supplements if you are healthy.

Breast-feeding:
- Don't take amino-acid supplements if you are healthy.

Effect on lab tests:
- None known.

Storage:
- Store in cool, dry place away from direct light, but don't freeze.
- Store safely out of reach of children.
- Don't store in bathroom medicine cabinet. Heat and moisture may change action of amino acid.

Others:
- Phenylalanine may cause high blood pressure to rise even higher.

Overdose/Toxicity

Signs and symptoms:
Unlikely to threaten life or cause significant symptoms.

What to do:
For symptoms of overdosage: Discontinue amino acid, and consult doctor. Also see *Adverse Reactions or Side Effects* section below.
For accidental overdosage (such as child taking entire bottle): Dial 911 (emergency), 0 for operator or your nearest Poison Control Center.

Adverse Reactions or Side Effects

Reaction or effect	What to do
Lowers blood pressure	Discontinue. Call doctor immediately.
Raises blood pressure	Discontinue. Call doctor immediately.
Migraine headaches	Discontinue. Call doctor immediately.

Interaction with Medicine, Vitamins or Minerals

Interacts with	Combined effect
Anti-depressant drugs (containing monamine oxidase inhibitors)	Dangerous or life-threatening blood-pressure elevation.
Tyrosine	Additive effect with phenylalanine greatly increases chance of undesirable side effects.

Interaction with Other Substances

None known

Taurine

Basic Information
Taurine is an amino acid.
Available from natural sources? Yes
Available from synthetic sources? Yes
Prescription required? No

Natural Sources

Eggs
Fish
Meat
Milk
Note: Not available from plant sources.

Reasons to Use

- May be helpful in treating epilepsy.
- Functions as building block for all proteins.
- Helps regulate nervous system.
- Helps regulate muscle system.

Unproved Speculated Benefits

- May be essential for growth of infants, children, adolescents.

Who Needs Additional Amounts?

- Single amino-acid deficiencies are unknown except in people on crash diets consisting of only a few foods.
- Amino-acid deficiencies appear more commonly as a result of total protein deficiency, which is rare in the United States and Canada.
- Anyone with inadequate caloric or nutritional dietary intake or increased nutritional requirements.
- Those with inadequate protein dietary intake.
- Children, pregnant or breast-feeding women who are vegan vegetarians.
- People with recent severe burns or injuries.
- Premature infants.

Deficiency Symptoms

In moderate deficiencies:
- Slowed growth in children
- Low levels of essential proteins in blood

In severe deficiencies:
- Apathy
- Depigmentation of hair
- Edema
- Lethargy
- Liver damage
- Loss of muscle and fat
- Skin lesions
- Weakness

Unproved Speculated Symptoms

- Vision problems

Lab Tests to Detect Deficiency

- None available, except for experimental purposes.

Dosage and Usage Information

Recommended Dietary Allowance (RDA):
No RDA has been established.

What this amino acid does:
- Provides part of all proteins.

Miscellaneous information:
- Taurine is synthesized from methionine and cystine.
- Supplements are not needed by healthy people who eat well-balanced diets.
- Poorly nourished people have a greater chance of adverse side effects from taking amino-acid supplements, including an amino-acid imbalance.
- The poorer the diet, the greater the chance of an amino-acid supplement creating a harmful combination.

Available as:
- Tablets: Swallow whole with full glass of liquid. Don't chew or crush. Take with meals or 1 to 1-1/2 hours after meals unless otherwise directed by your doctor.
- Capsules: Swallow whole with full glass of liquid. Don't chew or crush. Take with meals or 1 to 1-1/2 hours after meals unless otherwise directed by your doctor.

》▶

Warnings and Precautions

Don't take if you:
• Are allergic to any food protein such as eggs, milk, wheat.
• Are at risk of poor nutrition for any reason.

Consult your doctor if you have:
• Epilepsy.
• Eye problems.
• Self-prescribed taurine without medical supervision.

Over age 55:
• Don't take amino-acid supplements if you are healthy.

Pregnancy:
• Don't take amino-acid supplements if you are healthy.

Breast-feeding:
• Don't take amino-acid supplements if you are healthy.

Effect on lab tests:
• None known.

Storage:
• Store in cool, dry place away from direct light, but don't freeze.
• Store safely out of reach of children.
• Don't store in bathroom medicine cabinet. Heat and moisture may change action of amino acid.

 ## Overdose/Toxicity

Signs and symptoms:
Unlikely to threaten life or cause significant symptoms.

What to do:
For symptoms of overdosage: Discontinue amino acid, and consult doctor. Also see *Adverse Reactions or Side Effects* section below.
For accidental overdosage (such as child taking entire bottle): Dial 911 (emergency), 0 for operator or your nearest Poison Control Center.

 ## Adverse Reactions or Side Effects

Reaction or effect	What to do
Memory deficits	Discontinue. Call doctor when convenient.
May depress normal function of central nervous system	Discontinue. Call doctor immediately.

 ## Interaction with Medicine, Vitamins or Minerals

Interacts with	Combined effect
Anti-convulsants	May decrease frequency of seizures.

 ## Interaction with Other Substances

None known

AMINO ACID

Tryptophan

Basic Information
Tryptophan is an amino acid.
Brand names, see page 477.
Available from natural sources? Yes
Available from synthetic sources? Yes
Prescription required? No

Natural Sources

Bananas	Meat
Cottage cheese	Milk
Dried dates	Peanuts
Fish	Turkey

Reasons to Use

• Functions as building block of all proteins.

Unproved Speculated Benefits

• Is an effective sleep aid.
• Acts as an anti-depressant.
• Helps treat cocaine addiction.
• Treats mania and aggressive behavior.
• Decreases sensitivity to moderate pain.
• Suppresses appetite.

Who Needs Additional Amounts?

• Single amino-acid deficiencies are unknown except in people on crash diets consisting of only a few foods.
• Amino-acid deficiencies appear more commonly as a result of total protein deficiency, which is rare in the United States and Canada.
• Anyone with inadequate caloric or nutritional dietary intake or increased nutritional requirements.
• Those with inadequate protein dietary intake.
• Children, pregnant or breast-feeding women who are vegan vegetarians.
• People with recent severe burns or injuries.
• Premature infants.

Deficiency Symptoms

In moderate deficiencies:
• Slowed growth in children
• Low levels of essential proteins in blood

In severe deficiencies:
• Apathy
• Depigmentation of hair
• Edema
• Lethargy
• Liver damage
• Loss of muscle and fat
• Skin lesions
• Weakness

Unproved Speculated Symptoms

• None

Lab Tests to Detect Deficiency

• None available except for experimental purposes.

Dosage and Usage Information

Recommended Dietary Allowance (RDA):
No RDA has been established.

What this amino acid does:
• Provides part of all proteins.
• Participates in biosynthesis of a neurotransmitter called *serotonin*. Serotonin may be an inducer of certain stages of sleep.

Available as:
• Capsules or tablets: Swallow whole with full glass of liquid. Don't chew or crush. Take with or immediately after food to decrease stomach irritation.

Warnings and Precautions

Don't take if you:
- Are allergic to any food protein, such as eggs, milk, wheat.
- Are at risk of poor nutrition for any reason.
- Are severely depressed. Other medicines to treat severe depression are more effective.

Consult your doctor if you:
- Take medicines to induce sleep.

Over age 55:
- Don't take amino-acid supplements if you are healthy.

Pregnancy:
- Don't take amino-acid supplements if you are healthy.

Breast-feeding:
- Don't take amino-acid supplements if you are healthy.

Effect on lab tests:
- None known.

Storage:
- Store in cool, dry place away from direct light, but don't freeze.
- Store safely out of reach of children.
- Don't store in bathroom medicine cabinet. Heat and moisture may change action of amino acid.

Others:
- In experimental animal studies of animals with vitamin B-6 deficiency, large doses of tryptophan caused bladder cancer.

 ## Overdose/Toxicity

Signs and symptoms:
Unlikely to threaten life or cause significant symptoms.

What to do:
For symptoms of overdosage: Discontinue amino acid, and consult doctor. Also see *Adverse Reactions or Side Effects* section below.
For accidental overdosage (such as child taking entire bottle): Dial 911 (emergency), 0 for operator or your nearest Poison Control Center.

 ## Adverse Reactions or Side Effects

Reaction or effect	What to do
Fatigue	Discontinue. Call doctor when convenient.
Inertia	Discontinue. Call doctor when convenient.
Reduced vigor	Discontinue. Call doctor when convenient.

 ## Interaction with Medicine, Vitamins or Minerals

None known

 ## Interaction with Other Substances

None known

Tyrosine

Basic Information

Tyrosine is an amino acid.
Available from natural sources? Yes
Available from synthetic sources? Yes
Prescription required? No

Natural Sources

Almonds	Non-fat dried milk
Avocados	Peanuts
Bananas	Pickled herring
Cheese	Pumpkin seeds
Cottage cheese	Sesame seeds
Lima beans	

Reasons to Use

• Functions as building block of all proteins.
• Can induce significant short-term increases of blood levels of norepinephrine, dopamine and epinephrine. May be harmful at times and helpful at others. Don't take without medical supervision!

Unproved Speculated Benefits

• Treats mental depression.
• Improves memory.
• Diminishes pain.
• Increases mental alertness.
• Promotes sexual interest.
• Releases hormones that suppress appetite.
• Treats Parkinson's disease.

Who Needs Additional Amounts?

• Single amino-acid deficiencies are unknown except in people on crash diets consisting of only a few foods.
• Amino-acid deficiencies appear more commonly as a result of total protein deficiency, which is rare in the United States and Canada.
• Anyone with inadequate caloric or nutritional dietary intake or increased nutritional requirements.
• Those with inadequate protein dietary intake.
• Children, pregnant or breast-feeding women who are vegan vegetarians.
• People with recent severe burns or injuries.
• Premature infants.

Deficiency Symptoms

In moderate deficiencies:
• Slowed growth in children
• Low levels of essential proteins in blood
In severe deficiencies:
• Apathy
• Depigmentation of hair
• Edema
• Lethargy
• Liver damage
• Loss of muscle and fat
• Skin lesions
• Weakness

Unproved Speculated Symptoms

• Lack of sexual interest
• Impotence
• Poor memory
• Obesity

Lab Tests to Detect Deficiency

• None available, except for experimental purposes.

Dosage and Usage Information

Recommended Dietary Allowance (RDA):
No RDA has been established.

What this amino acid does:
• It is involved in production of dopamine and epinephrine, which affect transmission of impulses in the human brain and other parts of the nervous system.

Miscellaneous information:
• Supplements taken by healthy people will not make them healthier.
• Poorly nourished people have a greater chance of adverse side effects from taking amino-acid supplements, including an amino-acid imbalance.
• The poorer the diet, the greater the chance of an amino-acid supplement creating a harmful combination.

Available as:
• Tablets: Swallow whole with full glass of liquid. Don't chew or crush. Take with meals or 1 to 1-1/2 hours after meals unless otherwise directed by your doctor.

Tyrosine

Warnings and Precautions

Don't take if you:
- Are allergic to any food protein, such as eggs, milk, wheat.
- Are at risk of poor nutrition for any reason.
- Suffer from migraine headaches.
- Have phenylketonuria (PKU).
- Have pigmented malignant melanoma, a deadly form of skin cancer.
- Take any monamine oxidase inhibitor as an anti-depressant, including pargyline, isocarboxazid, phenelzine, procarbazine, tranylcypromine.

Consult your doctor if you have:
- High blood pressure.
- Self-medicated with tyrosine for any reason without medical supervision.

Over age 55:
- Don't take amino-acid supplements if you are healthy.

Pregnancy:
- Don't take amino-acid supplements if you are healthy.

Breast-feeding:
- Don't take amino-acid supplements if you are healthy.

Effect on lab tests:
- None known.

Storage:
- Store in cool, dry place away from direct light, but don't freeze.
- Store safely out of reach of children.
- Don't store in bathroom medicine cabinet. Heat and moisture may change action of amino acid.

Others:
- Tyrosine may cause high blood pressure to rise even higher at times.

Overdose/Toxicity

Signs and symptoms:
Unlikely to threaten life or cause significant symptoms.

What to do:
For symptoms of overdosage: Discontinue amino acid, and consult doctor. Also see *Adverse Reactions or Side Effects* section below.
For accidental overdosage (such as child taking entire bottle): Dial 911 (emergency), 0 for operator or your nearest Poison Control Center.

Adverse Reactions or Side Effects

Reaction or effect	What to do
Lowers blood pressure	Discontinue. Call doctor immediately.
Raises blood pressure	Discontinue. Call doctor immediately.
Migraine headaches	Discontinue. Call doctor immediately.

Interaction with Medicine, Vitamins or Minerals

Interacts with	Combined effect
Anti-depressant drugs (containing monamine oxidase inhibitors)	Dangerous or life-threatening blood-pressure elevation.
Phenylalanine	Additive effect with tyrosine greatly increases chance of undesirable side effects.

Interaction with Other Substances

None known

AMINO ACID

Other Supplements

These substances may play important roles in human health and nutrition, but they are neither vitamins nor minerals. The selected supplements discussed in this book include fats and lipids and miscellaneous supplements.

Fats, along with carbohydrates and proteins, make up our daily food. There has been much publicity regarding the potential dangers of fats in our diet, particularly the dangers of cholesterol, low-density lipoproteins and too much total fat intake. The discussion of all these important items is beyond the scope of this book and is not attempted. I've focussed on a few fatty acids that are making headlines. These include choline, lecithin, gamma-linolenic acid (evening primrose oil), inositol and omega-3 fatty acids (fish oil).

This section also describes what is known and not known about a number of supplements that are widely touted as important to human nutrition. These substances include superoxide dismutase, wheatgrass, barley grass and dietary fiber.

Acidophilus

Basic Information

Acidophilus is a bacterium found in yogurt, kefir and other products.
Chemical this supplements contains:
Enzymes, to aid digestion

Known Effects

- Helps maintain normal bacteria balance in lower intestines.
- Kills monilia, yeast or fungus on contact.

Miscellaneous information:
- Acidophilus is made by fermenting milk using *lactobaccillus acidophilus* and other bacteria.
- It is available as a liquid, in capsules or in milk products, such as yogurt or kefir.

Unproved Speculated Benefits

- Lowers cholesterol.
- Clears up skin problems.
- Helps prevent vaginal yeast infections in women who take antibiotics or who have diabetes.
- Extends life span.
- Helps digestion of milk and milk products in people with lactase deficiency.
- Enhances immunity.

Warnings and Precautions

Don't take if you:
- Have intestinal problems except under a physician's supervision.

Consult your doctor:
- Before you use acidophilus in vaginal area for yeast infections.
- If you take any medicinal drugs or herbs including aspirin, laxatives, cold and cough remedies, antacids, vitamins, minerals, amino acids, supplements, other prescription or non-prescription drugs.

Pregnancy:
- Problems in pregnant women taking small or usual amounts have not been proved. But the chance of problems does exist. Don't use unless prescribed by your doctor.

Breast-feeding:
- Problems in breast-fed infants of lactating mothers taking small or usual amounts have not been proved. But the chance of problems does exist. Don't use unless prescribed by your doctor.

Infants and children:
- Treating infants and children under 2 with any supplement is hazardous.

Storage:
- Keep cool and dry, but don't freeze. Store safely away from children.

Safe dosage:
- At present no "safe" dosage has been established.

Toxicity

Comparative-toxicity rating not available from standard references.

Adverse Reactions, Side Effects or Overdose Symptoms

None expected

Bee Pollen

Basic Information

Bee pollen is the microscopic male seed in flowering plants.
Chemicals this supplement contains:
Some vitamins, minerals and amino acids. Bee pollen is expensive and provides inadequate, uncertain quantities of nutrients.

Known Effects

- No beneficial effect in the body has been proved.

Miscellaneous information:
- Available in injectable form and capsules.

Unproved Speculated Benefits

- Acts as anti-aging agent.
- Energizes body.
- Regulates bowels.
- Treats prostate problems.
- Helps weight control.
- Renews skin.
- Reduces risk of heart disease and arthritis.
- Relieves stress.
- Boosts immunity.
- Inhibits cancer.
- Decreases allergy symptoms.

Warnings and Precautions

Don't take if you:
- Are pregnant, think you may be pregnant or plan pregnancy in the near future.

Consult your doctor if you:
- Take this herb for any medical problem that doesn't improve in 2 weeks. There may be safer, more-effective treatments.

Pregnancy:
- Problems in pregnant women taking small or usual amounts have not been proved. But the chance of problems does exist. Don't use unless prescribed by your doctor.

Breast-feeding:
- Problems in breast-fed infants of lactating mothers taking small or usual amounts have not been proved. But the chance of problems does exist. Don't use unless prescribed by your doctor.

Infants and children:
- Treating infants and children under 2 with any supplement is hazardous.

Storage:
- Keep cool and dry, but don't freeze. Store safely away from children.

Safe dosage:
- At present no "safe" dosage has been established.

Toxicity

Comparative-toxicity rating not available from standard references.

For symptoms of toxicity: See *Adverse Reactions, Side Effects or Overdose Symptoms* section below.

Adverse Reactions, Side Effects or Overdose Symptoms

Signs and symptoms	What to do
May cause allergic reactions in those sensitive to pollens. Mild allergic response is characterized by itching, pain at injection site and swelling occurring in 24-48 hours.	Discontinue. Call doctor immediately.
Life-threatening anaphylaxis may follow injection—symptoms include immediate severe itching, paleness, low blood pressure, loss of consciousness, coma	Yell for help. Don't leave victim. Begin CPR (cardiopulmonary resuscitation), mouth to mouth breathing and external cardiac massage. Have someone dial "O" (operator) or 911 (emergency). Don't stop CPR until help arrives.

Bioflavinoids (Vitamin P)

Basic Information

Bioflavinoids are a brightly colored, chemical constituent of pulp and rind of citrus fruits, green pepper, apricots, cherries, grapes, papaya, tomatoes, papaya, broccoli.
Chemicals this supplement contains:
*   *Hesperidin*
*   *Nobiletin*
*   *Rutin*
*   *Sinensetin*
*   *Tangeretin*

Known Effects:

*   Treats rare bioflavinoid deficiency characterized by fragile capillaries and unusual bleeding.
*   May act as an anti-oxidant, preventing vitamin C and adrenalin from being oxidized by copper-containing enzymes.

Miscellaneous information:

*   Bioflavinoids are sold under the brand names Rutin, Hesperiden, CVP, duo-CVP, Hesper capsules, Hesper bitabs and are included in numerous vitamin/mineral supplements.
*   Enough bioflavinoids are present in food to make supplements unnecessary in healthy humans.
*   Commercial products such as tablets or capsules often contain vitamin C.

Unproved Speculated Benefits

*   Increases effectiveness of vitamin C.
*   Prevents hemorrhoids.
*   Prevents miscarriages.
*   Prevents retinal bleeding in people with diabetes and hypertension.
*   Prevents capillary fragility.
*   Prevents nosebleed.
*   Prevents post-partem hemorrhage.
*   Prevents menstrual disorders.
*   Prevents blood clotting and platelet clumping.
*   Prevents easy bruising.

Warnings and Precautions

Don't take if you:

*   Have a bleeding problem until studies are done to diagnose the underlying disease.

Consult your doctor if you:

*   Self-medicate.
*   Take any medicinal drugs or herbs including aspirin, laxatives, cold and cough remedies, antacids, vitamins, minerals, amino acids, supplements, other prescription or non-prescription drugs.

Pregnancy:

*   Notify doctor if you take supplements.

Breast-feeding:

*   Notify doctor if you take supplements.

Others:

*   None expected if you are beyond childhood and under 45, basically healthy and take for only a short time.

Storage:

*   Keep cool and dry, but don't freeze. Store safely away from children.

Safe dosage:

*   At present no "safe" dosage has been established.

Toxicity

Comparative-toxicity rating not available from standard references.

Adverse Reactions, Side Effects or Overdose Symptoms

None expected

Basic Information

Non-leavening, with a slightly bitter taste. It is an excellent source of B vitamins, protein and minerals.

Chemicals this supplement contains:
B vitamins
DNA and RNA
Trace mineral, chromium

Known Effects

- Supplies B vitamins, protein and minerals.
- Provides bulk to prevent constipation.
- Good source of enzyme-producing vitamins.
- Chromium in brewer's yeast helps regulate sugar metabolism.

Miscellaneous information:

- Out of the can, the bitter taste of brewer's yeast may be unpleasant. Adding it to foods with a strong taste makes it tolerable.
- A good, inexpensive food supplement for aging adults and growing, developing children.
- Can be used in baking, soups, chili and casseroles to increase nutritional content.
- Available in powder, flakes and tablets.

Unproved Speculated Benefits

- Helps treat diabetes.
- Reduces risk of high cholesterol in blood.
- Treats contact dermatitis.

Warnings and Precautions

Don't take if you:

- Have intestinal disease.

Consult your doctor if you:

- Have an acute intestinal upset.
- Take any medicinal drugs or herbs including aspirin, laxatives, cold and cough remedies, antacids, vitamins, minerals, amino acids, supplements, other prescription or non-prescription drugs.

Pregnancy:

- Excellent, inexpensive source of nutrients. Don't overuse.

Breast-feeding:

- Excellent, inexpensive source of nutrients. Don't overuse.

Others:

- Quality and quantity of nutrients vary greatly among commercially available products.
- Brewer's yeast is usually non-toxic if you consume 1 tablespoon or less of the powder or equivalent amounts of tablets or flakes.

Storage:

- Keep cool and dry, but don't freeze. Store safely away from children.

Safe dosage:

- At present no "safe" dosage has been established.

Toxicity

Comparative-toxicity rating not available from standard references.

For symptoms of toxicity: See *Adverse Reactions, Side Effects or Overdose Symptoms* section below.

Adverse Reactions, Side Effects or Overdose Symptoms

Signs and symptoms	What to do
Diarrhea	Discontinue. Call doctor immediately.
Nausea	Discontinue. Call doctor immediately.

SUPPLEMENT

Choline

Basic Information
Available from natural sources? Yes
Available from synthetic sources? Yes
Prescription required? No

Natural Sources

Cabbage	Green beans
Calves' liver	Lentils
Cauliflower	Rice
Caviar	Soybeans
Eggs	Soy lecithin
Garbanzo beans	Split peas
(chickpeas)	

Found in all animal and plant products.

Reasons to Use

- Protects against damage to cells by oxidation.
- People taking niacin or nicotinic acid for treatment of high-serum cholesterol and triglycerides need lecithin or choline supplements because nicotinic acid and nicotinomide (vitamin B-3) can reduce normal amount of choline and lecithin available for basic body needs.

Unproved Speculated Benefits

- Protects against cardiovascular disease.
- Protects against memory loss.
- Prevents some diseases of the nervous system, such as Alzheimer's disease and tardive dyskinesia (involuntary, abnormal facial movements including grimacing, sticking out tongue and sucking movements).
- Treats Alzheimer's disease.
- Treats liver damage caused by alcoholism.
- Lowers cholesterol level in human serum.

Who Needs Additional Amounts?

- No one.

Deficiency Symptoms

- There are no specific deficiency symptoms in man, although some animals can suffer from lack of choline. Lecithin must be present for choline synthesis in the human body.

Unproved Speculated Symptoms

- Symptoms of heart or blood-vessel disease.
- Decreasing mental alertness.

Lab Tests to Detect Deficiency

- None available, except for experimental purposes.

Dosage and Usage Information

Recommended Dietary Allowance (RDA):
No RDA has been established.

What this supplement does:
- Choline is involved in production of acetylcholine. Acetylcholine must be present in the body for proper function of the nervous system, including mood, behavior, orientation, personality traits, judgment.

Miscellaneous information:
- Choline's major source is lecithin.
- It is used as a thickener in several foods, including mayonnaise, margarine, ice cream.

Available as:
- Capsules: Swallow whole with full glass of liquid. Don't chew or crush. Take with meals or 1 to 1-1/2 hours after meals unless otherwise directed by your doctor.

➤➤➤

Warnings and Precautions

Don't take if you:
• Are healthy and eat a well-balanced diet.

Consult your doctor if you have:
• Plans to use choline to treat Alzheimer's disease with lecithin/choline.

Over age 55:
• Don't take if you are healthy.

Pregnancy:
• Don't take if you are healthy. Check with your doctor if you have any questions.

Breast-feeding:
• Don't take if you are healthy. Check with your doctor if you have any questions.

Effect on lab tests:
• May cause inaccurate results in choline/sphingomyelin test as part of examination of amniotic fluid.

Storage:
• Store in cool, dry place away from direct light, but don't freeze.
• Store safely out of reach of children.
• Don't store in bathroom medicine cabinet. Heat and moisture may change action of supplement.

Others:
• Don't take more than 1 gram per day.

Overdose/Toxicity

Signs and symptoms:
Nausea, vomiting, dizziness.

What to do:
For symptoms of overdosage: Discontinue supplement, and consult doctor. Also see *Adverse Reactions or Side Effects* section below.
For accidental overdosage (such as child taking entire bottle): Dial 911 (emergency), 0 for operator or your nearest Poison Control Center.

Adverse Reactions or Side Effects

Reaction or effect	What to do
"Fishy" body odor	Discontinue. Call doctor when convenient.

Interaction with Medicine, Vitamins or Minerals

Interacts with	Combined effect
Nicotinic acid (nicotinamide, vitamin B-3)	Decreases choline effectiveness.

Interaction with Other Substances

None known

SUPPLEMENT

Chondroitin Sulfate

Basic Information
This substance is found in cartilages of most mammals.
Chemicals this supplement contains:
Complex protein molecules

Known Effects

• None proved.

Miscellaneous information:
• Available as capsules.

Unproved Speculated Benefits

• Lowers cholesterol levels.
• Lowers triglyceride levels.
• Prolongs clotting time.

Warnings and Precautions

Don't take if you:
• Have bleeding problems.
• Are pregnant, think you may be pregnant or plan pregnancy in the near future.

Consult your doctor if you:
• Take anti-coagulants.
• Take any medicinal drugs or herbs including aspirin, laxatives, cold and cough remedies, antacids, vitamins, minerals, amino acids, supplements, other prescription or non-prescription drugs.

Pregnancy:
• Problems in pregnant women taking small or usual amounts have not been proved. But the chance of problems does exist. Don't use unless prescribed by your doctor.

Breast-feeding:
• Problems in breast-fed infants of lactating mothers taking small or usual amounts have not been proved. But the chance of problems does exist. Don't use unless prescribed by your doctor.

Infants and children:
• Treating infants and children under 2 with any supplement is hazardous.

Others:
• None expected if you are beyond childhood and under 45, basically healthy and take for only a short time.

Storage:
• Keep cool and dry, but don't freeze. Store safely away from children.

Safe dosage:
• At present no "safe" dosage has been established.

Toxicity

Comparative-toxicity rating not available from standard references.

Adverse Reactions, Side Effects or Overdose Symptoms

None expected

Coenzyme Q (CoQ)

Basic Information

*Coenzyme Q is part of the mitochondria of cells
and is necessary for energy production.
Chemical this supplement contains:
Coenzyme Q10 (a nutrient); found in beef,
sardines, spinach, peanuts*

Known Effects

• Controls flow of oxygen within individual cells.

Miscellaneous information:
• Oral products are available, but most experts
do *not* recommend using them except under
medical supervision.

Unproved Speculated Benefits

• Improves heart-muscle metabolism.
• Treats chest pain caused by narrowed coronary
arteries (coronary insufficiency).
• Lowers blood pressure.
• Treats congestive heart failure by enhancing
pumping action of heart.

Warnings and Precautions

Don't take if you:
• Have heart disease, without consulting doctor.

Consult your doctor if you:
• Take any medicinal drugs or herbs including
aspirin, laxatives, cold and cough remedies,
antacids, vitamins, minerals, amino acids,
supplements, other prescription or non-
prescription drugs.

Pregnancy:
• Dangers outweigh any possible benefits. Don't
use.

Breast-feeding:
• Dangers outweigh any possible benefits. Don't
use.

Infants and children:
• Treating infants and children under 2 with any
supplement is hazardous.

Others:
• No contraindications if you are not pregnant
and do not take amounts larger than a
manufacturer's recommended dosage

Storage:
• Keep cool and dry, but don't freeze. Store
safely away from children.

Safe dosage:
• At present no "safe" dosage has been
established.

Toxicity

Comparative-toxicity rating not available from
standard references.

Adverse Reactions, Side Effects or Overdose Symptoms

None expected

Dessicated Liver

Basic Information

Dessicated liver is a concentrated form of dried liver and is available in tablets or powder.
Chemicals this supplement contains:
  Calcium
  Cholesterol
  Copper
  Iron
  Phosphorus
  Vitamins A, C, D

Known Effects

• Is a good source of vitamins A, C, D and iron, calcium, phosphorus, copper.

Miscellaneous information:
• Healthy people who eat a balanced diet probably do not need this supplement.

Unproved Speculated Benefits

• Acts as an anti-stress agent.
• Cures gum problems.

Warnings and Precautions

Don't take if you:
• Are pregnant, think you may be pregnant or plan pregnancy in the near future.

Consult your doctor if you:
• Take any medicinal drugs or herbs including aspirin, laxatives, cold and cough remedies, antacids, vitamins, minerals, amino acids, supplements, other prescription or non-prescription drugs.

Pregnancy:
• Problems in pregnant women taking small or usual amounts have not been proved. But the chance of problems does exist. Don't use unless prescribed by your doctor.

Breast-feeding:
• Problems in breast-fed infants of lactating mothers taking small or usual amounts have not been proved. But the chance of problems does exist. Don't use unless prescribed by your doctor.

Infants and children:
• Treating infants and children under 2 with any supplement is hazardous.

Others:
• None expected if you are beyond childhood and under 45, basically healthy and take for only a short time.

Storage:
• Keep cool and dry, but don't freeze. Store safely away from children.

Safe dosage:
• At present no "safe" dosage has been established.

Toxicity

Comparative-toxicity rating not available from standard references.

Adverse Reactions, Side Effects or Overdose Symptoms

None expected

Basic Information

Cell walls of plants are made of fiber that give a plant structure and stability. Fiber cannot be broken down by enzymes in the digestive tract, so fiber passes through without being absorbed.
Chemicals this supplement contains: Structured and non-structured substances in plant carbohydrate (starches)

Known Effects

- Absorbs many times its weight in water, causing bulkier stools and lessening chance of constipation.
- Helps control blood-sugar level in people with diabetes.
- Helps reduce cholesterol and triglycerides in blood.

Miscellaneous information:
- Best sources of dietary fiber include fresh fruits, vegetables, nuts, seeds, whole-grain products, potatoes.
- Available commercially in capsules, tablets, chewable tablets, oral suspension and flakes or wafers.

Unproved Speculated Benefits

- Reduces risk of heart disease.
- Reduces risk of cancer of colon and rectum.
- Reduces risk of diverticulitis.

Warnings and Precautions

Don't take if you:
- Have Crohn's disease.

Consult your doctor if you:
- Are pregnant, think you may be pregnant or plan pregnancy in the near future.

Pregnancy:
- Problems in pregnant women taking small or usual amounts have not been proved. But the chance of problems does exist. Don't use unless prescribed by your doctor.

Breast-feeding:
- Problems in breast-fed infants of lactating mothers taking small or usual amounts have not been proved. But the chance of problems does exist. Don't use unless prescribed by your doctor.

Infants and children:
- Treating infants and children under 2 with any supplement is hazardous.

Others:
- Intake of excessive amounts of fiber may decrease absorption of minerals, especially calcium, iron, zinc.

Storage:
- Keep cool and dry, but don't freeze. Store safely away from children.

Safe dosage:
- At present no "safe" dosage has been established. Most experts feel that increasing fiber is healthful, but no one knows for sure the optimal amount.

Toxicity

Comparative-toxicity rating not available from standard references.

For symptoms of toxicity: See *Adverse Reactions, Side Effects or Overdose Symptoms* section below.

Adverse Reactions, Side Effects or Overdose Symptoms

Signs and symptoms	What to do
Bloating of abdomen	Discontinue. Call doctor when convenient.
Excess flatulence	Discontinue. Call doctor when convenient.
Obstruction of large intestine. Rare, but more likely if there is pre-existing inflammatory disease. Symptoms of obstruction are tender, distended abdomen, abdominal pain, fever, no bowel movements.	Discontinue. Call doctor immediately.

Gamma-Linolenic Acid

Basic Information

Gamma-linolenic acid is found in a supplement called evening primrose oil.
Brand names, see page 477.
Available from natural sources? Yes
Available from synthetic sources? Yes
Prescription required? No

Natural Sources

Evening primrose (a plant)
Fish
Human mother's milk

Reasons to Use

• Helps inhibit coughing.
• Is an essential nutrient.
• Acts as an astringent.

Unproved Speculated Benefits

• May have an anti-clotting factor, which would make it useful in the prevention of heart attacks caused by thrombosis.
• Helps people suffering from atopic eczema or eczema due to allergy.
• Is used in external preparations to treat skin eruptions, such as psoriasis.
• Helps treat migraines.
• Helps treat asthma.
• Treats arthritis.
• Alleviates symptoms of premenstrual syndrome.
• Treats schizophrenia.
• Is effective against obesity.
• Makes fingernails stronger.
• Treats hangovers.
• General "cure-all" for many other disorders.

Who Needs Additional Amounts?

• Those on greatly restricted fat and oil intake.

Deficiency Symptoms

• None

Unproved Speculated Symptoms

• Eczema-like lesions
• Hair loss
• Reduced immunological response
• Kidney disease
• Inability of wounds to heal properly

Lab Tests to Detect Deficiency

• None available, except for experimental purposes.

Dosage and Usage Information

Recommended Dietary Allowance (RDA):
No RDA has been established.

What this supplement does:
• Functions as one of the sources of essential fatty acids.

Miscellaneous information:
• Evening primrose grows wild. A long spike of yellow flowers opens at night. Oil can be expressed from the tiny seeds of the flower.
• Linolenic acid, working with enzymes, becomes part of some prostaglandins. Prostaglandins sometimes *limit* inflammatory reactions in the body and sometimes *cause* inflammatory reactions. Taking evening primrose oil may cause unpredictable, harmful effects.
• Nutrition authorities do not recommend supplements in healthy people.

Available as:
• Capsules: Swallow whole with full glass of liquid. Don't chew or crush. Take with meals or 1 to 1-1/2 hours after meals unless otherwise directed by your doctor.

➤➤

Warnings and Precautions

Don't take if you:
• Are healthy and eat a well-balanced diet.

Consult your doctor if you have:
• Any illness.

Over age 55:
• Don't take if you are healthy.

Pregnancy:
• Don't take if you are healthy.

Breast-feeding:
• Don't take if you are healthy.

Effect on lab tests:
• None known.

Storage:
• Store in cool, dry place away from direct light, but don't freeze.
• Store safely out of reach of children.
• Don't store in bathroom medicine cabinet. Heat and moisture may change action of supplement.

Overdose/Toxicity

Signs and symptoms:
Unlikely to threaten life or cause significant symptoms.

What to do:
For symptoms of overdosage: Discontinue supplement, and consult doctor. Also see *Adverse Reactions or Side Effects* section below.
For accidental overdosage (such as child taking entire bottle): Dial 911 (emergency), 0 for operator or your nearest Poison Control Center.

Adverse Reactions or Side Effects

Reaction or effect	What to do
Can make symptoms of some problems, such as asthma, migraines, arthritis, worse	Don't take.

Interaction with Medicine, Vitamins or Minerals

None known

Interaction with Other Substances

None known

SUPPLEMENT

Gelatin (Plain)

Basic Information
Gelatin is a tasteless, odorless substance extracted by boiling bones, hoofs and animal tissues.
Chemical this supplement contains:
Proteins

Known Effects

• None except to serve as source of protein.

Miscellaneous information:
• Gelatin is available in capsule or powdered form.

Unproved Speculated Benefits

• Improves condition of broken, splitting, brittle nails.
• Prevents nosebleeds.

Warnings and Precautions

Don't take if you:
• Expect gelatin to cure anything.

Consult your doctor if:
• Conditions you take gelatin for don't improve by themselves.

Pregnancy:
• Problems in pregnant women taking small or usual amounts have not been proved. But the chance of problems does exist. Don't use unless prescribed by your doctor.

Breast-feeding:
• Problems in breast-fed infants of lactating mothers taking small or usual amounts have not been proved. But the chance of problems does exist. Don't use unless prescribed by your doctor.

Infants and children:
• Treating infants and children under 2 with any supplement is hazardous.

Others:
• None expected if you are beyond childhood and under 45, basically healthy and take for only a short time.

Storage:
• Keep cool and dry, but don't freeze. Store safely away from children.

Safe dosage:
• At present no "safe" dosage has been established.

Toxicity

Comparative-toxicity rating not available from standard references.

Adverse Reactions, Side Effects or Overdose Symptoms

None expected

Basic Information

*Glandulars are concentrated forms of various
animal glands including adrenals, thymus,
spleen, intestines.*
Chemicals this supplement contains:
Antibiotics
Enzymes
Herbicides
Hormones
Pesticides

Known Effects

• None ever proved or conclusively
demonstrated.

Miscellaneous information:
• Glandulars are available as extracts without
prescription. They are worthless and may be
harmful. Don't use them!

Unproved Speculated Benefits

• Increases sex drive.
• Cures infertility.
• Prevents aging.
• Reverses aging.
• Builds bodies.

Warnings and Precautions

Don't take:
• For any reason.

Consult your doctor if you have:
• Any problem these substances are advertised
to help.

Pregnancy:
• Danger outweighs benefits. Don't use.

Breast-feeding:
• Danger outweighs benefits. Don't use.

Infants and children:
• Treating infants and children under 2 with any
supplement is hazardous.

Others:
• Don't take these! If toxic symptoms occur,
discontinue. Call doctor immediately.

Storage:
• Keep cool and dry, but don't freeze. Store
safely away from children.

Safe dosage:
• At present no "safe" dosage has been
established.

Toxicity

Comparative-toxicity rating not available from
standard references.

For symptoms of toxicity: See *Adverse
Reactions, Side Effects or Overdose Symptoms*
section below.

Adverse Reactions, Side Effects or Overdose Symptoms

Signs and symptoms	What to do
Intestinal upsets, with nausea, vomiting, diarrhea	Discontinue. Call doctor immediately.
Numbness	Discontinue. Call doctor when convenient.
Tingling of feet and hands	Discontinue. Call doctor when convenient.

Inositol

Basic Information
Inositol is also called myo-inositol.
Available from natural sources? Yes
Available from synthetic sources? Yes
Prescription required? No

Natural Sources

Beans, dried	Nuts
Calves' liver	Oats
Cantaloupe	Pork
Citrus fruit, except	Rice
lemons	Veal
Garbanzo beans	Wheat germ
(chickpeas)	Whole-grain products
Lecithin granules	
Lentils	

Reasons to Use

• Plays a role similar to choline in helping move fats out of liver.

Unproved Speculated Benefits

• Protects against cardiovascular disease.
• Protects against peripheral neuritis associated with diabetes. (Some studies have shown promise for this use, but definitive, well-controlled studies have not been done.)
• Protects against hair loss.
• Helps maintain healthy hair.
• Functions as mild anti-anxiety agent.
• Helps control blood-cholesterol level.
• Promotes body's production of lecithin.
• Treats constipation with its stimulating effect on muscular action of alimentary canal.

Who Needs Additional Amounts?

• Heavy drinkers of coffee, tea, cocoa and other caffeine-containing substances.

Deficiency Symptoms

• Symptoms develop only in some animals; none are known in humans.

Unproved Speculated Symptoms

• Eczema
• Constipation
• Abnormalities of the eyes

Lab Tests to Detect Deficiency

• None available, except for experimental purposes.

Dosage and Usage Information

Recommended Dietary Allowance (RDA):
No RDA has been established.

What this supplement does:
• Inositol forms an important part of *phospholipids,* which are compounds manufactured in our bodies.

Miscellaneous information:
• Caffeine in large quantities may create an inositol shortage.

Available as:
• Capsules: Swallow whole with full glass of liquid. Don't chew or crush. Take with meals or 1 to 1-1/2 hours after meals unless otherwise directed by your doctor.

➤➤➤

Warnings and Precautions

Don't take if you:
• Are healthy.

Consult your doctor if you have:
• Diabetes with peripheral neuropathy—pain, numbness, tingling, alternating feelings of cold and hot in feet and hands. Medical supervision is necessary.

Over age 55:
• Don't take if you are healthy.

Pregnancy:
• Don't take if you are healthy.

Breast-feeding:
• Don't take if you are healthy.

Effect on lab tests:
• None known.

Storage:
• Store in cool, dry place away from direct light, but don't freeze.
• Store safely out of reach of children.
• Don't store in bathroom medicine cabinet. Heat and moisture may change action of supplement.

Overdose/Toxicity

Signs and symptoms:
Unlikely to threaten life or cause significant symptoms.

What to do:
For symptoms of overdosage: Discontinue supplement, and consult doctor.
For accidental overdosage (such as child taking entire bottle): Dial 911 (emergency), 0 for operator or your nearest Poison Control Center.

Adverse Reactions or Side Effects

None expected.

Interaction with Medicine, Vitamins or Minerals

None known

Interaction with Other Substances

Caffeine-containing foods and **beverages** may create inositol shortage in the body.

SUPPLEMENT

Jojoba (Coffeeberry, Goatnut)

Basic Information
Biological name is Simmondsia chinensis.
Chemical this supplement contains:
  Amino acids

Known Effects

• Acts as soothing ingredient in many shampoos, pre-electric-shave conditioners, after-shave preparations, skin lotion, makeup remover.

Miscellaneous information:
• Jojoba is unique among plants because its seeds contain a liquid wax oil.
• The plant grows in Arizona and is used as a medicinal herb among Southern Arizona Indians.

Unproved Speculated Benefits

• Suppresses appetite.
• Treats rheumatoid arthritis.
• Treats inflammation.
• Relieves swelling.
• Treats acne.
• Treats warts.
• Treats tuberculosis.

Warnings and Precautions

Don't take if you:
• No contraindications if you are not pregnant and do not take amounts larger than manufacturer's recommended dosage.

Consult your doctor if you:
• Are pregnant, think you may be pregnant or plan pregnancy in the near future.
• Take any medicinal drugs or herbs including aspirin, laxatives, cold and cough remedies, antacids, vitamins, minerals, amino acids, supplements, other prescription or non-prescription drugs.

Pregnancy:
• Problems in pregnant women taking small or usual amounts have not been proved. But the chance of problems does exist. Don't use unless prescribed by your doctor.

Breast-feeding:
• Problems in breast-fed infants of lactating mothers taking small or usual amounts have not been proved. But the chance of problems does exist. Don't use unless prescribed by your doctor.

Infants and children:
• Treating infants and children under 2 with any supplement is hazardous.

Others:
• No beneficial effects when taken by mouth have been proved.

Storage:
• Keep cool and dry, but don't freeze. Store safely away from children.

Safe dosage:
• At present no "safe" dosage has been established.

Toxicity

Comparative-toxicity rating not available from standard references.

Adverse Reactions, Side Effects or Overdose Symptoms

None expected

L-Carnitine

Basic Information
L-carnitine is synthesized in the body from the amino acids lysine and methionine.
Available from natural sources? Yes
Available from synthetic sources? Yes
Prescription required? No

Natural Sources

Avocados
Dairy products
Red meats, especially lamb and beef
Tempeh (fermented soybean product)

Reasons to Use

• Promotes normal growth and development.

Unproved Speculated Benefits

• Treats and possibly prevents some forms of cardiovascular disease.
• Protects against muscle disease.
• Helps build muscle.
• Protects against liver disease.
• Protects against diabetes.
• Protects against kidney disease.
• Aids in dieting. May make low-calorie diets easier to tolerate by reducing feelings of hunger and weakness.

Who Needs Additional Amounts?

• Anyone with deficient protein or amino acids in their diet because L-carnitine requires essential amino acids to be synthesized by the body.
• Children, pregnant or breast-feeding women who are vegan vegetarians.
• People with recent severe burns or injuries.
• Those on hemodialysis.
• Premature infants.

Deficiency Symptoms

• Muscle fatigue
• Cramps
• Changes in kidney-function chemistry following exercise

Unproved Speculated Symptoms

• Premature aging
• Heartbeat irregularities in someone who has had a heart attack
• Angina (See Glossary)

Lab Tests to Detect Deficiency

• None available, except for experimental purposes.

Dosage and Usage Information

Recommended Dietary Allowance (RDA):
No RDA has been established.

What this supplement does:
• Transports long-chain fatty acids into mitochondria, which are the metabolic furnaces of cells (particularly heart and kidney cells) where they may be oxidized to yield energy.

Miscellaneous information:
• L-carnitine is synthesized in human kidney and liver from the essential amino acids lysine and methionine, plus vitamins B-6, C and iron.

Available as:
• L-carnitine tablets: Swallow whole with full glass of liquid. Don't chew or crush. Take with meals or 1 to 1-1/2 hours after meals unless otherwise directed by your doctor. Avoid DL-carnitine tablets; they may be toxic. ⟫▸

SUPPLEMENT

L-Carnitine, Continued

Warnings and Precautions

Don't take if you:
• Are allergic to any food protein, such as eggs, milk, wheat.
• Are at risk of poor nutrition for any reason.
• Are pregnant, think you may be pregnant or plan pregnancy in the near future.

Consult your doctor if you have:
• Any liver or kidney problems.

Over age 55:
• No special problems expected.

Pregnancy:
• Problems in pregnant women taking small or usual amounts have not beem proved. But the chance of problems does exist. Don't use unless prescribed by your doctor.

Breast-feeding:
• Problems in breast-fed infants of lactating mothers taking small or usual amounts have not been proved. But the chance of problems does exist. Don't use unless prescribed by your doctor.

Effect on lab tests:
• None known.

Storage:
• Store in cool, dry place away from direct light, but don't freeze.
• Store safely out of reach of children.
• Don't store in bathroom medicine cabinet. Heat and moisture may change action of supplement.

Overdose/Toxicity

Signs and symptoms:
Muscle weakness.

What to do:
For symptoms of overdosage: Discontinue supplement, and consult doctor. Also see *Adverse Reactions or Side Effects* section below.
For accidental overdosage (such as child taking entire bottle): Dial 911 (emergency), 0 for operator or your nearest Poison Control Center.

Adverse Reactions or Side Effects

Reaction or effect	What to do
Symptoms of myasthenia (progressive weakness of certain muscle groups without evidence of atrophy or wasting) have been reported in kidney patients being maintained for prolonged periods on hemodialysis and supplemental DL-carnitine	Don't take supplements without doctor's prescription and supervision.

Interaction with Medicine, Vitamins or Minerals

None expected

Interaction with Other Substances

None known

Lecithin

Basic Information
Available from natural sources? Yes
Available from synthetic sources? Yes
Prescription required? No

Natural Sources

Cabbage	Green beans
Calves' liver	Lentils
Cauliflower	Rice
Caviar	Soy lecithin
Eggs	Soybeans
Garbanzo beans	Split peas
(chickpeas)	

Found in all animal and plant products.

Reasons to Use

- Protects against damage to cells by oxidation.
- Major source of the chemical nutrient choline. Choline's benefits are also lecithin's benefits. See Choline.
- People taking niacin or nicotinic acid for treatment of high-serum cholesterol and triglycerides need lecithin or choline supplements.
- May lower high blood cholesterol or triglycerides in some people.

Unproved Speculated Benefits

- Protects against cardiovascular disease.
- Protects against memory loss.
- Prevents some diseases of the nervous system, such as Alzheimer's disease and tardive dyskinesia (involuntary, abnormal facial movements including grimacing, sticking out tongue and sucking movements).
- Treats Alzheimer's disease.
- Treats liver damage caused by alcoholism.
- Lowers cholesterol level.

Who Needs Additional Amounts?

- No one.

Deficiency Symptoms

- There are no specific deficiency symptoms in man, although some animals can suffer from lack of choline. Lecithin must be present for choline synthesis in the human body.

Unproved Speculated Symptoms

- Symptoms of heart or blood-vessel disease
- Decreasing mental alertness

Lab Tests to Detect Deficiency

- None available, except for experimental purposes.

Dosage and Usage Information

Recommended Dietary Allowance (RDA):
No RDA has been established.

What this supplement does:
- Lecithin is a phospholipid composed of saturated, unsaturated and polyunsaturated fatty acids. It also contains glycerin, phosphorous and choline.
- It is found in chemicals that aid passage of many nutrients from the bloodstream into cells.

Miscellaneous information:
- Is used as a thickener in several foods, including mayonnaise, margarine, ice cream.
- Supplements are not needed by healthy people.

Available as:
- Tablets: Swallow whole with full glass of liquid. Don't chew or crush. Take with meals or 1 to 1-1/2 hours after meals unless otherwise directed by your doctor.
- Liquid: Dilute in at least 1/2 glass of water or other liquid. Take with meals or 1 to 1-1/2 hours after meals unless otherwise directed by your doctor.

SUPPLEMENT

Lecithin, Continued

Warnings and Precautions

Don't take if you:
• Are healthy and eat a well-balanced diet.

Consult your doctor if you have:
• Plans to treat Alzheimer's disease with lecithin/choline.

Over age 55:
• No problems expected.

Pregnancy:
• Supplements are not needed.

Breast-feeding:
• Supplements are not needed.

Effect on lab tests:
• May cause inaccurate results in lecithin/sphingomyelin test as part of examination of amniotic fluid.

Storage:
• Store in cool, dry place away from direct light, but don't freeze.
• Store safely out of reach of children.
• Don't store in bathroom medicine cabinet. Heat and moisture may change action of supplement.

Others:
• Don't take more than 1 gram per day.

Overdose/Toxicity

Signs and symptoms:
Nausea, vomiting, dizziness.

What to do:
For symptoms of overdosage: Discontinue supplement, and consult doctor. Also see *Adverse Reactions or Side Effects* section below.
For accidental overdosage (such as child taking entire bottle): Dial 911 (emergency), 0 for operator or your nearest Poison Control Center.

Adverse Reactions or Side Effects

Reaction or effect	What to do
"Fishy" body odor	Discontinue. Call doctor when convenient.

Interaction with Medicine, Vitamins or Minerals

Interacts with	Combined effect
Nicotinic acid (nicotinamide, vitamin B-3)	Decreases lecithin effectiveness.

Interaction with Other Substances

None known

Omega-3 Fatty Acids (Fish Oils)

Basic Information

This fatty acid is a dietary supplement available in capsules or oil. It is commercially advertised to protect against diseases of the heart and blood vessels.
Brand names, see page 477.
Chemicals this supplement contains:
Docosahexaenoic acid (DHA)
Eicosapentaenoic acid (EPA)

Known Effects

• Greenland Eskimos who eat foods high in omega-3 fatty acids have very low serum triglycerides and total cholesterol. The cholesterol they do have is mainly high-density-lipoprotein (HDL) cholesterol. These substances are known to protect against deposits of plaque, which can occlude critical blood vessels and cause heart attacks, strokes and other major health problems. Coastal Japanese people have similar diets and similar findings. We *assume* increasing omega-3 in our daily food may give us the same protection.

Miscellaneous information:
• Omega-3 fatty acids come from *cold-water fish*, particularly cod, tuna, salmon, halibut, shark, mackerel. Increasing "oily" fish in the diet may be safer than taking omega-3 fatty-acid supplements.

Unproved Speculated Benefits

• Protects against arthritis.
• Protects against arteriosclerosis.
• Protects against coronary-artery disease.
• Protects against strokes.
• Protects against kidney failure.

Warnings and Precautions

Don't take if you:
• Are pregnant, think you may be pregnant or plan pregnancy in the near future.

Consult your doctor if you:
• Take any medicinal drugs or herbs including aspirin, laxatives, cold and cough remedies, antacids, vitamins, minerals, amino acids, supplements, other prescription or non-prescription drugs.

Pregnancy:
• Problems in pregnant women taking small or usual amounts have not been proved. But the chance of problems does exist. Don't use unless prescribed by your doctor.

Breast-feeding:
• Problems in breast-fed infants of lactating mothers taking small or usual amounts have not been proved. But the chance of problems does exist. Don't use unless prescribed by your doctor.

Infants and children:
• Treating infants and children under 2 with any supplement is hazardous.

Others:
• This fat becomes rancid easily and quickly.
• No one knows how much is beneficial and non-toxic.

Storage:
• Keep cool and dry, but don't freeze. Store safely away from children.

Safe dosage:
• At present no "safe" dosage has been established.

Toxicity

Comparative-toxicity rating not available from standard references.

For symptoms of toxicity: See *Adverse Reactions, Side Effects or Overdose Symptoms* section below.

Adverse Reactions, Side Effects or Overdose Symptoms

Signs and symptoms	What to do
Large amounts may lead to bleeding problems, diminished immunity, predisposition to some malignancies	Discontinue. Call doctor immediately.

SUPPLEMENT

Para-Aminobenzoic Acid (PABA)

Basic Information
Brand names, see page 477.
Available from natural sources? Yes
Available from synthetic sources? Yes
Prescription required? No

 ## Natural Sources

Bran	Molasses
Brown rice	Sunflower seeds
Kidney	Wheat germ
Liver	Whole-grain products
	Yogurt

 ## Reasons to Use

- Shields skin from damage of ultraviolet radiation when used as a topical sunscreen.
- Treats vitiligo, a condition characterized by discoloration or depigmentation of some areas of the skin.

 ## Unproved Speculated Benefits

- Rejuvenates skin.
- Treats arthritis.
- Stops hair loss.
- Restores color to graying or white hair.
- Treats anemia.
- Treats constipation.
- Treats headaches.
- Treats skin disorders.

 ## Who Needs Additional Amounts?

- PABA is not an essential nutrient, so no nutritional deficiency has been documented.

 ## Deficiency Symptoms

- None

 ## Unproved Speculated Symptoms

- Eczema

 ## Lab Tests to Detect Deficiency

- None available, except for experimental purposes.

 ## Dosage and Usage Information

Recommended Dietary Allowance (RDA):
No RDA has been established.

What this supplement does:
- Stimulates intestinal bacteria, enabling them to produce folic acid, which aids in production of pantothenic acid.

Miscellaneous information:
- When used topically, PABA helps prevent sunburn.
- *Don't* take oral supplements without doctor's supervision.

Available as:
- A constituent of many multivitamin/mineral preparations.
- A constituent of many topical sunscreen products.

»➤

Para-Aminobenzoic Acid (PABA)

Warnings and Precautions

Don't take if you:
• Take any sulfonamide or antibiotic internally because PABA prevents them from exerting their full effect.

Consult your doctor if you:
• Are pregnant, think you may be pregnant or plan pregnancy in the near future.

Over age 55:
• No problems expected.

Pregnancy:
• Risks outweigh benefits. Don't take internally.
• No problems are expected if you use PABA topically.

Breast-feeding:
• Risks outweigh benefits. Don't take internally.
• No problems are expected if you use PABA topically.

Effect on lab tests:
• None expected.

Storage:
• Store in cool, dry place away from direct light, but don't freeze.
• Store safely out of reach of children.
• Don't store in bathroom medicine cabinet. Heat and moisture may change action of supplement.

Overdose/Toxicity

Signs and symptoms:
Liver disease evidenced by abnormal liver-function tests, jaundice (yellow skin and eyes), vomiting.

Possible consequences of overdose:
PABA is stored in the tissues. In continued high doses, it may prove toxic to the liver. Symptoms of toxicity are nausea and vomiting.

What to do:
For symptoms of overdosage: Discontinue supplement, and consult doctor. Also see *Adverse Reactions or Side Effects* section below.
For accidental overdosage (such as child taking entire bottle): Dial 911 (emergency), 0 for operator or your nearest Poison Control Center.

Adverse Reactions or Side Effects

Reaction or effect	What to do
Diarrhea	Discontinue. Call doctor immediately.
Nausea	Discontinue. Call doctor immediately.
Vomiting	Discontinue. Call doctor immediately.

Interaction with Medicine, Vitamins or Minerals

Interacts with	Combined effect
Antibiotics	Decreases effectiveness of antibiotics.
Folic acid	Increases effectiveness of folic acid.
Sulfonamide ("sulfa drugs")	Decreases effectiveness of sulfa drugs.
Vitamin-B complex	Increases effectiveness of vitamin-B complex.
Vitamin C	Increases effectiveness of vitamin C.

Interaction with Other Substances

None known

Royal Jelly

Basic Information

Royal jelly is a milky-white, gelatinous substance secreted by salivary glands of worker bees to stimulate growth and development of queen bees.
Chemicals this supplement contains:
Pantothenic acid (part of B-complex of vitamins)
10-hydroxydec-2-enoic acid

Known Effects

• None proved or conclusively demonstrated.

Miscellaneous information:
• This substance must be given by injection.

Unproved Speculated Benefits

• Extends life span.
• Treats bone and joint disorders, such as rheumatoid arthritis.
• Protects against leukemia.
• Contains antibiotic properties.

Warnings and Precautions

Don't take if you:
• Are pregnant, think you may be pregnant or plan pregnancy in the near future.

Consult your doctor if you:
• Take any medicinal drugs or herbs including aspirin, laxatives, cold and cough remedies, antacids, vitamins, minerals, amino acids, supplements, other prescription or non-prescription drugs.

Pregnancy:
• Dangers outweigh any possible benefits. Don't use.

Breast-feeding:
• Dangers outweigh any possible benefits. Don't use.

Infants and children:
• Treating infants and children under 2 with any supplement is hazardous.

Others:
• Dangers outweigh any possible benefits. Don't use.

Storage:
• Keep cool and dry, but don't freeze. Store safely away from children.

Safe dosage:
• At present no "safe" dosage has been established.

Toxicity

Comparative-toxicity rating not available from standard references.

For symptoms of toxicity: See *Adverse Reactions, Side Effects or Overdose Symptoms* section below.

Adverse Reactions, Side Effects or Overdose Symptoms

Signs and symptoms	What to do
Life-threatening anaphylaxis may follow injections— symptoms include immediate severe itching, paleness, low blood pressure, loss of consciousness, coma	Yell for help. Don't leave victim. Begin CPR (cardiopulmonary resuscitation), mouth to mouth breathing and external cardiac massage. Have someone dial "O" (operator) or 911 (emergency). Don't stop CPR until help arrives.

Basic Information

Biological names of spirulina are Spirulina geitler, Spirulina maxima, Spirulina platenis. *Chemicals this supplement contains:*
   B-complex vitamins
   Beta-carotene
   Gamma-linoleic acid (See Glossary)

 ## Known Effects

• None proved or conclusively demonstrated.

Miscellaneous information:
• Spirulina is expensive and tastes terrible.
• It is a blue-green microalgae that grows wild on the surface of brackish, alkaline lakes in the tropics. Spirulina is cultivated commercially in several places, including Mexico, Thailand, Japan and Southern California.
• It is available in powder and tablet forms.

 ## Unproven Speculated Benefits

• Treats obesity.
• Is used as tonic.
• Acts as energy booster.
• Treats diabetes mellitus.

 ## Warnings and Precautions

Don't take if you:
• Are pregnant, think you may be pregnant or plan pregnancy in the near future.

Consult your doctor if you:
• Take any medicinal drugs or herbs including aspirin, laxatives, cold and cough remedies, antacids, vitamins, minerals, amino acids, supplements, other prescription or non-prescription drugs.

Pregnancy:
• Problems in pregnant women taking small or usual amounts have not been proved. But the chance of problems does exist. Don't use unless prescribed by your doctor.

Breast-feeding:
• Problems in breast-fed infants of lactating mothers taking small or usual amounts have not been proved. But the chance of problems does exist. Don't use unless prescribed by your doctor.

Infants and children:
• Treating infants and children under 2 with any supplement is hazardous.

Others:
• No contraindications if you are not pregnant and do not take amounts larger than a reputable manufacturer recommends on the package.

Storage:
• Keep cool and dry, but don't freeze. Store safely away from children.

Safe dosage:
• At present no "safe" dosage has been established.

 ## Toxicity

Comparative-toxicity rating not available from standard references.

For symptoms of toxicity: See *Adverse Reactions, Side Effects or Overdose Symptoms* section below.

 ## Adverse Reactions, Side Effects or Overdose Symptoms

Signs and symptoms	What to do
Diarrhea	Discontinue. Call doctor immediately.
Nausea	Discontinue. Call doctor immediately.
Vomiting	Discontinue. Call doctor immediately.

SUPPLEMENT

Superoxide Dismutase

Basic Information

Superoxide dismutase is an enzyme associated with copper, zinc and manganese.
Chemical this supplement contains:
Enzyme that participates in utilization of copper, zinc and manganese by body cells

Known Effects

• Injectable forms may have anti-oxidant properties.

Miscellaneous information:

• Oral superoxide dismutase is destroyed in the intestines before being absorbed, so oral forms are worthless.
• Available in injectable forms, but these should be used *only* under close medical supervision.

Unproved Speculated Benefits

• Protects against free radicals. (See Glossary.)
• Treats arthritis.
• Treats cancer.
• Treats side effects of radiation.

Warnings and Precautions

Don't take if you:
• Have any medical problems.

Consult your doctor if you:
• Take any medicinal drugs or herbs including aspirin, laxatives, cold and cough remedies, antacids, vitamins, minerals, amino acids, supplements, other prescription or non-prescription drugs.

Pregnancy:
• Don't use without medical supervision.

Breast-feeding:
• Don't use without medical supervision.

Infants and children:
• Treating infants and children under 2 with any supplement is hazardous.

Others:
• Available oral forms are worthless. Injectable forms may cause anaphylaxis.

Storage:
• Keep cool and dry, but don't freeze. Store safely away from children.

Safe dosage:
• At present no "safe" dosage has been established.

Toxicity

Comparative-toxicity rating not available from standard references.

For symptoms of toxicity: See *Adverse Reactions, Side Effects or Overdose Symptoms* section below.

Adverse Reactions, Side Effects or Overdose Symptoms

Signs and symptoms	What to do
Life-threatening anaphylaxis may follow injections—symptoms include immediate severe itching, paleness, low blood pressure, loss of consciousness, coma	Yell for help. Don't leave victim. Begin CPR (cardiopulmonary resuscitation), mouth to mouth breathing and external cardiac massage. Have someone dial "O" (operator) or 911 (emergency). Don't stop CPR until help arrives.

Basic Information

Wheat germ is the embryo of the wheat grain located at the lower end. It is derived from both root and shoot.
Chemicals this supplement contains:
Calcium
Copper
Manganese
Magnesium
Most B vitamins
Octacosanol
Phosphorus
Vitamin E (one of the richest natural sources)

 ## Known Effects

• Excellent nutritional source of chemicals listed above.

Miscellaneous information:
• Wheat germ is available in food and as flakes to mix with other foods. It is also available as oil, which should be kept tightly covered and refrigerated.

 ## Unproved Speculated Benefits

• Treats muscular dystrophy.
• Improves physical stamina and performance.

 ## Warnings and Precautions

Don't take if you:
• No contraindications if you are not pregnant and do not take amounts larger than manufacturer's recommended dosage.

Consult your doctor if you:
• Are pregnant, think you may be pregnant or plan pregnancy in the near future.
• Take any medicinal drugs or herbs including aspirin, laxatives, cold and cough remedies, antacids, vitamins, minerals, amino acids, supplements, other prescription or non-prescription drugs.

Pregnancy:
• Problems in pregnant women taking small or usual amounts have not been proved. But the chance of problems does exist. Don't use unless prescribed by your doctor.

Breast-feeding:
• Problems in breast-fed infants of lactating mothers taking small or usual amounts have not been proved. But the chance of problems does exist. Don't use unless prescribed by your doctor.

Infants and children:
• Treating infants and children under 2 with any supplement is hazardous.

Others:
• None expected if you are beyond childhood and under 45, basically healthy and take for only a short time.

Storage:
• Keep cool and dry, but don't freeze. Store safely away from children.

Safe dosage:
• At present no "safe" dosage has been established.

 ## Toxicity

Comparative-toxicity rating not available from standard references.

 ## Adverse Reactions, Side Effects or Overdose Symptoms

None expected

SUPPLEMENT

Wheatgrass/Barley Grass, Green Plants (Cabbage, Broccoli, Brussels Sprouts)

Basic Information

These products are derived from roots and leaves.
Chemicals this supplement contains:
  Chlorophyll
  Superoxide-dismutase

Known Effects

• May function as an anti-oxidant.

Unproved Speculated Benefits

• Protects against pollutants.
• Protects against radiation damage to cells.
• "Detoxifies" body.
• "Purifies" blood.
• Protects against cancer.

Warnings and Precautions

Don't take if you:
• Are pregnant, think you may be pregnant or plan pregnancy in the near future.

Consult your doctor if you:
• Take this herb for any medical problem that doesn't improve in 2 weeks. There may be safer, more-effective treatments.
• Take any medicinal drugs or herbs including aspirin, laxatives, cold and cough remedies, antacids, vitamins, minerals, amino acids, supplements, other prescription or non-prescription drugs.

Pregnancy:
• Problems in pregnant women taking small or usual amounts have not been proved. But the chance of problems does exist. Don't use unless prescribed by your doctor.

Breast-feeding:
• Problems in breast-fed infants of lactating mothers taking small or usual amounts have not been proved. But the chance of problems does exist. Don't use unless prescribed by your doctor.

Infants and children:
• Treating infants and children under 2 with any supplement is hazardous.

Others:
• No contraindications if you are not pregnant and do not take amounts larger than manufacturer's recommended dosage.

Storage:
• Keep cool and dry.

Safe dosage:
• At present no "safe" dosage has been established.

Toxicity

Comparative-toxicity rating not available from standard references.

Adverse Reactions, Side Effects or Overdose Symptoms

None expected

Medicinal Herbs

The following section makes an attempt to provide you with enough knowledge to avoid toxicity if you choose to self-prescribe medicinal herbs. If you collect and use herbal medications, you must be an expert botanist or herbologist. If you buy them, you have the right to know everything about possible side effects, adverse reactions, toxicity and other dangers in using materials that are not subjected to rigid control procedures.

I am not qualified to extol the virtues of herbal medicines. But I feel qualified to point out some of the inherent dangers in using medicinal herbs. That is the main purpose of the following section.

The most important warnings I can give you are:

&. Don't use medicinal herbs for infants or children without guidance from an expert and your doctor's approval.

&. Don't use medicinal herbs at all unless you know enough to use them safely.

&. Don't use medicinal herbs if there is a safer, more effective medicine to use, whatever problem you have.

&. If you choose to use medicinal herbs, tell your doctor which ones you use and in what amounts when he or she asks if you take other medications. As you will see in the following charts, these substances could affect various medicines or courses of treatment your doctor may prescribe.

Aconite (Monkshood, Blue Rocket)

Basic Information
Biological name (genus and species):
  Aconitum napellus
Parts used for medicinal purposes:
  Roots
  Leaves
Chemicals this herb contains:
  Aconine
  Aconitine
  Benzoylamine
  Neopelline
  Picratonitine

Known Effects

- Small amounts stimulate central nervous system and peripheral nerves.
- Large amounts depress central nervous system and peripheral nerves.
- Normalizes heartbeat irregularities.
- A dose as low as 5ml (about 1 teaspoon) of the root can be lethal.

Miscellaneous information:
- Was used for centuries as arrow poison.

Unproved Speculated Benefits

- Decreases fever.
- Treats heartbeat irregularities.
- Increases sweating.
- Decreases blood pressure.
- Treats neuralgias.
- Treats other nerve disorders.

Warnings and Precautions

Don't take if you:
- Are pregnant, think you may be pregnant or plan pregnancy in the near future.
- Have any chronic disease of the gastrointestinal tract, such as stomach or duodenal ulcers, esophageal reflux (reflux esophagitis), ulcerative colitis, spastic colitis, diverticulosis, diverticulitis.

Consult your doctor if you:
- Take this herb for any medical problem that doesn't improve in 2 weeks. There may be safer, more-effective treatments.
- Take any medicinal drugs or herbs including aspirin, laxatives, cold and cough remedies, antacids, vitamins, minerals, amino acids, supplements, other prescription or non-prescription drugs.

Pregnancy:
- Dangers outweigh any possible benefits. Don't use.

Breast-feeding:
- Dangers outweigh any possible benefits. Don't use.

Infants and children:
- Treating infants and children under 2 with any herbal preparation is hazardous.

Others:
- Dangers outweigh any possible benefits. Don't use.

Storage:
- Keep cool and dry, but don't freeze. Store safely away from children.

Safe dosage:
- At present no "safe" dosage has been established.

Toxicity

Rated dangerous, particularly in children, persons over 55 and those who take larger than appropriate quantities for extended periods of time.

For symptoms of toxicity: See *Adverse Reactions, Side Effects or Overdose Symptoms* section below.

Adverse Reactions, Side Effects or Overdose Symptoms

Signs and symptoms:	What to do:
Burning tongue and lips	Discontinue. Call doctor immediately.
Difficulty swallowing	Discontinue. Call doctor immediately.
Irritability	Discontinue. Call doctor when convenient.
Nausea	Discontinue. Call doctor immediately.
Numbness of tongue and lips	Discontinue. Call doctor immediately.
Restlessness	Discontinue. Call doctor when convenient.
Speech difficulties	Discontinue. Call doctor immediately.
Vision doubled or blurred	Discontinue. Call doctor immediately.
Vomiting	Discontinue. Call doctor immediately.

Basic Information

Biological name (genus and species):
Agave lecheguilla
Parts used for medicinal purposes:
Roots
Leaves
Sap
Chemicals this herb contains:
Diosgenin
Photosensitizing pigment (See Glossary)
Steroidal chemicals (See Glossary)
Vitamin C

Known Effects

- Causes disintegration of red blood cells.
- Irritates skin.
- Irritates lining of gastrointestinal tract.
- Small amounts depress central nervous system.
- Damages cells, dissolves membranes of red blood cells and changes tissue permeability.

Miscellaneous information:
- Is used to make mescal, an alcoholic beverage.
- Fibers are used for rope.
- Sap is used as a syrup.
- Roots and leaves contain active chemicals.

Unproved Speculated Benefits

- Roots and leaves are used to relieve toothache.
- Provides nutrition.
- Is used as hormone replacement.
- Produces immunosuppressive effects on body.
- Causes abortion or miscarriage.
- Treats dysentery.
- Relieves pain of sprains.

Warnings and Precautions

Don't take if you:
- Are pregnant, think you may be pregnant or plan pregnancy in the near future.
- Have symptoms of a disease caused by a hormone deficiency.
- Have any chronic disease of the gastrointestinal tract, such as stomach or duodenal ulcers, esophageal reflux (reflux esophagitis), ulcerative colitis, spastic colitis, diverticulosis, diverticulitis.

Consult your doctor if you:
- Have stomach problems.
- Take cortisone, ACTH, testosterone, androgenic steroids.

- Are pregnant, think you may be pregnant or plan pregnancy in the near future.

Pregnancy:
- Problems in pregnant women taking small or usual amounts have not been proved. Don't use unless prescribed by your doctor.

Breast-feeding:
- Problems in breast-fed infants of lactating mothers taking small or usual amounts have not been proved. Don't use unless prescribed by your doctor.

Infants and children:
- Treating infants and children under 2 with any herbal preparation is hazardous.

Others:
- None expected if you are beyond childhood and under 45, basically healthy and take for only a short time.

Storage:
- Keep cool and dry, but don't freeze. Store safely away from children.

Safe dosage:
- At present no "safe" dosage established.

Toxicity

Comparative-toxicity rating not available from standard references.

For symptoms of toxicity: See below.

Adverse Reactions, Side Effects or Overdose Symptoms

Signs and symptoms:	What to do:
Abortion (remote possibility if taken in large amounts)	Seek emergency treatment.
Diarrhea	Discontinue. Call doctor immediately.
Increased sensitivity to sunlight	Discontinue. Call doctor when convenient.
Jaundice (yellow eyes and skin)	Discontinue. Call doctor immediately.
Nausea	Discontinue. Call doctor immediately.
Skin itching and rash	Discontinue. Call doctor when convenient.
Unusual bleeding	Discontinue. Call doctor immediately.
Vomiting	Discontinue. Call doctor immediately.

MEDICINAL HERB

Alder, Black (Alder Buckthorn)

Basic Information

Biological name (genus and species):
Rhamnus frangula, Frangula
Parts used for medicinal purposes:
Various parts of the entire plant, frequently differing by country and/or culture
Chemicals this herb contains:
Anthraquinone glycosides
Emodin
Rhamnose

Known Effects

- Irritates gastrointestinal tract.
- Causes vomiting.

Miscellaneous information:
- Is used in veterinary medicine for its cathartic properties.
- Emetic action (causing vomiting) is less when plant is dried for 1 year or more.

Unproved Speculated Benefits

- Temporarily relieves constipation.

Warnings and Precautions

Don't take if you:
- Are pregnant, think you may be pregnant or plan pregnancy in the near future.
- Have any chronic disease of gastrointestinal tract, such as stomach or duodenal ulcers, esophageal reflux (reflux esophagitis), ulcerative colitis, spastic colitis, diverticulosis, diverticulitis.

Consult your doctor if you:
- Take this herb for any medical problem that doesn't improve in 2 weeks. There may be safer, more-effective treatments.
- Take any medicinal drugs or herbs including aspirin, laxatives, cold and cough remedies, antacids, vitamins, minerals, amino acids, supplements, other prescription or non-prescription drugs.

Pregnancy:
- Problems in pregnant women taking small or usual amounts have not been proved. But the chance of problems does exist. Don't use unless prescribed by your doctor.

Breast-feeding:
- Problems in breast-fed infants of lactating mothers taking small or usual amounts have not been proved. But the chance of problems does exist. Don't use unless prescribed by your doctor.

Infants and children:
- Treating infants and children under 2 with any herbal preparation is hazardous.

Others:
- None expected if you are beyond childhood and under 45, basically healthy and take for only a short time.

Storage:
- Keep cool and dry, but don't freeze. Store safely away from children.

Safe dosage:
- At present no "safe" dosage has been established.

Toxicity

Rated slightly dangerous, particularly in children, persons over 55 and those who take larger than appropriate quantities for extended periods of time.

For symptoms of toxicity: See *Adverse Reactions, Side Effects or Overdose Symptoms* section below.

Adverse Reactions, Side Effects or Overdose Symptoms

Signs and symptoms:	What to do:
Abdominal cramps, severe	Discontinue. Call doctor immediately.
Abdominal pain	Discontinue. Call doctor when convenient.
Nausea	Discontinue. Call doctor immediately.
Vomiting	Discontinue. Call doctor immediately.

Basic Information

Biological name (genus and species):
Medicago sativa
Parts used for medicinal purposes:
Leaves
Petals/flower
Sprouts
Chemicals this herb contains:
Proteins
Vitamins A, B, D, K

Known Effects

• Provides useful proteins and vitamins for dietary use.
• Stimulates menstruation.
• Stimulates milk production in lactating women.

Miscellaneous information:
• Alfalfa is usually compressed into capsules or brewed as tea.

Unproved Speculated Benefits

• Treats arthritis.
• Treats unusual bleeding.
• Lowers cholesterol.

Warnings and Precautions

Don't take if you:
• Take anti-coagulants, such as warfarin sodium (Coumadin) or heparin.
• Have lupus erythematosus.

Consult your doctor if you:
• Have any bleeding disorder.

Pregnancy:
• Pregnant women should experience no problems taking usual amounts as part of a balanced diet. Other products extracted from this herb have not been proved to cause problems.

Breast-feeding:
• Breast-fed infants of lactating mothers should experience no problems when mother takes usual amounts as part of a balanced diet. Other products extracted from this herb have not been proved to cause problems.

Infants and children:
• Treating infants and children under 2 with any herbal preparation is hazardous.

Others:
• None expected if you are beyond childhood and under 45, basically healthy and take for only a short time.
• Alfalfa sprouts eaten in large amounts may cause one form of anemia.

Storage:
• Keep cool and dry, but don't freeze. Store safely away from children.

Safe dosage:
• At present no "safe" dosage has been established.

Toxicity

Generally regarded as safe when taken in appropriate quantities for short periods of time.

Adverse Reactions, Side Effects or Overdose Symptoms

None expected

Allspice (Jamaican Pepper, Clove Pepper)

Basic Information

Biological name (genus and species):
  Pimenta dioica
Parts used for medicinal purposes:
  Berries/fruits
Chemicals this herb contains:
  Acid-fixed oil
  Eugenol
  Resin (See Glossary)
  Tannic acid
  Volatile oils (See Glossary)

Known Effects

- Irritates mucous membranes including lining of gastrointestinal tract.
- Interferes with absorption of iron and other minerals when taken internally.

Miscellaneous information:
- Active chemicals are in allspice *berries.*
- Provides flavor in toothpaste and other products.
- Is used as an aromatic spice in foods.

Unproved Speculated Benefits

- Aids in expelling gas from intestines to relieve colic or griping.
- Relieves diarrhea.
- Relieves fatigue.

Warnings and Precautions

Don't take if you:
- Are pregnant, think you may be pregnant or plan pregnancy in the near future.
- Have any chronic disease of the gastrointestinal tract, such as stomach or duodenal ulcers, esophageal reflux (reflux esophagitis), ulcerative colitis, spastic colitis, diverticulosis, diverticulitis.

Consult your doctor if you:
- Take any medicinal drugs or herbs, including aspirin, laxatives, cold and cough remedies, antacids, vitamins, minerals, amino acids, supplements, other prescription or non-prescription drugs.

Pregnancy:
- Problems in pregnant women taking small or usual amounts have not been proved. But the chance of problems does exist. Don't use unless prescribed by your doctor.

Breast-feeding:
- Problems in breast-fed infants of lactating mothers taking small or usual amounts have not been proved. But the chance of problems does exist. Don't use unless prescribed by your doctor.

Infants and children:
- Treating infants and children under 2 with any herbal preparation is hazardous.

Others:
- None expected if you are beyond childhood and under 45, basically healthy and take for only a short time.

Storage:
- Keep cool and dry, but don't freeze. Store safely away from children.

Safe dosage:
- At present no "safe" dosage has been established.

Toxicity

Rated relatively safe when taken in appropriate quantities for short periods of time.

For symptoms of toxicity: See *Adverse Reactions, Side Effects or Overdose Symptoms* section below.

Adverse Reactions, Side Effects or Overdose Symptoms

Signs and symptoms:	What to do:
Excess of 5 ml (about 1 teaspoon) of eugenol (a volatile oil found in allspice) may cause convulsions, nausea, vomiting	Discontinue. Call doctor immediately.

Aloe (Mediterranean Aloe, Barbados Aloe, Curacao Aloe, Aloe Vera)

Basic Information

Biological name (genus and species):
Aloe vera, Aloe barbadensis, Aloe officinalis
Parts used for medicinal purposes:
Leaves
Chemicals this herb contains:
Barbaloin (not present in Aloe vera)
Beta-barbaloin (purgative)
Socaloin
Resin (See Glossary)
Tannins (See Glossary)

Known Effects

- Milky exudate (not dried preparations) from leaves helps reduce inflammation and hasten recovery in first- and second-degree burns.
- Acts as cathartic, but whether this is beneficial or dangerous depends on many factors.
- Treats X-ray or radiation burns.
- Interferes with absorption of iron and other minerals when taken internally.

Miscellaneous information:
- Not useful for clearing intestinal tract before surgery because only cleanses small intestine.
- Is used as an ingredient in many over-the-counter laxatives.
- Is used as an ingredient in some cosmetics.

Unproved Speculated Benefits

- Applied to skin, it kills *Pseudomonas aeruginosa,* a bacterium, but probably does not promote healing.
- Taken internally, treats amenorrhea (lack of menstrual periods).
- Is used as an aphrodisiac.
- Causes breast development to progress more quickly.
- Is applied to head to relieve headache.

Warnings and Precautions

Don't take if you:
- Have ulcers.
- Have small-bowel problems, such as regional enteritis.
- Have ulcerative colitis.
- Have diverticulosis or diverticulitis.
- Have proctitis or hemorrhoids.

Consult your doctor if you:
- Have any digestive disorder.
- Intend to take internally.

Pregnancy and breast-feeding:
- Problems in pregnant women or in breast-fed infants of lactating mothers taking small or usual amounts have not been proved. But the chance of problems does exist. Don't use unless prescribed by your doctor.

Infants and children:
- Treating infants and children under 2 with any herbal preparation is hazardous.

Others:
- Healing properties of aloe taken internally are still tentative and need more study.

Storage:
- Keep cool and dry, but don't freeze. Store safely away from children.

Safe dosage:
- At present no "safe" dosage has been established.

Toxicity

Generally regarded as safe when taken in appropriate quantities for short periods of time.

For symptoms of toxicity: See below.

Adverse Reactions, Side Effects or Overdose Symptoms

Signs and symptoms:	What to do:
Abdominal cramps	Discontinue. Call doctor when convenient.
Bowel irritation	Discontinue. Call doctor when convenient.
Diarrhea	Discontinue. Call doctor immediately.
High dose: bloody diarrhea, shock	Seek emergency treatment.
Minor skin irritation (for external applications)	Cleanse skin with clear water. Do not apply aloe again.
Nausea	Discontinue. Call doctor immediately.
Red urine	Discontinue. Call doctor when convenient.
Urinary frequency, backache, pain on urination with long, continued use	Discontinue. Call doctor immediately.
Vomiting	Discontinue. Call doctor immediately.

Alum Root (American Sanicle)

Basic Information

Biological name (genus and species):
  Heuchera
Parts used for medicinal purposes:
  Roots
Chemical this herb contains:
  Tannins (See Glossary)

Known Effects

- Shrinks tissues.
- Prevents secretion of fluids.

Miscellaneous information:
- Is used externally and internally by some tribes of North-American Indians for many disorders.
- Is used as a douche.

Unproved Speculated Benefits

- Treats heart disease.
- Prevents infection in injured skin.

Warnings and Precautions

Don't take if you:
- Have liver or kidney disease.
- Are pregnant, think you may be pregnant or plan pregnancy in the near future.
- Have any chronic disease of the gastrointestinal tract, such as stomach or duodenal ulcers, esophageal reflux (reflux esophagitis), ulcerative colitis, spastic colitis, diverticulosis, diverticulitis.

Consult your doctor if you:
- Take this herb for any medical problem that doesn't improve in 2 weeks. There may be safer, more-effective treatments.
- Take any medicinal drugs or herbs including aspirin, laxatives, cold and cough remedies, antacids, vitamins, minerals, amino acids, supplements, other prescription or non-prescription drugs.

Pregnancy:
- Problems in pregnant women taking small or usual amounts have not been proved. But the chance of problems does exist. Don't use unless prescribed by your doctor.

Breast-feeding:
- Problems in breast-fed infants of lactating mothers taking small or usual amounts have not been proved. But the chance of problems does exist. Don't use unless prescribed by your doctor.

Infants and children:
- Treating infants and children under 2 with any herbal preparation is hazardous.

Others:
- Toxic effects greatly outweigh any possible benefits. *Don't take this herb internally!*

Storage:
- Keep cool and dry, but don't freeze. Store safely away from children.

Safe dosage:
- At present no "safe" dosage has been established.

Toxicity

Comparative-toxicity rating not available from standard references. However, it is believed toxic effects greatly outweigh any possible benefits.

For symptoms of toxicity: See *Adverse Reactions, Side Effects or Overdose Symptoms* section below.

Adverse Reactions, Side Effects or Overdose Symptoms

Signs and symptoms:	What to do:
Burning indigestion	Discontinue. Call doctor when convenient.
Edema (swelling of hands and feet)	Discontinue. Call doctor when convenient.
Jaundice (yellow eyes and skin)	Discontinue. Call doctor immediately.
Nausea	Discontinue. Call doctor immediately.
Vomiting	Discontinue. Call doctor immediately.

American Dogwood (Dogwood, American Boxwood)

Basic Information

Biological name (genus and species):
  Cornus florida
Parts used for medicinal purposes:
  Bark
Chemicals this herb contains:
  Betulic acid
  Cornin

 ## Known Effects

- Irritates gastrointestinal tract and acts as a cathartic.
- Causes uterine contractions.

 ## Unproved Speculated Benefits

- Reduces fever.
- Kills bacteria in boils, carbuncles, infected skin rashes, insect bites.

 ## Warnings and Precautions

Don't take if you:
- Are pregnant. It may cause miscarriage.

Consult your doctor if you:
- Take this herb for any medical problem that doesn't improve in 2 weeks. There may be safer, more-effective treatments.

Pregnancy:
- Dangers outweigh any possible benefits. Don't use.

Breast-feeding:
- Dangers outweigh any possible benefits. Don't use.

Infants and children:
- Treating infants and children under 2 with any herbal preparation is hazardous.

Storage:
- Keep cool and dry, but don't freeze. Store safely away from children.

Safe dosage:
- At present no "safe" dosage has been established.

 ## Toxicity

Rated relatively safe when taken in appropriate quantities for short periods of time.

For symptoms of toxicity: See *Adverse Reactions, Side Effects or Overdose Symptoms* section below.

 ## Adverse Reactions, Side Effects or Overdose Symptoms

Signs and symptoms:	What to do:
Abortion	Seek emergency treatment.
Dermatitis	Discontinue. Call doctor when convenient.

MEDICINAL HERB

263

Angelica (Garden Angelica, European Angelica)

Basic Information

Biological name (genus and species):
Angelica archangelica
Parts used for medicinal purposes:
Entire plant
Chemicals this herb contains:
Angelic acid
Resin (See Glossary)
Volatile oils (See Glossary)

Known Effects

Volatile oil gives angelica the following effects:
• Decreases thickness and increases fluidity of mucus from lungs and bronchial tubes.
• Increases perspiration.

Unproved Speculated Benefits

• Seeds and roots are used to reduce odor and volume of intestinal gases.
• Brings on menstruation.

Warnings and Precautions

Don't take if you:
• Are pregnant, think you may be pregnant or plan pregnancy in the near future.
• Have any chronic disease of the gastrointestinal tract, such as stomach or duodenal ulcers, esophageal reflux (reflux esophagitis), ulcerative colitis, spastic colitis, diverticulosis, diverticulitis.

Consult your doctor if you:
• Take this herb for any medical problem that doesn't improve in 2 weeks. There may be safer, more-effective treatments.
• Take any medicinal drugs or herbs, including aspirin, laxatives, cold and cough remedies, antacids, vitamins, minerals, amino acids, supplements, other prescription or non-prescription drugs.

Pregnancy:
• Dangers outweigh any possible benefits. Don't use.

Breast-feeding:
• Dangers outweigh any possible benefits. Don't use.

Infants and children:
• Treating infants and children under 2 with any herbal preparation is hazardous.

Others:
• None expected if you are healthy, take it for a short time and do not exceed manufacturer's recommended dosage.

Storage:
• Keep cool and dry, but don't freeze. Store safely away from children.

Safe dosage:
• At present no "safe" dosage has been established.

Toxicity

Rated relatively safe when taken in appropriate quantities for short periods of time.

Adverse Reactions, Side Effects or Overdose Symptoms

None expected

Basic Information

Biological name (genus and species):
Pimpinella anisum
Parts used for medicinal purposes:
Seeds
Chemicals this herb contains:
Anethole
Essential oils (See Glossary)

 ## Known Effects

- Aids in expelling gas from intestinal tract.
- Helps body dispose of excess fluid by increasing amount of urine produced.
- Increases perspiration.
- Decreases thickness and increases fluidity of mucus from lungs and bronchial tubes.
- Causes hallucinations.

Miscellaneous information:
- Anise is also used in perfumes, soaps, beverages, baked goods, liqueur and as a flavoring.

 ## Unproved Speculated Benefits

- Increases sex drive.
- Decreases colic.
- Treats asthma.
- Kills body lice when applied externally.
- Treats bronchitis.

 ## Warnings and Precautions

Don't take if you:
- Are pregnant, think you may be pregnant or plan pregnancy in the near future.
- Have any chronic disease of the gastrointestinal tract, such as stomach or duodenal ulcers, esophageal reflux (reflux esophagitis), ulcerative colitis, spastic colitis, diverticulosis, diverticulitis.

Consult your doctor if you:
- Take this herb for any medical problem that doesn't improve in 2 weeks. There may be safer, more-effective treatments.
- Take any medicinal drugs or herbs including aspirin, laxatives, cold and cough remedies, antacids, vitamins, minerals, amino acids, supplements, other prescription or non-prescription drugs.

Pregnancy:
- Dangers outweigh any possible benefits. Don't use.

Breast-feeding:
- Dangers outweigh any possible benefits. Don't use.

Storage:
- Keep cool and dry, but don't freeze. Store safely away from children.

Safe dosage:
- At present no "safe" dosage has been established.

 ## Toxicity

Rated relatively safe when taken in appropriate quantities for short periods of time.

For symptoms of toxicity: See *Adverse Reactions, Side Effects or Overdose Symptoms* section below.

 ## Adverse Reactions, Side Effects or Overdose Symptoms

Signs and symptoms:	What to do:
Oil may cause:	
Difficulty breathing	Seek emergency treatment.
Nausea	Discontinue. Call doctor immediately.
Seizures	Seek emergency treatment.
Skin irritation when applied to skin	Discontinue. Call doctor when convenient.
Vomiting	Discontinue. Call doctor immediately.

Asafetida (Devil's Dung)

Basic Information
Biological name (genus and species):
Ferula assafoetida, Ferula foetida
Parts used for medicinal purposes:
Roots
Chemicals this herb contains:
Gum (See Glossary)
Volatile oils (See Glossary)
Resin (See Glossary)

Known Effects

• Irritates lining of gastrointestinal tract and produces laxative effect.

Miscellaneous information:
• Introduced by Arab physicians to European medical practitioners.
• Has garlic-like odor and bitter taste. May have good placebo effect because it is so disagreeable.
• Is used in sack around the neck by some people to repel evil.
• Is used as a condiment.
• Provides flavor as an ingredient in Worcestershire sauce.

Unproved Speculated Benefits

• Decreases thickness and increases fluidity of mucus from lungs and bronchial tubes.
• Treats colic (See Glossary).
• Temporarily relieves constipation.
• Treats nerve disorders.

Warnings and Precautions

Don't take if you:
• Are pregnant, think you may be pregnant or plan pregnancy in the near future.
• Have any chronic disease of the gastrointestinal tract, such as stomach or duodenal ulcers, esophageal reflux (reflux esophagitis), ulcerative colitis, spastic colitis, diverticulosis, diverticulitis.

Consult your doctor if you:
• Take this herb for any medical problem that doesn't improve in 2 weeks. There may be safer, more-effective treatments.
• Take any medicinal drugs or herbs including aspirin, laxatives, cold and cough remedies, antacids, vitamins, minerals, amino acids, supplements, other prescription or non-prescription drugs.

Pregnancy:
• Dangers outweigh any possible benefits. Don't use.

Breast-feeding:
• Dangers outweigh any possible benefits. Don't use.

Infants and children:
• Treating infants and children under 2 with any herbal preparation is hazardous.

Others:
• No contraindications if you are not pregnant and do not exceed manufacturer's recommended dosage.

Storage:
• Keep cool and dry, but don't freeze. Store safely away from children.

Safe dosage:
• At present no "safe" dosage has been established.

Toxicity

Rated relatively safe when taken in appropriate quantities for short periods of time.

For symptoms of toxicity: See *Adverse Reactions, Side Effects or Overdose Symptoms* section below.

Adverse Reactions, Side Effects or Overdose Symptoms

Signs and symptoms:	What to do:
Diarrhea	Discontinue. Call doctor immediately.

Barberry (European Barberry)

Basic Information
Biological name (genus and species):
  Berberis vulgaris
Parts used for medicinal purposes:
  Berries/fruits
  Rootbark
Chemicals this herb contains:
  Berbamine
  Berberine
  Berberrubine
  Columbamine
  Hydrastine
  Jatrorrhizine
  Oxycanthine
  Palmatine

Known Effects

- Dilates blood vessels.
- Decreases heart rate.
- Depresses breathing.
- Stimulates intestinal movement.
- Reduces bronchial constriction.
- Kills bacteria on skin.

Miscellaneous information:
- Fruit is made into jelly.
- Roots are used to dye wool.
- Benefits are mainly based in folklore.

Unproved Speculated Benefits

- Treats diarrhea.
- Treats dyspepsia.
- Treats skin infections.

Warnings and Precautions

Don't take if you:
- Are pregnant, think you may be pregnant or plan pregnancy in the near future.
- Take this herb for any medical problem that doesn't improve in 2 weeks. There may be safer, more-effective treatments.

Consult your doctor if you:
- Take any medicinal drugs or herbs including aspirin, laxatives, cold and cough remedies, antacids, vitamins, minerals, amino acids, supplements, other prescription or non-prescription drugs.

Pregnancy:
- Dangers outweigh any possible benefits. Don't use.

Breast-feeding:
- Dangers outweigh any possible benefits. Don't use.

Infants and children:
- Treating infants and children under 2 with any herbal preparation is hazardous.

Others:
- None expected if you are beyond childhood and under 45, not pregnant, basically healthy and take for only a short time.

Storage:
- Keep cool and dry, but don't freeze. Store safely away from children.

Safe dosage:
- At present no "safe" dosage has been established.

Toxicity

Rated slightly dangerous, particularly in children, persons over 55 and those who take larger than appropriate quantities for extended periods of time.

Adverse Reactions, Side Effects or Overdose Symptoms

None expected

Barley

Basic Information

Biological name (genus and species):
Hordeum distichon, Hordeum spp

Parts used for medicinal purposes:
Various parts of the entire plant, frequently differing by country and/or culture

Chemicals this herb contains:
Ash
Cellulose
Hordenine
Invert sugar
Lignin
Malt
Nitrogen
Pectin
Pentosan
Protein
Starch
Sucrose

Known Effects

• Provides nutrition to body.

Miscellaneous information:
• Barley is a grain and primarily contains nutrients.

Unproved Speculated Benefits

• Is used as a "restorative" following stomach and intestinal irritation.
• Protects scraped tissues.

Warnings and Precautions

Don't take if you:
• Are allergic or sensitive to barley or gluten.

Consult your doctor if you:
• Take this herb for any medical problem that doesn't improve in 2 weeks. There may be safer, more-effective treatments.
• Take any medicinal drugs or herbs including aspirin, laxatives, cold and cough remedies, antacids, vitamins, minerals, amino acids, supplements, other prescription or non-prescription drugs.

Pregnancy:
• Problems in pregnant women taking small or usual amounts have not been proved. But the chance of problems does exist. Don't use unless prescribed by your doctor.

Breast-feeding:
• Problems in breast-fed infants of lactating mothers taking small or usual amounts have not been proved. But the chance of problems does exist. Don't use unless prescribed by your doctor.

Infants and children:
• Treating infants and children under 2 with any herbal preparation is hazardous.

Others:
• Barley infested with fungus can cause poisoning in animals.

Storage:
• Keep cool and dry, but don't freeze. Store safely away from children.

Safe dosage:
• At present no "safe" dosage has been established.

Toxicity

Comparative-toxicity rating not available from standard references.

Adverse Reactions, Side Effects or Overdose Symptoms

None expected

Basic Information

Biological name (genus and species):
  Myrica cerifera
Parts used for medicinal purposes:
  Bark
  Berries/fruits
  Leaves
Chemicals this herb contains:
  Gallic acid
  Mycricic acid containing palmitin
  Myricinic acid, related to saponin
  Resin (See Glossary)
  Tannic acid

 ## Known Effects

- Shrinks tissues.
- Prevents secretion of fluids.
- Interferes with absorption of iron and other minerals when taken internally.

Miscellaneous information:
- Injections of bark extract have caused cancer in laboratory animals.
- Frequently used as a basic ingredient in cosmetics, pharmaceuticals, candle making.

 ## Unproved Speculated Benefits

Internal use:
- Causes vomiting.
- Treats the common cold.
- Treats diarrhea.
- Treats jaundice.

External use:
- Heals ulcers.
- Treats gum problems.

 ## Warnings and Precautions

Don't take if you:
- Are pregnant, think you may be pregnant or plan pregnancy in the near future.

Consult your doctor if you:
- Take this herb for any medical problem that doesn't improve in 2 weeks. There may be safer, more-effective treatments.
- Take any medicinal drugs or herbs including aspirin, laxatives, cold and cough remedies, antacids, vitamins, minerals, amino acids, supplements, other prescription or non-prescription drugs.

Pregnancy:
- Problems in pregnant women taking small or usual amounts have not been proved. But the chance of problems does exist. Don't use unless prescribed by your doctor.

Breast-feeding:
- Problems in breast-fed infants of lactating mothers taking small or usual amounts have not been proved. But the chance of problems does exist. Don't use unless prescribed by your doctor.

Infants and children:
- Treating infants and children under 2 with any herbal preparation is hazardous.

Others:
- None expected if you are beyond childhood and under 45, basically healthy and take for only a short time.

Storage:
- Keep cool and dry, but don't freeze. Store safely away from children.

Safe dosage:
- At present no "safe" dosage has been established.

 ## Toxicity

Rated relatively safe when taken in appropriate quantities for short periods of time.

 ## Adverse Reactions, Side Effects or Overdose Symptoms

None expected

Bearberry (Uva-ursi)

Basic Information

Biological name (genus and species):
  Arctostaphylos uva-ursi
Parts used for medicinal purposes:
  Leaves
Chemicals this herb contains:
  Arbutin
  Ericolin
  Gallic acid
  Hydroquinolone
  Malic acid
  Quercetin
  Tannins (See Glossary)
  Ursolic acid
  Volatile oils (See Glossary)

Known Effects

- Shrinks urinary tissues.
- Prevents secretion of fluids.
- Relieves urinary pain.
- Helps body dispose of excess fluid by increasing amount of urine produced.
- Interferes with absorption of iron and other minerals when taken internally.

Miscellaneous information:
- Bearberry turns urine green.

Unproved Speculated Benefits

Boiled, bruised leaves:
- Act as a sedative.
- Relieve nausea.
- Decrease ringing in ears.
- Treat breathing problems.

Warnings and Precautions

Don't take if you:
- Are pregnant, think you may be pregnant or plan pregnancy in the near future.

Consult your doctor if you:
- Take this herb for any medical problem that doesn't improve in 2 weeks. There may be safer, more-effective treatments.
- Take any medicinal drugs or herbs including aspirin, laxatives, cold and cough remedies, antacids, vitamins, minerals, amino acids, supplements, other prescription or non-prescription drugs.

Pregnancy:
- Problems in pregnant women taking small or usual amounts have not been proved. But the chance of problems does exist. Don't use unless prescribed by your doctor.

Breast-feeding:
- Problems in breast-fed infants of lactating mothers taking small or usual amounts have not been proved. But the chance of problems does exist. Don't use unless prescribed by your doctor.

Infants and children:
- Treating infants and children under 2 with any herbal preparation is hazardous.

Others:
- No contraindications if you are not pregnant and do not take amounts larger than manufacturer's recommended dosage.

Storage:
- Keep cool and dry, but don't freeze. Store safely away from children.

Safe dosage:
- At present no "safe" dosage has been established.

Toxicity

Rated relatively safe when taken in appropriate quantities for short periods of time.

Adverse Reactions, Side Effects or Overdose Symptoms

None expected

Birch

Basic Information
Biological name (genus and species):
  Betula alba, B. lenta
Parts used for medicinal purposes:
  Bark
  Leaves
Chemicals this herb contains:
  Betulin in bark
  Methyl salicylate (similar to aspirin) in bark
  Resin in shoots and leaves (See Glossary)
  Tar (creosol, phenol, creosote, guaiacol) in
    bark

 Known Effects

- Provides counterirritation when applied to skin overlying an inflamed or irritated joint.
- Decreases inflammation in tissues.

Miscellaneous information:
- Leaves have agreeable aromatic odor but bitter taste.

 Unproved Speculated Benefits

- Is steeped to extract its medicinal properties for rheumatism and congestive heart failure.
- Treats skin disorders when applied topically.
- Shrinks tissues.
- Treats arthritis.
- Prevents secretion of fluids.

 Warnings and Precautions

Don't take if you:
- Are pregnant, think you may be pregnant or plan pregnancy in the near future.

Consult your doctor if you:
- Take this herb for any medical problem that doesn't improve in 2 weeks. There may be safer, more-effective treatments.
- Take any medicinal drugs or herbs including aspirin, laxatives, cold and cough remedies, antacids, vitamins, minerals, amino acids, supplements, other prescription or non-prescription drugs.

Pregnancy:
- Problems in pregnant women taking small or usual amounts have not been proved. But the chance of problems does exist. Don't use unless prescribed by your doctor.

Breast-feeding:
- Problems in breast-fed infants of lactating mothers taking small or usual amounts have not been proved. But the chance of problems does exist. Don't use unless prescribed by your doctor.

Infants and children:
- Treating infants and children under 2 with any herbal preparation is hazardous.

Others:
- No contraindications if you are not pregnant and do not take amounts larger than a reputable manufacturer recommends on package.

Storage:
- Keep cool and dry, but don't freeze. Store safely away from children.

Safe dosage:
- At present no "safe" dosage has been established.

 Toxicity

Comparative-toxicity rating not available from standard references.

 Adverse Reactions, Side Effects or Overdose Symptoms

None expected

Birthroot (Bethroot)

Basic Information

Biological name (genus and species):
Trillium erectum, T. pendulum
Parts used for medicinal purposes:
Various parts of the entire plant, frequently differing by country and/or culture
Chemicals this herb contains:
Resin (See Glossary)
Saponin (See Glossary)
Starch
Tannins (See Glossary)
Volatile oils (See Glossary)

Known Effects

• Irritates mucous membranes.

Miscellaneous information:
• Name *birthroot* resulted from pioneers using this herb to stop bleeding after childbirth.

Unproved Speculated Benefits

• Is used as an aphrodisiac by Indians in southeastern United States.
• Treats gastrointestinal upsets.
• Decreases heartbeat irregularities.
• Controls skin infections.
• Stops excessive bleeding.
• Treats menstrual irregularity or increased menstrual frequency.
• Shrinks tissues.
• Prevents secretion of fluids.
• Decreases thickness and increases fluidity of mucus from lungs and bronchial tubes.
• Is used as an astringent poultice.

Warnings and Precautions

Don't take if you:
• Are pregnant, think you may be pregnant or plan pregnancy in the near future.

Consult your doctor if you:
• Take this herb for any medical problem that doesn't improve in 2 weeks. There may be safer, more-effective treatments.
• Take any medicinal drugs or herbs including aspirin, laxatives, cold and cough remedies, antacids, vitamins, minerals, amino acids, supplements, other prescription or non-prescription drugs.

Pregnancy:
• Problems in pregnant women taking small or usual amounts have not been proved. But the chance of problems does exist. Don't use unless prescribed by your doctor.

Breast-feeding:
• Problems in breast-fed infants of lactating mothers taking small or usual amounts have not been proved. But the chance of problems does exist. Don't use unless prescribed by your doctor.

Infants and children:
• Treating infants and children under 2 with any herbal preparation is hazardous.

Others:
• No contraindications if you are not pregnant and do not take amounts larger than manufacturer's recommended dosage.

Storage:
• Keep cool and dry, but don't freeze. Store safely away from children.

Safe dosage:
• At present no "safe" dosage has been established.

Toxicity

Comparative-toxicity rating not available from standard references.

Adverse Reactions, Side Effects or Overdose Symptoms

None expected

Bistort (Snakeweed)

Basic Information

Biological name (genus and species):
Polygonum bistorta
Parts used for medicinal purposes:
Various parts of the entire plant, frequently differing by country and/or culture
Chemical this herb contains:
Tannins (See Glossary)

Known Effects

- Precipitates proteins.
- Shrinks tissues.
- Prevents secretion of fluids.
- Interferes with absorption of iron and other minerals.

Unproved Speculated Benefits

Roots:
- Are used for astringent gargle.
- Treat unusual bleeding.
- Cause vomiting.
- Treat cavities in teeth.

Warnings and Precautions

Don't take if you:
- Are pregnant, think you may be pregnant or plan pregnancy in the near future.
- Have any chronic disease of gastrointestinal tract, such as stomach or duodenal ulcers, esophageal reflux (reflux esophagitis), ulcerative colitis, spastic colitis, diverticulosis, diverticulitis.

Consult your doctor if you:
- Take this herb for any medical problem that doesn't improve in 2 weeks. There may be safer, more-effective treatments.
- Take any medicinal drugs or herbs including aspirin, laxatives, cold and cough remedies, antacids, vitamins, minerals, amino acids, supplements, other prescription or non-prescription drugs.

Pregnancy:
- Dangers outweigh any possible benefits. Don't use.

Breast-feeding:
- Dangers outweigh any possible benefits. Don't use.

Infants and children:
- Treating infants and children under 2 with any herbal preparation is hazardous.

Others:
- No contraindications if you are not pregnant and do not take amounts larger than manufacturer's recommended dosage.

Storage:
- Keep cool and dry, but don't freeze. Store safely away from children.

Safe dosage:
- At present no "safe" dosage has been established.

Toxicity

Comparative-toxicity rating not available from standard references.

For symptoms of toxicity: See *Adverse Reactions, Side Effects or Overdose Symptoms* section below.

Adverse Reactions, Side Effects or Overdose Symptoms

Signs and symptoms:	What to do:
Bleeding from stomach characterized by vomiting bright-red blood or material that looks like coffee grounds	Discontinue. Call doctor immediately.
Kidney damage characterized by blood in urine, decreased urine flow, swelling of hands and feet.	Seek emergency treatment.
Nausea	Discontinue. Call doctor immediately.
Vomiting	Discontinue. Call doctor immediately.

Bitter Lettuce

Basic Information

Biological name (genus and species):
Lactuca virosa, L. sativa, L. scariola
Parts used for medicinal purposes:
Latex, which exudes from stem of flower stalks
Chemicals this herb contains:
Caoutchouc
Hyoscyamine
Lactucerol
Latucic acid
Lactucin
Mannite
Nitrates
Volatile oils (See Glossary)

Known Effects

• Depresses central nervous system.

Unproved Speculated Benefits

• Acts as a sedative to relieve anxiety or nervous disorders.
• Treats coughs.
• Treats chest pain due to coronary artery disease (angina).
• Causes a "high" when smoked.

Warnings and Precautions

Don't take if you:
• Are pregnant, think you may be pregnant or plan pregnancy in the near future.

Consult your doctor if you:
• Take this herb for any medical problem that doesn't improve in 2 weeks. There may be safer, more-effective treatments.
• Take any medicinal drugs or herbs including aspirin, laxatives, cold and cough remedies, antacids, vitamins, minerals, amino acids, supplements, other prescription or non-prescription drugs.

Pregnancy:
• Dangers outweigh any possible benefits. Don't use.

Breast-feeding:
• Dangers outweigh any possible benefits. Don't use.

Infants and children:
• Treating infants and children under 2 with any herbal preparation is hazardous.

Others:
• Dangers outweigh any possible benefits. Don't use.

Storage:
• Keep cool and dry, but don't freeze. Store safely away from children.

Safe dosage:
• At present no "safe" dosage has been established.

Toxicity

Rated relatively safe when taken in appropriate quantities for short periods of time.

For symptoms of toxicity: See *Adverse Reactions, Side Effects or Overdose Symptoms* section below.

Adverse Reactions, Side Effects or Overdose Symptoms

Signs and symptoms:	What to do:
Breathing difficulties	Seek emergency treatment.

Bitter Root (Wild Ipecac, Spreading Dogbane, Rheumatism Weed)

Basic Information

Biological name (genus and species):
Apocynum androsaemifolium

Parts used for medicinal purposes:
Roots, Bark, Petals/flower

Chemicals this herb contains:

Apocynein	Cymarin
Apocynin	Saponin (See Glossary)

 ## Known Effects

- Slows heartbeat.
- Helps body dispose of excess fluid by increasing amount of urine produced.
- Causes vomiting.

Miscellaneous information:

- Bitter root has marked effect on the heart. Prescribed FDA-approved digitalis preparations are far superior in treating heart disorders such as congestive heart failure and heartbeat irregularities.
- Many plants of varying potency and toxicity are called by this name. Be sure you know what you buy and take.
- You will need increased potassium if you take this herb. Take potassium supplement or eat more food high in potassium, such as apricots, citrus fruits, bananas.

 ## Unproved Speculated Benefits

- Treats congestive heart failure.
- Treats palpitations.
- Treats gallstones.
- "Corrects" bile flow.
- Roots and rhizomes are used to make a medicinal preparation to restore normal tone to tissues or to stimulate appetite.

 ## Warnings and Precautions

Don't take if you:

- Are pregnant, think you may be pregnant or plan pregnancy in the near future.
- Have any chronic disease of gastrointestinal tract, such as stomach or duodenal ulcers, esophageal reflux (reflux esophagitis), ulcerative colitis, spastic colitis, diverticulosis, diverticulitis.

Consult your doctor if you:

- Take this herb for any medical problem that doesn't improve in 2 weeks. There may be safer, more-effective treatments.
- Take any medicinal drugs or herbs including aspirin, laxatives, cold and cough remedies, antacids, vitamins, minerals, amino acids, supplements, other prescription or non-prescription drugs.

Pregnancy:

- Problems in pregnant women taking small or usual amounts have not been proved. But the chance of problems does exist. Don't use unless prescribed by your doctor.

Breast-feeding:

- Problems in breast-fed infants of lactating mothers taking small or usual amounts have not been proved. But the chance of problems does exist. Don't use unless prescribed by your doctor.

Infants and children:

- Treating infants and children under 2 with any herbal preparation is hazardous.

Others:

- Use only under medical supervision.

Storage:

- Keep cool and dry, but don't freeze. Store safely away from children.

Safe dosage:

- At present no "safe" dosage has been established.

 ## Toxicity

Rated slightly dangerous, particularly in children, persons over 55 and those who take larger than appropriate quantities for extended periods of time.

For symptoms of toxicity: See below.

 ## Adverse Reactions, Side Effects or Overdose Symptoms

Signs and symptoms:	What to do:
Precipitous blood-pressure drop—symptoms include, faintness, cold sweat, paleness, rapid pulse.	Seek emergency treatment.
Gastritis	Discontinue. Call doctor when convenient.
Heartbeat irregularities	Seek emergency treatment.
Vomiting	Discontinue. Call doctor immediately.

Bittersweet (European Bittersweet, Bitter Nightshade, Felonwood)

Basic Information
Biological name (genus and species):
  Solanum dulcamara
Parts used for medicinal purposes:
  Leaves
  Roots
Chemicals this herb contains:
  Dulcamarin
  Saponin (See Glossary)
  Solanine
  Solanidine

Known Effects

• Depresses central nervous system.

Miscellaneous information:
• Bittersweet is a potentially dangerous herb. Toxic amounts depress nervous system and cause drowsiness. Berries are poisonous.

Unproved Speculated Benefits

• Treats glandular problems of thyroid, pancreas, ovaries.
• Is used as a lymphatic medicine.
• Treats eczema (See Glossary).
• Kills pain.
• Treats arthritis.
• Is used as an aphrodisiac.
• Treats skin diseases.

Warnings and Precautions

Don't take if you:
• Are pregnant, think you may be pregnant or plan pregnancy in the near future.
• Have any chronic disease of the gastrointestinal tract, such as stomach or duodenal ulcers, esophageal reflux (reflux esophagitis), ulcerative colitis, spastic colitis, diverticulosis, diverticulitis.

Consult your doctor if you:
• Take this herb for any medical problem that doesn't improve in 2 weeks. There may be safer, more-effective treatments.
• Take any medicinal drugs or herbs including aspirin, laxatives, cold and cough remedies, antacids, vitamins, minerals, amino acids, supplements, other prescription or non-prescription drugs.

Pregnancy:
• Dangers outweigh any possible benefits. Don't use.

Breast-feeding:
• Dangers outweigh any possible benefits. Don't use.

Infants and children:
• Treating infants and children under 2 with any herbal preparation is hazardous.

Others:
• Dangers outweigh any possible benefits. Don't use.

Storage:
• Keep cool and dry, but don't freeze. Store safely away from children.

Safe dosage:
• At present no "safe" dosage has been established.

Toxicity

Rated slightly dangerous, particularly in children, persons over 55 and those who take larger than appropriate quantities for extended periods of time.

For symptoms of toxicity: See *Adverse Reactions, Side Effects or Overdose Symptoms* section below.

Adverse Reactions, Side Effects or Overdose Symptoms

Signs and symptoms:	What to do:
Toxins are mostly in unripe fruit, which cause the following symptoms:	
Burning throat	Discontinue. Call doctor when convenient.
Coma	Seek emergency treatment.
Dilated pupils	Discontinue. Call doctor immediately.
Dizziness	Discontinue. Call doctor immediately.
Headache	Discontinue. Call doctor when convenient.
Muscle weakness	Discontinue. Call doctor immediately.
Nausea	Discontinue. Call doctor immediately.
Slow pulse	Seek emergency treatment.
Vomiting	Discontinue. Call doctor immediately.

Basic Information

Biological name (genus and species):
Juglans nigra
Parts used for medicinal purposes:
Husks
Inner bark
Leaves
Chemicals this herb contains:
Ellagic acid
Juglone
Nucin

Known Effects

- Shrinks tissues.
- Prevents secretion of fluids.

Miscellaneous information:
- Nut husks yield brown dye for hair and clothing.

Unproved Speculated Benefits

- Leaves, bark and cut-open ends of nut husks are used to treat fungal infections of skin.

Warnings and Precautions

Don't take if you:
- Are pregnant, think you may be pregnant or plan pregnancy in the near future.
- Have any chronic disease of the gastrointestinal tract, such as stomach or duodenal ulcers, esophageal reflux (reflux esophagitis), ulcerative colitis, spastic colitis, diverticulosis, diverticulitis.

Consult your doctor if you:
- Take this herb for any medical problem that doesn't improve in 2 weeks. There may be safer, more-effective treatments.
- Take any medicinal drugs or herbs including aspirin, laxatives, cold and cough remedies, antacids, vitamins, minerals, amino acids, supplements, other prescription or non-prescription drugs.

Pregnancy:
- Problems in pregnant women taking small or usual amounts have not been proved. But the chance of problems does exist. Don't use unless prescribed by your doctor.

Breast-feeding:
- Problems in breast-fed infants of lactating mothers taking small or usual amounts have not been proved. But the chance of problems does exist. Don't use unless prescribed by your doctor.

Infants and children:
- Treating infants and children under 2 with any herbal preparation is hazardous.

Others:
- None expected if you are under 45, not pregnant, basically healthy, take it for a short time and do not exceed manufacturer's recommended dosage.

Storage:
- Keep cool and dry, but don't freeze. Store safely away from children.

Safe dosage:
- At present no "safe" dosage has been established.

Toxicity

Comparative-toxicity rating not available from standard references.

For symptoms of toxicity: See *Adverse Reactions, Side Effects or Overdose Symptoms* section below.

Adverse Reactions, Side Effects or Overdose Symptoms

Signs and symptoms:	What to do:
Nausea	Discontinue. Call doctor immediately.
Upper-abdominal pain	Discontinue. Call doctor when convenient.

Bladderwrack

Basic Information

Biological name (genus and species):
Fuycus vesiculosus
Parts used for medicinal purposes:
Various parts of the entire plant, frequently differing by country and/or culture
Chemicals this herb contains:
Alginic acid
Bromine iodine
Fucodin
Laminarin

Known Effects

• Absorbs water in intestines to form bulk.

Unproved Speculated Benefits

• Treats obesity.
• Increases thyroid activity.
• Kills intestinal parasites.

Warnings and Precautions

Don't take if you:
• Are pregnant, think you may be pregnant or plan pregnancy in the near future.

Consult your doctor if you:
• Take this herb for any medical problem that doesn't improve in 2 weeks. There may be safer, more-effective treatments.
• Take any medicinal drugs or herbs including aspirin, laxatives, cold and cough remedies, antacids, vitamins, minerals, amino acids, supplements, other prescription or non-prescription drugs.

Pregnancy:
• Problems in pregnant women taking small or usual amounts have not been proved. But the chance of problems does exist. Don't use unless prescribed by your doctor.

Breast-feeding:
• Problems in breast-fed infants of lactating mothers taking small or usual amounts have not been proved. But the chance of problems does exist. Don't use unless prescribed by your doctor.

Infants and children:
• Treating infants and children under 2 with any herbal preparation is hazardous.

Others:
• None expected if you are under 45, not pregnant, basically healthy, take it for only a short time and do not exceed manufacturer's recommended dosage.

Storage:
• Keep cool and dry, but don't freeze. Store safely away from children.

Safe dosage:
• At present no "safe" dosage has been established.

Toxicity

Comparative-toxicity rating not available from standard references.

Adverse Reactions, Side Effects or Overdose Symptoms

None expected

Basic Information

Biological name (genus and species):
  Cincus benedictus
Parts used for medicinal purposes:
  Various parts of the entire plant, frequently
  differing by country and/or culture
Chemicals this herb contains:
  Cincin
  Volatile oils (See Glossary)

 ## Known Effects

- Stimulates secretions from stomach.
- Irritates mucous membranes.

Miscellaneous information:
- Is applied to skin overlying a joint to cause an irritant to relieve another irritant.
- Effects have not been studied to any great extent.
- Careful handling is necessary to avoid toxic effects on skin.

 ## Unproved Speculated Benefits

- Increases stomach secretions.
- Increases appetite.

 ## Warnings and Precautions

Don't take if you:
- Are pregnant, think you may be pregnant or plan pregnancy in the near future.
- Have any chronic disease of the gastrointestinal tract, such as stomach or duodenal ulcers, esophageal reflux (reflux esophagitis), ulcerative colitis, spastic colitis, diverticulosis, diverticulitis.

Consult your doctor if you:
- Take this herb for any medical problem that doesn't improve in 2 weeks. There may be safer, more-effective treatments.
- Take any medicinal drugs or herbs including aspirin, laxatives, cold and cough remedies, antacids, vitamins, minerals, amino acids, supplements, other prescription or non-prescription drugs.

Pregnancy:
- Problems in pregnant women taking small or usual amounts have not been proved. But the chance of problems does exist. Don't use unless prescribed by your doctor.

Breast-feeding:
- Problems in breast-fed infants of lactating mothers taking small or usual amounts have not been proved. But the chance of problems does exist. Don't use unless prescribed by your doctor.

Infants and children:
- Treating infants and children under 2 with any herbal preparation is hazardous.

Others:
- None expected if you are beyond childhood and under 45, basically healthy and take for only a short time.

Storage:
- Keep cool and dry, but don't freeze. Store safely away from children.

Safe dosage:
- At present no "safe" dosage has been established.

 ## Toxicity

Comparative-toxicity rating not available from standard references.

For symptoms of toxicity: See *Adverse Reactions, Side Effects or Overdose Symptoms* section below.

 ## Adverse Reactions, Side Effects or Overdose Symptoms

Signs and symptoms:	What to do:
Vomiting	Discontinue. Call doctor immediately.

Blueberry

Basic Information

Biological name (genus and species):
Vaccinum spp
Parts used for medicinal purposes:
Leaves
Stems
Chemicals this herb contains:
Fatty acids (See Glossary)
Hydroquinone
Loeanolic acid
Neomyrtillin
Tannins (See Glossary)
Ursolic acid

Known Effects

- Decreases blood sugar.
- Interferes with absorption of iron and other minerals when taken internally.

Unproved Speculated Benefits

- Treats diarrhea.
- Treats gastroenteritis.
- Helps body dispose of excess fluid by increasing amount of urine produced.
- Treats and prevents scurvy.

Warnings and Precautions

Don't take if you:
- Are pregnant, think you may be pregnant or plan pregnancy in the near future.

Consult your doctor if you:
- Take this herb for any medical problem that doesn't improve in 2 weeks. There may be safer, more-effective treatments.
- Take any medicinal drugs or herbs including aspirin, laxatives, cold and cough remedies, antacids, vitamins, minerals, amino acids, supplements, other prescription or non-prescription drugs.

Pregnancy:
- Problems in pregnant women taking small or usual amounts have not been proved. But the chance of problems does exist. Don't use unless prescribed by your doctor.

Breast-feeding:
- Problems in breast-fed infants of lactating mothers taking small or usual amounts have not been proved. But the chance of problems does exist. Don't use unless prescribed by your doctor.

Infants and children:
- Treating infants and children under 2 with any herbal preparation is hazardous.

Others:
- None expected if you are beyond childhood and under 45, basically healthy and take for only a short time.

Storage:
- Keep cool and dry, but don't freeze. Store safely away from children.

Safe dosage:
- At present no "safe" dosage has been established.

Toxicity

Comparative-toxicity rating not available from standard references.

Adverse Reactions, Side Effects or Overdose Symptoms

None expected

Boneset (Richweed, White Snakeroot, Ague Weed)

Basic Information

Biological name (genus and species):
Eupatorium perfoliatum, E. rugosum
Parts used for medicinal purposes:
Leaves
Petals/flower
Chemicals this herb contains:
Eupatroin
Resin (See Glossary)
Sugar
Tremetrol
Volatile oils (See Glossary)
Wax (See Glossary)

Known Effects

- Irritates gastrointestinal tract.
- Can produce "milk sickness" in humans, an acute disease characterized by trembling, vomiting and severe abdominal pain. It is caused by eating dairy products or beef from cattle poisoned by eating boneset.
- Increases perspiration.
- Causes vomiting.

Miscellaneous information:
- Tremetrol can accumulate slowly in animal bodies and cause toxic symptoms. It may do the same in humans.

Unproved Speculated Benefits

- Decreases blood sugar.
- Treats malaria.
- Treats fever.

Warnings and Precautions

Don't take if you:
- Are pregnant, think you may be pregnant or plan pregnancy in the near future.
- Have any chronic disease of the gastrointestinal tract, such as stomach or duodenal ulcers, esophageal reflux (reflux esophagitis), ulcerative colitis, spastic colitis, diverticulosis, diverticulitis.

Consult your doctor if you:
- Take this herb for any medical problem that doesn't improve in 2 weeks. There may be safer, more-effective treatments.

- Take any medicinal drugs or herbs including aspirin, laxatives, cold and cough remedies, antacids, vitamins, minerals, amino acids, supplements, other prescription or non-prescription drugs.

Pregnancy:
- Dangers outweigh any possible benefits. Don't use.

Breast-feeding:
- Dangers outweigh any possible benefits. Don't use.

Infants and children:
- Treating infants and children under 2 with any herbal preparation is hazardous.

Others:
- Dangers outweigh any possible benefits. Don't use.

Storage:
- Keep cool and dry, but don't freeze. Store safely away from children.

Safe dosage:
- At present no "safe" dosage has been established.

Toxicity

Comparative-toxicity rating not available from standard references.

For symptoms of toxicity: See below.

Adverse Reactions, Side Effects or Overdose Symptoms

Signs and symptoms:	What to do:
Breathing difficulties	Seek emergency treatment.
Coma	Seek emergency treatment.
Drooling	Discontinue. Call doctor when convenient.
Muscle trembling	Discontinue. Call doctor immediately.
Nausea	Discontinue. Call doctor immediately.
Stiffness	Discontinue. Call doctor when convenient.
Vomiting	Discontinue. Call doctor immediately.
Weakness	Discontinue. Call doctor when convenient.

Buchu (Honey Buchu, Short-Leaf Mountain Buchu)

Basic Information

Biological name (genus and species):
Barosma betulina
Parts used for medicinal purposes:
Leaves
Chemicals this herb contains:
Diasmin
Hesperidin
l-enthone
Mucilage (See Glossary)
Resin (See Glossary)
Volatile oils (See Glossary)

Known Effects

- Helps body dispose of excess fluid by increasing amount of urine produced.
- Works as a urinary antiseptic.
- Aids in expelling gas from intestinal tract.

Miscellaneous information:
- Has peppermint-like odor.
- This herb is no longer used by medical profession.

Unproved Speculated Benefits

- Treats kidney stones.
- Treats chronic prostatitis.
- Treats bladder irritation.
- Treats urethral irritation.
- Stimulates central nervous system.
- Increases perspiration.

Warnings and Precautions

Don't take if you:
- Are pregnant, think you may be pregnant or plan pregnancy in the near future.
- Have any chronic disease of the gastrointestinal tract, such as stomach or duodenal ulcers, esophageal reflux (reflux esophagitis), ulcerative colitis, spastic colitis, diverticulosis, diverticulitis.

Consult your doctor if you:
- Take this herb for any medical problem that doesn't improve in 2 weeks. There may be safer, more-effective treatments.
- Take any medicinal drugs or herbs including aspirin, laxatives, cold and cough remedies, antacids, vitamins, minerals, amino acids, supplements, other prescription or non-prescription drugs.

Pregnancy:
- Dangers outweigh any possible benefits. Don't use.

Breast-feeding:
- Dangers outweigh any possible benefits. Don't use.

Infants and children:
- Treating infants and children under 2 with any herbal preparation is hazardous.

Others:
- None expected if you are under 45, not pregnant, basically healthy, take it for only a short time and do not exceed small doses.

Storage:
- Keep cool and dry, but don't freeze. Store safely away from children.

Safe dosage:
- At present no "safe" dosage has been established.

Toxicity

Rated relatively safe when taken in appropriate quantities for short periods of time.

For symptoms of toxicity: See *Adverse Reactions, Side Effects or Overdose Symptoms* section below.

Adverse Reactions, Side Effects or Overdose Symptoms

Signs and symptoms:	What to do:
Nausea	Discontinue. Call doctor immediately.
Vomiting	Discontinue. Call doctor immediately.

Basic Information

Biological name (genus and species):
Rhamnus cathartica
Parts used for medicinal purposes:
Bark
Berries/fruits
Chemicals this herb contains:
Anthra-quinone
Emodin

Known Effects

- Irritates gastrointestinal tract and causes watery, explosive bowel movements.

Miscellaneous information:
- Several dyes are made from juice of berries. Children can have toxic symptoms after eating as few as 20 berries.
- Syrup of buckthorn is made from berries

Unproved Speculated Benefits

- Is used as laxative.

Warnings and Precautions

Don't take if you:
- Are pregnant, think you may be pregnant or plan pregnancy in the near future.
- Have any chronic disease of the gastrointestinal tract, such as stomach or duodenal ulcers, esophageal reflux (reflux esophagitis), ulcerative colitis, spastic colitis, diverticulosis, diverticulitis.

Consult your doctor if you:
- Take this herb for any medical problem that doesn't improve in 2 weeks. There may be safer, more-effective treatments.
- Take any medicinal drugs or herbs including aspirin, laxatives, cold and cough remedies, antacids, vitamins, minerals, amino acids, supplements, other prescription or non-prescription drugs.

Pregnancy:
- Dangers outweigh any possible benefits. Don't use.

Breast-feeding:
- Dangers outweigh any possible benefits. Don't use.

Infants and children:
- Treating infants and children under 2 with any herbal preparation is hazardous.

Others:
- None expected if you are under 45, not pregnant, basically healthy, take it for only a short time and do not exceed manufacturer's recommended dosage.

Storage:
- Keep cool and dry, but don't freeze. Store safely away from children.

Safe dosage:
- At present no "safe" dosage has been established.

Toxicity

Comparative-toxicity rating not available from standard references.

For symptoms of toxicity: See *Adverse Reactions, Side Effects or Overdose Symptoms* section below.

Adverse Reactions, Side Effects or Overdose Symptoms

Signs and symptoms:	What to do:
Diarrhea, severe and watery	Discontinue. Call doctor immediately.
Kidney damage with large amounts over long period of time; characterized by blood in urine, decreased urine flow, swelling of hands and feet	Seek emergency treatment.
Nausea	Discontinue. Call doctor immediately.
Vomiting	Discontinue. Call doctor immediately.

Burdock (Edible Burdock, Lappa, Great Burdock)

Basic Information

Biological name (genus and species):
  Arctium lappa
Parts used for medicinal purposes:
  Roots
  Seeds
Chemicals this herb contains:
  Arctiin
  Inulin
  Tannins (See Glossary)
  Volatile oils (See Glossary)

Known Effects

- Burdock contains no pharmacologically active chemicals, but it may be contaminated by atropine-like chemicals that can be poisonous.
- Interferes with absorption of iron and other minerals when taken internally.

Unproved Speculated Benefits

- Treats skin disorders.
- Treats gout.
- Stimulates body's defenses against disease.

Warnings and Precautions

Don't take if you:
- Are pregnant, think you may be pregnant or plan pregnancy in the near future.

Consult your doctor if you:
- Take this herb for any medical problem that doesn't improve in 2 weeks. There may be safer, more-effective treatments.
- Take any medicinal drugs or herbs including aspirin, laxatives, cold and cough remedies, antacids, vitamins, minerals, amino acids, supplements, other prescription or non-prescription drugs.

Pregnancy:
- Dangers outweigh any possible benefits. Don't use.

Breast-feeding:
- Dangers outweigh any possible benefits. Don't use.

Infants and children:
- Treating infants and children under 2 with any herbal preparation is hazardous.

Storage:
- Keep cool and dry, but don't freeze. Store safely away from children.

Safe dosage:
- At present no "safe" dosage has been established.

Toxicity

Rated relatively safe when taken in appropriate quantities for short periods of time.

For symptoms of toxicity: See *Adverse Reactions, Side Effects or Overdose Symptoms* section below.

Adverse Reactions, Side Effects or Overdose Symptoms

Signs and symptoms:	What to do:
Dilated pupils	Discontinue. Call doctor immediately.
Dry mouth	Discontinue. Call doctor when convenient.
Hallucinations	Seek emergency treatment.

Calamus Root (Sweet Root, Acore, Rat Root, Sweet Flag, Sweet Myrtle, Sweet Cane, Sweet Sedge, Flagroot, Calamus)

Basic Information

Biological name (genus and species):
Acorus calamus

Parts used for medicinal purposes:
Roots

Chemicals this herb contains:
Asarone
Beta-asarone
Camphene
Caryophyllene
Eugenol
Pinene
Volatile oils (See Glossary)

Known Effects

- Aids in expelling gas from the intestinal tract.
- Depresses central nervous system.
- Causes hallucinations.

Miscellaneous information:
- Used primarily in India for many illnesses.
- Essential oil extracted from the root causes cancer in rats. The FDA has banned all varieties of this plant for human use.

Unproved Speculated Benefits

- Treats asthma.
- Treats coughs.
- Treats dyspepsia.
- Treats convulsions.
- Treats epilepsy.
- Treats hysteria.
- Treats insanity.
- Treats intestinal parasites.
- Is used as an aphrodisiac.
- Reduces fever.

Warnings and Precautions

Don't take if you:
- Are pregnant, think you may be pregnant or plan pregnancy in the near future.

Consult your doctor if you:
- Take this herb for any medical problem that doesn't improve in 2 weeks. There may be safer, more-effective treatments.
- Take any medicinal drugs or herbs including aspirin, laxatives, cold and cough remedies, antacids, vitamins, minerals, amino acids, supplements, other prescription or non-prescription drugs.

Pregnancy:
- Problems in pregnant women taking small or usual amounts have not been proved. But the chance of problems does exist. Don't use unless prescribed by your doctor.

Breast-feeding:
- Problems in breast-fed infants of lactating mothers taking small or usual amounts have not been proved. But the chance of problems does exist. Don't use unless prescribed by your doctor.

Infants and children:
- Treating infants and children under 2 with any herbal preparation is hazardous.

Others:
- None expected if you are under 45, not pregnant, basically healthy, take it for only a short time and do not exceed manufacturer's recommended dosage.

Storage:
- Keep cool and dry, but don't freeze. Store safely away from children.

Safe dosage:
- At present no "safe" dosage has been established.

Toxicity

Rated dangerous, particularly in children, persons over 55 and those who take larger than appropriate quantities for extended periods of time.

For symptoms of toxicity: See *Adverse Reactions, Side Effects or Overdose Symptoms* section below.

Adverse Reactions, Side Effects or Overdose Symptoms

Signs and symptoms:	What to do:
Drowsiness	Discontinue. Call doctor when convenient.
Hallucinations	Seek emergency treatment.

California Poppy

Basic Information

Biological name (genus and species):
Eschoscholtzia californica
Parts used for medicinal purposes:
Entire plant, except roots
Chemicals this herb contains:
Coptisine
Sanguinarine

 Known Effects

- Feeble narcotic action.
- Increases perspiration.
- Depresses central nervous system.

Miscellaneous information:
- Does not contain any narcotic derivatives, such as morphine or codeine. The poppy plant that has narcotic properties is different from this one.

 Unproved Speculated Benefits

- Is used by drug abusers for sedative or mind-altering effects.

 Warnings and Precautions

Don't take if you:
- Are pregnant, think you may be pregnant or plan pregnancy in the near future.

Consult your doctor if you:
- Take this herb for any medical problem that doesn't improve in 2 weeks. There may be safer, more-effective treatments.
- Take any medicinal drugs or herbs including aspirin, laxatives, cold and cough remedies, antacids, vitamins, minerals, amino acids, supplements, other prescription or non-prescription drugs.

Pregnancy:
- Problems in pregnant women taking small or usual amounts have not been proved. But the chance of problems does exist. Don't use unless prescribed by your doctor.

Breast-feeding:
- Problems in breast-fed infants of lactating mothers taking small or usual amounts have not been proved. But the chance of problems does exist. Don't use unless prescribed by your doctor.

Infants and children:
- Treating infants and children under 2 with any herbal preparation is hazardous.

Others:
- None expected if you are beyond childhood and under 45, basically healthy and take for only a short time.

Storage:
- Keep cool and dry, but don't freeze. Store safely away from children.

Safe dosage:
- At present no "safe" dosage has been established.

 Toxicity

Rated slightly dangerous, particularly in children, persons over 55 and those who take larger than appropriate quantities for extended periods of time.

 Adverse Reactions, Side Effects or Overdose Symptoms

None expected

Capsicum (Red-Hot Pepper, Hot Pepper, Cayenne, Chili Pepper, Africa Pepper, American Pepper, Red Pepper, Spanish Pepper)

Basic Information

Biological name (genus and species):
  Capsicum frutescens, Capsicum annum
Parts used for medicinal purposes:
  Berries/fruits
Chemicals this herb contains:
  Apsaicine
  Capsacutin
  Capsico
  Capsaicin

Known Effects

• Provides counterirritation when applied to skin overlying an inflamed or irritated joint.
• No effects are expected on the body, either good or bad, when herb is used in very small amounts to enhance the flavor of food.

Miscellaneous information:
• Available in powder form.
• Available as fresh food.
• Is used in small amounts as a condiment.

Unproved Speculated Benefits

• Reduces incidence of clotting in blood vessels (thromboembolism).
• Relieves toothache.
• Wards off infections.
• Settles "upset stomach."
• Treats intestinal disorders.
• Is used as external rub or poultice.

Warnings and Precautions

Don't take if you:
• Are pregnant, think you may be pregnant or plan pregnancy in the near future.
• Have a bleeding problem.
• Have any chronic disease of the gastrointestinal tract, such as stomach or duodenal ulcers, esophageal reflux (reflux esophagitis), ulcerative colitis, spastic colitis, diverticulosis, diverticulitis.

Consult your doctor if you:
• Take this herb for any medical problem that doesn't improve in 2 weeks. There may be safer, more-effective treatments.
• Take any medicinal drugs or herbs including aspirin, laxatives, cold and cough remedies, antacids, vitamins, minerals, amino acids, supplements, other prescription or non-prescription drugs.

Pregnancy:
• Problems in pregnant women taking small or usual amounts have not been proved. But the chance of problems does exist. Don't use unless prescribed by your doctor.

Breast-feeding:
• Problems in breast-fed infants of lactating mothers taking small or usual amounts have not been proved. But the chance of problems does exist. Don't use unless prescribed by your doctor.

Infants and children:
• Treating infants and children under 2 with any herbal preparation is hazardous.

Storage:
• Keep cool and dry, but don't freeze. Store safely away from children.

Safe dosage:
• At present no "safe" dosage has been established.

Toxicity

Rated relatively safe when taken in appropriate quantities for short periods of time.

For symptoms of toxicity: See *Adverse Reactions, Side Effects or Overdose Symptoms* section below.

Adverse Reactions, Side Effects or Overdose Symptoms

Signs and symptoms:	What to do:
Diarrhea, regular or bloody	Discontinue. Call doctor immediately.
Nausea	Discontinue. Call doctor immediately.
Vomiting	Discontinue. Call doctor immediately.
Vomiting blood	Seek emergency treatment.

Caraway

Basic Information
Biological name (genus and species):
Carum carvi
Parts used for medicinal purposes:
Leaves
Seeds
Chemicals this herb contains:
Calcium oxalate
Carveo
Carvone, as volatile oils (See Glossary)
Dihydrocarvone
Fatty acids (See Glossary)
Proteins

Known Effects

• Acts as an aromatic (See Glossary).
• Aids in expelling gas from intestinal tract.
• No effects are expected on the body, either good or bad, when herb is used in very small amounts to enhance the flavor of food.

Miscellaneous information:
• Is used as a flavoring agent in baking.
• Oil is used in making ice cream.

Unproved Speculated Benefits

• Reduces flatulence in infants.
• Treats abdominal cramping.
• Treats nausea.
• Treats scabies.

Warnings and Precautions

Don't take if you:
• Are pregnant, think you may be pregnant or plan pregnancy in the near future.
• Have any chronic disease of the gastrointestinal tract, such as stomach or duodenal ulcers, esophageal reflux (reflux esophagitis), ulcerative colitis, spastic colitis, diverticulosis, diverticulitis.

Consult your doctor if you:
• Take this herb for any medical problem that doesn't improve in 2 weeks. There may be safer, more-effective treatments.
• Take any medicinal drugs or herbs including aspirin, laxatives, cold and cough remedies, antacids, vitamins, minerals, amino acids, supplements, other prescription or non-prescription drugs.

Pregnancy:
• Problems in pregnant women taking small or usual amounts have not been proved. But the chance of problems does exist. Don't use unless prescribed by your doctor.

Breast-feeding:
• Problems in breast-fed infants of lactating mothers taking small or usual amounts have not been proved. But the chance of problems does exist. Don't use unless prescribed by your doctor.

Infants and children:
• Treating infants and children under 2 with any herbal preparation is hazardous.

Others:
• None expected if you are under 45, not pregnant, basically healthy, take it for only a short time and do not exceed manufacturer's recommended dosage.

Storage:
• Keep cool and dry, but don't freeze. Store safely away from children.

Safe dosage:
• At present no "safe" dosage has been established.

Toxicity

Comparative-toxicity rating not available from standard references.

For symptoms of toxicity: See *Adverse Reactions, Side Effects or Overdose Symptoms* section below.

Adverse Reactions, Side Effects or Overdose Symptoms

Signs and symptoms:	What to do:
In very large amounts only:	
Central-nervous-system depression	Seek emergency treatment.
Nausea	Discontinue. Call doctor immediately.
Vomiting	Discontinue. Call doctor immediately.

Cardamom Seed

Basic Information

Biological name (genus and species):
Ellettaria cardamonum, Amonum cardamonum

Parts used for medicinal purposes:
Seeds

Chemicals this herb contains:
Dipentene
Fixed oil (See Glossary)
Gum (See Glossary)
Limonene
Terpene alcohol
Terpinene
Starch
Volatile oils (See Glossary)
Yellow coloring

Known Effects

- Aids in expelling gas from intestinal tract.
- No effects are expected on the body, either good or bad, when herb is used in very small amounts to enhance the flavor of food.

Miscellaneous information:
- Provides flavor.

Unproved Speculated Benefits

- Acts as vigorous laxative.
- Causes explosive, watery diarrhea.

Warnings and Precautions

Don't take if you:
- Are pregnant, think you may be pregnant or plan pregnancy in the near future.
- Have any chronic disease of the gastrointestinal tract, such as stomach or duodenal ulcers, esophageal reflux (reflux esophagitis), ulcerative colitis, spastic colitis, diverticulosis, diverticulitis.

Consult your doctor if you:
- Take this herb for any medical problem that doesn't improve in 2 weeks. There may be safer, more-effective treatments.
- Take any medicinal drugs or herbs including aspirin, laxatives, cold and cough remedies, antacids, vitamins, minerals, amino acids, supplements, other prescription or non-prescription drugs.

Pregnancy:
- Problems in pregnant women taking small or usual amounts have not been proved. But the chance of problems does exist. Don't use unless prescribed by your doctor.

Breast-feeding:
- Problems in breast-fed infants of lactating mothers taking small or usual amounts have not been proved. But the chance of problems does exist. Don't use unless prescribed by your doctor.

Infants and children:
- Treating infants and children under 2 with any herbal preparation is hazardous.

Others:
- None expected if you are beyond childhood and under 45, basically healthy and take for only a short time.

Storage:
- Keep cool and dry, but don't freeze. Store safely away from children.

Safe dosage:
- At present no "safe" dosage has been established.

Toxicity

Comparative-toxicity rating not available from standard references.

For symptoms of toxicity: See *Adverse Reactions, Side Effects or Overdose Symptoms* section below.

Adverse Reactions, Side Effects or Overdose Symptoms

Signs and symptoms:	What to do:
Diarrhea	Discontinue. Call doctor immediately.
Nausea	Discontinue. Call doctor immediately.
Vomiting	Discontinue. Call doctor immediately.

MEDICINAL HERB

Cascara Sagrada (Cascara Buckthorn)

Basic Information

Biological name (genus and species):
Rhamnus purshiana
Parts used for medicinal purposes:
Bark
Chemicals this herb contains:
Anthraquinone
Cascarosides

Known Effects

- Causes irritation to gastrointestinal tract and can cause watery, explosive diarrhea.

Miscellaneous information:
- Not recommended for prolonged use.
- This is a standard medicinal product listed in the *United States Pharmacopeia.*

Unproved Speculated Benefits

- Treats chronic constipation.

Warnings and Precautions

Don't take if you:
- Are pregnant, think you may be pregnant or plan pregnancy in the near future.
- Have any chronic disease of the gastrointestinal tract, such as stomach or duodenal ulcers, esophageal reflux (reflux esophagitis), ulcerative colitis, spastic colitis, diverticulosis, diverticulitis.

Consult your doctor if you:
- Take this herb for any medical problem that doesn't improve in 2 weeks. There may be safer, more-effective treatments.
- Take any medicinal drugs or herbs including aspirin, laxatives, cold and cough remedies, antacids, vitamins, minerals, amino acids, supplements, other prescription or non-prescription drugs.

Pregnancy:
- Dangers outweigh any possible benefits. Don't use.

Breast-feeding:
- Dangers outweigh any possible benefits. Don't use.

Infants and children:
- Treating infants and children under 2 with any herbal preparation is hazardous.

Others:
- None expected if you are under 45, not pregnant, basically healthy, take it for only a short time and do not exceed small doses.

Storage:
- Keep cool and dry, but don't freeze. Store safely away from children.

Safe dosage:
- At present no "safe" dosage has been established.

Toxicity

Rated slightly dangerous, particularly in children, persons over 55 and those who take larger than appropriate quantities for extended periods of time.

For symptoms of toxicity: See *Adverse Reactions, Side Effects or Overdose Symptoms* section below.

Adverse Reactions, Side Effects or Overdose Symptoms

Signs and symptoms:	What to do:
With excessive dosage:	
Diarrhea, violent and watery	Discontinue. Call doctor immediately.
Nausea	Discontinue. Call doctor immediately.
Vomiting	Discontinue. Call doctor immediately.

Basic Information

Biological name (genus and species):
Catalpa bignonioides
Parts used for medicinal purposes:
Various parts of the entire plant, frequently differing by country and/or culture
Chemicals this herb contains:
Catalpin
Catalposide

Known Effects

• Irritates gastrointestinal tract.

Unproved Speculated Benefits

• Treats asthma.

Warnings and Precautions

Don't take if you:
• Are pregnant, think you may be pregnant or plan pregnancy in the near future.
• Have any chronic disease of the gastrointestinal tract, such as stomach or duodenal ulcers, esophageal reflux (reflux esophagitis), ulcerative colitis, spastic colitis, diverticulosis, diverticulitis.

Consult your doctor if you:
• Take this herb for any medical problem that doesn't improve in 2 weeks. There may be safer, more-effective treatments.
• Take any medicinal drugs or herbs including aspirin, laxatives, cold and cough remedies, antacids, vitamins, minerals, amino acids, supplements, other prescription or non-prescription drugs.

Pregnancy:
• Dangers outweigh any possible benefits. Don't use.

Breast-feeding:
• Dangers outweigh any possible benefits. Don't use.

Infants and children:
• Treating infants and children under 2 with any herbal preparation is hazardous.

Others:
• Dangers outweigh any possible benefits. Don't use.

Storage:
• Keep cool and dry, but don't freeze. Store safely away from children.

Safe dosage:
• At present no "safe" dosage has been established.

Toxicity

Comparative-toxicity rating not available from standard references.

For symptoms of toxicity: See *Adverse Reactions, Side Effects or Overdose* section below.

Adverse Reactions, Side Effects or Overdose Symptoms

Signs and symptoms:	What to do:
Precipitous blood-pressure drop—symptoms include, faintness, cold sweat, paleness, rapid pulse	Seek emergency treatment.
Cold, clammy skin	Discontinue. Call doctor when convenient.
Diarrhea	Discontinue. Call doctor immediately.
Nausea	Discontinue. Call doctor immediately.
Rapid, weak pulse	Seek emergency treatment.
Vomiting	Discontinue. Call doctor immediately.

MEDICINAL HERB

291

Catechu, Black

Basic Information

Biological name (genus and species):
  Acacia catechu
Parts used for medicinal purposes:
  Various parts of the entire plant, frequently
  differing by country and/or culture
Chemical this herb contains:
  Tannins (See Glossary)

Known Effects

- Shrinks tissues.
- Prevents secretion of fluids.
- Interferes with absorption of iron and other minerals when taken internally.

Unproved Speculated Benefits

- Decreases unusual bleeding.
- Treats chronic diarrhea.
- Is used as gargle for sore throat.

Warnings and Precautions

Don't take if you:
- Are pregnant, think you may be pregnant or plan pregnancy in the near future.
- Have any chronic disease of the gastrointestinal tract, such as stomach or duodenal ulcers, esophageal reflux (reflux esophagitis), ulcerative colitis, spastic colitis, diverticulosis, diverticulitis.

Consult your doctor if you:
- Take this herb for any medical problem that doesn't improve in 2 weeks. There may be safer, more-effective treatments.
- Take any medicinal drugs or herbs including aspirin, laxatives, cold and cough remedies, antacids, vitamins, minerals, amino acids, supplements, other prescription or non-prescription drugs

Pregnancy:
- Dangers outweigh any possible benefits. Don't use.

Breast-feeding:
- Dangers outweigh any possible benefits. Don't use.

Infants and children:
- Treating infants and children under 2 with any herbal preparation is hazardous.

Others:
- None expected if you are under 45, not pregnant, basically healthy, take it for only a short time and do not exceed manufacturer's recommended dosage.

Storage:
- Keep cool and dry, but don't freeze. Store safely away from children.

Safe dosage:
- At present no "safe" dosage has been established.

Toxicity

Comparative-toxicity rating not available from standard references.

For symptoms of toxicity: See *Adverse Reactions, Side Effects or Overdose Symptoms* section below.

Adverse Reactions, Side Effects or Overdose Symptoms

Signs and symptoms:	What to do:
Diarrhea	Discontinue. Call doctor immediately.
Kidney damage characterized by blood in urine, decreased urine flow, swelling of hands and feet	Seek emergency treatment.
Vomiting	Discontinue. Call doctor immediately.

Catha (Khat plant)

Basic Information
Biological name (genus and species):
  Catha edulis
Parts used for medicinal purposes:
  Leaves
Chemicals this herb contains:
  Cathidine
  Cathine (a form of ephedrine)
  Celastrin
  Choline
  Ratine
  Tannins (See Glossary)

Known Effects

- Stimulates brain and spinal cord through synapses.
- Interferes with absorption of iron and other minerals when taken internally.

Miscellaneous information:
- Can be habit forming. Addicts become talkative then depressed and apathetic.

Unproved Speculated Benefits

- Leaves chewed or steeped to make tea to treat fatigue.
- Suppresses appetite.

Warnings and Precautions

Don't take if you:
- Are pregnant, think you may be pregnant or plan pregnancy in the near future.
- Have heart trouble.
- Have high blood pressure.

Consult your doctor if you:
- Take this herb for any medical problem that doesn't improve in 2 weeks. There may be safer, more-effective treatments.
- Take any medicinal drugs or herbs including aspirin, laxatives, cold and cough remedies, antacids, vitamins, minerals, amino acids, supplements, other prescription or non-prescription drugs.

Pregnancy:
- Dangers outweigh any possible benefits. Don't use.

Breast-feeding:
- Dangers outweigh any possible benefits. Don't use.

Infants and children:
- Treating infants and children under 2 with any herbal preparation is hazardous.

Storage:
- Keep cool and dry, but don't freeze. Store safely away from children.

Safe dosage:
- At present no "safe" dosage has been established.

Toxicity

Rated slightly dangerous, particularly in children, persons over 55 and those who take larger than appropriate quantities for extended periods of time.

For symptoms of toxicity: See *Adverse Reactions, Side Effects or Overdose Symptoms* section below.

Adverse Reactions, Side Effects or Overdose Symptoms

Signs and symptoms:	What to do:
Large amounts:	
Breathing difficulties	Seek emergency treatment.
Depression	Discontinue. Call doctor when convenient.
Euphoria	Discontinue. Call doctor when convenient.
Increased blood pressure	Discontinue. Call doctor immediately.
Increased heart rate	Seek emergency treatment.
Paralysis	Seek emergency treatment.
Stomach irritation, with bleeding	Discontinue. Call doctor immediately.

Catnip (Catnep, Catmint)

Basic Information

Biological name (genus and species):
  Nepeta cataria
Parts used for medicinal purposes:
  Leaves
Chemicals this herb contains:
  Acetic acid
  Buteric acid
  Citral
  Dipentene
  Lifronella
  Limonene
  Nepetalic acid
  Tannins (See Glossary)
  Terpene
  Valeric acid
  Volatile oils (See Glossary)

Known Effects

- Stimulates central nervous system.
- Relieves spasm in skeletal or smooth muscle.
- Interferes with absorption of iron and other minerals when taken internally.

Miscellaneous information:
- Catnip is not a psychedelic or euphoria-producing drug, despite several reports to the contrary.

Unproved Speculated Benefits

- Steeped leaves produce increased sweating for reducing fevers.
- Leaves used as snuff to treat colic. (See Glossary.)
- Treats insomnia.

Warnings and Precautions

Don't take if you:
- Are pregnant, think you may be pregnant or plan pregnancy in the near future.

Consult your doctor if you:
- Take this herb for any medical problem that doesn't improve in 2 weeks. There may be safer, more-effective treatments.
- Take any medicinal drugs or herbs including aspirin, laxatives, cold and cough remedies, antacids, vitamins, minerals, amino acids, supplements, other prescription or non-. prescription drugs.

Pregnancy:
- Problems in pregnant women taking small or usual amounts have not been proved. But the chance of problems does exist. Don't use unless prescribed by your doctor.

Breast-feeding:
- Problems in breast-fed infants of lactating mothers taking small or usual amounts have not been proved. But the chance of problems does exist. Don't use unless prescribed by your doctor.

Infants and children:
- Treating infants and children under 2 with any herbal preparation is hazardous.

Others:
- None expected if you are beyond childhood and under 45, basically healthy and take for only a short time.

Storage:
- Keep cool and dry, but don't freeze. Store safely away from children.

Safe dosage:
- At present no "safe" dosage has been established.

Toxicity

Generally regarded as safe when taken in appropriate quantities for short periods of time.

Adverse Reactions, Side Effects or Overdose Symptoms

None expected

Basic Information

Biological name (genus and species):
  Apium graveolens
Parts used for medicinal purposes:
  Juice
  Roots
  Seeds
Chemicals this herb contains:
  D-limonene
  Nitrates
  Resin (See Glossary)
  Sedanoloid
  Sedanonic anhydrides
  Volatile oils (See Glossary)

Known Effects

- Relieves spasm in skeletal or smooth muscle.
- Causes uterine contractions, whether pregnant or not.
- Celery juice reduces blood pressure.
- Reduces gas in gastrointestinal tract.

Miscellaneous information:

- No effects are expected on the body, either good or bad, when herb is used in very small amounts to enhance the flavor of food.
- When eaten as a common food, no problems are expected for anyone.
- Workers in celery fields may develop skin rashes.

Unproved Speculated Benefits

- Seeds act as anti-oxidant.
- Acts as a sedative.
- Treats dysmenorrhea (menstrual cramps).
- Treats arthritis.
- Roots act as aphrodisiac.

Warnings and Precautions

Don't take if you:
- Are in your third trimester of a pregnancy.

Consult your doctor if you:
- Take this herb for any medical problem that doesn't improve in 2 weeks. There may be safer, more-effective treatments.
- Take any medicinal drugs or herbs including aspirin, laxatives, cold and cough remedies, antacids, vitamins, minerals, amino acids, supplements, other prescription or non-prescription drugs.

Pregnancy:
- Pregnant women should experience no problems taking usual amounts as part of a balanced diet. Don't drink large quantities of celery juice.

Breast-feeding:
- Breast-fed infants of lactating mothers should experience no problems when mother takes usual amounts as part of a balanced diet. Other products extracted from this herb have not been proved to cause problems.

Infants and children:
- Treating infants and children under 2 with any herbal preparation is hazardous.

Others:
- None expected if you are beyond childhood and under 45, not pregnant, basically healthy and take for only a short time.

Storage:
- Keep cool and dry, but don't freeze. Store safely away from children.

Safe dosage:
- At present no "safe" dosage has been established.

Toxicity

Rated relatively safe when taken in appropriate quantities for short periods of time.

For symptoms of toxicity: See *Adverse Reactions, Side Effects or Overdose Symptoms* section below.

Adverse Reactions, Side Effects or Overdose Symptoms

Signs and symptoms:	What to do:
Deep sedation with large amounts	Seek emergency treatment.
Premature labor	Seek emergency treatment.

Centuary (Minor Centuary)

Basic Information

Biological name (genus and species):
Centaurium erythraea, C. umbellatum
Parts used for medicinal purposes:
Petals/flower
Chemicals this herb contains:
Amarogentin
Erytaurin
Erythrocentaurin
Gentiopicrin
Gentisin

 Known Effects

• Increases stomach secretions.

 Unproved Speculated Benefits

• Treats malaria.
• Reduces fever.

 Warnings and Precautions

Don't take if you:
• Are pregnant, think you may be pregnant or plan pregnancy in the near future.
• Have any chronic disease of the gastrointestinal tract, such as stomach or duodenal ulcers, esophageal reflux (reflux esophagitis), ulcerative colitis, spastic colitis, diverticulosis, diverticulitis.

Consult your doctor if you:
• Take this herb for any medical problem that doesn't improve in 2 weeks. There may be safer, more-effective treatments.
• Take any medicinal drugs or herbs including aspirin, laxatives, cold and cough remedies, antacids, vitamins, minerals, amino acids, supplements, other prescription or non-prescription drugs.

Pregnancy:
• Problems in pregnant women taking small or usual amounts have not been proved. But the chance of problems does exist. Don't use unless prescribed by your doctor.

Breast-feeding:
• Problems in breast-fed infants of lactating mothers taking small or usual amounts have not been proved. But the chance of problems does exist. Don't use unless prescribed by your doctor.

Infants and children:
• Treating infants and children under 2 with any herbal preparation is hazardous.

Others:
• None expected if you are beyond childhood and under 45, basically healthy and take for only a short time.

Storage:
• Keep cool and dry, but don't freeze. Store safely away from children.

Safe dosage:
• At present no "safe" dosage has been established.

 Toxicity

Comparative-toxicity rating not available from standard references.

For symptoms of toxicity: See *Adverse Reactions, Side Effects or Overdose Symptoms* section below.

 Adverse Reactions, Side Effects or Overdose Symptoms

Signs and symptoms:	What to do:
Only with very large amounts or accidental overdose:	
Nausea	Discontinue. Call doctor immediately.
Vomiting	Discontinue. Call doctor immediately.

Basic Information

Biological name (genus and species):
   Anthemis flores, A. nobilis
Parts used for medicinal purposes:
   Various parts of the entire plant, frequently differing by country and/or culture
Chemicals this herb contains:

Antheme	Resin (See Glossary)
Anthemic acid	Tannic acid
Anthesterol	Tiglic acid
Apigenin	Volatile oils (See Glossary)
Chamazulene	

 ## Known Effects

- Used as an aromatic. (See Glossary.)
- Irritates mucous membranes.
- Decreases spasm of smooth or skeletal muscle.
- Reduces inflammation.
- Interferes with absorption of iron and other minerals when taken internally.
- Kills bacteria on skin.

Miscellaneous information:
- Flowers are used to make extract and herbal tea.

 ## Unproved Speculated Benefits

Internal use:
- Treats minor infections.
- Treats diarrhea.
- Treats indigestion.
- Relieves cramps.
- Decreases intestinal gas.

External use:
- Is used as a poultice. (See Glossary.) It is occasionally used as a way to apply medication to skin abscesses.

 ## Warnings and Precautions

Don't take if you:
- Are pregnant, think you may be pregnant or plan pregnancy in the near future.
- Have any chronic disease of the gastrointestinal tract, such as stomach or duodenal ulcers, esophageal reflux (reflux esophagitis), ulcerative colitis, spastic colitis, diverticulosis, diverticulitis.

Consult your doctor if you:
- Take this herb for any medical problem that doesn't improve in 2 weeks. There may be safer, more-effective treatments.

- Take any medicinal drugs or herbs including aspirin, laxatives, cold and cough remedies, antacids, vitamins, minerals, amino acids, supplements, other prescription or non-prescription drugs.

Pregnancy:
- Dangers outweigh any benefits. Don't use.

Breast-feeding:
- Dangers outweigh any benefits. Don't use.

Infants and children:
- Treating infants and children under 2 with any herbal preparation is hazardous.

Others:
- Dangers outweigh any possible benefits. Don't use.

Storage:
- Keep cool and dry, but don't freeze. Store safely away from children.

Safe dosage:
- At present no "safe" dosage has been established.

 ## Toxicity

Comparative-toxicity rating not available from standard references.

For symptoms of toxicity: See below.

 ## Adverse Reactions, Side Effects or Overdose Symptoms

Signs and symptoms:	What to do:
Allergic reactions in individuals who are sensitized to ragweed pollens (rare)	Discontinue. Call doctor immediately.
Life-threatening anaphylaxis may follow injections— symptoms include: immediate, severe itching, paleness, low blood pressure, loss of consciousness, coma	Yell for help. Don't leave victim. Begin CPR (cardiopulmonary resuscitation), mouth to mouth breathing and external cardiac massage. Have someone dial "O" (operator) or 911 (emergency). Don't stop CPR until help arrives.
Skin irritation	Discontinue. Call doctor when convenient.
Vomiting	Discontinue. Call doctor immediately.

MEDICINAL HERB

Chickweed

Basic Information

Biological name (genus and species):
    Stellaria media
Parts used for medicinal purposes:
    Various parts of the entire plant, frequently
    differing by country and/or culture
Chemicals this herb contains:
    Ascorbic acid (vitamin C)
    Potash salts
    Rutin

Known Effects

- Reduces thickness of mucus in lungs.
- Increases urine production.

Miscellaneous information:
- Chickweed has been proved ineffective for medicinal purposes.

Unproved Speculated Benefits

Internal use:
- Treats asthma
- Protects scraped tissues.
- Treats gastrointestinal disorders.
- Is used as a vitamin-C supplement.
- Relieves constipation

External use:
- Is used as an ointment for rashes and sores.

Warnings and Precautions

Don't take if you:
- Are pregnant, think you may be pregnant or plan pregnancy in the near future.

Consult your doctor if you:
- Take this herb for any medical problem that doesn't improve in 2 weeks. There may be safer, more-effective treatments.
- Take any medicinal drugs or herbs including aspirin, laxatives, cold and cough remedies, antacids, vitamins, minerals, amino acids, supplements, other prescription or non-prescription drugs.

Pregnancy:
- Problems in pregnant women taking small or usual amounts have not been proved. But the chance of problems does exist. Don't use unless prescribed by your doctor.

Breast-feeding:
- Problems in breast-fed infants of lactating mothers taking small or usual amounts have not been proved. But the chance of problems does exist. Don't use unless prescribed by your doctor.

Infants and children:
- Treating infants and children under 2 with any herbal preparation is hazardous.

Others:
- None expected if you are beyond childhood and under 45, basically healthy and take for only a short time.

Storage:
- Keep cool and dry, but don't freeze. Store safely away from children.

Safe dosage:
- At present no "safe" dosage has been established.

Toxicity

Rated relatively safe when taken in appropriate quantities for short periods of time.

For symptoms of toxicity: See *Adverse Reactions, Side Effects or Overdose Symptoms* section below.

Adverse Reactions, Side Effects or Overdose Symptoms

Signs and symptoms:	What to do:
Temporary paralysis (large amounts only)	Seek emergency treatment.

Basic Information
Biological name (genus and species):
 Cichorium intybus
Parts used for medicinal purposes:
 Roots
Chemicals this herb contains:
 Ascorbic acid (vitamin C)
 Inulin
 Vitamin A

Known Effects

- Reduces kidney inflammation.
- Helps body dispose of excess fluid by increasing amount of urine produced.

Unproved Speculated Benefits

- Treats dyspepsia.

Warnings and Precautions

Don't take if you:
- Are pregnant, think you may be pregnant or plan pregnancy in the near future.

Consult your doctor if you:
- Take this herb for any medical problem that doesn't improve in 2 weeks. There may be safer, more-effective treatments.
- Take any medicinal drugs or herbs including aspirin, laxatives, cold and cough remedies, antacids, vitamins, minerals, amino acids, supplements, other prescription or non-prescription drugs.

Pregnancy:
- Problems in pregnant women taking small or usual amounts have not been proved. But the chance of problems does exist. Don't use unless prescribed by your doctor.

Breast-feeding:
- Problems in breast-fed infants of lactating mothers taking small or usual amounts have not been proved. But the chance of problems does exist. Don't use unless prescribed by your doctor.

Infants and children:
- Treating infants and children under 2 with any herbal preparation is hazardous.

Storage:
- Keep cool and dry, but don't freeze. Store safely away from children.

Safe dosage:
- At present no "safe" dosage has been established.

Toxicity

Comparative-toxicity rating not available from standard references.

Adverse Reactions, Side Effects or Overdose Symptoms

None expected

MEDICINAL HERB

299

Chinese Rhubarb (Canton Rhubarb, Shensi Rhubarb)

Basic Information

Biological name (genus and species):
Rheum officinalis, R. palmatum
Parts used for medicinal purposes:
Roots
Chemicals this herb contains:
Aloe-emodin
Anthraquinone
Chrysophanol
Emodin
Tannins (See Glossary)

Known Effects

- Shrinks tissues.
- Prevents secretion of fluids.
- Irritates mucous membranes of intestinal tract.
- Interferes with absorption of iron and other minerals when taken internally.

Miscellaneous information:
- This is *not* the garden variety of rhubarb.

Unproved Speculated Benefits

- Relieves diarrhea (in small amounts).

Warnings and Precautions

Don't take if you:
- Are pregnant, think you may be pregnant or plan pregnancy in the near future.
- Have any chronic disease of the gastrointestinal tract, such as stomach or duodenal ulcers, esophageal reflux (reflux esophagitis), ulcerative colitis, spastic colitis, diverticulosis, diverticulitis.

Consult your doctor if you:
- Take this herb for any medical problem that doesn't improve in 2 weeks. There may be safer, more-effective treatments.
- Take any medicinal drugs or herbs including aspirin, laxatives, cold and cough remedies, antacids, vitamins, minerals, amino acids, supplements, other prescription or non-prescription drugs.

Pregnancy:
- Avoid overeating this herb.

Breast-feeding:
- Avoid overeating this herb.

Infants and children:
- Treating infants and children under 2 with any herbal preparation is hazardous.

Others:
- None expected if you are under 45, not pregnant, basically healthy, take it for only a short time and do not exceed manufacturer's recommended dosage.

Storage:
- Keep cool and dry, but don't freeze. Store safely away from children.

Safe dosage:
- At present no "safe" dosage has been established.

Toxicity

Rated relatively safe when taken in appropriate quantities for short periods of time.

For symptoms of toxicity: See *Adverse Reactions, Side Effects or Overdose Symptoms* section below.

Adverse Reactions, Side Effects or Overdose Symptoms

Signs and symptoms:	What to do:
Cramping, abdominal pain	Discontinue. Call doctor immediately.
Explosive, watery diarrhea	Discontinue. Call doctor immediately.

Cinnamon (Camphor, Hon-Sho)

Basic Information
Biological name (genus and species):
Cinnamonum camphora
Parts used for medicinal purposes:
Leaves
Roots
Chemicals this herb contains:

Camphor oil	Limonene
Cineol	Mannitol
Cinnamaldehyde	Safrole
Fatty acids	Tannins (See Glossary)
Gum (See Glossary)	Oils

Known Effects

- Aids in expelling gas from intestinal tract.
- Shrinks tissues.
- Prevents secretion of fluids.
- Provides flavor.
- Safrole is a possible carcinogen.
- Interferes with absorption of iron and other minerals when taken internally.

Miscellaneous information:
- Sometimes cinnamon is mixed with marijuana then smoked.
- Used as placticizer to make celluloid, explosives and other chemicals.

Unproved Speculated Benefits

- None

Warnings and Precautions

Don't take if you:
- Are pregnant, think you may be pregnant or plan pregnancy in the near future.
- Have any chronic disease of the gastrointestinal tract, such as stomach or duodenal ulcers, esophageal reflux (reflux esophagitis), ulcerative colitis, spastic colitis, diverticulosis, diverticulitis.

Consult your doctor if you:
- Take this herb for any medical problem that doesn't improve in 2 weeks. There may be safer, more-effective treatments.
- Take any medicinal drugs or herbs including aspirin, laxatives, cold and cough remedies, antacids, vitamins, minerals, amino acids, supplements, other prescription or non-prescription drugs.

Pregnancy:
- Dangers outweigh any possible benefits. Don't use.

Breast-feeding:
- Dangers outweight any possible benefits. Don't use.

Infants and children:
- Treating infants and children under 2 with any herbal preparation is hazardous.

Others:
- None expected if you are beyond childhood and under 45, not pregnant, basically healthy and take for only a short time.

Storage:
- Keep cool and dry, but don't freeze. Store safely away from children.

Safe dosage:
- At present no "safe" dosage has been established.

Toxicity

Rated dangerous, particularly in children, persons over 55 and those who take larger than appropriate quantities for extended periods of time.

For symptoms of toxicity: See below.

Adverse Reactions, Side Effects or Overdose Symptoms

Signs and symptoms:	What to do:
Convulsions	Seek emergency treatment.
Dizziness	Discontinue. Call doctor immediately.
Hallucinations	Seek emergency treatment.
Large overdose (0.5ml/kg body weight) can cause coma or kidney damage	Seek emergency treatment.
Nausea	Discontinue. Call doctor immediately.
Skin contact with oil can cause redness and burning sensation	Discontinue. Call doctor when convenient.
Vomiting	Discontinue. Call doctor immediately.

Coconut

Basic Information

Biological name (genus and species):
  Cocus nucifera
Parts used for medicinal purposes:
  Oil from seeds
Chemicals this herb contains:
  Fixed oil (See Glossary)
  Tannins (See Glossary)
  Trilaurin
  Trimyristin
  Triolein
  Tripalmatic acid
  Tripalmatin
  Tristearin

Known Effects

- Shrinks tissues.
- Prevents secretion of fluids.
- Interferes with absorption of iron and other minerals when taken internally.

Miscellaneous information:
- Is used in making soaps, scalp applications, hand creams, some foodstuffs.
- Coconut-oil-based soaps are useful for marine purposes because they are not easily precipitated by saltwater or salty solutions.

Unproved Speculated Benefits

- Kills intestinal parasites.
- Relieves toothache.

Warnings and Precautions

Don't take if you:
- Have any chronic disease of the gastrointestinal tract, such as stomach or duodenal ulcers, esophageal reflux (reflux esophagitis), ulcerative colitis, spastic colitis, diverticulosis, diverticulitis.

Consult your doctor if you:
- Take this herb for any medical problem that doesn't improve in 2 weeks. There may be safer, more-effective treatments.
- Take any medicinal drugs or herbs including aspirin, laxatives, cold and cough remedies, antacids, vitamins, minerals, amino acids, supplements, other prescription or non-prescription drugs.

Pregnancy:
- Pregnant women should experience no problems taking usual amounts as part of a balanced diet. Other products extracted from this herb have not been proved to cause problems.

Breast-feeding:
- Breast-fed infants of lactating mothers should experience no problems when mother takes usual amounts as part of a balanced diet. Other products extracted from this herb have not been proved to cause problems.

Infants and children:
- Treating infants and children under 2 with any herbal preparation is hazardous.

Others:
- None expected if you are beyond childhood and under 45, basically healthy and take for only a short time.

Storage:
- Keep cool and dry, but don't freeze. Store safely away from children.

Safe dosage:
- At present no "safe" dosage has been established.

Toxicity

Comparative-toxicity rating not available from standard references.

For symptoms of toxicity: See *Adverse Reactions, Side Effects or Overdose Symptoms* section below.

Adverse Reactions, Side Effects or Overdose Symptoms

Signs and symptoms:	What to do:
Diarrhea	Discontinue. Call doctor immediately.

Cohosh, Black (Black Snakeroot, Squaw Root, Rattle Root)

Basic Information

Biological name (genus and species):
  Cimicifuga spp
Parts used for medicinal purposes:
  Rhizomes
  Roots
Chemicals this herb contains:
  Cimicifugin
  Isoferulic acid
  Oleic acid
  Palmitic acid
  Tannins (See Glossary)

Known Effects

- Irritates gastrointestinal system.
- Roots and rhizomes impair digestive function and cause an uncomfortable feeling of indigestion.
- Interferes with absorption of iron and other minerals when taken internally.

Unproved Speculated Benefits

- Treats arthritis.
- Treats diarrhea.
- Treats coughs.
- Is used as an antidote for rattlesnake poison.

Warnings and Precautions

Don't take if you:
- Are pregnant, think you may be pregnant or plan pregnancy in the near future.
- Have any chronic disease of the gastrointestinal tract, such as stomach or duodenal ulcers, esophageal reflux (reflux esophagitis), ulcerative colitis, spastic colitis, diverticulosis, diverticulitis.

Consult your doctor if you:
- Take this herb for any medical problem that doesn't improve in 2 weeks. There may be safer, more-effective treatments.
- Take any medicinal drugs or herbs including aspirin, laxatives, cold and cough remedies, antacids, vitamins, minerals, amino acids, supplements, other prescription or non-prescription drugs.

Pregnancy:
- Problems in pregnant women taking small or usual amounts have not been proved. But the chance of problems does exist. Don't use unless prescribed by your doctor.

Breast-feeding:
- Problems in breast-fed infants of lactating mothers taking small or usual amounts have not been proved. But the chance of problems does exist. Don't use unless prescribed by your doctor.

Infants and children:
- Treating infants and children under 2 with any herbal preparation is hazardous.

Others:
- Dangers outweigh any possible benefits. Don't use.

Storage:
- Keep cool and dry, but don't freeze. Store safely away from children.

Safe dosage:
- At present no "safe" dosage has been established.

Toxicity

Rated slightly dangerous, particularly in children, persons over 55 and those who take larger than appropriate quantities for extended periods of time.

For symptoms of toxicity: See *Adverse Reactions, Side Effects or Overdose Symptoms* section below.

Adverse Reactions, Side Effects or Overdose Symptoms

Signs and symptoms:	What to do:
Gastroenteritis, characterized by stomach pain, nausea, diarrhea	Discontinue. Call doctor immediately.
Nausea	Discontinue. Call doctor immediately.
Vomiting	Discontinue. Call doctor immediately.

Cohosh, Blue (Papoose Root, Squaw Root)

Basic Information

Biological name (genus and species):
  Caulophyllum thalictroides
Parts used for medicinal purposes:
  Roots
Chemicals this herb contains:
  Leontin (a saponin)
  Methylcystine
  Coulosaponin

Known Effects

- Stimulates contraction of smooth muscle (blood vessels including small muscles surrounding certain arteries and muscle fibers in the uterus).
- Raises blood pressure.

Unproved Speculated Benefits

- Treats menstrual problems.
- Stimulates uterine contractions during labor.
- Elevates blood pressure.

Warnings and Precautions

Don't take if you:
- Are pregnant, think you may be pregnant or plan pregnancy in the near future.
- Have any chronic disease of the gastrointestinal tract, such as stomach or duodenal ulcers, esophageal reflux (reflux esophagitis), ulcerative colitis, spastic colitis, diverticulosis, diverticulitis.

Consult your doctor if you:
- Take this herb for any medical problem that doesn't improve in 2 weeks. There may be safer, more-effective treatments.
- Take any medicinal drugs or herbs including aspirin, laxatives, cold and cough remedies, antacids, vitamins, minerals, amino acids, supplements, other prescription or non-prescription drugs.

Pregnancy:
- Dangers outweigh any possible benefits. Don't use.

Breast-feeding:
- Dangers outweigh any possible benefits. Don't use.

Infants and children
- Treating infants and children under 2 with any herbal preparation is hazardous.

Others:
- Don't self-medicate for *any* purpose. May cause toxic symptoms.

Storage:
- Keep cool and dry, but don't freeze. Store safely away from children.

Safe dosage:
- At present no "safe" dosage has been established.

Toxicity

Rated slightly dangerous, particularly in children, persons over 55 and those who take larger than appropriate quantities for extended periods of time.

For symptoms of toxicity: See *Adverse Reactions, Side Effects or Overdose Symptoms* section below.

Adverse Reactions, Side Effects or Overdose Symptoms

Signs and symptoms:	What to do:
Chest pain	Seek emergency treatment.
Convulsions	Seek emergency treatment.
Dilated pupils	Discontinue. Call doctor immediately.
Headache	Discontinue. Call doctor immediately.
Nausea	Discontinue. Call doctor immediately.
Stomach irritation, with possible bleeding	Discontinue. Call doctor immediately.
Thirst	Discontinue. Call doctor when convenient.
Vomiting	Discontinue. Call doctor immediately.
Weakness	Discontinue. Call doctor immediately.

Basic Information

Biological name (genus and species):
Actaea alba, A. arguta
Parts used for medicinal purposes:
Various parts of the entire plant, frequently differing by country and/or culture
Chemicals this herb contains:
Glycosides (See Glossary)
Protoanemonin
Volatile oils (See Glossary)

Known Effects

• Irritates mucous membranes.

Unproved Speculated Benefits

• Acts as mild sedative to relieve anxiety.
• Helps bring on menstruation.

Warnings and Precautions

Don't take if you:
• Are pregnant, think you may be pregnant or plan pregnancy in the near future.
• Have any chronic disease of the gastrointestinal tract, such as stomach or duodenal ulcers, esophageal reflux (reflux esophagitis), ulcerative colitis, spastic colitis, diverticulosis, diverticulitis.

Consult your doctor if you:
• Take this herb for any medical problem that doesn't improve in 2 weeks. There may be safer, more-effective treatments.
• Take any medicinal drugs or herbs including aspirin, laxatives, cold and cough remedies, antacids, vitamins, minerals, amino acids, supplements, other prescription or non-prescription drugs.

Pregnancy:
• Dangers outweigh any possible benefits. Don't use.

Breast-feeding:
• Dangers outweigh any possible benefits. Don't use.

Infants and children:
• Treating infants and children under 2 with any herbal preparation is hazardous.

Others:
• This product will *not* help you and may cause toxic symptoms.

Storage:
• Keep cool and dry, but don't freeze. Store safely away from children.

Safe dosage:
• At present no "safe" dosage has been established.

Toxicity

Comparative-toxicity rating not available from standard references.

For symptoms of toxicity: See *Adverse Reactions, Side Effects or Overdose Symptoms* section below.

Adverse Reactions, Side Effects or Overdose Symptoms

Signs and symptoms:	What to do:
Diarrhea (sometimes bloody)	Discontinue. Call doctor immediately.
Hallucinations	Seek emergency treatment.
Nausea	Discontinue. Call doctor immediately.
Skin rashes or eye irritation, if used on skin or in eye	Discontinue. Call doctor immediately.
Vomiting	Discontinue. Call doctor immediately.

Coltsfoot (Coughwort, Horse-Hoof)

Basic Information

Biological name (genus and species):
  Tussilago farfara
Parts used for medicinal purposes:
  Berries/fruits
  Leaves
Chemicals this herb contains:
  Caoutchouc
  Pectin
  Resin (See Glossary)
  Tannins (See Glossary)
  Volatile oils (See Glossary)

Known Effects

• Shrinks tissues.
• Prevents secretion of fluids.
• Interferes with absorption of iron and other minerals when taken internally.

Miscellaneous information:
• Has been found to have carcinogenic properties.

Unproved Speculated Benefits

Internal use:
• Treats persistent cough.
External use:
• Soothes various skin disorders.

Warnings and Precautions

Don't take if you:
• Are pregnant, think you may be pregnant or plan pregnancy in the near future.

Consult your doctor if you:
• Take this herb for any medical problem that doesn't improve in 2 weeks. There may be safer, more-effective treatments.
• Take any medicinal drugs or herbs including aspirin, laxatives, cold and cough remedies, antacids, vitamins, minerals, amino acids, supplements, other prescription or non-prescription drugs.

Pregnancy:
• Problems in pregnant women taking small or usual amounts have not been proved. But the chance of problems does exist. Don't use unless prescribed by your doctor.

Breast-feeding:
• Problems in breast-fed infants of lactating mothers taking small or usual amounts have not been proved. But the chance of problems does exist. Don't use unless prescribed by your doctor.

Infants and children:
• Treating infants and children under 2 with any herbal preparation is hazardous.

Others:
• None expected if you are under 45, not pregnant, basically healthy, take it for only a short time and do not exceed manufacturer's recommended dosage.

Storage:
• Keep cool and dry, but don't freeze. Store safely away from children.

Safe dosage:
• At present no "safe" dosage has been established.

Toxicity

Rated relatively safe when taken in small quantities for short periods of time. However, cumulative effects may produce malignant growths.

Adverse Reactions, Side Effects or Overdose Symptoms

None expected

Comfrey (Knitbone)

Basic Information
Biological name (genus and species):
   Symphytum officinale
Parts used for medicinal purposes:
   Leaves
   Roots
Chemicals this herb contains:
   Allantoin
   Consolidine
   Mucilage (See Glossary)
   Phosphorous
   Potassium
   Pyrrolizidine
   Starch
   Symphytocynglossine
   Tannins (See Glossary)
   Vitamins A and C

Known Effects

- Shrinks tissues.
- Prevents secretion of fluids.
- Interferes with absorption of iron and other minerals when taken internally.

Unproved Speculated Benefits

- Roots and leaves are used in poultices to heal wounds and ulcers. (See Glossary.)
- Protects scraped tissues.
- Is used as a laxative by providing bulk.
- Helps body dispose of excess fluid by increasing the amount of urine produced.

Warnings and Precautions

Don't take if you:
- Are pregnant, think you may be pregnant or plan pregnancy in the near future.
- Need to restrict potassium in your diet.

Consult your doctor if you:
- Take this herb for any medical problem that doesn't improve in 2 weeks. There may be safer, more-effective treatments.
- Take any medicinal drugs or herbs including aspirin, laxatives, cold and cough remedies, antacids, vitamins, minerals, amino acids, supplements, other prescription or non-prescription drugs.

Pregnancy:
- Problems in pregnant women taking small or usual amounts have not been proved. But the chance of problems does exist. Don't use unless prescribed by your doctor.

Breast-feeding:
- Problems in breast-fed infants of lactating mothers taking small or usual amounts have not been proved. But the chance of problems does exist. Don't use unless prescribed by your doctor.

Infants and children:
- Treating infants and children under 2 with any herbal preparation is hazardous.

Others:
- None expected if you are beyond childhood and under 45, basically healthy and take for only a short time.

Storage:
- Keep cool and dry, but don't freeze. Store safely away from children.

Safe dosage:
- At present no "safe" dosage has been established.

Toxicity

Rated relatively safe when taken in appropriate quantities for short periods of time.

For symptoms of toxicity: See *Adverse Reactions, Side Effects or Overdose Symptoms* section below.

Adverse Reactions, Side Effects or Overdose Symptoms

Signs and symptoms:	What to do:
Coma	Seek emergency treatment.
Drowsiness	Discontinue. Call doctor immediately.
Lethargy	Discontinue. Call doctor immediately.

Cottonwood (Balm of Gilead)

Basic Information
Biological name (genus and species):
   Populus deltoides, P. candicans, P. spp
Parts used for medicinal purposes:
   Roots
Chemicals this herb contains:
   Salacin

Known Effects

• Acts as an anti-inflammatory.
• Reduces pain.
• Reduces fever.

Miscellaneous information:
• Used extensively by North-American Indians for many disorders.

Unproved Speculated Benefits

• Relieves toothache.
• Treats arthritis.
• Treats heart diseases.
• Treats any illness accompanied by fever, pain or inflammation.

Warnings and Precautions

Don't take if you:
• Are pregnant, think you may be pregnant or plan pregnancy in the near future.

Consult your doctor if you:
• Take this herb for any medical problem that doesn't improve in 2 weeks. There may be safer, more-effective treatments.
• Take any medicinal drugs or herbs including aspirin, laxatives, cold and cough remedies, antacids, vitamins, minerals, amino acids, supplements, other prescription or non-prescription drugs.

Pregnancy:
• Problems in pregnant women taking small or usual amounts have not been proved. But the chance of problems does exist. Don't use unless prescribed by your doctor.

Breast-feeding:
• Problems in breast-fed infants of lactating mothers taking small or usual amounts have not been proved. But the chance of problems does exist. Don't use unless prescribed by your doctor.

Infants and children:
• Treating infants and children under 2 with any herbal preparation is hazardous.

Others:
• None expected if you are beyond childhood and under 45, basically healthy and take for only a short time.

Storage:
• Keep cool and dry, but don't freeze. Store safely away from children.

Safe dosage:
• At present no "safe" dosage has been established.

Toxicity

Comparative-toxicity rating not available from standard references.

For symptoms of toxicity: See *Adverse Reactions, Side Effects or Overdose Symptoms* section below.

Adverse Reactions, Side Effects or Overdose Symptoms

Signs and symptoms:	What to do:
Coma	Seek emergency treatment.
Confusion	Discontinue. Call doctor immediately.
Convulsions	Seek emergency treatment.

Couchgrass (Dog Grass, Triticum)

Basic Information

Biological name (genus and species):
Agropyrum repens
Parts used for medicinal purposes:
Roots
Chemicals this herb contains:

Dextrose (simple sugar) Levulose (simple sugar)
Gum (See Glossary) Mannite
Inosite Silica
Lactic acid Vannilin

Known Effects

- Helps body dispose of excess fluid by increasing amount of urine produced.
- If contaminated with ergot, causes constriction of blood vessels and muscular spasm of uterus.

Miscellaneous information:
- Frequently contaminated with a poisonous fungus containing ergot. Discard *any* grass that has a black coating.

Unproved Speculated Benefits

- Protects scraped tissues.
- Is used as a nutrient.
- Treats bladder infections.
- Treats arthritis.

Warnings and Precautions

Don't take if you:
- Are pregnant, think you may be pregnant or plan pregnancy in the near future.
- Have liver disease.
- Have any chronic disease of the gastrointestinal tract, such as stomach or duodenal ulcers, esophageal reflux (reflux esophagitis), ulcerative colitis, spastic colitis, diverticulosis, diverticulitis.

Consult your doctor if you:
- Take this herb for any medical problem that doesn't improve in 2 weeks. There may be safer, more-effective treatments.
- Take any medicinal drugs or herbs including aspirin, laxatives, cold and cough remedies, antacids, vitamins, minerals, amino acids, supplements, other prescription or non-prescription drugs.

Pregnancy:
- Dangers outweigh any possible benefits. Don't use.

Breast-feeding:
- Dangers outweigh any possible benefits. Don't use.

Infants and children:
- Treating infants and children under 2 with any herbal preparation is hazardous.

Others:
- None expected if you are beyond childhood and under 45, basically healthy take it for only a short time and do not exceed manufacturer's recommended dosage.

Storage:
- Keep cool and dry, but don't freeze. Store safely away from children.

Safe dosage:
- At present no "safe" dosage has been established.

Toxicity

Comparative-toxicity rating not available from standard references.

For symptoms of toxicity: See *Adverse Reactions, Side Effects or Overdose Symptoms* section below.

Adverse Reactions, Side Effects or Overdose Symptoms

Signs and symptoms:	What to do:
Only if contaminated with ergot:	
Coma	Seek emergency treatment.
Diarrhea	Discontinue. Call doctor when convenient.
Rapid, weak pulse	Seek emergency treatment.
Tingling, itching	Discontinue. Call doctor when convenient.
Unquenchable thirst	Discontinue. Call doctor immediately.
Vomiting	Discontinue. Call doctor immediately.

Cow Parsnip (Hogweed, Keck)

Basic Information

Biological name (genus and species):
Heracleum lanatum
Parts used for medicinal purposes:
Fruit
Leaves
Roots
Seeds
Chemicals this herb contains:
Volatile oils (See Glossary)

Known Effects

- Decreases thickness and increases fluidity of mucus from lungs and bronchial tubes.
- Depresses central nervous system.
- Decreases spasm of smooth muscle or skeletal muscle.

Miscellaneous information:
- Young plants may look like hemlock, which is poisonous.

Unproved Speculated Benefits

- Fruits and leaves are used as a sedative.

Warnings and Precautions

Don't take if you:
- Are pregnant, think you may be pregnant or plan pregnancy in the near future.

Consult your doctor if you:
- Take this herb for any medical problem that doesn't improve in 2 weeks. There may be safer, more-effective treatments.
- Take any medicinal drugs or herbs including aspirin, laxatives, cold and cough remedies, antacids, vitamins, minerals, amino acids, supplements, other prescription or non-prescription drugs.

Pregnancy:
- Dangers outweigh any possible benefits. Don't use.

Breast-feeding:
- Dangers outweigh any possible benefits. Don't use.

Infants and children:
- Treating infants and children under 2 with any herbal preparation is hazardous.

Others:
- None expected if you are beyond childhood and under 45, basically healthy and take for only a short time.

Storage:
- Keep cool and dry, but don't freeze. Store safely away from children.

Safe dosage:
- At present no "safe" dosage has been established.

Toxicity

Comparative-toxicity rating not available from standard references.

Adverse Reactions, Side Effects or Overdose Symptoms

None reported

Basic Information

Biological name (genus and species):
Geranium maculatum
Parts used for medicinal purposes:
Roots
Leaves
Chemicals this herb contains:
Coloring materials
Gallic acid   Starch
Gum (See Glossary)   Sugar
Pectin   Tannins (See Glossary)

Known Effects

- Produces puckering.
- Shrinks tissues.
- Prevents secretion of fluids.
- May increase blood clotting.
- Interferes with absorption of iron and other minerals when taken internally.

Miscellaneous information:
- Is used as a mouthwash.
- Is used as a gargle for sore throat.
- Used in traps to kill Japanese beetles which are attracted to it. They die when they eat cranesbill leaves.

Unproved Speculated Benefits

- Acts as an astringent.
- Decreases nosebleeds.
- Treats bleeding from stomach, mouth, intestines.
- Treats diarrhea.
- Is used as a poultice. (See Glossary.) Occasionally used as a means of applying medications.

Warnings and Precautions

Don't take if you:
- Are pregnant, think you may be pregnant or plan pregnancy in the near future.
- Have any chronic disease of the gastrointestinal tract, such as stomach or duodenal ulcers, esophageal reflux (reflux esophagitis), ulcerative colitis, spastic colitis, diverticulosis, diverticulitis.

Consult your doctor if you:
- Take this herb for any medical problem that doesn't improve in 2 weeks. There may be safer, more-effective treatments.

- Take any medicinal drugs or herbs including aspirin, laxatives, cold and cough remedies, antacids, vitamins, minerals, amino acids, supplements, other prescription or non-prescription drugs.

Pregnancy:
- Problems in pregnant women taking small or usual amounts have not been proved. But the chance of problems does exist. Don't use unless prescribed by your doctor.

Breast-feeding:
- Problems in breast-fed infants of lactating mothers taking small or usual amounts have not been proved. But the chance of problems does exist. Don't use unless prescribed by your doctor.

Infants and children:
- Treating infants and children under 2 with any herbal preparation is hazardous.

Others:
- None expected if you are beyond childhood and under 45, basically healthy and take for only a short time.

Storage:
- Keep cool and dry, but don't freeze. Store safely away from children.

Safe dosage:
- At present no "safe" dosage has been established.

Toxicity

Comparative-toxicity rating not available from standard references.

For symptoms of toxicity: See below.

Adverse Reactions, Side Effects or Overdose Symptoms

Signs and symptoms:	What to do:
Diarrhea	Discontinue. Call doctor immediately.
Kidney damage characterized by blood in urine, decreased urine flow, swelling of hands and feet	Seek emergency treatment.
Nausea	Discontinue. Call doctor immediately.
Vomiting	Discontinue. Call doctor immediately.

Cubeb (Tailed Pepper, Java Pepper)

Basic Information

Biological name (genus and species):
Piper cubeba
Parts used for medicinal purposes:
Berries/fruits
Chemicals this herb contains:
Cubebic acid
Cubebin
Fixed oil (See Glossary)
Gum (See Glossary)
Resin (See Glossary)
Sesquiterpene alcohol (cubeb camphor)
Terpenes
Volatile oils (See Glossary)

Known Effects

• Cubebic acid irritates the ureter, bladder and urethra.

Miscellaneous information:
• Active chemicals are in fully grown, *unripe* fruit.

Unproved Speculated Benefits

• Helps body dispose of excess fluid by increasing amount of urine produced.
• Is used as urinary antiseptic.
• Decreases thickness and increases fluidity of mucus from lungs and bronchial tubes.
• Aids in expelling gas from intestinal tract.

Warnings and Precautions

Don't take if you:
• Are pregnant, think you may be pregnant or plan pregnancy in the near future.
• Have any chronic disease of the gastrointestinal tract, such as stomach or duodenal ulcers, esophageal reflux (reflux esophagitis), ulcerative colitis, spastic colitis, diverticulosis, diverticulitis.

Consult your doctor if you:
• Take this herb for any medical problem that doesn't improve in 2 weeks. There may be safer, more-effective treatments.
• Have chronic intestinal disease. Cubeb may make it worse.

Pregnancy:
• Problems in pregnant women taking small or usual amounts have not been proved. But the chance of problems does exist. Don't use unless prescribed by your doctor.

Breast-feeding:
• Problems in breast-fed infants of lactating mothers taking small or usual amounts have not been proved. But the chance of problems does exist. Don't use unless prescribed by your doctor.

Infants and children:
• Treating infants and children under 2 with any herbal preparation is hazardous.

Others:
• None expected if you are under 45, not pregnant, basically healthy, take it for only a short time and do not exceed manufacturer's recommded dosage.

Storage:
• Keep cool and dry, but don't freeze. Store safely away from children.

Safe dosage:
• At present no "safe" dosage has been established.

Toxicity

Comparative-toxicity rating not available from standard references.

For symptoms of toxicity: See *Adverse Reactions, Side Effects or Overdose Symptoms* section below.

Adverse Reactions, Side Effects or Overdose Symptoms

Signs and symptoms:	What to do:
Nausea	Discontinue. Call doctor immediately.
Vomiting	Discontinue. Call doctor immediately.

Basic Information

Biological name (genus and species):
  Turnera diffusa
Parts used for medicinal purposes:
  Leaves
Chemicals this herb contains:
  Arbutin
  Chlorophyll
  Damianian
  Resin (See Glossary)
  Starch
  Sugar
  Tannins (See Glossary)
  Volatile oils (See Glossary)

Known Effects

- Stimulates muscular contractions of intestinal tract.
- Interferes with absorption of iron and other minerals when taken internally.

Miscellaneous Information:
- Tastes very bitter.

Unproved Speculated Benefits

- Acts as a purgative.
- Is used as an aphrodisiac.
- Is used as a headache remedy.
- Decreases or cures bedwetting.

Warnings and Precautions

Don't take if you:
- Are pregnant, think you may be pregnant or plan pregnancy in the near future.
- Have any chronic disease of the gastrointestinal tract, such as stomach or duodenal ulcers, esophageal reflux (reflux esophagitis), ulcerative colitis, spastic colitis, diverticulosis, diverticulitis.
- Have kidney or urinary-tract disease.

Consult your doctor if you:
- Take this herb for any medical problem that doesn't improve in 2 weeks. There may be safer, more-effective treatments.
- Take any medicinal drugs or herbs including aspirin, laxatives, cold and cough remedies, antacids, vitamins, minerals, amino acids, supplements, other prescription or non-prescription drugs.

Pregnancy:
- Dangers outweigh any possible benefits. Don't use.

Breast-feeding:
- Dangers outweigh any possible benefits. Don't use.

Infants and children:
- Treating infants and children under 2 with any herbal preparation is hazardous.

Others:
- None expected if you are under 45, not pregnant, basically healthy, take it for only a short time and do not exceed manufacturer's recommended dosage.

Storage:
- Keep cool and dry, but don't freeze. Store safely away from children.

Safe dosage:
- At present no "safe" dosage has been established.

Toxicity

Rated relatively safe when taken in appropriate quantities for short periods of time.

For symptoms of toxicity: See *Adverse Reactions, Side Effects or Overdose Symptoms* section below.

Adverse Reactions, Side Effects or Overdose Symptoms

Signs and symptoms:	What to do:
No documented cases reported. Theoretically:	
Diarrhea	Discontinue. Call doctor immediately.
Nausea	Discontinue. Call doctor immediately.
Urinary frequency	Discontinue. Call doctor when convenient.
Vomiting	Discontinue. Call doctor immediately.

Dandelion

Basic Information

Biological name (genus and species):
  Taraxacum officinale
Parts used for medicinal purposes:
  Leaves
  Roots
  Young tops
Chemicals this herb contains:
  Bitters (See Glossary)
  Fats
  Gluten
  Gum (See Glossary)
  Inulin
  Iron
  Niacin
  Potash
  Proteins
  Resin (See Glossary)
  Teraxacerin
  Vitamins A and C

Known Effects

- Helps body dispose of excess fluid by increasing amount of urine produced.
- Stimulates stomach secretions.

Miscellaneous information:
- Is a source of vitamins A and C.

Unproved Speculated Benefits

- Treats dyspepsia.
- Treats constipation.

Warnings and Precautions

Don't take if you:
- Are pregnant, think you may be pregnant or plan pregnancy in the near future.

Consult your doctor if you:
- Take this herb for any medical problem that doesn't improve in 2 weeks. There may be safer, more-effective treatments.
- Take any medicinal drugs or herbs including aspirin, laxatives, cold and cough remedies, antacids, vitamins, minerals, amino acids, supplements, other prescription or non-prescription drugs.

Pregnancy:
- Problems in pregnant women taking small or usual amounts have not been proved. But the chance of problems does exist. Don't use unless prescribed by your doctor.

Breast-feeding:
- Problems in breast-fed infants of lactating mothers taking small or usual amounts have not been proved. But the chance of problems does exist. Don't use unless prescribed by your doctor.

Infants and children:
- Treating infants and children under 2 with any herbal preparation is hazardous.

Others:
- None expected if you are beyond childhood and under 45, basically healthy and take for only a short time.

Storage:
- Keep cool and dry, but don't freeze. Store safely away from children.

Safe dosage:
- At present no "safe" dosage has been established.

Toxicity

Generally regarded as safe when taken in appropriate quantities for short periods of time.

Adverse Reactions, Side Effects or Overdose Symptoms

None expected

Basic Information

Biological name (genus and species):
Sambucus canadensis

Parts used for medicinal purposes:
Bark
Berries/fruits
Inner bark
Leaves

Chemicals this herb contains:

Albumin	Tannic acid
Cyanide	Tyrosin
Itydrocyanic aid	Viburnic acid
Resin (See Glossary)	Vitamin C
Rutin	Volatile oils (See Glossary)
Sambucine	Wax (See Glossary)

Sambunigrin—found in stem; breaks down
to cyanide

Known Effects

- Bark, berries and leaves irritate the gastrointestinal tract and act as a laxative and purgative.
- Causes vomiting (sometimes).
- Helps body dispose of excess fluid by increasing amount of urine produced.
- Increases perspiration.

Miscellaneous information:
- Stems contain cyanide and can be *extremely* toxic.

Unproved Speculated Benefits

- Treats headache.
- Treats arthritis.
- Treats gout.
- Treats the common cold.
- Treats fevers.
- Treats sore throat.
- Treats abdominal pain.
- Aids discomfort of menstrual cramps.
- Poultices promote healing of bruises and sprains. (See Glossary.)

Warnings and Precautions

Don't take if you:
- Are pregnant, think you may be pregnant or plan pregnancy in the near future.
- Have any chronic disease of the gastrointestinal tract, such as stomach or duodenal ulcers, esophageal reflux (reflux esophagitis), ulcerative colitis, spastic colitis, diverticulosis, diverticulitis.

Consult your doctor if you:
- Take this herb for any medical problem that doesn't improve in 2 weeks. There may be safer, more-effective treatments.
- Take any medicinal drugs or herbs including aspirin, laxatives, cold and cough remedies, antacids, vitamins, minerals, amino acids, supplements, other prescription or non-prescription drugs.

Pregnancy:
- Dangers outweigh any possible benefits. Don't use.

Breast-feeding:
- Dangers outweigh any possible benefits. Don't use.

Infants and children:
- Treating infants and children under 2 with any herbal preparation is hazardous.

Others:
- Ripe berries are probably non-toxic.
- Beware of stems. Enough cyanide from them could cause death.

Storage:
- Keep cool and dry, but don't freeze. Store safely away from children.

Safe dosage:
- At present no "safe" dosage has been established.

Toxicity

Rated slightly dangerous, particularly in children, persons over 55 and those who take larger than appropriate quantities for extended periods of time.

For symptoms of toxicity: See *Adverse Reactions, Side Effects or Overdose Symptoms* section below.

Adverse Reactions, Side Effects or Overdose Symptoms

Signs and symptoms:	What to do:
Abdominal pain	Discontinue. Call doctor immediately.
Diarrhea	Discontinue. Call doctor immediately.
Nausea	Discontinue. Call doctor immediately.
Vomiting	Discontinue. Call doctor immediately.

MEDICINAL HERB

Eyebright

Basic Information

Biological name (genus and species):
  Euphrasia officinalis
Parts used for medicinal purposes:
  Entire plant, except roots
Chemicals this herb contains:
  Bitters (See Glossary)
  Tannins (See Glossary)
  Volatile oils (See Glossary)

Known Effects

- Shrinks tissues.
- Prevents secretion of fluids.
- Interferes with absorption of iron and other minerals when taken internally.

Miscellaneous information:
- No proved benefits.

Unproved Speculated Benefits

- Is used as an eyewash to relieve discomfort caused from eyestrain or minor irritation.
- Is used internally for many alleged benefits.

Warnings and Precautions

Don't take if you:
- Are pregnant, think you may be pregnant or plan pregnancy in the near future.

Consult your doctor if you:
- Take this herb for any medical problem that doesn't improve in 2 weeks. There may be safer, more-effective treatments.
- Take any medicinal drugs or herbs including aspirin, laxatives, cold and cough remedies, antacids, vitamins, minerals, amino acids, supplements, other prescription or non-prescription drugs.

Pregnancy:
- Problems in pregnant women taking small or usual amounts have not been proved. But the chance of problems does exist. Don't use unless prescribed by your doctor.

Breast-feeding:
- Problems in breast-fed infants of lactating mothers taking small or usual amounts have not been proved. But the chance of problems does exist. Don't use unless prescribed by your doctor.

Infants and children:
- Treating infants and children under 2 with any herbal preparation is hazardous.

Others:
- None expected if you are under 45, not pregnant, basically healthy, take it for only a short time and do not exceed manufacturer's recommended dosage.

Storage:
- Keep cool and dry, but don't freeze. Store safely away from children.

Safe dosage:
- At present no ''safe'' dosage has been established.

Toxicity

Rated relatively safe when taken in appropriate quantities for short periods of time.

Adverse Reactions, Side Effects or Overdose Symptoms

None expected

Fennel (Finocchio)

Basic Information

Biological name (genus and species):
  Foeniculum vulgare
Parts used for medicinal purposes:
  Berries/fruits
  Roots
  Stems
Chemicals this herb contains:
  Anethole
  Fixed oil (See Glossary)
  Volatile oils (See Glossary)

Known Effects

• Aids in expelling gas from intestinal tract.
• Stimulates respiration.
• Increases stomach acidity.

Miscellaneous information:
• Provides flavor.

Unproved Speculated Benefits

• Treats dyspepsia.
• Is used for common colds.
• Is used for coughs.

Warnings and Precautions

Don't take if you:
• Are pregnant, think you may be pregnant or plan pregnancy in the near future.
• Have any chronic disease of the gastrointestinal tract, such as stomach or duodenal ulcers, esophageal reflux (reflux esophagitis), ulcerative colitis, spastic colitis, diverticulosis, diverticulitis.

Consult your doctor if you:
• Take this herb for any medical problem that doesn't improve in 2 weeks. There may be safer, more-effective treatments.
• Take any medicinal drugs or herbs including aspirin, laxatives, cold and cough remedies, antacids, vitamins, minerals, amino acids, supplements, other prescription or non-prescription drugs.

Pregnancy:
• Dangers outweigh any possible benefits. Don't use.

Breast-feeding:
• Dangers outweigh any possible benefits. Don't use.

Infants and children:
• Treating infants and children under 2 with any herbal preparation is hazardous.

Others:
• If you stay away from the oil extract, none expected if you are beyond childhood and under 45, basically healthy and take for only a short time.

Storage:
• Keep cool and dry, but don't freeze. Store safely away from children.

Safe dosage:
• At present no "safe" dosage has been established.

Toxicity

Generally regarded as safe when taken in appropriate quantities for short periods of time.

For symptoms of toxicity: See *Adverse Reactions, Side Effects or Overdose Symptoms* section below.

Adverse Reactions, Side Effects or Overdose Symptoms

Signs and symptoms:	What to do:
Oil extracted from fennel may cause:	
Congestive heart failure	Seek emergency treatment.
Nausea	Discontinue. Call doctor immediately.
Seizures	Seek emergency treatment.
Vomiting	Discontinue. Call doctor immediately.

Fenugreek

Basic Information

Biological name (genus and species):
　Trigonella foenum-graecum
Parts used for medicinal purposes:
　Seeds
Chemicals this herb contains:
　Choline
　Fixed oil (See Glossary)
　Iron
　Lecithin
　Mucilage (See Glossary)
　Phosphates (See Glossary)
　Protein
　Trigonelline
　Trimethylamine
　Volatile oils (See Glossary)

 Known Effects

• Increases stomach acidity.

Miscellaneous information:
• Fenugreek has a disagreeable odor and bitter taste.
• Prescribed frequently by veterinarians, particularly for horses.

 Unproved Speculated Benefits

• Seeds act as a bulk laxative.
• Protects scraped tissues.

 Warnings and Precautions

Don't take if you:
• Are pregnant, think you may be pregnant or plan pregnancy in the near future.

Consult your doctor if you:
• Take this herb for any medical problem that doesn't improve in 2 weeks. There may be safer, more-effective treatments.
• Take any medicinal drugs or herbs including aspirin, laxatives, cold and cough remedies, antacids, vitamins, minerals, amino acids, supplements, other prescription or non-prescription drugs.

Pregnancy:
• Problems in pregnant women taking small or usual amounts have not been proved. But the chance of problems does exist. Don't use unless prescribed by your doctor.

Breast-feeding:
• Problems in breast-fed infants of lactating mothers taking small or usual amounts have not been proved. But the chance of problems does exist. Don't use unless prescribed by your doctor.

Infants and children:
• Treating infants and children under 2 with any herbal preparation is hazardous.

Storage:
• Keep cool and dry, but don't freeze. Store safely away from children.

Safe dosage:
• At present no "safe" dosage has been established.

 Toxicity

Rated relatively safe when taken in appropriate quantities for short periods of time.

 Adverse Reactions, Side Effects or Overdose Symptoms

None expected

Feverfew (Bachelor's Buttons, Altamisa)

Basic Information

Biological name (genus and species):
Chrysanthemum parthenium
Parts used for medicinal purposes:
Bark
Dried flowers
Leaves
Chemicals this herb contains:
Parthenolide
Pyrethrins
Santamarin

Known Effects

- Kills insects.
- Decreases thickness and increases fluidity of mucus from lungs and bronchial tubes.
- Stimulates uterine contractions.

Unproved Speculated Benefits

Leaves:
- Treat menstrual disorders.
- Treat common cold.
- Treat indigestion and diarrhea.
- Stimulate appetite.
- Decrease edema.
- Make childbirth easier.

Dried flowers:
- Treat intestinal parasites (worms).
- Start menstrual flow.
- Cause abortion.
- Stimulate appetite.
- Aid in expelling gas from intestinal tract.

Warnings and Precautions

Don't take if you:
- Are allergic to pyrethrins.
- Are pregnant, think you may be pregnant or plan pregnancy in the near future.

Consult your doctor if you:
- Take this herb for any medical problem that doesn't improve in 2 weeks. There may be safer, more-effective treatments.
- Take any medicinal drugs or herbs including aspirin, laxatives, cold and cough remedies, antacids, vitamins, minerals, amino acids, supplements, other prescription or non-prescription drugs.

Pregnancy:
- Dangers outweigh any possible benefits. Don't use.

Breast-feeding:
- Dangers outweigh any possible benefits. Don't use.

Infants and children:
- Treating infants and children under 2 with any herbal preparation is hazardous.

Others:
- Feverfew has no proven usefulness. It may be dangerous.

Storage:
- Keep cool and dry, but don't freeze. Store safely away from children.

Safe dosage:
- At present no "safe" dosage has been established.

Toxicity

Generally regarded as safe when taken in very small quantities for short periods of time.

For symptoms of toxicity: See *Adverse Reactions, Side Effects or Overdose Symptoms* section below.

Adverse Reactions, Side Effects or Overdose Symptoms

Signs and symptoms:	What to do:
Life-threatening anaphylaxis may follow injections—symptoms include immediate severe itching, paleness, low blood pressure, loss of consciousness, coma	Yell for help. Don't leave victim. Begin CPR (cardiopulmonary resuscitation), mouth to mouth breathing and external cardiac massage. Have someone dial "O" (operator) or 911 (emergency). Don't stop CPR until help arrives.

Flaxseed (Linseed)

Basic Information

Biological name (genus and species):
  Linum usitatissimum
Parts used for medicinal purposes:
  Seeds
Chemicals this herb contains:
  Gum (See Glossary)
  Fixed oil (See Glossary)
  Linamarin
  Mucilage (See Glossary)
  Protein
  Tannins (See Glossary)
  Wax (See Glossary)

Known Effects

- Protects scraped tissues.
- Forms bulk in intestinal tract.
- Interferes with absorption of iron and other minerals when taken internally.

Miscellaneous information:
- All parts of flax can contain toxic chemicals, but immature seeds grown in warm climates may have higher toxic concentrations.
- Linamarin may be converted to cyanide in the body.

Unproved Speculated Benefits

- Soothes coughs.
- Oil softens or smoothes skin.
- Is used as a bulk-forming laxative.
- Is used for poultices to apply to chest for colds and coughs. (See Glossary.)

Warnings and Precautions

Don't take if you:
- Are pregnant, think you may be pregnant or plan pregnancy in the near future.

Consult your doctor if you:
- Take this herb for any medical problem that doesn't improve in 2 weeks. There may be safer, more-effective treatments.
- Take any medicinal drugs or herbs including aspirin, laxatives, cold and cough remedies, antacids, vitamins, minerals, amino acids, supplements, other prescription or non-prescription drugs.

Pregnancy:
- Dangers outweigh any possible benefits. Don't use.

Breast-feeding:
- Dangers outweigh any possible benefits. Don't use.

Infants and children:
- Treating infants and children under 2 with any herbal preparation is hazardous.

Others:
- None expected if you are beyond childhood and under 45, basically healthy and take for only a short time.

Storage:
- Keep cool and dry, but don't freeze. Store safely away from children.

Safe dosage:
- At present no "safe" dosage has been established.

Toxicity

Comparative-toxicity rating not available from standard references.

For symptoms of toxicity: See *Adverse Reactions, Side Effects or Overdose Symptoms* section below.

Adverse Reactions, Side Effects or Overdose Symptoms

Signs and symptoms:	What to do:
Convulsions	Seek emergency treatment.
Fast breathing	Discontinue. Call doctor immediately.
Paralysis	Seek emergency treatment.
Unusual excitement	Discontinue. Call doctor immediately.
Weakness	Discontinue. Call doctor immediately.

Basic Information

Biological name (genus and species):
Fritillia vericillia, F. meleagris
Parts used for medicinal purposes:
Roots
Chemicals this herb contains:
Frimitime
Fritilline
Peimine
Peiminine
Verticine
Verticilline
(Peimine and peiminine may resemble steroid hormones.)

Known Effects

- Peimine and peiminine may affect the electrical system of the heart.
- Decreases blood pressure.
- Increases blood sugar.

Miscellaneous information:
- Only roots have medicinal properties.

Unproved Speculated Benefits

- Reduces fevers.
- Decreases thickness and increases fluidity of mucus from lungs and bronchial tubes.
- Increases flow of breast milk in lactating women.

Warnings and Precautions

Don't take if you:
- Are pregnant, think you may be pregnant or plan pregnancy in the near future.
- Have heart disease.

Consult your doctor if you:
- Take this herb for any medical problem that doesn't improve in 2 weeks. There may be safer, more-effective treatments.
- Take any medicinal drugs or herbs including aspirin, laxatives, cold and cough remedies, antacids, vitamins, minerals, amino acids, supplements, other prescription or non-prescription drugs.

Pregnancy:
- Dangers outweigh any possible benefits. Don't use.

Breast-feeding:
- Dangers outweigh any possible benefits. Don't use.

Infants and children:
- Treating infants and children under 2 with any herbal preparation is hazardous.

Others:
- None expected if you are beyond childhood and under 45, not pregnant, basically healthy and take for only a short time.

Storage:
- Keep cool and dry, but don't freeze. Store safely away from children.

Safe dosage:
- At present no "safe" dosage has been established.

Toxicity

Comparative-toxicity rating not available from standard references.

For symptoms of toxicity: See *Adverse Reactions, Side Effects or Overdose Symptoms* section below.

Adverse Reactions, Side Effects or Overdose Symptoms

Signs and symptoms:	What to do:
Heart block characterized by slow heart rate (below 50)	Seek emergency treatment.
Heartbeat irregularity	Discontinue. Call doctor immediately.

MEDICINAL HERB

Galanga Major & Minor (India Root, Chinese Ginger)

Basic Information

Biological name (genus and species):
Alpinia galanga, Alpinia officinarum
Parts used for medicinal purposes:
Various parts of the entire plant, frequently differing by country and/or culture
Chemicals this herb contains:
Cineloe
Galangin
Galangol
Kaempferid
Resin (See Glossary)
Volatile oils (See Glossary)

 ## Known Effects

- Anti-bacterial effect acts against bacterial germs, such as streptococci, staphylococci and coliform bacteria.

Miscellaneous information:
- Related botanically and pharmacologically to ginger.
- Known and used by ancient Greeks and Arabs.

 ## Unproved Speculated Benefits

- Aids in expelling gas from intestinal tract.
- Treats impotence.
- Reduces excess phlegm caused by allergies.
- Treats painful teeth and gums.
- Stimulates respiration.

 ## Warnings and Precautions

Don't take if you:
- Are pregnant, think you may be pregnant or plan pregnancy in the near future.
- Have any chronic disease of the gastrointestinal tract, such as stomach or duodenal ulcers, esophageal reflux (reflux esophagitis), ulcerative colitis, spastic colitis, diverticulosis, diverticulitis.

Consult your doctor if you:
- Take this herb for any medical problem that doesn't improve in 2 weeks. There may be safer, more-effective treatments.
- Take any medicinal drugs or herbs including aspirin, laxatives, cold and cough remedies, antacids, vitamins, minerals, amino acids, supplements, other prescription or non-prescription drugs.

Pregnancy:
- Problems in pregnant women taking small or usual amounts have not been proved. But the chance of problems does exist. Don't use unless prescribed by your doctor.

Breast-feeding:
- Problems in breast-fed infants of lactating mothers taking small or usual amounts have not been proved. But the chance of problems does exist. Don't use unless prescribed by your doctor.

Infants and children:
- Treating infants and children under 2 with any herbal preparation is hazardous.

Others:
- None expected if you are under 45, not pregnant, basically healthy, take it for only a short time and do not exceed manufacturer's recommended dosage.

Storage:
- Keep cool and dry, but don't freeze. Store safely away from children.

Safe dosage:
- At present no "safe" dosage has been established.

 ## Toxicity

Comparative-toxicity rating not available from standard references.

For symptoms of toxicity: See *Adverse Reactions, Side Effects or Overdose Symptoms* section below.

 ## Adverse Reactions, Side Effects or Overdose Symptoms

Signs and symptoms:	What to do:
Diarrhea	Discontinue. Call doctor immediately.
Nausea	Discontinue. Call doctor immediately.
Vomiting	Discontinue. Call doctor immediately.

Basic Information

Biological name (genus and species):
Galega officinalis
Parts used for medicinal purposes:
Various parts of the entire plant, frequently differing by country and/or culture
Chemicals this herb contains:
Bitters (See Glossary)
Galegine
Tannins (See Glossary)

Known Effects

• Reduces blood sugar.
• Interferes with absorption of iron and other minerals when taken internally.

Miscellaneous information:
• Plant smells bad when it is bruised.

Unproved Speculated Benefits

• Treats diabetes.
• Increases flow of breast milk in lactating women.

Warnings and Precautions

Don't take if you:
• Are pregnant, think you may be pregnant or plan pregnancy in the near future.

Consult your doctor if you:
• Take this herb for any medical problem that doesn't improve in 2 weeks. There may be safer, more-effective treatments.
• Take any medicinal drugs or herbs including aspirin, laxatives, cold and cough remedies, antacids, vitamins, minerals, amino acids, supplements, other prescription or non-prescription drugs.

Pregnancy:
• Problems in pregnant women taking small or usual amounts have not been proved. But the chance of problems does exist. Don't use unless prescribed by your doctor.

Breast-feeding:
• Problems in breast-fed infants of lactating mothers taking small or usual amounts have not been proved. But the chance of problems does exist. Don't use unless prescribed by your doctor.

Infants and children:
• Treating infants and children under 2 with any herbal preparation is hazardous.

Others:
• None expected if you are under 45, not pregnant, basically healthy, take it for only a short time and do not exceed manufacturer's recommended dosage.

Storage:
• Keep cool and dry, but don't freeze. Store safely away from children.

Safe dosage:
• At present no "safe" dosage has been established.

Toxicity

Comparative-toxicity rating not available from standard references.

For symptoms of toxicity: See *Adverse Reactions, Side Effects or Overdose Symptoms* section below.

Adverse Reactions, Side Effects or Overdose Symptoms

Signs and symptoms:	What to do:
Headache	Discontinue. Call doctor when convenient.
Jitteriness	Discontinue. Call doctor when convenient.
Weakness	Discontinue. Call doctor immediately.

MEDICINAL HERB

Gambier (Pale Catechu, Gambir)

Basic Information

Biological name (genus and species):
   Uncaria gambier
Parts used for medicinal purposes:
   Leaves
   Twigs
Chemicals this herb contains:
   Catechin
   Catechutannic acid
   Tannins (See Glossary)

 ## Known Effects

- Shrinks tissues.
- Prevents secretion of fluids.
- Interferes with absorption of iron and other minerals when taken internally.

 ## Unproved Speculated Benefits

- Decreases unusual bleeding.
- Treats chronic diarrhea.
- Is used as gargle for sore throats.

 ## Warnings and Precautions

Don't take if you:
- Have any chronic disease of the gastrointestinal tract, such as stomach or duodenal ulcers, esophageal reflux (reflux esophagitis), ulcerative colitis, spastic colitis, diverticulosis, diverticulitis.

Consult your doctor if you:
- Take this herb for any medical problem that doesn't improve in 2 weeks. There may be safer, more-effective treatments.
- Take any medicinal drugs or herbs including aspirin, laxatives, cold and cough remedies, antacids, vitamins, minerals, amino acids, supplements, other prescription or non-prescription drugs.

Pregnancy:
- Dangers outweigh any possible benefits. Don't use.

Breast-feeding:
- Dangers outweigh any possible benefits. Don't use.

Infants and children:
- Treating infants and children under 2 with any herbal preparation is hazardous.

Storage:
- Keep cool and dry, but don't freeze. Store safely away from children.

Safe dosage:
- At present no "safe" dosage has been established.

 ## Toxicity

Rated relatively safe when taken in appropriate quantities for short periods of time.

For symptoms of toxicity: See *Adverse Reactions, Side Effects or Overdose Symptoms* section below.

 ## Adverse Reactions, Side Effects or Overdose Symptoms

Signs and symptoms:	What to do:
Diarrhea	Discontinue. Call doctor immediately.
Kidney damage characterized by blood in urine, decreased urine flow, swelling of hands and feet	Seek emergency treatment.
Vomiting	Discontinue. Call doctor immediately.

Basic Information

Biological name (genus and species):
Allium sativum
Parts used for medicinal purposes:
Bulb
Chemicals this herb contains:
Allicin
Allyl disulfides
Phytoncides
Unsaturated aldehydes
Volatile oils (See Glossary)

Known Effects

- Kills larvae.
- Stops germs from reproducing.
- Decreases thickness and increases fluidity of mucus from lungs and bronchial tubes.
- Stimulates perspiration.
- Helps body dispose of excess fluid by increasing amount of urine produced.
- No effects are expected on the body, either good or bad, when herb is used in very small amounts to enhance the flavor of food.

Miscellaneous information:
- Is used as a condiment.
- Avoid using garlic as a medicinal herb in *any amount* with children! It is acceptable to use garlic as a flavoring in children's food.

Unproved Speculated Benefits

- Treats hypertension.
- Treats cramping abdominal pain in adults.
- Lowers high blood fats (hyperlipemia, hypercholesterolemia).
- Reddens skin by increasing blood flow to it.

Warnings and Precautions

Don't take if you:
- Have any medical problem that your doctor is treating you for. Consult him or her first.

Consult your doctor if you:
- Take this herb for any medical problem that doesn't improve in 2 weeks. There may be safer, more-effective treatments.
- Take any medicinal drugs or herbs including aspirin, laxatives, cold and cough remedies, antacids, vitamins, minerals, amino acids, supplements, other prescription or non-prescription drugs.

Pregnancy:
- Problems in pregnant women taking small or usual amounts have not been proved. Don't take unless prescribed by your doctor.

Breast-feeding:
- Problems in breast-fed infants of lactating mothers taking small or usual amounts have not been proved. Don't take unless prescribed by your doctor

Infants and children:
- Treating infants and children under 2 with any herbal preparation is hazardous.

Others:
- None expected if you are beyond childhood and under 45, basically healthy and take small amounts for only a short time.

Storage:
- Keep cool and dry, but don't freeze. Store safely away from children.

Safe dosage:
- At present no "safe" dosage has been established.

Toxicity

Comparative-toxicity rating not available from standard references.

For symptoms of toxicity: See *Adverse Reactions, Side Effects or Overdose Symptoms* section below.

Adverse Reactions, Side Effects or Overdose Symptoms

Signs and symptoms:	What to do:
Precipitous blood-pressure drop—symptoms include, faintness, cold sweat, paleness, rapid pulse	Seek emergency treatment.
Increased number of circulating white blood cells as determined by laboratory studies	Discontinue. Call doctor immediately.
Skin eruptions	Discontinue. Call doctor when convenient.

Gentian (Yellow Gentian)

Basic Information

Biological name (genus and species):
  Gentiana lutea
Parts used for medicinal purposes:
  Roots
Chemicals this herb contains:
  Gentiamarin
  Gentiin
  Gentiopicrin
  Gentisin
  Mesogentioigenin
  Protogentiogenin
  Sugar
  Xanthone pigment

Known Effects

- Irritates mucous membranes.
- Kills plasmodium, which causes malaria.

Miscellaneous information:
- Has been known and used since ancient times in Greece.

Unproved Speculated Benefits

- Increase contractions of stomach muscles.
- Stimulates gastric secretions.
- Is used as a tonic to stimulate appetite.
- Aids digestion.

Warnings and Precautions

Don't take if you:
- Are pregnant, think you may be pregnant or plan pregnancy in the near future.
- Have any chronic disease of the gastrointestinal tract, such as stomach or duodenal ulcers, esophageal reflux (reflux esophagitis), ulcerative colitis, spastic colitis, diverticulosis, diverticulitis.

Consult your doctor if you:
- Take this herb for any medical problem that doesn't improve in 2 weeks. There may be safer, more-effective treatments.
- Take any medicinal drugs or herbs including aspirin, laxatives, cold and cough remedies, antacids, vitamins, minerals, amino acids, supplements, other prescription or non-prescription drugs.

Pregnancy:
- Problems in pregnant women taking small or usual amounts have not been proved. But the chance of problems does exist. Don't use unless prescribed by your doctor.

Breast-feeding:
- Problems in breast-fed infants of lactating mothers taking small or usual amounts have not been proved. But the chance of problems does exist. Don't use unless prescribed by your doctor.

Infants and children:
- Treating infants and children under 2 with any herbal preparation is hazardous.

Others:
- None expected if you are under 45, not pregnant, basically healthy, take it for only a short time and do not exceed manufacturer's recommended dosage.

Storage:
- Keep cool and dry, but don't freeze. Store safely away from children.

Safe dosage:
- At present no "safe" dosage has been established.

Toxicity

Rated relatively safe when taken in appropriate quantities for short periods of time.

For symptoms of toxicity: See *Adverse Reactions, Side Effects or Overdose Symptoms* section below.

Adverse Reactions, Side Effects or Overdose Symptoms

Signs and symptoms:	What to do:
Nausea	Discontinue. Call doctor immediately.
Vomiting	Discontinue. Call doctor immediately.

German Chamomile (Mazanilla, Matricaria, Hungarian Chamomile)

Basic Information

Biological name (genus and species):
   Matricaria chamomilla
Parts used for medicinal purposes:
   Petals/flower
Chemicals this herb contains:
   Alphabisabolol
   Azulene
   Fatty acid
   Furfural
   Paraffin hydrocarbons
   Sesquiterpene
   Sesquiterpene alcohol
   Tannins (See Glossary)

Known Effects

- Acts as an anti-inflammatory.
- Weakens muscles.
- Interferes with absorption of iron and other minerals when taken internally.

Miscellaneous information:
- Ice cream, candy and liqueur manufacturers use small, non-toxic amounts for flavoring.

Unproved Speculated Benefits

- Relieves spasms in skeletal or smooth muscle.
- Is used as a tonic.
- Is used as a sedative.
- Aids in expelling gas from intestinal tract.

Warnings and Precautions

Don't take if you:
- Are pregnant, think you may be pregnant or plan pregnancy in the near future.
- Have any chronic disease of the gastrointestinal tract, such as stomach or duodenal ulcers, esophageal reflux (reflux esophagitis), ulcerative colitis, spastic colitis, diverticulosis, diverticulitis.

Consult your doctor if you:
- Take this herb for any medical problem that doesn't improve in 2 weeks. There may be safer, more-effective treatments.
- Take any medicinal drugs or herbs including aspirin, laxatives, cold and cough remedies, antacids, vitamins, minerals, amino acids, supplements, other prescription or non-prescription drugs.

Pregnancy:
- Problems in pregnant women taking small or usual amounts have not been proved. But the chance of problems does exist. Don't use unless prescribed by your doctor.

Breast-feeding:
- Problems in breast-fed infants of lactating mothers taking small or usual amounts have not been proved. But the chance of problems does exist. Don't use unless prescribed by your doctor.

Infants and children:
- Treating infants and children under 2 with any herbal preparation is hazardous.

Others:
- None expected if you are beyond childhood and under 45, basically healthy and take for only a short time.

Storage:
- Keep cool and dry, but don't freeze. Store safely away from children.

Safe dosage:
- At present no "safe" dosage has been established.

Toxicity

Generally regarded as safe when taken in appropriate quantities for short periods of time.

For symptoms of toxicity: See *Adverse Reactions, Side Effects or Overdose Symptoms* section below.

Adverse Reactions, Side Effects or Overdose Symptoms

Signs and symptoms:	What to do:
Diarrhea	Discontinue. Call doctor immediately.
Excess sedation	Discontinue. Call doctor immediately.
Nausea	Discontinue. Call doctor immediately.
Skin eruptions	Discontinue. Call doctor when convenient.
Vomiting	Discontinue. Call doctor immediately.

Ginger

Basic Information
Biological name (genus and species):
  Zingiber
Parts used for medicinal purposes:
  Roots
Chemicals this herb contains:
  Bisabolene
  Borneal
  Cineole
  Citral
  Sequiterpene
  Volatile oils (See Glossary)
  Zingerone
  Zingiberene

Known Effects

- Aids in expelling gas from intestinal tract.
- Provides counterirritation when applied to skin overlying an inflamed or irritated joint.
- No effects are expected on the body, either good or bad, when herb is used in very small amounts to enhance the flavor of food.

Miscellaneous information:
- Is used as a flavoring agent.

Unproved Speculated Benefits

- Treats indigestion.
- Treats abdominal discomfort.

Warnings and Precautions

Don't take if you:
- Are pregnant, think you may be pregnant or plan pregnancy in the near future.
- Have any chronic disease of the gastrointestinal tract, such as stomach or duodenal ulcers, esophageal reflux (reflux esophagitis), ulcerative colitis, spastic colitis, diverticulosis, diverticulitis.

Consult your doctor if you:
- Take this herb for any medical problem that doesn't improve in 2 weeks. There may be safer, more-effective treatments.
- Take any medicinal drugs or herbs including aspirin, laxatives, cold and cough remedies, antacids, vitamins, minerals, amino acids, supplements, other prescription or non-prescription drugs.
- Have stomach or intestinal diseases.

Pregnancy:
- Problems in pregnant women taking small or usual amounts have not been proved. But the chance of problems does exist. Don't use unless prescribed by your doctor.

Breast-feeding:
- Problems in breast-fed infants of lactating mothers taking small or usual amounts have not been proved. But the chance of problems does exist. Don't use unless prescribed by your doctor.

Infants and children:
- Treating infants and children under 2 with any herbal preparation is hazardous.

Others:
- None expected if you are under 45, not pregnant, basically healthy, take it for only a short time and do not exceed manufacturer's recommended dosage.

Storage:
- Keep cool and dry, but don't freeze. Store safely away from children.

Safe dosage:
- At present no "safe" dosage has been established.

Toxicity

Comparative-toxicity rating not available from standard references.

For symptoms of toxicity: See *Adverse Reactions, Side Effects or Overdose Symptoms* section below.

Adverse Reactions, Side Effects or Overdose Symptoms

Signs and symptoms:	What to do:
Diarrhea	Discontinue. Call doctor immediately.
Nausea	Discontinue. Call doctor immediately.
Vomiting	Discontinue. Call doctor immediately.

Ginseng (Sang)

Basic Information

Biological name (genus and species):
Panax quinquefolium
Parts used for medicinal purposes:
Roots
Chemicals this herb contains:

Arabinose	Mucilage (See Glossary)
Camphor	Panaxosides
Gineosides	Resin (See Glossary)
	Saponin (See Glossary)
	Starch

Known Effects

- Stimulates brain, heart, blood vessels.
- Decreases blood sugar.
- Increases secretion of histamine.
- Decreases eosinophils in blood.
- Increases corticosteroid content of blood.

Miscellaneous information:
- A favorite Chinese remedy used for almost everything.
- A native plant in the state of Georgia, USA.

Unproved Speculated Benefits

- Treats biological "stress."
- Is used as an aphrodisiac.
- Increases mental and physical efficiency.
- Treats impotence.
- Treats anemia.
- Treats hardening of the arteries.
- Reduces depression.
- Treats diabetes.
- Treats ulcers.
- Treats edema.

Warnings and Precautions

Don't take if you:
- Are pregnant, think you may be pregnant or plan pregnancy in the near future.
- Have any chronic disease of the gastrointestinal tract, such as stomach or duodenal ulcers, esophageal reflux (reflux esophagitis), ulcerative colitis, spastic colitis, diverticulosis, diverticulitis.

Consult your doctor if you:
- Take this herb for any medical problem that doesn't improve in 2 weeks. There may be safer, more-effective treatments.
- Take any medicinal drugs or herbs including aspirin, laxatives, cold and cough remedies, antacids, vitamins, minerals, amino acids, supplements, other prescription or non-prescription drugs.

Pregnancy:
- Problems in pregnant women taking small or usual amounts have not been proved. But the chance of problems does exist. Don't use unless prescribed by your doctor.

Breast-feeding:
- Problems in breast-fed infants of lactating mothers taking small or usual amounts have not been proved. But the chance of problems does exist. Don't use unless prescribed by your doctor.

Infants and children:
- Treating infants and children under 2 with any herbal preparation is hazardous.

Others:
- None expected if you are under 45, not pregnant, basically healthy, take it for only a short time and do not exceed manufacturer's recommended dosage.

Storage:
- Keep cool and dry, but don't freeze. Store safely away from children.

Safe dosage:
- At present no "safe" dosage has been established.

Toxicity

Generally regarded as safe when taken in appropriate quantities for short periods of time.

For symptoms of toxicity: See *Adverse Reactions, Side Effects or Overdose Symptoms* section below.

Adverse Reactions, Side Effects or Overdose Symptoms

Signs and symptoms:	What to do:
Diarrhea	Discontinue. Call doctor immediately.
Insomnia	Discontinue. Call doctor when convenient.
Nervousness	Discontinue. Call doctor when convenient.
Nausea	Discontinue. Call doctor immediately.
Vomiting	Discontinue. Call doctor immediately.

Goldenseal

Basic Information

Biological name (genus and species):
Hydrastis canadensis
Parts used for medicinal purposes:
Rhizomes
Roots
Chemicals this herb contains:

Albumin	Lignin
Berberine	Resin (See Glossary)
Candine	Starch
Fats	Sugar
Hydrastine	Volatile oils (See Glossary)

Known Effects

- Decreases uterine bleeding.
- Large amounts stimulate central nervous system.
- Large amounts given intravenously can reduce blood pressure.
- Depresses muscle tone of small blood vessels.

Miscellaneous information:
- Has a very bitter taste.

Unproved Speculated Benefits

- Treats dyspepsia.
- Increases appetite.

Warnings and Precautions

Don't take if you:
- Are pregnant, think you may be pregnant or plan pregnancy in the near future.
- Have any chronic disease of the gastrointestinal tract, such as stomach or duodenal ulcers, esophageal reflux (reflux esophagitis), ulcerative colitis, spastic colitis, diverticulosis, diverticulitis.

Consult your doctor if you:
- Take this herb for any medical problem that doesn't improve in 2 weeks. There may be safer, more-effective treatments.
- Take any medicinal drugs or herbs including aspirin, laxatives, cold and cough remedies, antacids, vitamins, minerals, amino acids, supplements, other prescription or non-prescription drugs.

Pregnancy:
- Dangers outweigh any possible benefits. Don't use.

Breast-feeding:
- Dangers outweigh any possible benefits. Don't use.

Infants and children:
- Treating infants and children under 2 with any herbal preparation is hazardous.

Others:
- None expected if you are under 45, not pregnant, basically healthy, take it for only a short time and do not exceed manufacturer's recommended dosage.

Storage:
- Keep cool and dry, but don't freeze. Store safely away from children.

Safe dosage:
- At present no "safe" dosage has been established.

Toxicity

Rated slightly dangerous, particularly in children, persons over 55 and those who take larger than appropriate quantities for extended periods of time.

For symptoms of toxicity: See *Adverse Reactions, Side Effects or Overdose Symptoms* section below.

Adverse Reactions, Side Effects or Overdose Symptoms

Signs and symptoms:	What to do:
Breathing difficulties	Seek emergency treatment.
Diarrhea	Discontinue. Call doctor immediately.
Mouth and throat irritation	Discontinue. Call doctor immediately.
Nausea	Discontinue. Call doctor immediately.
Numbness of hands and feet	Discontinue. Call doctor immediately.
Vomiting	Discontinue. Call doctor immediately.
Weakness leading to paralysis of muscles	Seek emergency treatment.

Gotu Cola (Kola, Gbanja Kola)

Basic Information

Biological name (genus and species):
Cola nitida

Parts used for medicinal purposes:
Seeds/nuts

Chemicals this herb contains:
Caffeine
Catechol
Epicatechol
Theobromine

 Known Effects

- Stimulates central nervous system.
- Helps body dispose of excess fluid by increasing amount of urine produced.
- Acts as an astringent.
- Shrinks tissues.
- Prevents secretion of fluids.
- No effects are expected on the body, either good or bad, when herb is used in very small amounts to enhance the flavor of food.

 Unproved Speculated Benefits

- Decreases fatigue.
- Increases sex drive.
- Treats high blood pressure and congestive heart failure.

 Warnings and Precautions

Don't take if you:
- Are pregnant, think you may be pregnant or plan pregnancy in the near future.
- Have any chronic disease of the gastrointestinal tract, such as stomach or duodenal ulcers, esophageal reflux (reflux esophagitis), ulcerative colitis, spastic colitis, diverticulosis, diverticulitis.

Consult your doctor if you:
- Take this herb for any medical problem that doesn't improve in 2 weeks. There may be safer, more-effective treatments.
- Take any medicinal drugs or herbs including aspirin, laxatives, cold and cough remedies, antacids, vitamins, minerals, amino acids, supplements, other prescription or non-prescription drugs.

Pregnancy:
- Dangers outweigh any possible benefits. Don't use.

Breast-feeding:
- Dangers outweigh any possible benefits. Don't use.

Infants and children:
- Treating infants and children under 2 with any herbal preparation is hazardous.

Others:
- None expected if you are under 45, not pregnant, basically healthy, take it for only a short time and do not exceed manufacturer's recommended dosage.

Storage:
- Keep cool and dry, but don't freeze. Store safely away from children.

Safe dosage:
- At present no "safe" dosage has been established.

 Toxicity

Rated relatively safe when taken in appropriate quantities for short periods of time.

For symptoms of toxicity: See *Adverse Reactions, Side Effects or Overdose Symptoms* section below.

 Adverse Reactions, Side Effects or Overdose Symptoms

Signs and symptoms:	What to do:
Aggravates peptic ulcers in stomach, duodenum or esophagus	Discontinue. Call doctor immediately.
Inability to sleep	Discontinue. Call doctor when convenient.
Nervousness	Discontinue. Call doctor when convenient.

Grape Hyacinth

Basic Information

Biological name (genus and species):
 Muscari racemonsum, M. comosum
Parts used for medicinal purposes:
 Bulb
Chemicals this herb contains:
 Cosmisic acid
 Saponin (See Glossary)

Known Effects

• Irritates gastrointestinal tract.

Unproved Speculated Benefits

• Treats constipation.
• Stimulates central nervous system.
• Helps body dispose of excess fluid by increasing amount of urine produced.

Warnings and Precautions

Don't take if you:
• Are pregnant, think you may be pregnant or plan pregnancy in the near future.
• Have any chronic disease of the gastrointestinal tract, such as stomach or duodenal ulcers, esophageal reflux (reflux esophagitis), ulcerative colitis, spastic colitis, diverticulosis, diverticulitis.

Consult your doctor if you:
• Take this herb for any medical problem that doesn't improve in 2 weeks. There may be safer, more-effective treatments.
• Take any medicinal drugs or herbs including aspirin, laxatives, cold and cough remedies, antacids, vitamins, minerals, amino acids, supplements, other prescription or non-prescription drugs.

Pregnancy:
• Dangers outweigh any possible benefits. Don't use.

Breast-feeding:
• Dangers outweigh any possible benefits. Don't use.

Infants and children:
• Treating infants and children under 2 with any herbal preparation is hazardous.

Others:
• No evidence of any useful therapeutic effect. Don't use.

Storage:
• Keep cool and dry, but don't freeze. Store safely away from children.

Safe dosage:
• At present no "safe" dosage has been established.

Toxicity

Comparative-toxicity rating not available from standard references.

For symptoms of toxicity: See *Adverse Reactions, Side Effects or Overdose Symptoms* section below.

Adverse Reactions, Side Effects or Overdose Symptoms

Signs and symptoms:	What to do:
Diarrhea	Discontinue. Call doctor immediately.
Nausea	Discontinue. Call doctor immediately.
Vomiting	Discontinue. Call doctor immediately.

Grindelia (Gumweed, Rosinweed)

Basic Information

Biological name (genus and species):
Grindelia camporum, G. humilus, G. squarrosa
Parts used for medicinal purposes:
Leaves
Chemicals this herb contains:
Balsamic resin
Grindelol
Robustic acid
Saponins (See Glossary)
Tannins (See Glossary)
Volatile oils (See Glossary)

Known Effects

- Depresses central nervous system in high amounts.
- Dilates pupils of eyes.
- Decreases heart rate.
- Increases blood pressure.
- Interferes with absorption of iron and other minerals when taken internally.

Unproved Speculated Benefits

- Decreases thickness and increases fluidity of mucus from lungs and bronchial tubes.
- Acts as a sedative.
- Treats asthma.
- Treats bronchitis.
- Soothes and heals burns when applied topically.
- Treats vaginitis.
- Is used in poultices as a means of applying medications (See Glossary.)

Warnings and Precautions

Don't take if you:
- Are pregnant, think you may be pregnant or plan pregnancy in the near future.

Consult your doctor if you:
- Take this herb for any medical problem that doesn't improve in 2 weeks. There may be safer, more-effective treatments.
- Take any medicinal drugs or herbs including aspirin, laxatives, cold and cough remedies, antacids, vitamins, minerals, amino acids, supplements, other prescription or non-prescription drugs.

Pregnancy:
- Dangers outweigh any possible benefits. Don't use.

Breast-feeding:
- Dangers outweigh any possible benefits. Don't use.

Infants and children:
- Treating infants and children under 2 with any herbal preparation is hazardous.

Storage:
- Keep cool and dry, but don't freeze. Store safely away from children.

Safe dosage:
- At present no "safe" dosage has been established.

Toxicity

Rated slightly dangerous, particularly in children, persons over 55 and those who take larger than appropriate quantities for extended periods of time.

For symptoms of toxicity: See *Adverse Reactions, Side Effects or Overdose Symptoms* section below.

Adverse Reactions, Side Effects or Overdose Symptoms

Signs and symptoms:	What to do:
Kidney damage characterized by blood in urine, decreased urine flow, swelling of hands and feet	Seek emergency treatment.

Guaiac

Basic Information

Biological name (genus and species):
  Guaiacum officinale, G. sanctum
Parts used for medicinal purposes:
  Stems
Chemicals this herb contains:
  Guaiaconic acid
  Guaiaretic acid
  Resin (See Glossary)
  Saponin (See Glossary)
  Vanillin

Known Effects

• Irritates gastrointestinal tract.
• Increases perspiration.
• Tests for oxidizing enzymes to detect blood in stool or urine.

Miscellaneous information:

• When added to a stool specimen, hydrogen peroxide and guaiac establish the presence or absence of blood. This test is a useful screening procedure to detect malignant and non-malignant disorders of the intestinal tract.

Unproved Speculated Benefits

• Treats arthritis.
• Treats scrofula.
• Treats constipation.
• Reduces edema.

Warnings and Precautions

Don't take if you:

• Are pregnant, think you may be pregnant or plan pregnancy in the near future.
• Have any chronic disease of the gastrointestinal tract, such as stomach or duodenal ulcers, esophageal reflux (reflux esophagitis), ulcerative colitis, spastic colitis, diverticulosis, diverticulitis.

Consult your doctor if you:

• Take this herb for any medical problem that doesn't improve in 2 weeks. There may be safer, more-effective treatments.
• Take any medicinal drugs or herbs including aspirin, laxatives, cold and cough remedies, antacids, vitamins, minerals, amino acids, supplements, other prescription or non-prescription drugs.

Pregnancy:

• Dangers outweigh any possible benefits. Don't use.

Breast-feeding:

• Dangers outweigh any possible benefits. Don't use.

Infants and children:

• Treating infants and children under 2 with any herbal preparation is hazardous.

Others:

• None expected if you are under 45, not pregnant, basically healthy, take it for only a short time and do not exceed manufacturer's recommended dosage.

Storage:

• Keep cool and dry, but don't freeze. Store safely away from children.

Safe dosage:

• At present no "safe" dosage has been established.

Toxicity

Comparative-toxicity rating not available from standard references.

For symptoms of toxicity: See *Adverse Reactions, Side Effects or Overdose Symptoms* section below.

Adverse Reactions, Side Effects or Overdose Symptoms

Signs and symptoms:	What to do:
Nausea	Discontinue. Call doctor immediately.
Vomiting	Discontinue. Call doctor immediately.

Harmel (Wild Rue, African Rue, Syrian Rue)

Basic Information

Biological name (genus and species):
  Peganum harmala
Parts used for medicinal purposes:
  Various parts of the entire plant, frequently
  differing by country and/or culture
Chemicals this herb contains:
  Harmaline
  Harmalol
  Harmine
  Peganine

Known Effects

- Causes hallucinations.
- Destroys bacteria (germs) or suppresses their growth or reproduction.

Miscellaneous information:
- Wild rue is often abused where it grows in Arizona, New Mexico and Texas.

Unproved Speculated Benefits

- Destroys intestinal worms.
- Decreases pain.

Warnings and Precautions

Don't take if you:
- Are pregnant, think you may be pregnant or plan pregnancy in the near future.

Consult your doctor if you:
- Take this herb for any medical problem that doesn't improve in 2 weeks. There may be safer, more-effective treatments.
- Take any medicinal drugs or herbs including aspirin, laxatives, cold and cough remedies, antacids, vitamins, minerals, amino acids, supplements, other prescription or non-prescription drugs.

Pregnancy:
- Dangers outweigh any possible benefits. Don't use.

Breast-feeding:
- Dangers outweigh any possible benefits. Don't use.

Infants and children:
- Treating infants and children under 2 with any herbal preparation is hazardous.

Others:
- Dangers outweigh any possible benefits. Don't use.

Storage:
- Keep cool and dry, but don't freeze. Store safely away from children.

Safe dosage:
- At present no "safe" dosage has been established.

Toxicity

Rated slightly dangerous, particularly in children, persons over 55 and those who take larger than appropriate quantities for extended periods of time.

For symptoms of toxicity: See *Adverse Reactions, Side Effects or Overdose Symptoms* section below.

Adverse Reactions, Side Effects or Overdose Symptoms

Signs and symptoms:	What to do:
Hallucinations	Seek emergency treatment.
Muscle weakness	Discontinue. Call doctor immediately.

Hawthorn

Basic Information

Biological name (genus and species):
Crataegus oxyacantha
Parts used for medicinal purposes:
Berries/fruits
Leaves
Chemicals this herb contains:
Anthocyanin-type pigments
Cratagolic acid
Flavinonoid
Glycosides (See Glossary)
Purines
Saponins (See Glossary)

Known Effects

- Depresses respiration.
- Depresses heart rate.
- Causes irregular heartbeats.
- Causes congestive heart failure.
- Relaxes smooth muscle of uterus and intestines.
- Constricts bronchial tubes.

Unproved Speculated Benefits

- Treats high blood pressure.
- May have value in other cardiovascular disorders, but studies need to be completed before doses, efficacy and safety can be established. Do *not* self-medicate!

Warnings and Precautions

Don't take if you:
- Are pregnant, think you may be pregnant or plan pregnancy in the near future.
- Have heart disease.

Consult your doctor if you:
- Take this herb for any medical problem that doesn't improve in 2 weeks. There may be safer, more-effective treatments.
- Take any medicinal drugs or herbs including aspirin, laxatives, cold and cough remedies, antacids, vitamins, minerals, amino acids, supplements, other prescription or non-prescription drugs.

Pregnancy:
- Dangers outweigh any possible benefits. Don't use.

Breast-feeding:
- Dangers outweigh any possible benefits. Don't use.

Infants and children:
- Treating infants and children under 2 with any herbal preparation is hazardous.

Others:
- Any possible beneficial uses have been supplanted by other chemicals that are safer, more-effective and easier to control.

Storage:
- Keep cool and dry, but don't freeze. Store safely away from children.

Safe dosage:
- At present no "safe" dosage has been established.

Toxicity

Comparative-toxicity rating not available from standard references.

For symptoms of toxicity: See *Adverse Reactions, Side Effects or Overdose Symptoms* section below.

Adverse Reactions, Side Effects or Overdose Symptoms

Signs and symptoms:	What to do:
Breathing difficulties	Seek emergency treatment.
Heartbeat irregularities	Seek emergency treatment.

Basic Information

Biological name (genus and species):
   Heliotropium europaem
Parts used for medicinal purposes:
   Juice
   Leaves
   Seeds
Chemicals this herb contains:
   Heliotrine
   Lassiocarpine

Known Effects

• Kills liver cells.
• Stimulates production of bile.

Miscellaneous information:
• Heliotrope is a common weed.

Unproved Speculated Benefits

• Leaves and juice are used to treat ulcers, warts polyps and tumors.

Warnings and Precautions

Don't take if you:
• Are pregnant, think you may be pregnant or plan pregnancy in the near future.

Consult your doctor if you:
• Take this herb for any medical problem that doesn't improve in 2 weeks. There may be safer, more-effective treatments.
• Take any medicinal drugs or herbs including aspirin, laxatives, cold and cough remedies, antacids, vitamins, minerals, amino acids, supplements, other prescription or non-prescription drugs.

Pregnancy:
• Dangers outweigh any possible benefits. Don't use.

Breast-feeding:
• Dangers outweigh any possible benefits. Don't use.

Infants and children:
• Treating infants and children under 2 with any herbal preparation is hazardous.

Storage:
• Keep cool and dry, but don't freeze. Store safely away from children.

Safe dosage:
• At present no "safe" dosage has been established.

Toxicity

Rated slightly dangerous, particularly in children, persons over 55 and those who take larger than appropriate quantities for extended periods of time.

For symptoms of toxicity: See *Adverse Reactions, Side Effects or Overdose Symptoms* section below.

Adverse Reactions, Side Effects or Overdose Symptoms

Signs and symptoms:	What to do:
Jaundice (yellow eyes and skin)	Discontinue. Call doctor immediately.

MEDICINAL HERB

Hellebore (American Hellebore, Green Hellebore, Liliaceae)

Basic Information

Biological name (genus and species):
Veratrum viride
Parts used for medicinal purposes:
Rhizome
Root
Chemicals this herb contains:
Germidine
Germitrine
Jervine
Pseudojervine
Rubijervine
Veratrum alkaloids

Known Effects

- Decreases blood pressure.
- Decreases heart rate.
- Depresses central nervous system.

Unproved Speculated Benefits

- Treats hypertension.
- Treats toxemia of pregnancy.
- Irritates gastrointestinal system.

Warnings and Precautions

Don't take if you:
- Are pregnant, think you may be pregnant or plan pregnancy in the near future.
- Have any chronic disease of the gastrointestinal tract, such as stomach or duodenal ulcers, esophageal reflux (reflux esophagitis), ulcerative colitis, spastic colitis, diverticulosis, diverticulitis.

Consult your doctor if you:
- Take this herb for any medical problem that doesn't improve in 2 weeks. There may be safer, more-effective treatments.
- Take any medicinal drugs or herbs including aspirin, laxatives, cold and cough remedies, antacids, vitamins, minerals, amino acids, supplements, other prescription or non-prescription drugs.

Pregnancy:
- Dangers outweigh any possible benefits. Don't use.

Breast-feeding:
- Dangers outweigh any possible benefits. Don't use.

Infants and children:
- Treating infants and children under 2 with any herbal preparation is hazardous.

Others:
- *All* parts of the plant may be toxic.

Storage:
- Keep cool and dry, but don't freeze. Store safely away from children.

Safe dosage:
- At present no "safe" dosage has been established.

Toxicity

Rated dangerous, particularly in children, persons over 55 and those who take larger than appropriate quantities for extended periods of time.

For symptoms of toxicity: See *Adverse Reactions, Side Effects or Overdose Symptoms* section below.

Adverse Reactions, Side Effects or Overdose Symptoms

Signs and symptoms:	What to do:
Abdominal pain	Discontinue. Call doctor immediately.
Burning sensation in mouth	Discontinue. Call doctor when convenient.
Diarrhea	Discontinue. Call doctor immediately.
Headache	Discontinue. Call doctor when convenient.
Nausea	Discontinue. Call doctor immediately.
Precipitous blood-pressure drop—symptoms include, faintness, cold sweat, paleness, rapid pulse	Seek emergency treatment.
Vomiting	Discontinue. Call doctor immediately.

Helonias (False Unicorn Root, Fairy Wand)

Basic Information

Biological name (genus and species):
  Chamaelirium luteum
Parts used for medicinal purposes:
  Roots
Chemicals this herb contains:
  Chamaelirin
  Saponin (See Glossary)

Known Effects

- Irritates gastrointestinal system.
- Helps body dispose of excess fluid by increasing amount of urine produced.
- Produces puckering.

Unproved Speculated Benefits

- Prevents miscarriage.
- Treats menopause symptoms.
- Increases appetite.
- Acts as a vigorous laxative.

Warnings and Precautions

Don't take if you:
- Are pregnant, think you may be pregnant or plan pregnancy in the near future.
- Have any chronic disease of the gastrointestinal tract, such as stomach or duodenal ulcers, esophageal reflux (reflux esophagitis), ulcerative colitis, spastic colitis, diverticulosis, diverticulitis.

Consult your doctor if you:
- Take this herb for any medical problem that doesn't improve in 2 weeks. There may be safer, more-effective treatments.
- Take any medicinal drugs or herbs including aspirin, laxatives, cold and cough remedies, antacids, vitamins, minerals, amino acids, supplements, other prescription or non-prescription drugs.

Pregnancy:
- Dangers outweigh any possible benefits. Don't use.

Breast-feeding:
- Dangers outweigh any possible benefits. Don't use.

Infants and children:
- Treating infants and children under 2 with any herbal preparation is hazardous.

Others:
- None expected if you are beyond childhood and under 45, basically healthy and take for only a short time.

Storage:
- Keep cool and dry, but don't freeze. Store safely away from children.

Safe dosage:
- At present no "safe" dosage has been established.

Toxicity

Comparative-toxicity rating not available from standard references.

For symptoms of toxicity: See *Adverse Reactions, Side Effects or Overdose Symptoms* section below.

Adverse Reactions, Side Effects or Overdose Symptoms

Signs and symptoms:	What to do:
Diarrhea	Discontinue. Call doctor immediately.
Nausea	Discontinue. Call doctor immediately.

MEDICINAL HERB

Henbane (Hyoscyamus)

Basic Information

Biological name (genus and species):
  Hyoscyamus niger
Parts used for medicinal purposes:
  Berries/fruits
  Leaves
  Roots
Chemicals this herb contains:
  Hyoscyamine
  Scopolamine

Known Effects

- Blocks effects of parasympathetic nervous system causing increased heart rate, dilated pupils, dry mouth, hallucinations, urinary retention, reduced contractions of gastrointestinal tract.

Miscellaneous information:
- Henbane is poisonous, especially to children!

Unproved Speculated Benefits

- Treats whooping cough.
- Treats asthma.
- Is used as a mouthwash.
- Acts as a sedative.
- Is used as a pain killer.

Warnings and Precautions

Don't take if you:
- Are pregnant, think you may be pregnant or plan pregnancy in the near future.

Consult your doctor if you:
- Take this herb for any medical problem that doesn't improve in 2 weeks. There may be safer, more-effective treatments.
- Take any medicinal drugs or herbs including aspirin, laxatives, cold and cough remedies, antacids, vitamins, minerals, amino acids, supplements, other prescription or non-prescription drugs.

Pregnancy:
- Dangers outweigh any possible benefits. Don't use.

Breast-feeding:
- Dangers outweigh any possible benefits. Don't use.

Infants and children:
- Treating infants and children under 2 with any herbal preparation is hazardous.

Others:
- Must be obtained from a reliable source. Unpredictable concentrations can be dangerous.

Storage:
- Keep cool and dry, but don't freeze. Store safely away from children.

Safe dosage:
- At present no "safe" dosage has been established.

Toxicity

Rated dangerous, particularly in children, persons over 55 and those who take larger than appropriate quantities for extended periods of time.

For symptoms of toxicity: See *Adverse Reactions, Side Effects or Overdose Symptoms* section below.

Adverse Reactions, Side Effects or Overdose Symptoms

Signs and symptoms:	What to do:
Delirium	Seek emergency treatment.
Hallucinations	Seek emergency treatment.
Rapid heartbeat	Seek emergency treatment.

Basic Information

Biological name (genus and species):
Humulus lupulus
Parts used for medicinal purposes:
Berries/fruits
Chemicals this herb contains:
Humulene
Lupulinic acid
Lupulon

Known Effects

- Inhibits growth and development of germs.
- Depresses central nervous system.

Miscellaneous information:
- If fruit is not fresh, it smells bad.
- Hops are used extensively in brewing industry.
- Produces odors because it evaporates at room temperature.

Unproved Speculated Benefits

- Is used as a tonic.
- Treats dyspepsia.
- Helps body dispose of excess fluid by increasing amount of urine produced.
- Treats insomnia.
- Causes hallucinations.

Warnings and Precautions

Don't take if you:
- Are pregnant, think you may be pregnant or plan pregnancy in the near future.

Consult your doctor if you:
- Take this herb for any medical problem that doesn't improve in 2 weeks. There may be safer, more-effective treatments.
- Take any medicinal drugs or herbs including aspirin, laxatives, cold and cough remedies, antacids, vitamins, minerals, amino acids, supplements, other prescription or non-prescription drugs.

Pregnancy:
- Problems in pregnant women taking small or usual amounts have not been proved. But the chance of problems does exist. Don't use unless prescribed by your doctor.

Breast-feeding:
- Problems in breast-fed infants of lactating mothers taking small or usual amounts have not been proved. But the chance of problems does exist. Don't use unless prescribed by your doctor.

Infants and children:
- Treating infants and children under 2 with any herbal preparation is hazardous.

Others:
- None expected if you are beyond childhood and under 45, basically healthy and take for only a short time.

Storage:
- Keep cool and dry, but don't freeze. Store safely away from children.

Safe dosage:
- At present no "safe" dosage has been established.

Toxicity

Rated relatively safe when taken in appropriate quantities for short periods of time.

Adverse Reactions, Side Effects or Overdose Symptoms

None expected

Horehound

Basic Information

Biological name (genus and species):
  Marrubium vulgare
Parts used for medicinal purposes:
  Flowers
  Leaves
Chemicals this herb contains:
  Marrubiin
  Resin (See Glossary)
  Tannins (See Glossary)
  Volatile oils (See Glossary)

Known Effects

- Aids in expelling gas from intestinal tract.
- Decreases thickness and increases fluidity of mucus from lungs and bronchial tubes.
- Increases stomach secretions.
- Interferes with absorption of iron and other minerals when taken internally.

Miscellaneous information:
- Leaves and flowers are used to make tincture.

Unproved Speculated Benefits

Is used as a cough and cold remedy.
- Relieves various symptoms.
- Increases perspiration.

Warnings and Precautions

Don't take if you:
- Are pregnant, think you may be pregnant or plan pregnancy in the near future.
- Have any chronic disease of the gastrointestinal tract, such as stomach or duodenal ulcers, esophageal reflux (reflux esophagitis), ulcerative colitis, spastic colitis, diverticulosis, diverticulitis.

Consult your doctor if you:
- Take this herb for any medical problem that doesn't improve in 2 weeks. There may be safer, more-effective treatments.
- Take any medicinal drugs or herbs including aspirin, laxatives, cold and cough remedies, antacids, vitamins, minerals, amino acids, supplements, other prescription or non-prescription drugs.

Pregnancy:
- Problems in pregnant women taking small or usual amounts have not been proved. But the chance of problems does exist. Don't use unless prescribed by your doctor.

Breast-feeding:
- Problems in breast-fed infants of lactating mothers taking small or usual amounts have not been proved. But the chance of problems does exist. Don't use unless prescribed by your doctor.

Infants and children:
- Treating infants and children under 2 with any herbal preparation is hazardous.

Others:
- None expected if you are beyond childhood and under 45, basically healthy and take for only a short time.

Storage:
- Keep cool and dry, but don't freeze. Store safely away from children.

Safe dosage:
- At present no "safe" dosage has been established.

Toxicity

Comparative-toxicity rating not available from standard references.

For symptoms of toxicity: See *Adverse Reactions, Side Effects or Overdose Symptoms* section below.

Adverse Reactions, Side Effects or Overdose Symptoms

Signs and symptoms:	What to do:
Diarrhea	Discontinue. Call doctor immediately.
Nausea	Discontinue. Call doctor immediately.
Vomiting	Discontinue. Call doctor immediately.

Basic Information

Biological name (genus and species):
Aesculus hippocastanum
Parts used for medicinal purposes:
Bark
Leaves
Seeds/nuts
Chemicals this herb contains:
Aesculin
Argyroscin
Capsuloescinic acid
Escin

Known Effects

- Increases bleeding time (a laboratory test for blood clotting).
- Irritates mucous membrane.

Miscellaneous information:
- There are more reliable, safer anti-coagulants approved by FDA.
- Eating even a few nuts can cause toxic symptoms.

Unproved Speculated Benefits

- Is used as anti-coagulant.
- 4% solution is used as sunscreen.

Warnings and Precautions

Don't take if you:
- Are pregnant, think you may be pregnant or plan pregnancy in the near future.
- Have any chronic disease of the gastrointestinal tract, such as stomach or duodenal ulcers, esophageal reflux (reflux esophagitis), ulcerative colitis, spastic colitis, diverticulosis, diverticulitis.

Consult your doctor if you:
- Take this herb for any medical problem that doesn't improve in 2 weeks. There may be safer, more-effective treatments.
- Take any medicinal drugs or herbs including aspirin, laxatives, cold and cough remedies, antacids, vitamins, minerals, amino acids, supplements, other prescription or non-prescription drugs.

Pregnancy:
- Dangers outweigh any possible benefits. Don't use.

Breast-feeding:
- Dangers outweigh any possible benefits. Don't use.

Infants and children:
- Treating infants and children under 2 with any herbal preparation is hazardous.

Storage:
- Keep cool and dry, but don't freeze. Store safely away from children.

Safe dosage:
- At present no "safe" dosage has been established.

Toxicity

Rated slightly dangerous, particularly in children, persons over 55 and those who take larger than appropriate quantities for extended periods of time.

For symptoms of toxicity: See *Adverse Reactions, Side Effects or Overdose Symptoms* section below.

Adverse Reactions, Side Effects or Overdose Symptoms

Signs and symptoms:	What to do:
Lack of coordination	Discontinue. Call doctor immediately.
Nausea	Discontinue. Call doctor immediately.
Unusual bleeding	Discontinue. Call doctor immediately.
Vomiting	Discontinue. Call doctor immediately.

MEDICINAL HERB

Horsemint

Basic Information
Biological name (genus and species):
  Monarda punctata
Parts used for medicinal purposes:
  Leaves
  Stems
Chemicals this herb contains:
  Carvacrol
  Cyemene
  d-limonene
  Hydrothymoquinone
  Linalool
  Monarda oil
  Thymol

Known Effects

- Irritates tissues and mucous membranes.
- Kills germs when used on the skin for external infections.

Unproved Speculated Benefits

Internal use:
- Kills intestinal parasites.
- Aids in expelling gas from intestinal tract.
- Treats abdominal cramps.
- Treats nausea.

External use:
- Kills fungus infections on skin.
- Kills bacterial infections on skin.

Warnings and Precautions

Don't take if you:
- Are pregnant, think you may be pregnant or plan pregnancy in the near future.
- Have any chronic disease of the gastrointestinal tract, such as stomach or duodenal ulcers, esophageal reflux (reflux esophagitis), ulcerative colitis, spastic colitis, diverticulosis, diverticulitis.

Consult your doctor if you:
- Take this herb for any medical problem that doesn't improve in 2 weeks. There may be safer, more-effective treatments.
- Take any medicinal drugs or herbs including aspirin, laxatives, cold and cough remedies, antacids, vitamins, minerals, amino acids, supplements, other prescription or non-prescription drugs.

Pregnancy:
- Problems in pregnant women taking small or usual amounts have not been proved. But the chance of problems does exist. Don't use unless prescribed by your doctor.

Breast-feeding:
- Problems in breast-fed infants of lactating mothers taking small or usual amounts have not been proved. But the chance of problems does exist. Don't use unless prescribed by your doctor.

Infants and children:
- Treating infants and children under 2 with any herbal preparation is hazardous.

Others:
- None expected if you are under 45, not pregnant, basically healthy, take it for only a short time and do not exceed manufacturer's recommended dosage.

Storage:
- Keep cool and dry, but don't freeze. Store safely away from children.

Safe dosage:
- At present no "safe" dosage has been established.

Toxicity

Comparative-toxicity rating not available from standard references.

For symptoms of toxicity: See *Adverse Reactions, Side Effects or Overdose Symptoms* section below.

Adverse Reactions, Side Effects or Overdose Symptoms

Signs and symptoms:	What to do:
Diarrhea	Discontinue. Call doctor immediately.
Nausea	Discontinue. Call doctor immediately.
Skin rash when used on skin	Discontinue. Call doctor when convenient.
Vomiting	Discontinue. Call doctor immediately.

Basic Information

Biological name (genus and species):
Armoraciae radix, Cochlearia armoracia
Parts used for medicinal purposes:
Roots
Chemicals this herb contains:
Allyl isothiocyanate
Sinigriu

 ## Known Effects

External:
• Irritates skin.
• Blisters skin.
Internal:
• Irritates gastrointestinal tract.

Miscellaneous information:
• Eating large amounts of raw root can be toxic.
• Horseradish is used to add flavor to foods.
• No effects are expected on the body, either good or bad, when herb is used in very small amounts to enhance the flavor of food.

 ## Unproved Speculated Benefits

• Stimulates appetite.

 ## Warnings and Precautions

Don't take if you:
• Are pregnant, think you may be pregnant or plan pregnancy in the near future.
• Have any chronic disease of the gastrointestinal tract, such as stomach or duodenal ulcers, esophageal reflux (reflux esophagitis), ulcerative colitis, spastic colitis, diverticulosis, diverticulitis.

Consult your doctor if you:
• Take this herb for any medical problem that doesn't improve in 2 weeks. There may be safer, more-effective treatments.
• Take any medicinal drugs or herbs including aspirin, laxatives, cold and cough remedies, antacids, vitamins, minerals, amino acids, supplements, other prescription or non-prescription drugs.

Pregnancy:
• Problems in pregnant women taking small or usual amounts have not been proved. But the chance of problems does exist. Don't use unless prescribed by your doctor.

Breast-feeding:
• Problems in breast-fed infants of lactating mothers taking small or usual amounts have not been proved. But the chance of problems does exist. Don't use unless prescribed by your doctor.

Infants and children:
• Treating infants and children under 2 with any herbal preparation is hazardous.

Others:
• None expected if you are beyond childhood and under 45, basically healthy and take for only a short time.
• Eating large amounts of raw root can be toxic.

Storage:
• Keep cool and dry, but don't freeze. Store safely away from children.

Safe dosage:
• At present no "safe" dosage has been established.

 ## Toxicity

Comparative-toxicity rating not available from standard references.

For symptoms of toxicity: See *Adverse Reactions, Side Effects or Overdose Symptoms* section below.

 ## Adverse Reactions, Side Effects or Overdose Symptoms

Signs and symptoms:	What to do:
Diarrhea, with blood	Discontinue. Call doctor immediately.
Nausea	Discontinue. Call doctor immediately.
Vomiting	Discontinue. Call doctor immediately.
Vomiting, with blood	Seek emergency treatment.

Horsetails (Shave Grass, Bottle Brush, Field Horsetail)

Basic Information

Biological name (genus and species):
  Equisetum arvense
Parts used for medicinal purposes:
  Stems
Chemicals this herb contains:
  Aconitic acid
  Equisitine
  Fatty acids
  Nicotine
  Silica
  Starch

Known Effects

- Shrinks tissues.
- Prevents secretion of fluids.

Unproved Speculated Benefits

- Treats diarrhea.
- Treats dyspepsia.
- Helps body dispose of excess fluid by increasing amount of urine produced.
- Helps heal sores on skin.

Warnings and Precautions

Don't take if you:
- Are pregnant, think you may be pregnant or plan pregnancy in the near future.
- Have heart disease.

Consult your doctor if you:
- Take this herb for any medical problem that doesn't improve in 2 weeks. There may be safer, more-effective treatments.
- Take any medicinal drugs or herbs including aspirin, laxatives, cold and cough remedies, antacids, vitamins, minerals, amino acids, supplements, other prescription or non-prescription drugs.

Pregnancy:
- Dangers outweigh any possible benefits. Don't use.

Breast-feeding:
- Dangers outweigh any possible benefits. Don't use.

Infants and children:
- Treating infants and children under 2 with any herbal preparation is hazardous.

Others:
- None expected if you are beyond childhood and under 45, basically healthy and take for only a short time.

Storage:
- Keep cool and dry, but don't freeze. Store safely away from children.

Safe dosage:
- At present no "safe" dosage has been established.

Toxicity

Rated slightly dangerous, particularly in children, persons over 55 and those who take larger than appropriate quantities for extended periods of time.

For symptoms of toxicity: See *Adverse Reactions, Side Effects or Overdose Symptoms* section below.

Adverse Reactions, Side Effects or Overdose Symptoms

Signs and symptoms:	What to do:
Cold hands and feet	Discontinue. Call doctor when convenient.
Fever	Discontinue. Call doctor immediately.
Gait disturbances	Discontinue. Call doctor immediately.
Heartbeat irregularities	Seek emergency treatment.
Muscle weakness	Discontinue. Call doctor immediately.
Weight loss	Discontinue. Call doctor when convenient.

Houseleek (Jupiter's Eye, Thor's Beard)

Basic Information

Biological name (genus and species):
Sempervivum tectorum, Sempervivum
Parts used for medicinal purposes:
Leaves
Chemical this herb contains:
Malic acid

Known Effects

- Shrinks tissues.
- Prevents secretion of fluids.

Unproved Speculated Benefits

- Helps body dispose of excess fluid by increasing amount of urine produced.
- As a poultice is used to treat insect bites, burns, bruises, skin disease. (See Glossary)

Warnings and Precautions

Don't take if you:
- Are pregnant, think you may be pregnant or plan pregnancy in the near future.
- Have any chronic disease of the gastrointestinal tract, such as stomach or duodenal ulcers, esophageal reflux (reflux esophagitis), ulcerative colitis, spastic colitis, diverticulosis, diverticulitis.

Consult your doctor if you:
- Take this herb for any medical problem that doesn't improve in 2 weeks. There may be safer, more-effective treatments.
- Take any medicinal drugs or herbs including aspirin, laxatives, cold and cough remedies, antacids, vitamins, minerals, amino acids, supplements, other prescription or non-prescription drugs.

Pregnancy:
- Problems in pregnant women taking small or usual amounts have not been proved. But the chance of problems does exist. Don't use unless prescribed by your doctor.

Breast-feeding:
- Problems in breast-fed infants of lactating mothers taking small or usual amounts have not been proved. But the chance of problems does exist. Don't use unless prescribed by your doctor.

Infants and children:
- Treating infants and children under 2 with any herbal preparation is hazardous.

Others:
- None expected if you are beyond childhood and under 45, basically healthy and take for only a short time.

Storage:
- Keep cool and dry, but don't freeze. Store safely away from children.

Safe dosage:
- At present no "safe" dosage has been established.

Toxicity

Comparative-toxicity rating not available from standard references.

For symptoms of toxicity: See *Adverse Reactions, Side Effects or Overdose Symptoms* section below.

Adverse Reactions, Side Effects or Overdose Symptoms

Signs and symptoms:	What to do:
Vomiting	Discontinue. Call doctor immediately.
Watery, explosive diarrhea	Discontinue. Call doctor immediately.

Huckleberry

Basic Information

Biological name (genus and species):
Vaccinum myrtillus
Parts used for medicinal purposes:
Entire plant
Chemicals this herb contains:
Fatty acids
Hydroquinone
Loeanolic acid
Neomyrtillin
Tannins (See Glossary)
Ursolic acid

Known Effects

- Decreases blood sugar.
- Helps body dispose of excess fluid by increasing amount of urine produced.
- Interferes with absorption of iron and other minerals when taken internally.

Unproved Speculated Benefits

- Treats diarrhea.
- Treats gastroenteritis.
- Treats and prevents scurvy.

Warnings and Precautions

Don't take if you:
- Are allergic to blueberries or huckleberries.

Consult your doctor if you:
- Take this herb for any medical problem that doesn't improve in 2 weeks. There may be safer, more-effective treatments.
- Take any medicinal drugs or herbs including aspirin, laxatives, cold and cough remedies, antacids, vitamins, minerals, amino acids, supplements, other prescription or non-prescription drugs.

Pregnancy:
- Pregnant women should experience no problems taking usual amounts as part of a balanced diet. Other products extracted from this herb have not been proved to cause problems.

Breast-feeding:
- Breast-fed infants of lactating mothers should experience no problems when mother takes usual amounts as part of a balanced diet. Other products extracted from this herb have not been proved to cause problems.

Infants and children:
- Treating infants and children under 2 with any herbal preparation is hazardous.

Others:
- None expected if you are beyond childhood and under 45, basically healthy and take for only a short time.

Storage:
- Keep cool and dry, but don't freeze. Store safely away from children.

Safe dosage:
- At present no "safe" dosage has been established.

Toxicity

Generally regarded as safe when taken in appropriate quantities for short periods of time.

Adverse Reactions, Side Effects or Overdose Symptoms

None expected

Hydrangea (Seven Barks, Peegee)

Basic Information

Biological name (genus and species):
Hydrangea paniculata
Parts used for medicinal purposes:
Roots
Chemicals this herb contains:
Hydrangin (can change to cyanide)
Resin (See Glossary)
Saponin (See Glossary)
Volatile oils (See Glossary)

 ## Known Effects

- Aids in expelling gas from intestinal tract.
- Shrinks tissues.
- Prevents secretion of fluids.

Miscellaneous information:
- Leaves contain cyanide. Smoking can cause mind-altering effects and toxicity.

 ## Unproved Speculated Benefits

- Treats cystitis.
- Treats bladder stones.
- Treats dyspepsia.

Warnings and Precautions

Don't take if you:
- Are pregnant, think you may be pregnant or plan pregnancy in the near future.
- Have any chronic disease of the gastrointestinal tract, such as stomach or duodenal ulcers, esophageal reflux (reflux esophagitis), ulcerative colitis, spastic colitis, diverticulosis, diverticulitis.

Consult your doctor if you:
- Take this herb for any medical problem that doesn't improve in 2 weeks. There may be safer, more-effective treatments.
- Take any medicinal drugs or herbs including aspirin, laxatives, cold and cough remedies, antacids, vitamins, minerals, amino acids, supplements, other prescription or non-prescription drugs.

Pregnancy:
- Dangers outweigh any possible benefits. Don't use.

Breast-feeding:
- Dangers outweigh any possible benefits. Don't use.

Infants and children:
- Treating infants and children under 2 with any herbal preparation is hazardous.

Storage:
- Keep cool and dry, but don't freeze. Store safely away from children.

Safe dosage:
- At present no "safe" dosage has been established.

 ## Toxicity

Rated relatively safe when taken in appropriate quantities for short periods of time.

For symptoms of toxicity: See *Adverse Reactions, Side Effects or Overdose Symptoms* section below.

 ## Adverse Reactions, Side Effects or Overdose Symptoms

Signs and symptoms:	What to do:
Dizziness	Discontinue. Call doctor immediately.
Heavy feeling in chest	Discontinue. Call doctor immediately.
Nausea	Discontinue. Call doctor immediately.
Vomiting	Discontinue. Call doctor immediately.

Indian Nettle (Kuppi, Mercury Weed, Indian Acalypha, Hierba de Cancer)

Basic Information

Biological name (genus and species):
  Acalypha indica, A. virginica
Parts used for medicinal purposes:
  Leaves
Chemicals this herb contains:
  Acalyphine
  Cyanogenic glycoside (See Glossary)
  Inositol methylether
  Resin (See Glossary)
  Triacetomamine
  Volatile oils (See Glossary)

Known Effects

- Irritates stomach lining.
- Decreases thickness and increases fluidity of mucus from lungs and bronchial tubes.
- Causes vomiting.

Miscellaneous information:
- Basic ingredients are similar to ipecac.

Unproved Speculated Benefits

- Stimulates bowel movements.
- Is used as a poultice. (See Glossary)
- Is used as a mouthwash.

Warnings and Precautions

Don't take if you:
- Are pregnant, think you may be pregnant or plan pregnancy in the near future.

Consult your doctor if you:
- Take this herb for any medical problem that doesn't improve in 2 weeks. There may be safer, more-effective treatments.
- Take any medicinal drugs or herbs including aspirin, laxatives, cold and cough remedies, antacids, vitamins, minerals, amino acids, supplements, other prescription or non-prescription drugs.

Pregnancy:
- Dangers outweigh any possible benefits. Don't use.

Breast-feeding:
- Dangers outweigh any possible benefits. Don't use.

Infants and children:
- Treating infants and children under 2 with any herbal preparation is hazardous.

Others:
- None expected if you are under 45, not pregnant, basically healthy, take it for only a short time and do not exceed manufacturer's recommended dosage.

Storage:
- Keep cool and dry, but don't freeze. Store safely away from children.

Safe dosage:
- At present no "safe" dosage has been established.

Toxicity

Rated relatively safe when taken in appropriate quantities for short periods of time.

For symptoms of toxicity: See *Adverse Reactions, Side Effects or Overdose Symptoms* section below.

Adverse Reactions, Side Effects or Overdose Symptoms

Signs and symptoms:	What to do:
Diarrhea	Discontinue. Call doctor immediately.
Nausea	Discontinue. Call doctor immediately.
Vomiting	Discontinue. Call doctor immediately.

Indian Tobacco (Lobelia, Asthma Weed)

Basic Information

Biological name (genus and species):
Lobelia inflata
Parts used for medicinal purposes:
Leaves
Seeds
Chemicals this herb contains:
Isolobenine
Lobelanine
Lobelidine
Lobeline
Nor-lobelaine

Known Effects

• Large amounts stimulate central nervous system.
• Small amounts depress central nervous system as blood level drops.
• Activates vomiting center in people not accustomed to lobelia.

Miscellaneous information:
• Sometimes advertised as "legal grass." Do not be misled! Toxic effects can be dangerous.

Unproved Speculated Benefits

• Treats asthma.
• Decreases thickness and increases fluidity of mucus from lungs and bronchial tubes.
• Promotes weight loss.

Warnings and Precautions

Don't take if you:
• Are pregnant, think you may be pregnant or plan pregnancy in the near future.
• Have any chronic disease of the gastrointestinal tract, such as stomach or duodenal ulcers, esophageal reflux (reflux esophagitis), ulcerative colitis, spastic colitis, diverticulosis, diverticulitis.

Consult your doctor if you:
• Take this herb for any medical problem that doesn't improve in 2 weeks. There may be safer, more-effective treatments.
• Take any medicinal drugs or herbs including aspirin, laxatives, cold and cough remedies, antacids, vitamins, minerals, amino acids, supplements, other prescription or non-prescription drugs.

Pregnancy:
• Dangers outweigh any possible benefits. Don't use.

Breast-feeding:
• Dangers outweigh any possible benefits. Don't use.

Infants and children:
• Treating infants and children under 2 with any herbal preparation is hazardous.

Others:
• Dangers outweigh any possible benefits. Don't use.

Storage:
• Keep cool and dry, but don't freeze. Store safely away from children.

Safe dosage:
• At present no "safe" dosage has been established.

Toxicity

Rated slightly dangerous, particularly in children, persons over 55 and those who take larger than appropriate quantities for extended periods of time.

For symptoms of toxicity: See *Adverse Reactions, Side Effects or Overdose Symptoms* section below.

Adverse Reactions, Side Effects or Overdose Symptoms

Signs and symptoms:	What to do:
Coma	Seek emergency treatment.
Diarrhea	Discontinue. Call doctor immediately.
Excess salivation	Discontinue. Call doctor when convenient.
Excess tear formation	Discontinue. Call doctor when convenient.
Giddiness	Discontinue. Call doctor when convenient.
Headache	Discontinue. Call doctor when convenient.
Nausea	Discontinue. Call doctor immediately.
Stupor	Seek emergency treatment.
Tremors	Discontinue. Call doctor immediately.
Vomiting	Discontinue. Call doctor immediately.

Indigo, Wild

Basic Information

Biological name (genus and species):
Baptisia tinctoria
Parts used for medicinal purposes:
Roots
Chemicals this herb contains:
Baptisin
Baptisine
Bapitoxine
Cystisine
Quinolizidine

Known Effects

- Irritates gastrointestinal-lining membrane.
- Causes watery, explosive bowel movements.
- Causes vomiting.

Miscellaneous information:
- Blue dye in wild indigo is inferior to domestically grown indigo.

Unproved Speculated Benefits

- Treats typhoid fever.
- Treats amebiasis.

Warnings and Precautions

Don't take if you:
- Are pregnant, think you may be pregnant or plan pregnancy in the near future.
- Have any chronic disease of the gastrointestinal tract, such as stomach or duodenal ulcers, esophageal reflux (reflux esophagitis), ulcerative colitis, spastic colitis, diverticulosis, diverticulitis.

Consult your doctor if you:
- Take this herb for any medical problem that doesn't improve in 2 weeks. There may be safer, more-effective treatments.
- Take any medicinal drugs or herbs including aspirin, laxatives, cold and cough remedies, antacids, vitamins, minerals, amino acids, supplements, other prescription or non-prescription drugs.

Pregnancy:
- Dangers outweigh any possible benefits. Don't use.

Breast-feeding:
- Dangers outweigh any possible benefits. Don't use.

Infants and children:
- Treating infants and children under 2 with any herbal preparation is hazardous.

Others:
- None expected if you are under 45, not pregnant, basically healthy, take it for only a short time and do not exceed manufacturer's recommended dosage.

Storage:
- Keep cool and dry, but don't freeze. Store safely away from children.

Safe dosage:
- At present no "safe" dosage has been established.

Toxicity

Comparative-toxicity rating not available from standard references.

For symptoms of toxicity: See *Adverse Reactions, Side Effects or Overdose Symptoms* section below.

Adverse Reactions, Side Effects or Overdose Symptoms

Signs and symptoms:	What to do:
Diarrhea	Discontinue. Call doctor immediately.
Nausea	Discontinue. Call doctor immediately.
Vomiting	Discontinue. Call doctor immediately.

Irish Moss

Basic Information

Biological name (genus and species):
Chondrus crispus, Gigartina mamillosa
Parts used for medicinal purposes:
Entire plant
Chemicals this herb contains:
Bromine
Calcium
Carrageenan
Chlorine
Protein
Sodium

Known Effects

• Protects scraped tissues.
• Interferes with blood-clotting mechanism.

Miscellaneous information:
• Used for hand lotions and as substitute for gelatin in jellies.
• Chemically similar to agar, a substance used in laboratories as a base for growing germ cultures.

Unproved Speculated Benefits

• Forms bulky stools.
• Treats coughs.
• Treats diarrhea.

Warnings and Precautions

Don't take if you:
• Are pregnant, think you may be pregnant or plan pregnancy in the near future.
• Have any chronic disease of the gastrointestinal tract, such as stomach or duodenal ulcers, esophageal reflux (reflux esophagitis), ulcerative colitis, spastic colitis, diverticulosis, diverticulitis.
• Take anti-coagulants.

Consult your doctor if you:
• Take this herb for any medical problem that doesn't improve in 2 weeks. There may be safer, more-effective treatments.
• Take any medicinal drugs or herbs including aspirin, laxatives, cold and cough remedies, antacids, vitamins, minerals, amino acids, supplements, other prescription or non-prescription drugs.

Pregnancy:
• Problems in pregnant women taking small or usual amounts have not been proved. But the chance of problems does exist. Don't use unless prescribed by your doctor.

Breast-feeding:
• Problems in breast-fed infants of lactating mothers taking small or usual amounts have not been proved. But the chance of problems does exist. Don't use unless prescribed by your doctor.

Infants and children:
• Treating infants and children under 2 with any herbal preparation is hazardous.

Others:
• None expected if you are under 45, not pregnant, basically healthy, take it for only a short time and do not exceed manufacturer's recommended dosage.

Storage:
• Keep cool and dry, but don't freeze. Store safely away from children.

Safe dosage:
• At present no "safe" dosage has been established.

Toxicity

Comparative-toxicity rating not available from standard references.

For symptoms of toxicity: See *Adverse Reactions, Side Effects or Overdose Symptoms* section below.

Adverse Reactions, Side Effects or Overdose Symptoms

Signs and symptoms:	What to do:
May interact with other anti-coagulants to increase anti-coagulant effect	Discontinue. Call doctor immediately.
Nausea	Discontinue. Call doctor immediately.

Jalap Root (Conqueror Root, High John Root, Ipomea, Turpeth)

Basic Information

Biological name (genus and species):
 Exagonium purga
Parts used for medicinal purposes:
 Roots
Chemicals this herb contains:
 Convolvulin
 Gum *(See Glossary)*
 Jalapin
 Jalapinolic acid
 Starch
 Sugar
 Volatile oils *(See Glossary)*

 ## Known Effects

• Irritates the gastrointestinal system.

 ## Unproved Speculated Benefits

• Treats constipation.

 ## Warnings and Precautions

Don't take if you:
• Are pregnant, think you may be pregnant or plan pregnancy in the near future.
• Have any chronic disease of the gastrointestinal tract, such as stomach or duodenal ulcers, esophageal reflux (reflux esophagitis), ulcerative colitis, spastic colitis, diverticulosis, diverticulitis.

Consult your doctor if you:
• Take this herb for any medical problem that doesn't improve in 2 weeks. There may be safer, more-effective treatments.
• Take any medicinal drugs or herbs including aspirin, laxatives, cold and cough remedies, antacids, vitamins, minerals, amino acids, supplements, other prescription or non-prescription drugs.

Pregnancy:
• Dangers outweigh any possible benefits. Don't use.

Breast-feeding:
• Dangers outweigh any possible benefits. Don't use.

Infants and children:
• Treating infants and children under 2 with any herbal preparation is hazardous.

Others:
• None expected if you are under 45, not pregnant, basically healthy, take it for only a short time and do not exceed manufacturer's recommended dosage.

Storage:
• Keep cool and dry, but don't freeze. Store safely away from children.

Safe dosage:
• At present no "safe" dosage has been established.

 ## Toxicity

Comparative-toxicity rating not available from standard references.

For symptoms of toxicity: See *Adverse Reactions, Side Effects or Overdose Symptoms* section below.

 ## Adverse Reactions, Side Effects or Overdose Symptoms

Signs and symptoms:	What to do:
Explosive, watery diarrhea, with possible fluid and electrolyte depletion leading to weakness and possible heartbeat irregularities	Discontinue. Call doctor immediately.

Jamaican Dogwood (Fish-Poison Tree)

Basic Information

Biological name (genus and species):
Piscidia piscipula
Parts used for medicinal purposes:
Bark
Chemicals this herb contains:
Piscidin
Rotenone

Known Effects

- Causes hallucinations.
- Treats painful conditions.
- Depresses uterine contractions.

Miscellaneous information:
- Poisonous to fish.
- Active chemicals in bark have odor similar to opium.

Unproved Speculated Benefits

- Produces euphoria.
- Treats dysmenorrhea (painful menstruation).

Warnings and Precautions

Don't take if you:
- Are pregnant, think you may be pregnant or plan pregnancy in the near future.

Consult your doctor if you:
- Take this herb for any medical problem that doesn't improve in 2 weeks. There may be safer, more-effective treatments.
- Take any medicinal drugs or herbs including aspirin, laxatives, cold and cough remedies, antacids, vitamins, minerals, amino acids, supplements, other prescription or non-prescription drugs.

Pregnancy:
- Problems in pregnant women taking small or usual amounts have not been proved. But the chance of problems does exist. Don't use unless prescribed by your doctor.

Breast-feeding:
- Problems in breast-fed infants of lactating mothers taking small or usual amounts have not been proved. But the chance of problems does exist. Don't use unless prescribed by your doctor.

Infants and children:
- Treating infants and children under 2 with any herbal preparation is hazardous.

Others:
- None expected if you are beyond childhood and under 45, basically healthy and take for only a short time.

Storage:
- Keep cool and dry, but don't freeze. Store safely away from children.

Safe dosage:
- At present no "safe" dosage has been established.

Toxicity

Rated slightly dangerous, particularly in children, persons over 55 and those who take larger than appropriate quantities for extended periods of time.

For symptoms of toxicity: See *Adverse Reactions, Side Effects or Overdose Symptoms* section below.

Adverse Reactions, Side Effects or Overdose Symptoms

Signs and symptoms:	What to do:
Hallucinations	Seek emergency treatment.

Jequirity Bean (Crab's Eyes, Indian Licorice, Rosary Pea)

Basic Information

Biological name (genus and species):
Abrus precatorius
Parts used for medicinal purposes:
Seeds/beans
Chemicals this herb contains:
Abric acid
Abrive
Glycyrrhizin
Hemoglutin
N-methyltryptophan
Toxalbumin abrin

Known Effects

- Toxalbumin in seed causes cell destruction.

Miscellaneous information:
- No longer used therapeutically.
- Causes toxic reactions with ingestion.
- Common weed in Florida, Central America and South America.

Unproved Speculated Benefits

- Is used as drops for eye problems.

Warnings and Precautions

Don't take if you:
- Are pregnant, think you may be pregnant or plan pregnancy in the near future.
- Have any chronic disease of the gastrointestinal tract, such as stomach or duodenal ulcers, esophageal reflux (reflux esophagitis), ulcerative colitis, spastic colitis, diverticulosis, diverticulitis.

Consult your doctor if you:
- Take this herb for any medical problem that doesn't improve in 2 weeks. There may be safer, more-effective treatments.
- Take any medicinal drugs or herbs including aspirin, laxatives, cold and cough remedies, antacids, vitamins, minerals, amino acids, supplements, other prescription or non-prescription drugs.

Pregnancy:
- Dangers outweigh any possible benefits. Don't use.

Breast-feeding:
- Dangers outweigh any possible benefits. Don't use.

Infants and children:
- Treating infants and children under 2 with any herbal preparation is hazardous.

Others:
- Dangers outweigh any possible benefits for *anyone.* Don't use. Swallowing even one bean can cause toxic symptoms hours or even days after eating.

Storage:
- Keep cool and dry, but don't freeze. Store safely away from children.

Safe dosage:
- At present no "safe" dosage has been established.

Toxicity

Rated dangerous, particularly in children, persons over 55 and those who take larger than appropriate quantities for extended periods of time.

For symptoms of toxicity: See *Adverse Reactions, Side Effects or Overdose Symptoms* section below.

Adverse Reactions, Side Effects or Overdose Symptoms

Signs and symptoms:	What to do:
Convulsions	Seek emergency treatment.
Diarrhea	Discontinue. Call doctor immediately.
Increased heart rate	Discontinue. Call doctor immediately.
Kidney damage characterized by blood in urine, decreased urine flow, swelling of hands and feet	Seek emergency treatment.
Nausea	Discontinue. Call doctor immediately.
Vomiting	Discontinue. Call doctor immediately.

Basic Information

Biological name (genus and species):
Ceanothus americanus
Parts used for medicinal purposes:
Roots
Chemicals this herb contains:
Ceanothic acid
Malonic acid
Orthophosphoric acid
Oxalic acid
Pyrophosphoric acid
Resin (See Glossary)
Succinic acid
Tannins (See Glossary)

Known Effects

- Shrinks tissues.
- Prevents secretion of fluids.
- Increases blood clotting.
- Interferes with absorption of iron and other minerals when taken internally.

Unproved Speculated Benefits

- Treats syphilis (archaic).
- Is used as a "spleen" remedy.
- Stops mild bleeding from broken capillaries in skin.
- Decreases thickness and increases fluidity of mucus from lungs and bronchial tubes.
- Acts as a sedative.
- Relieves spasm in skeletal muscle or smooth muscle.
- Treats depression.

Warnings and Precautions

Don't take if you:
- Are pregnant, think you may be pregnant or plan pregnancy in the near future.

Consult your doctor if you:
- Take this herb for any medical problem that doesn't improve in 2 weeks. There may be safer, more-effective treatments.
- Take any medicinal drugs or herbs including aspirin, laxatives, cold and cough remedies, antacids, vitamins, minerals, amino acids, supplements, other prescription or non-prescription drugs.

Pregnancy:
- Problems in pregnant women taking small or usual amounts have not been proved. But the chance of problems does exist. Don't use unless prescribed by your doctor.

Breast-feeding:
- Problems in breast-fed infants of lactating mothers taking small or usual amounts have not been proved. But the chance of problems does exist. Don't use unless prescribed by your doctor.

Infants and children:
- Treating infants and children under 2 with any herbal preparation is hazardous.

Others:
- None expected if you are under 45, not pregnant, basically healthy, take it for only a short time and do not exceed manufacturer's recommended dosage.

Storage:
- Keep cool and dry, but don't freeze. Store safely away from children.

Safe dosage:
- At present no "safe" dosage has been established.

Toxicity

Comparative-toxicity rating not available from standard references.

For symptoms of toxicity: See *Adverse Reactions, Side Effects or Overdose Symptoms* section below.

Adverse Reactions, Side Effects or Overdose Symptoms

Signs and symptoms:	What to do:
Prolonged minor bleeding	Discontinue. Call doctor immediately.

Jimson Weed (Sacred Datura, Thorn Apple, Stramonium)

Basic Information

Biological name (genus and species):
  Datura stramonium
Parts used for medicinal purposes:
  Leaves
  Seeds
Chemicals this herb contains:
  Atropine
  Hyoscyamine
  Scopolamine

Known Effects

• Negates normal activity of acetylcholine, an important chemical at the synapses (connection between nerve cells) of heart, brain, smooth muscles and glands.

Miscellaneous information:
• There are more-refined, predictable sources for active chemicals.
• Highest concentration of toxins are in seeds but may be in all parts of the plant.

Unproved Speculated Benefits

• Treats asthma.
• Treats gastrointestinal problems.
• Produces hallucinations.
• Acts as a sedative.

Warnings and Precautions

Don't take if you:
• Are pregnant, think you may be pregnant or plan pregnancy in the near future.
• Have heart disease.

Consult your doctor if you:
• Take this herb for any medical problem that doesn't improve in 2 weeks. There may be safer, more-effective treatments.
• Take any medicinal drugs or herbs including aspirin, laxatives, cold and cough remedies, antacids, vitamins, minerals, amino acids, supplements, other prescription or non-prescription drugs.

Pregnancy:
• Dangers outweigh any possible benefits. Don't use.

Breast-feeding:
• Dangers outweigh any possible benefits. Don't use.

Infants and children:
• Treating infants and children under 2 with any herbal preparation is hazardous.

Others:
• Dangers outweigh any possible benefits. Don't use.

Storage:
• Keep cool and dry, but don't freeze. Store safely away from children.

Safe dosage:
• At present no "safe" dosage has been established.

Toxicity

Rated dangerous, particularly in children, persons over 55 and those who take larger than appropriate quantities for extended periods of time.

For symptoms of toxicity: See *Adverse Reactions, Side Effects or Overdose Symptoms* section below.

Adverse Reactions, Side Effects or Overdose Symptoms

Signs and symptoms:	What to do:
Convulsions	Seek emergency treatment.
Dilated pupils	Discontinue. Call doctor immediately.
Dry mouth	Discontinue. Call doctor when convenient.
Extremely fast heart rate	Seek emergency treatment.
Flushing	Discontinue. Call doctor when convenient.
Hallucinations	Seek emergency treatment.
Increased blood pressure	Discontinue. Call doctor immediately.
Unconsciousness	Seek emergency treatment.

Basic Information

Biological name (genus and species):
Juniperus communis or J. depressa
Parts used for medicinal purposes:
Berries/fruits
Chemicals this herb contains:

Alcohols	Sabinal
Alpha-pinene	Sugar
Cadinene	Tannins (See Glossary)
Camphene	Terpinene
Flavone	Volatile oils (See Glossary)
Resin (See Glossary)	

Known Effects

- Irritates kidneys.
- May cause hallucinations.
- Interferes with absorption of iron and other minerals when taken internally.

Miscellaneous information:
- Provides flavor in gin.

Unproved Speculated Benefits

- Treats chronic kidney disorders.
- Aids in expelling gas from intestinal tract.
- Causes abortions (miscarriages).
- Treats colic.
- Treats flatulence.
- Helps body dispose of excess fluid by increasing amount of urine produced.

Warnings and Precautions

Don't take if you:
- Are pregnant, think you may be pregnant or plan pregnancy in the near future.
- Have kidney disease.
- Have any chronic disease of the gastrointestinal tract, such as stomach or duodenal ulcers, esophageal reflux (reflux esophagitis), ulcerative colitis, spastic colitis, diverticulosis, diverticulitis.

Consult your doctor if you:
- Take this herb for any medical problem that doesn't improve in 2 weeks. There may be safer, more-effective treatments.
- Take any medicinal drugs or herbs including aspirin, laxatives, cold and cough remedies, antacids, vitamins, minerals, amino acids, supplements, other prescription or non-prescription drugs.

Pregnancy:
- Dangers outweigh any possible benefits. Don't use.

Breast-feeding:
- Dangers outweigh any possible benefits. Don't use.

Infants and children:
- Treating infants and children under 2 with any herbal preparation is hazardous.

Others:
- None expected if you are under 45, not pregnant, basically healthy, take it for only a short time and do not exceed manufacturer's recommended dosage.

Storage:
- Keep cool and dry, but don't freeze. Store safely away from children.

Safe dosage:
- At present no "safe" dosage has been established.

Toxicity

Rated slightly dangerous, particularly in children, persons over 55 and those who take larger than appropriate quantities for extended periods of time.

For symptoms of toxicity: See *Adverse Reactions, Side Effects or Overdose Symptoms* section below.

Adverse Reactions, Side Effects or Overdose Symptoms

Signs and symptoms:	What to do:
Single dose:	
Diarrhea, watery, explosive	Discontinue. Call doctor immediately.
Small, repeated doses:	
Convulsions	Seek emergency treatment.
Hallucinations	Seek emergency treatment.
Kidney damage	Discontinue. Call doctor immediately.
Personality changes	Discontinue. Call doctor immediately.

Kava-Kava

Basic Information
Biological name (genus and species):
Piper methysticum
Parts used for medicinal purposes:
Roots
Chemicals this herb contains:
Demethoxyyangonin
Dihydrokawin
Dihydromethysticin
Flavorawin A
Kawain
Methysticin
Starch
Yangonin

 ## Known Effects

- Depresses the central nervous system.
- Produces skin pigmentation.

Miscellaneous information:
- Is used to make a fermented liquor.
- Sedative effect is very mild.

 ## Unproved Speculated Benefits

- Is used as a sedative for anxiety disorders.
- Induces restful sleep.
- Treats fatigue.
- Helps body dispose of excess fluid by increasing amount of urine produced.
- Is used as a genito-urinary antiseptic.
- Treats coughs.
- Is used as a douche.

 ## Warnings and Precautions

Don't take if you:
- Are pregnant, think you may be pregnant or plan pregnancy in the near future.

Consult your doctor if you:
- Take this herb for any medical problem that doesn't improve in 2 weeks. There may be safer, more-effective treatments.
- Take any medicinal drugs or herbs including aspirin, laxatives, cold and cough remedies, antacids, vitamins, minerals, amino acids, supplements, other prescription or non-prescription drugs.

Pregnancy:
- Problems in pregnant women taking small or usual amounts have not been proved. But the chance of problems does exist. Don't use unless prescribed by your doctor.

Breast-feeding:
- Problems in breast-fed infants of lactating mothers taking small or usual amounts have not been proved. But the chance of problems does exist. Don't use unless prescribed by your doctor.

Infants and children:
- Treating infants and children under 2 with any herbal preparation is hazardous.

Others:
- None expected if you are beyond childhood and under 45, basically healthy and take for only a short time.

Storage:
- Keep cool and dry, but don't freeze. Store safely away from children.

Safe dosage:
- At present no "safe" dosage has been established.

 ## Toxicity

Rated slightly dangerous, particularly in children, persons over 55 and those who take larger than appropriate quantities for extended periods of time.

For symptoms of toxicity: See *Adverse Reactions, Side Effects or Overdose Symptoms* section below.

 ## Adverse Reactions, Side Effects or Overdose Symptoms

Signs and symptoms:	What to do:
Oversedation	Discontinue. Call doctor immediately.
Repeated small amounts may lead to undesirable skin coloring, inflammation of the body and eyes	Discontinue. Call doctor immediately.

Basic Information

Biological name (genus and species):
Laminaria, Fucus, Sargassum
Parts used for medicinal purposes:
Leaves
Chemicals this herb contains:
Alginic acid
Bromine
Iodine
Potassium
Sodium

Known Effects

• Provides bulk for bowel movements.

Miscellaneous information:
• Iodine can interfere with normal thyroid function.

Unproved Speculated Benefits

• Treats chronic constipation without catharsis.
• Softens stools.
• Treats ulcers.
• Controls obesity.

Warnings and Precautions

Don't take if you:
• Are pregnant, think you may be pregnant or plan pregnancy in the near future.
• Are allergic to iodine in any form, particularly if you have had an allergic reaction to injected dye used for X-ray studies of the kidney or other organs.

Consult your doctor if you:
• Take this herb for any medical problem that doesn't improve in 2 weeks. There may be safer, more-effective treatments.
• Take any medicinal drugs or herbs including aspirin, laxatives, cold and cough remedies, antacids, vitamins, minerals, amino acids, supplements, other prescription or non-prescription drugs.

Pregnancy:
• Problems in pregnant women taking small or usual amounts have not been proved. But the chance of problems does exist. Don't use unless prescribed by your doctor.

Breast-feeding:
• Problems in breast-fed infants of lactating mothers taking small or usual amounts have not been proved. But the chance of problems does exist. Don't use unless prescribed by your doctor.

Infants and children:
• Treating infants and children under 2 with any herbal preparation is hazardous.

Others:
• None expected if you are under 45, not pregnant, basically healthy, take it for only a short time and do not exceed manufacturer's recommended dosage.

Storage:
• Keep cool and dry, but don't freeze. Store safely away from children.

Safe dosage:
• At present no "safe" dosage has been established.

Toxicity

Comparative-toxicity rating not available from standard references.

Adverse Reactions, Side Effects or Overdose Symptoms

None expected

Lemongrass

Basic Information
Biological name (genus and species):
  Cymbopogon citracus
Parts used for medicinal purposes:
  Various parts of the entire plant, frequently
  differing by country and/or culture
Chemicals this herb contains:
  Citronellal
  Methylneptenone
  Terpene
  Terpene alcohol

Known Effects

• In insecticides kills insects, but less efficiently
  than malathion or parathione.

Miscellaneous information:
• Is used in perfumes.
• Sometimes used as an insect repellent.

Unproved Speculated Benefits

• Treats constipation.

Warnings and Precautions

Don't take if you:
• Are pregnant, think you may be pregnant or
  plan pregnancy in the near future.
• Have any chronic disease of the
  gastrointestinal tract, such as stomach or
  duodenal ulcers, esophageal reflux (reflux
  esophagitis), ulcerative colitis, spastic colitis,
  diverticulosis, diverticulitis.

Consult your doctor if you:
• Take this herb for any medical problem that
  doesn't improve in 2 weeks. There may be
  safer, more-effective treatments.
• Take any medicinal drugs or herbs including
  aspirin, laxatives, cold and cough remedies,
  antacids, vitamins, minerals, amino acids,
  supplements, other prescription or non-
  prescription drugs.

Pregnancy:
• Problems in pregnant women taking small or
  usual amounts have not been proved. But the
  chance of problems does exist. Don't use
  unless prescribed by your doctor.

Breast-feeding:
• Problems in breast-fed infants of lactating
  mothers taking small or usual amounts have
  not been proved. But the chance of problems
  does exist. Don't use unless prescribed by your
  doctor.

Infants and children:
• Treating infants and children under 2 with any
  herbal preparation is hazardous.

Others:
• None expected if you are beyond childhood
  and under 45, basically healthy and take for
  only a short time.

Storage:
• Keep cool and dry, but don't freeze. Store
  safely away from children.

Safe dosage:
• At present no "safe" dosage has been
  established.

Toxicity

Comparative-toxicity rating not available from
standard references.

For symptoms of toxicity: See *Adverse
Reactions, Side Effects or Overdose Symptoms*
section below.

Adverse Reactions, Side Effects or Overdose Symptoms

Signs and symptoms:	What to do:
Diarrhea	Discontinue. Call doctor immediately.
Nausea	Discontinue. Call doctor immediately.
Vomiting	Discontinue. Call doctor immediately.

Licorice, Common (Licorice Root, Spanish Licorice Root)

Basic Information

Biological name (genus and species):
  Glycyrrhiza glabra
Parts used for medicinal purposes:
  Roots
Chemicals this herb contains:
  Asparagine
  Fat
  Glycyrrhizin
  Gum (See Glossary)
  Pentacyclic terpenes
  Protein
  Sugar
  Yellow dye

Known Effects

- Decreases inflammation.
- Provides estrogen-like hormone effects.
- Decreases spasm of smooth muscle or skeletal muscle.
- Decreases thickness and increases fluidity of mucus from lungs and bronchial tubes.

Miscellaneous information;
- *Warning:* Consuming large amounts of licorice may lead to high blood pressure.

Unproved Speculated Benefits

- Protects scraped tissues.
- Softens or soothes skin.
- Treats coughs.

Warnings and Precautions

Don't take if you:
- Are pregnant, think you may be pregnant or plan pregnancy in the near future.
- Have heart disease.
- Take diuretics.

Consult your doctor if you:
- Take this herb for any medical problem that doesn't improve in 2 weeks. There may be safer, more-effective treatments.
- Take any medicinal drugs or herbs including aspirin, laxatives, cold and cough remedies, antacids, vitamins, minerals, amino acids, supplements, other prescription or non-prescription drugs.

Pregnancy:
- Dangers outweigh any possible benefits. Don't use.

Breast-feeding:
- Dangers outweigh any possible benefits. Don't use.

Infants and children:
- Treating infants and children under 2 with any herbal preparation is hazardous.

Others:
- None expected if you are under 45, not pregnant, basically healthy, take it for only a short time and do not exceed manufacturer's recommended dosage.

Storage:
- Keep cool and dry, but don't freeze. Store safely away from children.

Safe dosage:
- At present no "safe" dosage has been established.

Toxicity

Rated slightly dangerous, particularly in children, persons over 55 and those who take larger than appropriate quantities for extended periods of time.

For symptoms of toxicity: See *Adverse Reactions, Side Effects or Overdose Symptoms* section below.

Adverse Reactions, Side Effects or Overdose Symptoms

Signs and symptoms:	What to do:
Causes sodium retention in blood, which may lead to edema, lung congestion	Discontinue. Call doctor immediately.
Depletes sodium from cells; may cause weakness, nausea, heartbeat irregularities	Discontinue. Call doctor immediately.
High blood pressure	Discontinue. Call doctor immediately.

Liferoot (Golden Groundsel, Squaw Weed)

Basic Information

Biological name (genus and species):
Senecio vulgaris, S. aureus
Parts used for medicinal purposes:
Roots
Chemicals this herb contains:
Pyrrolizidine (has high potential for causing liver disorders, including cancer)

Known Effects

• Increases blood pressure.
• Stimulates uterine contractions.

Unproved Speculated Benefits

• Treats menstrual irregularities.
• Treats dysmenorrhea (painful menstruation).
• Treats excessive menstrual bleeding (menorrhagia).
• Helps relieve vaginal discharge.
• Causes overdue labor to begin.

Warnings and Precautions

Don't take if you:
• Are pregnant, think you may be pregnant or plan pregnancy in the near future.

Consult your doctor if you:
• Take this herb for any medical problem that doesn't improve in 2 weeks. There may be safer, more-effective treatments.
• Take any medicinal drugs or herbs including aspirin, laxatives, cold and cough remedies, antacids, vitamins, minerals, amino acids, supplements, other prescription or non-prescription drugs.

Pregnancy:
• Dangers outweigh any possible benefits. Don't use.

Breast-feeding:
• Dangers outweigh any possible benefits. Don't use.

Infants and children:
• Treating infants and children under 2 with any herbal preparation is hazardous.

Others:
• None expected if you are beyond childhood and under 45, basically healthy and take for only a short time.

Storage:
• Keep cool and dry, but don't freeze. Store safely away from children.

Safe dosage:
• At present no "safe" dosage has been established.

Toxicity

Rated slightly dangerous, particularly in children, persons over 55 and those who take larger than appropriate quantities for extended periods of time.

For symptoms of toxicity: See *Adverse Reactions, Side Effects or Overdose Symptoms* section below.

Adverse Reactions, Side Effects or Overdose Symptoms

Signs and symptoms:	What to do:
Abnormal liver function tests	Discontinue. Call doctor immediately.
Jaundice (yellow eyes and skin)	Discontinue. Call doctor immediately.

Basic Information

Biological name (genus and species):
  Convallaria majalis
Parts used for medicinal purposes:
  Berries/fruits
  Petals/flower
  Roots
Chemicals this herb contains:
  Convallamarin
  Convallarin
  Convallatoxin (highly toxic)

Known Effects

- Increases efficiency of heart-muscle contraction.
- Helps body dispose of excess fluid by increasing amount of urine produced.

Miscellaneous information:

- Although lily-of-the-valley has similar action to digitalis, there are safer, less-expensive, more-reliable products to use.

Unproved Speculated Benefits

- Treats congestive heart failure.
- Treats heartbeat irregularities.
- Improves circulation.

Warnings and Precautions

Don't take if you:

- Are pregnant, think you may be pregnant or plan pregnancy in the near future.
- Have any chronic disease of the gastrointestinal tract, such as stomach or duodenal ulcers, esophageal reflux (reflux esophagitis), ulcerative colitis, spastic colitis, diverticulosis, diverticulitis.
- Have heart disease.

Consult your doctor if you:

- Take this herb for any medical problem that doesn't improve in 2 weeks. There may be safer, more-effective treatments.
- Take any medicinal drugs or herbs including aspirin, laxatives, cold and cough remedies, antacids, vitamins, minerals, amino acids, supplements, other prescription or non-prescription drugs.

Pregnancy:

- Dangers outweigh any possible benefits. Don't use.

Breast-feeding:

- Dangers outweigh any possible benefits. Don't use.

Infants and children:

- Treating infants and children under 2 with any herbal preparation is hazardous.

Others:

- Dangers outweigh any possible benefits. Don't use.

Storage:

- Keep cool and dry, but don't freeze. Store safely away from children.

Safe dosage:

- At present no "safe" dosage has been established.

Toxicity

Rated dangerous, particularly in children, persons over 55 and those who take larger than appropriate quantities for extended periods of time.

For symptoms of toxicity: See *Adverse Reactions, Side Effects or Overdose Symptoms* section below.

Adverse Reactions, Side Effects or Overdose Symptoms

Signs and symptoms:	What to do:
Heartbeat irregularities	Seek emergency treatment.
Nausea	Discontinue. Call doctor immediately.
Vomiting	Discontinue. Call doctor immediately.

Linden Tree (American, Lime Tree in Europe)

Basic Information

Biological name (genus and species):
Tilia europea
Parts used for medicinal purposes:
Petals/flower
Chemicals this herb contains:
Tannins (See Glossary)
Volatile oils (See Glossary)

Known Effects

- Decreases spasm of smooth or skeletal muscle.
- Shrinks tissues.
- Prevents secretion of fluids.
- Increases perspiration.
- Interferes with absorption of iron and other minerals when taken internally.

Unproved Speculated Benefits

- Treats coughs.
- Decreases thickness and increases fluidity of mucus from lungs and bronchial tubes.
- Reduces fever.

Warnings and Precautions

Don't take if you:
- Are pregnant, think you may be pregnant or plan pregnancy in the near future.

Consult your doctor if you:
- Take this herb for any medical problem that doesn't improve in 2 weeks. There may be safer, more-effective treatments.
- Take any medicinal drugs or herbs including aspirin, laxatives, cold and cough remedies, antacids, vitamins, minerals, amino acids, supplements, other prescription or non-prescription drugs.

Pregnancy:
- Problems in pregnant women taking small or usual amounts have not been proved. But the chance of problems does exist. Don't use unless prescribed by your doctor.

Breast-feeding:
- Problems in breast-fed infants of lactating mothers taking small or usual amounts have not been proved. But the chance of problems does exist. Don't use unless prescribed by your doctor.

Infants and children:
- Treating infants and children under 2 with any herbal preparation is hazardous.

Others:
- None expected if you are under 45, not pregnant, basically healthy, take it for only a short time and do not exceed manufacturer's recommended dosage.

Storage:
- Keep cool and dry, but don't freeze. Store safely away from children.

Safe dosage:
- At present no "safe" dosage has been established.

Toxicity

Rated relatively safe when taken in appropriate quantities for short periods of time.

For symptoms of toxicity: See *Adverse Reactions, Side Effects or Overdose Symptoms* section below.

Adverse Reactions, Side Effects or Overdose Symptoms

Signs and symptoms:	What to do:
Drowsiness	Discontinue. Call doctor when convenient.

Basic Information

Biological name (genus and species):
Myristica fragrans

Parts used for medicinal purposes:
Fibrous covering,
Seeds

Chemicals this herb contains:

Elemicin	Methyleugenol
Eugenol	Methylisoeugenol
Fixed oil (See Glossary)	Myristicin
Isoeugenol	Protein
Methoxyeugenol	Starch

Known Effects

- Stimulates muscular movement of intestinal tract.
- Stimulates central nervous system.

Miscellaneous information:
- May produce hallucinations.
- No effects are expected on the body, either good or bad, when herb is used in very small amounts to enhance the flavor of food.
- Nutmeg is the seed. Mace is the fibrous covering.

Unproved Speculated Benefits

- Alters mood.
- Treats digestive disorders.
- Treats cholera.
- Causes abortion.
- Triggers menstruation to begin.
- Arouses or enhances instinctive sexual desire.

Warnings and Precautions

Don't take if you:
- Are pregnant, think you may be pregnant or plan pregnancy in the near future.
- Have any chronic disease of the gastrointestinal tract, such as stomach or duodenal ulcers, esophageal reflux (reflux esophagitis), ulcerative colitis, spastic colitis, diverticulosis, diverticulitis.

Consult your doctor if you:
- Take this herb for any medical problem that doesn't improve in 2 weeks. There may be safer, more-effective treatments.
- Take any medicinal drugs or herbs including aspirin, laxatives, cold and cough remedies, antacids, vitamins, minerals, amino acids, supplements, other prescription or non-prescription drugs.

Pregnancy:
- Dangers outweigh any possible benefits. Don't use.

Breast-feeding:
- Dangers outweigh any possible benefits. Don't use.

Infants and children:
- Treating infants and children under 2 with any herbal preparation is hazardous.

Others:
- Mind-altering and hallucinogenic effects are unpleasant. Do not use nutmeg for these purposes.

Storage:
- Keep cool and dry, but don't freeze. Store safely away from children.

Safe dosage:
- At present no "safe" dosage has been established.

Toxicity

Rated slightly dangerous, particularly in children, persons over 55 and those who take larger than appropriate quantities for extended periods of time.

For symptoms of toxicity: See *Adverse Reactions, Side Effects or Overdose Symptoms* section below.

Adverse Reactions, Side Effects or Overdose Symptoms

Signs and symptoms:	What to do:
Diarrhea	Discontinue. Call doctor immediately.
Drowsiness	Discontinue. Call doctor when convenient.
Hallucinations	Seek emergency treatment.
Nausea	Discontinue. Call doctor immediately.
Reduced body temperature	Discontinue. Call doctor immediately.
Vomiting	Discontinue. Call doctor immediately.
Weak, thready, rapid pulse	Seek emergency treatment.

Malabar Nut (Adotodai, Parettia, Vasaka)

Basic Information

Biological name (genus and species):
Adhatoda vasica
Parts used for medicinal purposes:
Leaves
Chemicals this herb contains:
Adhatodic acid
Peganine
Vasicine

Known Effects

* Dilates bronchial tubes.
* Decreases thickness and increases fluidity of mucus from lungs and bronchial tubes.

Unproved Speculated Benefits

* Treats coughs and colds.
* Treats bronchitis.
* Treats asthma.

Warnings and Precautions

Don't take if you:
* Are pregnant, think you may be pregnant or plan pregnancy in the near future.
* Have any chronic disease of the gastrointestinal tract, such as stomach or duodenal ulcers, esophageal reflux (reflux esophagitis), ulcerative colitis, spastic colitis, diverticulosis, diverticulitis.

Consult your doctor if you:
* Take this herb for any medical problem that doesn't improve in 2 weeks. There may be safer, more-effective treatments.
* Take any medicinal drugs or herbs including aspirin, laxatives, cold and cough remedies, antacids, vitamins, minerals, amino acids, supplements, other prescription or non-prescription drugs.

Pregnancy:
* Problems in pregnant women taking small or usual amounts have not been proved. But the chance of problems does exist. Don't use unless prescribed by your doctor.

Breast-feeding:
* Problems in breast-fed infants of lactating mothers taking small or usual amounts have not been proved. But the chance of problems does exist. Don't use unless prescribed by your doctor.

Infants and children:
* Treating infants and children under 2 with any herbal preparation is hazardous.

Others:
* None expected if you are under 45, not pregnant, basically healthy, take it for only a short time and do not exceed manufacturer's recommended dosage.

Storage:
* Keep cool and dry, but don't freeze. Store safely away from children.

Safe dosage:
* At present no "safe" dosage has been established.

Toxicity

Rated slightly dangerous, particularly in children, persons over 55 and those who take larger than appropriate quantities for extended periods of time.

For symptoms of toxicity: See *Adverse Reactions, Side Effects or Overdose Symptoms* section below.

Adverse Reactions, Side Effects or Overdose Symptoms

Signs and symptoms:	What to do:
Diarrhea	Discontinue. Call doctor immediately.
Nausea	Discontinue. Call doctor immediately.
Vomiting	Discontinue. Call doctor immediately.

Male Fern (Aspidium)

Basic Information

Biological name (genus and species):
Dryopteris filix-mass, Dryopteris

Parts used for medicinal purposes:
Leaves, Roots

Chemicals this herb contains:

Albaspadin	Resin (See Glossary)
Aspidin	Starch
Aspidinol	Sugar
Filicic acid	Tannins (See Glossary)
Filmaron acid	Volatile oils (See Glossary)
Flavaspininic acid	Wax

Known Effects

- Destroys intestinal worms.
- Decreases normal muscle function.
- Interferes with absorption of iron and other minerals when taken internally.

Unproved Speculated Benefits

- None

Warnings and Precautions

Don't take if you:
- Are pregnant, think you may be pregnant or plan pregnancy in the near future.
- Have any chronic disease of the gastrointestinal tract, such as stomach or duodenal ulcers, esophageal reflux (reflux esophagitis), ulcerative colitis, spastic colitis, diverticulosis, diverticulitis.
- Are over age 55.
- Have heart disease.
- Have kidney disease.

Consult your doctor if you:
- Take this herb for any medical problem that doesn't improve in 2 weeks. There may be safer, more-effective treatments.
- Take any medicinal drugs or herbs including aspirin, laxatives, cold and cough remedies, antacids, vitamins, minerals, amino acids, supplements, other prescription or non-prescription drugs.

Pregnancy:
- Dangers outweigh any possible benefits. Don't use.

Breast-feeding:
- Dangers outweigh any possible benefits. Don't use.

Infants and children:
- Treating infants and children under 2 with any herbal preparation is hazardous.

Others:
- Dangers outweigh any possible benefits. Don't use.

Storage:
- Keep cool and dry, but don't freeze. Store safely away from children.

Safe dosage:
- At present no "safe" dosage has been established.

Toxicity

Rated dangerous, particularly in children, persons over 55 and those who take larger than appropriate quantities for extended periods of time.

For symptoms of toxicity: See *Adverse Reactions, Side Effects or Overdose Symptoms* section below.

Adverse Reactions, Side Effects or Overdose Symptoms

Signs and symptoms:	What to do:
Abdominal cramping	Discontinue. Call doctor when convenient.
Breathing difficulty	Seek emergency treatment.
Coma	Seek emergency treatment.
Convulsions	Seek emergency treatment.
Diarrhea	Discontinue. Call doctor immediately.
Headache	Discontinue. Call doctor when convenient.
Heartbeat irregularities	Seek emergency treatment.
Impaired vision	Discontinue. Call doctor when convenient.
Nausea	Discontinue. Call doctor immediately.
Vomiting	Discontinue. Call doctor immediately.

MEDICINAL HERB

Mandrake (Love Apple, Satan's Apple)

Basic Information

Biological name (genus and species):
Mandragora officanarum
Parts used for medicinal purposes:
Roots
Chemicals this herb contains:
Hyoscyamine
Mandragorin
Scopolamine

Known Effects

- Increases heart rate.
- Dilates pupils.
- Causes dry mouth.
- Causes urinary retention.
- Causes hallucinations.
- Reduces muscular movements of intestinal tract.

Unproved Speculated Benefits

- Relieves pain.
- Acts as a sedative.
- Is used as an aphrodisiac.
- Treats ulcers.
- Treats skin diseases.
- Treats hemorrhoids.
- Destroys or repels demons.
- Causes explosive, watery diarrhea.
- Is used as an anesthetic.

Warnings and Precautions

Don't take if you:
- Are pregnant, think you may be pregnant or plan pregnancy in the near future.
- Have heart disease.

Consult your doctor if you:
- Take this herb for any medical problem that doesn't improve in 2 weeks. There may be safer, more-effective treatments.
- Take any medicinal drugs or herbs including aspirin, laxatives, cold and cough remedies, antacids, vitamins, minerals, amino acids, supplements, other prescription or non-prescription drugs.

Pregnancy:
- Dangers outweigh any possible benefits. Don't use.

Breast-feeding:
- Dangers outweigh any possible benefits. Don't use.

Infants and children:
- Treating infants and children under 2 with any herbal preparation is hazardous.

Others:
- None expected if you are beyond childhood and under 45, basically healthy and take for only a short time.

Storage:
- Keep cool and dry, but don't freeze. Store safely away from children.

Safe dosage:
- At present no "safe" dosage has been established.

Toxicity

Rated dangerous, particularly in children, persons over 55 and those who take larger than appropriate quantities for extended periods of time.

For symptoms of toxicity: See *Adverse Reactions, Side Effects or Overdose Symptoms* section below.

Adverse Reactions, Side Effects or Overdose Symptoms

Signs and symptoms:	What to do:
Coma	Seek emergency treatment.
Confusion	Discontinue. Call doctor immediately.
Irregular heartbeat	Seek emergency treatment.

Basic Information

Biological name (genus and species):
Althea officinalis
Parts used for medicinal purposes:
Leaves
Roots
Chemicals this herb contains:
Asparagine
Fat
Mucilage (See Glossary)
Pectin
Starch
Sugar

Known Effects

• Softens or soothes skin.

Miscellaneous information:
• Marshmallow plant is used as a "filler" in a variety of pills.

Unproved Speculated Benefits

• Protects injured or scraped skin or mucous membranes.
• Used as a poultice for applying medications. (See Glossary.)

Warnings and Precautions

Don't take if you:
• Are pregnant, think you may be pregnant or plan pregnancy in the near future.
• Have any chronic disease of the gastrointestinal tract, such as stomach or duodenal ulcers, esophageal reflux (reflux esophagitis), ulcerative colitis, spastic colitis, diverticulosis, diverticulitis.

Consult your doctor if you:
• Take this herb for any medical problem that doesn't improve in 2 weeks. There may be safer, more-effective treatments.
• Take any medicinal drugs or herbs including aspirin, laxatives, cold and cough remedies, antacids, vitamins, minerals, amino acids, supplements, other prescription or non-prescription drugs.

Pregnancy:
• Problems in pregnant women taking small or usual amounts have not been proved. But the chance of problems does exist. Don't use unless prescribed by your doctor.

Breast-feeding:
• Problems in breast-fed infants of lactating mothers taking small or usual amounts have not been proved. But the chance of problems does exist. Don't use unless prescribed by your doctor.

Infants and children:
• Treating infants and children under 2 with any herbal preparation is hazardous.

Others:
• None expected if you are beyond childhood and under 45, basically healthy and take for only a short time.

Storage:
• Keep cool and dry, but don't freeze. Store safely away from children.

Safe dosage:
• At present no "safe" dosage has been established.

Toxicity

Comparative-toxicity rating not available from standard references.

Adverse Reactions, Side Effects or Overdose Symptoms

None expected

MEDICINAL HERB

Mayapple (American Mandrake)

Basic Information

Biological name (genus and species):
 Podophyllum peltatum
Parts used for medicinal purposes:
 Roots
Chemicals this herb contains:
 Alpha-peltatin
 Beta-peltatin
 Podophyllotoxin

Known Effects

• Inhibits or prevents cell division.
• Stimulates gastrointestinal tract.

Unproved Speculated Benefits

• Treats constipation.
• Treats recurrent fecal impactions.
• Is used as topical application for virus infections of skin around genitals.

Warnings and Precautions

Don't take if you:
• Are pregnant, think you may be pregnant or plan pregnancy in the near future.
• Have any chronic disease of the gastrointestinal tract, such as stomach or duodenal ulcers, esophageal reflux (reflux esophagitis), ulcerative colitis, spastic colitis, diverticulosis, diverticulitis.

Consult your doctor if you:
• Take this herb for any medical problem that doesn't improve in 2 weeks. There may be safer, more-effective treatments.
• Take any medicinal drugs or herbs including aspirin, laxatives, cold and cough remedies, antacids, vitamins, minerals, amino acids, supplements, other prescription or non-prescription drugs.

Pregnancy:
• Problems in pregnant women taking small or usual amounts have not been proved. But the chance of problems does exist. Don't use unless prescribed by your doctor.

Breast-feeding:
• Problems in breast-fed infants of lactating mothers taking small or usual amounts have not been proved. But the chance of problems does exist. Don't use unless prescribed by your doctor.

Infants and children:
• Treating infants and children under 2 with any herbal preparation is hazardous.

Others:
• None expected if you are beyond childhood and under 45, basically healthy and take for only a short time.

Storage:
• Keep cool and dry, but don't freeze. Store safely away from children.

Safe dosage:
• At present no "safe" dosage has been established.

Toxicity

Rated slightly dangerous, particularly in children, persons over 55 and those who take larger than appropriate quantities for extended periods of time.

For symptoms of toxicity: See *Adverse Reactions, Side Effects or Overdose Symptoms* section below.

Adverse Reactions, Side Effects or Overdose Symptoms

Signs and symptoms:	What to do:
Diarrhea	Discontinue. Call doctor immediately.
Drowsiness	Discontinue. Call doctor when convenient.
Lethargy	Discontinue. Call doctor when convenient.
Nausea	Discontinue. Call doctor immediately.
Unconsciousness	Seek emergency treatment.
Vomiting	Discontinue. Call doctor immediately.

Meadowsweet (Spirea, Queen-of-the-Meadow)

Basic Information

Biological name (genus and species):
  Filipendula ulmaria
Parts used for medicinal purposes:
  Petals/flower
  Roots
Chemicals this herb contains:
  Gallic acid
  Methyl salicylate
  Salicylic acid
  Salicylic aldehyde
  Tannic acid
  Volatile oils (See Glossary)

Known Effects

• Shrinks tissues.
• Prevents secretion of fluids.

Unproved Speculated Benefits

• Helps body dispose of excess fluid by increasing amount of urine produced.
• Treats diarrhea.
• Reduces pain.

Warnings and Precautions

Don't take if you:
• Are pregnant, think you may be pregnant or plan pregnancy in the near future.
• Have chronic kidney problems.

Consult your doctor if you:
• Take this herb for any medical problem that doesn't improve in 2 weeks. There may be safer, more-effective treatments.
• Take any medicinal drugs or herbs including aspirin, laxatives, cold and cough remedies, antacids, vitamins, minerals, amino acids, supplements, other prescription or non-prescription drugs.

Pregnancy:
• Problems in pregnant women taking small or usual amounts have not been proved. But the chance of problems does exist. Don't use unless prescribed by your doctor.

Breast-feeding:
• Problems in breast-fed infants of lactating mothers taking small or usual amounts have not been proved. But the chance of problems does exist. Don't use unless prescribed by your doctor.

Infants and children:
• Treating infants and children under 2 with any herbal preparation is hazardous.

Others:
• None expected if you are under 45, not pregnant, basically healthy, take it for only a short time and do not exceed manufacturer's recommended dosage.

Storage:
• Keep cool and dry, but don't freeze. Store safely away from children.

Safe dosage:
• At present no "safe" dosage has been established.

Toxicity

Generally regarded as safe when taken in appropriate quantities for short periods of time.

For symptoms of toxicity: See *Adverse Reactions, Side Effects or Overdose Symptoms* section below.

Adverse Reactions, Side Effects or Overdose Symptoms

Signs and symptoms:	What to do:
Coma	Seek emergency treatment.
Kidney damage characterized by blood in urine, decreased urine flow, swelling of hands and feet	Seek emergency treatment.
Lethargy	Discontinue. Call doctor when convenient.
Unconsciousness	Seek emergency treatment.

Mexican Sarsaparilla

Basic Information

Biological name (genus and species):
   Smilax aristolochiaefolia, S. regelii,
     S. febrifuga, S. ornata
Parts used for medicinal purposes:
   Bark
   Berries
   Roots
Chemicals this herb contains:
   Resin (See Glossary)
   Sarsasapogenin
   Smilagenin
   Starch
   Stigmasterol
   Volatile oils (See Glossary)

Known Effects

- Depresses central nervous system.
- Irritates mucous membranes.
- Irritates gastrointestinal tract.
- Helps body dispose of excess fluid by increasing amount of urine produced.

Miscellaneous information:
- Berries are edible.
- Berries, bark and other parts of plant are used to make the soft drink of the same name.

Unproved Speculated Benefits

- Relieves toothache.
- Increases sexual potency.
- Treats psoriasis.
- Temporarily relieves constipation.

Warnings and Precautions

Don't take if you:
- Are pregnant, think you may be pregnant or plan pregnancy in the near future.

Consult your doctor if you:
- Take this herb for any medical problem that doesn't improve in 2 weeks. There may be safer, more-effective treatments.
- Take any medicinal drugs or herbs including aspirin, laxatives, cold and cough remedies, antacids, vitamins, minerals, amino acids, supplements, other prescription or non-prescription drugs.

Pregnancy:
- Problems in pregnant women taking small or usual amounts have not been proved. But the chance of problems does exist. Don't use unless prescribed by your doctor.

Breast-feeding:
- Problems in breast-fed infants of lactating mothers taking small or usual amounts have not been proved. But the chance of problems does exist. Don't use unless prescribed by your doctor.

Infants and children:
- Treating infants and children under 2 with any herbal preparation is hazardous.

Others:
- None expected if you are under 45, not pregnant, basically healthy, take it for only a short time and do not exceed manufacturer's recommended dosage.

Storage:
- Keep cool and dry, but don't freeze. Store safely away from children.

Safe dosage:
- At present no "safe" dosage has been established.

Toxicity

Rated relatively safe when taken in appropriate quantities for short periods of time.

Adverse Reactions, Side Effects or Overdose Symptoms

None expected

Milkweed, Common (Blood-flower)

Basic Information

Biological name (genus and species):
Asclepias syriaca
Parts used for medicinal purposes:
Roots
Chemicals this herb contains:
Asclepiadin   Galitoxin
Asclepion (a bitter)

Known Effects

- Irritates and stimulates gastrointestinal tract.
- Decreases thickness and increases fluidity of mucus from lungs and bronchial tubes.
- Increases perspiration.

Miscellaneous information:
- *All parts* of milkweed plant may be toxic.

Unproved Speculated Benefits

- Treats bronchitis.
- Treats arthritis.

Warnings and Precautions

Don't take if you:
- Are pregnant, think you may be pregnant or plan pregnancy in the near future.
- Have any chronic disease of the gastrointestinal tract, such as stomach or duodenal ulcers, esophageal reflux (reflux esophagitis), ulcerative colitis, spastic colitis, diverticulosis, diverticulitis.

Consult your doctor if you:
- Take this herb for any medical problem that doesn't improve in 2 weeks. There may be safer, more-effective treatments.
- Take any medicinal drugs or herbs including aspirin, laxatives, cold and cough remedies, antacids, vitamins, minerals, amino acids, supplements, other prescription or non-prescription drugs.

Pregnancy:
- Dangers outweigh any possible benefits. Don't use.

Breast-feeding:
- Dangers outweigh any possible benefits. Don't use.

Infants and children:
- Treating infants and children under 2 with any herbal preparation is hazardous.

Others:
- Dangers outweigh any possible benefits. Don't use.

Storage:
- Keep cool and dry, but don't freeze. Store safely away from children.

Safe dosage:
- At present no "safe" dosage has been established.

Toxicity

Rated slightly dangerous, particularly in children, persons over 55 and those who take larger than appropriate quantities for extended periods of time.

For symptoms of toxicity: See *Adverse Reactions, Side Effects or Overdose Symptoms* section below.

Adverse Reactions, Side Effects or Overdose Symptoms

Signs and symptoms:	What to do:
Coma	Seek emergency treatment.
Diarrhea	Discontinue. Call doctor immediately.
Drowsiness	Discontinue. Call doctor when convenient.
Jaundice (yellow skin and eyes)	Discontinue. Call doctor immediately.
Kidney damage characterized by blood in urine, decreased urine flow, swelling of hands and feet	Seek emergency treatment.
Lethargy	Discontinue. Call doctor when convenient.
Loss of appetite	Discontinue. Call doctor when convenient.
Nausea	Discontinue. Call doctor immediately.
Seizures	Seek emergency treatment.
Unsteady gait	Discontinue. Call doctor immediately.
Vomiting	Discontinue. Call doctor immediately.

Milkwort

Basic Information
Biological name (genus and species):
Polygala vulgaris, P. senega
Parts used for medicinal purposes:
Roots
Chemical this herb contains:
Saponins (See Glossary)

Known Effects

- Causes increased secretions from bronchial tubes.
- Irritates intestinal tract.
- Decreases thickness and increases fluidity of mucus from lungs and bronchial tubes.
- Helps body dispose of excess fluid by increasing amount of urine produced.
- Increases perspiration.

Unproved Speculated Benefits

- Treats croup.
- Treats arthritis.
- Treats hives.
- Treats gout.
- Treats pleurisy.
- Treats constipation.
- Increases milk production in lactating women.

Warnings and Precautions

Don't take if you:
- Are pregnant, think you may be pregnant or plan pregnancy in the near future.
- Have any chronic disease of the gastrointestinal tract, such as stomach or duodenal ulcers, esophageal reflux (reflux esophagitis), ulcerative colitis, spastic colitis, diverticulosis, diverticulitis.

Consult your doctor if you:
- Take this herb for any medical problem that doesn't improve in 2 weeks. There may be safer, more-effective treatments.
- Take any medicinal drugs or herbs including aspirin, laxatives, cold and cough remedies, antacids, vitamins, minerals, amino acids, supplements, other prescription or non-prescription drugs.

Pregnancy:
- Dangers outweigh any possible benefits. Don't use.

Breast-feeding:
- Dangers outweigh any possible benefits. Don't use.

Infants and children:
- Treating infants and children under 2 with any herbal preparation is hazardous.

Others:
- None expected if you are under 45, not pregnant, basically healthy, take it for only a short time and do not exceed manufacturer's recommended dosage.

Storage:
- Keep cool and dry, but don't freeze. Store safely away from children.

Safe dosage:
- At present no "safe" dosage has been established.

Toxicity

Comparative-toxicity rating not available from standard references.

For symptoms of toxicity: See *Adverse Reactions, Side Effects or Overdose Symptoms* section below.

Adverse Reactions, Side Effects or Overdose Symptoms

Signs and symptoms:	What to do:
Coma	Seek emergency treatment.
Diarrhea	Discontinue. Call doctor immediately.
Drowsiness	Discontinue. Call doctor when convenient.
Lethargy	Discontinue. Call doctor when convenient.
Nausea, violent	Discontinue. Call doctor immediately.
Vomiting	Discontinue. Call doctor immediately.

Mistletoe

Basic Information

Biological name (genus and species):
  Phoradendron serotinum
Parts used for medicinal purposes:
  Berries/fruits
  Leaves
  Stems
Chemicals this herb contains:
  Beta phenylethylamine
  Tyramine

 ## Known Effects

- Stimulates central nervous system.
- Increases blood pressure.
- Causes contraction of smooth muscle in intestines or uterus.

Miscellaneous information:
- Mistletoe is particularly dangerous for people taking monamine-oxidase medications to treat high blood pressure.

 ## Unproved Speculated Benefits

- Controls excessive bleeding after childbirth.
- Treats cholera.
- Calms nervousness.

 ## Warnings and Precautions

Don't take if you:
- Are pregnant, think you may be pregnant or plan pregnancy in the near future.
- Have any chronic disease of the gastrointestinal tract, such as stomach or duodenal ulcers, esophageal reflux (reflux esophagitis), ulcerative colitis, spastic colitis, diverticulosis, diverticulitis.

Consult your doctor if you:
- Take this herb for any medical problem that doesn't improve in 2 weeks. There may be safer, more-effective treatments.
- Take any medicinal drugs or herbs including aspirin, laxatives, cold and cough remedies, antacids, vitamins, minerals, amino acids, supplements, other prescription or non-prescription drugs.

Pregnancy:
- Dangers outweigh any possible benefits. Don't use.

Breast-feeding:
- Dangers outweigh any possible benefits. Don't use.

Infants and children:
- Treating infants and children under 2 with any herbal preparation is hazardous.

Others:
- Do not allow children to eat berries of this popular Christmas plant. As few as one or two berries may cause toxic symptoms.

Storage:
- Keep cool and dry, but don't freeze. Store safely away from children.

Safe dosage:
- At present no "safe" dosage has been established.

 ## Toxicity

Rated slightly dangerous, particularly in children, persons over 55 and those who take larger than appropriate quantities for extended periods of time.

For symptoms of toxicity: See *Adverse Reactions, Side Effects or Overdose Symptoms* section below.

 ## Adverse Reactions, Side Effects or Overdose Symptoms

Signs and symptoms:	What to do:
Convulsions	Seek emergency treatment.
Diarrhea	Discontinue. Call doctor immediately.
Hallucinations	Seek emergency treatment.
Increased blood pressure	Discontinue. Call doctor immediately.
Nausea	Discontinue. Call doctor immediately.
Slow heartbeat	Seek emergency treatment.
Vomiting	Discontinue. Call doctor immediately.

Mormon Tea (Nevada Jointfir)

Basic Information

Biological name (genus and species):
Ephedra nevadensis, E. trifurca
Parts used for medicinal purposes:
Stems
Chemical this herb contains:
Ephedrine

 ## Known Effects

- Stimulates central nervous system.
- Increases blood pressure.
- Increases heart rate.
- Helps body dispose of excess fluid by increasing amount of urine produced.

 ## Unproved Speculated Benefits

- Elevates mood.
- Treats congestive heart failure, kidney failure, liver failure.
- Decreases appetite.
- Stimulates energy.
- Treats fatigue.

 ## Warnings and Precautions

Don't take if you:
- Are pregnant, think you may be pregnant or plan pregnancy in the near future.
- Have diabetes mellitus. It may make control with diet or insulin more difficult.
- Have heart disease.

Consult your doctor if you:
- Take this herb for any medical problem that doesn't improve in 2 weeks. There may be safer, more-effective treatments.
- Take any medicinal drugs or herbs including aspirin, laxatives, cold and cough remedies, antacids, vitamins, minerals, amino acids, supplements, other prescription or non-prescription drugs.

Pregnancy:
- Dangers outweigh any possible benefits. Don't use.

Breast-feeding:
- Dangers outweigh any possible benefits. Don't use.

Infants and children:
- Treating infants and children under 2 with any herbal preparation is hazardous.

Others:
- None expected if you are beyond childhood and under 45, not pregnant, basically healthy and take for only a short time.

Storage:
- Keep cool and dry, but don't freeze. Store safely away from children.

Safe dosage:
- At present no "safe" dosage has been established.

 ## Toxicity

Rated slightly dangerous, particularly in children, persons over 55 and those who take larger than appropriate quantities for extended periods of time.

For symptoms of toxicity: See *Adverse Reactions, Side Effects or Overdose Symptoms* section below.

 ## Adverse Reactions, Side Effects or Overdose Symptoms

Signs and symptoms:	What to do:
Excessively high blood pressure	Seek emergency treatment.
Irregular heartbeat	Seek emergency treatment.
Rapid heartbeat	Discontinue. Call doctor immediately.

Basic Information
Biological name (genus and species):
Ipomoea purpurea
Parts used for medicinal purposes:
Seeds
Chemicals this herb contains:
Cetyl alcohol
Dihydroxycinnamic acid
Lysergic acid
Scopoletin

Known Effects

- Stimulates central nervous system.
- Stimulates gastrointestinal tract.

Miscellaneous information.
- May cause hallucinations.

Unproved Speculated Benefits

- Is used as purgative for constipation.
- Elevates mood.

Warnings and Precautions

Don't take if you:
- Are pregnant, think you may be pregnant or plan pregnancy in the near future.
- Have any chronic disease of the gastrointestinal tract, such as stomach or duodenal ulcers, esophageal reflux (reflux esophagitis), ulcerative colitis, spastic colitis, diverticulosis, diverticulitis.

Consult your doctor if you:
- Take this herb for any medical problem that doesn't improve in 2 weeks. There may be safer, more-effective treatments.
- Take any medicinal drugs or herbs including aspirin, laxatives, cold and cough remedies, antacids, vitamins, minerals, amino acids, supplements, other prescription or non-prescription drugs.

Pregnancy:
- Problems in pregnant women taking small or usual amounts have not been proved. But the chance of problems does exist. Don't use unless prescribed by your doctor.

Breast-feeding:
- Problems in breast-fed infants of lactating mothers taking small or usual amounts have not been proved. But the chance of problems does exist. Don't use unless prescribed by your doctor.

Infants and children:
- Treating infants and children under 2 with any herbal preparation is hazardous.

Others:
- None expected if you are beyond childhood and under 45, basically healthy and take for only a short time.

Storage:
- Keep cool and dry, but don't freeze. Store safely away from children.

Safe dosage:
- At present no "safe" dosage has been established.

Toxicity

Rated slightly dangerous, particularly in children, persons over 55 and those who take larger than appropriate quantities for extended periods of time.

For symptoms of toxicity: See *Adverse Reactions, Side Effects or Overdose Symptoms* section below.

Adverse Reactions, Side Effects or Overdose Symptoms

Signs and symptoms:	What to do:
Confusion	Discontinue. Call doctor immediately.
Diarrhea, explosive and watery	Discontinue. Call doctor immediately.
Disturbed vision	Discontinue. Call doctor immediately.
Hallucinations	Seek emergency treatment.
Nausea	Discontinue. Call doctor immediately.
Vomiting	Discontinue. Call doctor immediately.

MEDICINAL HERB

Mountain Ash (Rowan Tree)

Basic Information

Biological name (genus and species):
  Sorbus aucuparia
Parts used for medicinal purposes:
  Berries/fruits
  Seeds
Chemicals this herb contains:
  Fixed oil (See Glossary)
  Malic acid
  Sorbic acid
  Sorbitol
  Sorbose

Known Effects

- Irritates and stimulates gastrointestinal tract.
- Helps body dispose of excess fluid by increasing amount of urine produced.

Miscellaneous information:
- Is used as a sweetener.

Unproved Speculated Benefits

- Prevents scurvy.
- Treats hemorrhoids.
- Treats stomach and duodenal ulcers.

Warnings and Precautions

Don't take if you:
- Are pregnant, think you may be pregnant or plan pregnancy in the near future.

Consult your doctor if you:
- Take this herb for any medical problem that doesn't improve in 2 weeks. There may be safer, more-effective treatments.
- Take any medicinal drugs or herbs including aspirin, laxatives, cold and cough remedies, antacids, vitamins, minerals, amino acids, supplements, other prescription or non-prescription drugs.

Pregnancy:
- Problems in pregnant women taking small or usual amounts have not been proved. But the chance of problems does exist. Don't use unless prescribed by your doctor.

Breast-feeding:
- Problems in breast-fed infants of lactating mothers taking small or usual amounts have not been proved. But the chance of problems does exist. Don't use unless prescribed by your doctor.

Infants and children:
- Treating infants and children under 2 with any herbal preparation is hazardous.

Others:
- None expected if you are beyond childhood and under 45, basically healthy and take for only a short time.

Storage:
- Keep cool and dry, but don't freeze. Store safely away from children.

Safe dosage:
- At present no "safe" dosage has been established.

Toxicity

Comparative-toxicity rating not available from standard references.

For symptoms of toxicity: See *Adverse Reactions, Side Effects or Overdose Symptoms* section below.

Adverse Reactions, Side Effects or Overdose Symptoms

Signs and symptoms:	What to do:
Diarrhea	Discontinue. Call doctor immediately.

Mountain Tobacco (Leopard's Bane, Wolf's Bane)

Basic Information

Biological name (genus and species):
  Arnica montana
Parts used for medicinal purposes:
  Petals/flower
Chemicals this herb contains:
  Angelic acid
  Arnidendiol (also found in dandelion flowers).
  Choline
  Fatty acids
  Formic acid
  Thymohydroquinone

Known Effects

- Provides counterirritation when applied to skin overlying an inflamed or irritated joint.
- Depresses central nervous system.
- Irritates gastrointestinal tract.

Unproved Speculated Benefits

- Relieves discomfort of sprains, strains, bruises when applied to skin over injury.

Warnings and Precautions

Don't take if you:
- Are pregnant, think you may be pregnant or plan pregnancy in the near future.
- Have any chronic disease of the gastrointestinal tract, such as stomach or duodenal ulcers, esophageal reflux (reflux esophagitis), ulcerative colitis, spastic colitis, diverticulosis, diverticulitis.

Consult your doctor if you:
- Take this herb for any medical problem that doesn't improve in 2 weeks. There may be safer, more-effective treatments.
- Take any medicinal drugs or herbs including aspirin, laxatives, cold and cough remedies, antacids, vitamins, minerals, amino acids, supplements, other prescription or non-prescription drugs.

Pregnancy:
- Dangers outweigh any possible benefits. Don't use.

Breast-feeding:
- Dangers outweigh any possible benefits. Don't use.

Infants and children:
- Treating infants and children under 2 with any herbal preparation is hazardous.

Others:
- Don't take internally. Probably safe for application to skin.

Storage:
- Keep cool and dry, but don't freeze. Store safely away from children.

Safe dosage:
- At present no "safe" dosage has been established.

Toxicity

Rated slightly dangerous, particularly in children, persons over 55 and those who take larger than appropriate quantities for extended periods of time.

For symptoms of toxicity: See *Adverse Reactions, Side Effects or Overdose Symptoms* section below.

Adverse Reactions, Side Effects or Overdose Symptoms

Signs and symptoms:	What to do:
Explosive, watery diarrhea	Discontinue. Call doctor immediately.
Heartbeat irregularities	Seek emergency treatment.
Muscle weakness	Discontinue. Call doctor immediately.
Nausea	Discontinue. Call doctor immediately.
Precipitous blood-pressure drop—symptoms include, faintness, cold sweat, paleness, rapid pulse	Seek emergency treatment.
Vomiting	Discontinue. Call doctor immediately.

Mulberry

Basic Information

Biological name (genus and species):
  Morus rubra
Parts used for medicinal purposes:
  Bark
  Berries/fruits
Chemicals this herb contains:
  Unidentified

Known Effects

- Stimulates gastrointestinal tract.
- Depresses central nervous system.

Unproved Speculated Benefits

- Reduces fever.
- Induces drowsiness.
- Acts as a mild laxative.

Warnings and Precautions

Don't take if you:
- Are pregnant, think you may be pregnant or plan pregnancy in the near future.
- Have any chronic disease of the gastrointestinal tract, such as stomach or duodenal ulcers, esophageal reflux (reflux esophagitis), ulcerative colitis, spastic colitis, diverticulosis, diverticulitis.

Consult your doctor if you:
- Take this herb for any medical problem that doesn't improve in 2 weeks. There may be safer, more-effective treatments.
- Take any medicinal drugs or herbs including aspirin, laxatives, cold and cough remedies, antacids, vitamins, minerals, amino acids, supplements, other prescription or non-prescription drugs.

Pregnancy:
- Dangers outweigh any possible benefits. Don't use.

Breast-feeding:
- Dangers outweigh any possible benefits. Don't use.

Infants and children:
- Treating infants and children under 2 with any herbal preparation is hazardous.

Others:
- This product will not help you and may cause toxic symptoms.

Storage:
- Keep cool and dry, but don't freeze. Store safely away from children.

Safe dosage:
- At present no "safe" dosage has been established.

Toxicity

Comparative-toxicity rating not available from standard references.

For symptoms of toxicity: See *Adverse Reactions, Side Effects or Overdose Symptoms* section below.

Adverse Reactions, Side Effects or Overdose Symptoms

Signs and symptoms:	What to do:
Diarrhea	Discontinue. Call doctor immediately.
Hallucinations	Seek emergency treatment.
Nausea	Discontinue. Call doctor immediately.
Vomiting	Discontinue. Call doctor immediately.

Basic Information

Biological name (genus and species):
  Verbascum thapsiforme, V. phlomoides,
  V. thapsus
Parts used for medicinal purposes:
  Leaves
Chemical this herb contains:
  Saponins (See Glossary)

 ## Known Effects

• Covers and protects scraped tissues.
• Softens and soothes irritated skin.
• Shrinks tissues.
• Prevents secretion of fluids.

Miscellaneous information:
• Action of mullein when taken orally is probably
  too weak to be effective.

 ## Unproved Speculated Benefits

• Smoking mullein relieves bronchial irritation.
• Topical applications treat sunburn,
  hemorrhoids, injured skin and mucous
  membranes.

 ## Warnings and Precautions

Don't take if you:
• Are pregnant, think you may be pregnant or
  plan pregnancy in the near future.

Consult your doctor if you:
• Take this herb for any medical problem that
  doesn't improve in 2 weeks. There may be
  safer, more-effective treatments.
• Take any medicinal drugs or herbs including
  aspirin, laxatives, cold and cough remedies,
  antacids, vitamins, minerals, amino acids,
  supplements, other prescription or non-
  prescription drugs.

Pregnancy:
• Problems in pregnant women taking small or
  usual amounts have not been proved. But the
  chance of problems does exist. Don't use
  unless prescribed by your doctor.

Breast-feeding:
• Problems in breast-fed infants of lactating
  mothers taking small or usual amounts have
  not been proved. But the chance of problems
  does exist. Don't use unless prescribed by your
  doctor.

Infants and children:
• Treating infants and children under 2 with any
  herbal preparation is hazardous.

Others:
• None expected if you are beyond childhood
  and under 45, basically healthy and take for
  only a short time.

Storage:
• Keep cool and dry, but don't freeze. Store
  safely away from children.

Safe dosage:
• At present no "safe" dosage has been
  established.

 ## Toxicity

Comparative-toxicity rating not available from
standard references.

 ## Adverse Reactions, Side Effects or Overdose Symptoms

None expected

Myrrh

Basic Information

Biological name (genus and species):
Commiphora molmol
Parts used for medicinal purposes:
Leaves
Resin from stems
Chemicals this herb contains:
Acetic acid
Formic acid
Myrrholic acids
Resin (See Glossary)
Volatile oils (See Glossary)

 ## Known Effects

- Stimulates muscular movements of intestines.
- Shrinks tissues.
- Prevents secretion of fluids to mucous membranes.
- Stimulates gastrointestinal tract.
- Aids in expelling gas from intestinal tract.

Miscellaneous information:
- Primary use of myrrh is in perfumes and incense.

 ## Unproved Speculated Benefits

- Causes watery, explosive bowel movements.
- Treats dyspepsia.
- Is used as mouthwash.

 ## Warnings and Precautions

Don't take if you:
- Are pregnant, think you may be pregnant or plan pregnancy in the near future.

Consult your doctor if you:
- Take this herb for any medical problem that doesn't improve in 2 weeks. There may be safer, more-effective treatments.
- Take any medicinal drugs or herbs including aspirin, laxatives, cold and cough remedies, antacids, vitamins, minerals, amino acids, supplements, other prescription or non-prescription drugs.

Pregnancy:
- Problems in pregnant women taking small or usual amounts have not been proved. But the chance of problems does exist. Don't use unless prescribed by your doctor.

Breast-feeding:
- Problems in breast-fed infants of lactating mothers taking small or usual amounts have not been proved. But the chance of problems does exist. Don't use unless prescribed by your doctor.

Infants and children:
- Treating infants and children under 2 with any herbal preparation is hazardous.

Others:
- None expected if you are under 45, not pregnant, basically healthy, take it for only a short time and do not exceed manufacturer's recommended dosage.

Storage:
- Keep cool and dry, but don't freeze. Store safely away from children.

Safe dosage:
- At present no "safe" dosage has been established.

 ## Toxicity

Comparative-toxicity rating not available from standard references.

For symptoms of toxicity: See *Adverse Reactions, Side Effects or Overdose Symptoms* section below.

 ## Adverse Reactions, Side Effects or Overdose Symptoms

Signs and symptoms:	What to do:
Convulsions	Seek emergency treatment.
Drowsiness	Discontinue. Call doctor when convenient.
Lethargy	Discontinue. Call doctor when convenient.

Basic Information

Biological name (genus and species):
  Myrtus communis
Parts used for medicinal purposes:
  Leaves
Chemicals this herb contains:
  d-pinene
  Eucalyptol
  Myrol

Known Effects

- Irritates mucous membranes.
- Large amounts may depress central nervous system.

Miscellaneous information:
- Myrtle is used as a condiment, flavoring and perfume essence.

Unproved Speculated Benefits

- Is used as gargle.
- Treats stomach irritations.
- Treats bronchitis.
- Treats cystitis.

Warnings and Precautions

Don't take if you:
- Are pregnant, think you may be pregnant or plan pregnancy in the near future.
- Have chronic kidney disease.

Consult your doctor if you:
- Take this herb for any medical problem that doesn't improve in 2 weeks. There may be safer, more-effective treatments.
- Take any medicinal drugs or herbs including aspirin, laxatives, cold and cough remedies, antacids, vitamins, minerals, amino acids, supplements, other prescription or non-prescription drugs.

Pregnancy:
- Problems in pregnant women taking small or usual amounts have not been proved. But the chance of problems does exist. Don't use unless prescribed by your doctor.

Breast-feeding:
- Problems in breast-fed infants of lactating mothers taking small or usual amounts have not been proved. But the chance of problems does exist. Don't use unless prescribed by your doctor.

Infants and children:
- Treating infants and children under 2 with any herbal preparation is hazardous.

Others:
- None expected if you are beyond childhood and under 45, basically healthy and take for only a short time.

Storage:
- Keep cool and dry, but don't freeze. Store safely away from children.

Safe dosage:
- At present no "safe" dosage has been established.

Toxicity

Comparative-toxicity rating not available from standard references.

For symptoms of toxicity: See *Adverse Reactions, Side Effects or Overdose Symptoms* section below.

Adverse Reactions, Side Effects or Overdose Symptoms

Signs and symptoms:	What to do:
Coma	Seek emergency treatment.
Convulsions	Seek emergency treatment.
Kidney damage characterized by blood in urine, decreased urine flow, swelling of hands and feet	Seek emergency treatment.

MEDICINAL HERB

Oak Bark

Basic Information

Biological name (genus and species):
  Quercus
Parts used for medicinal purposes:
  Bark
  Seeds
Chemical this herb contains:
  Quercitannic acid

 Known Effects

- Shrinks tissues.
- Prevents secretion of fluids.
- Causes protein molecules to clump together.

 Unproved Speculated Benefits

- Treats hemorrhoids.
- Treats diarrhea.
- Is used as gargle for sore throats.

 Warnings and Precautions

Don't take if you:
- Are pregnant, think you may be pregnant or plan pregnancy in the near future.
- Have any chronic disease of the gastrointestinal tract, such as stomach or duodenal ulcers, esophageal reflux (reflux esophagitis), ulcerative colitis, spastic colitis, diverticulosis, diverticulitis.

Consult your doctor if you:
- Take this herb for any medical problem that doesn't improve in 2 weeks. There may be safer, more-effective treatments.
- Take any medicinal drugs or herbs including aspirin, laxatives, cold and cough remedies, antacids, vitamins, minerals, amino acids, supplements, other prescription or non-prescription drugs.

Pregnancy:
- Dangers outweigh any possible benefits. Don't use.

Breast-feeding:
- Dangers outweigh any possible benefits. Don't use.

Infants and children:
- Treating infants and children under 2 with any herbal preparation is hazardous.

Others:
- None expected if you are beyond childhood and under 45, basically healthy and take for only a short time.

Storage:
- Keep cool and dry, but don't freeze. Store safely away from children.

Safe dosage:
- At present no "safe" dosage has been established.

 Toxicity

Comparative-toxicity rating not available from standard references.

For symptoms of toxicity: See *Adverse Reactions, Side Effects or Overdose Symptoms* section below.

 Adverse Reactions, Side Effects or Overdose Symptoms

Signs and symptoms:	What to do:
Constipation	Discontinue. Call doctor when convenient.
Dry mouth	Discontinue. Call doctor when convenient.
Increased urination	Discontinue. Call doctor when convenient.
Jaundice (yellow skin and eyes)	Discontinue. Call doctor immediately.
Kidney damage characterized by blood in urine, decreased urine flow, swelling of hands and feet	Seek emergency treatment.
Skin eruptions	Discontinue. Call doctor when convenient.
Thirst	Discontinue. Call doctor when convenient.

Basic Information

Biological name (genus and species):
 Avena sativa
Parts used for medicinal purposes:
 Seeds
Chemicals this herb contains:
 Albumin
 Gluten
 Gum oil
 Protein compound
 Salts
 Saponin (See Glossary)
 Starch
 Sugar

Known Effects

- Stimulates muscular contractions.
- Coats and protects scraped tissues.
- Stimulates central nervous system.

Miscellaneous information:

- "Feeling his oats" refers to the stimulant effect of this herb on some animals, particularly horses.

Unproved Speculated Benefits

- Is a satisfactory food source.
- Is used as a tonic.
- Decreases depression.
- Decreases dependence on nicotine and narcotics.

Warnings and Precautions

Don't take if you:

- No absolute contraindications.

Consult your doctor if you:

- Take this herb for any medical problem that doesn't improve in 2 weeks. There may be safer, more-effective treatments.
- Take any medicinal drugs or herbs including aspirin, laxatives, cold and cough remedies, antacids, vitamins, minerals, amino acids, supplements, other prescription or non-prescription drugs.

Pregnancy:

- Pregnant women should experience no problems taking usual amounts as part of a balanced diet. Other products extracted from this substance have not been proved to cause problems.

Breast-feeding:

- Breast-fed infants of lactating mothers should experience no problems when mother takes usual amounts as part of a balanced diet. Other products extracted from this substance have not been proved to cause problems.

Infants and children:

- Treating infants and children under 2 with any herbal preparation is hazardous.

Others:

- None expected if you are beyond childhood and under 45, basically healthy and take for only a short time.

Storage:

- Keep cool and dry, but don't freeze. Store safely away from children.

Safe dosage:

- At present no "safe" dosage has been established.

Toxicity

Comparative-toxicity rating not available from standard references.

Adverse Reactions, Side Effects or Overdose Symptoms

None expected

Orris Root (Black Flag)

Basic Information
Biological name (genus and species):
  Iris versicolor, Iris spp
Parts used for medicinal purposes:
  Roots
Chemicals this herb contains:
  Gum (See Glossary)
  Oleoresin (See Glossary)
  Tannins (See Glossary)

Known Effects

• Depresses central nervous system.
• Causes vomiting.
• Interferes with absorption of iron and other minerals when taken internally.

Unproved Speculated Benefits

• Treats skin disorders.
• Treats arthritis.
• Treats tumors.

Warnings and Precautions

Don't take if you:
• Are pregnant, think you may be pregnant or plan pregnancy in the near future.
• Have any chronic disease of the gastrointestinal tract, such as stomach or duodenal ulcers, esophageal reflux (reflux esophagitis), ulcerative colitis, spastic colitis, diverticulosis, diverticulitis.

Consult your doctor if you:
• Take this herb for any medical problem that doesn't improve in 2 weeks. There may be safer, more-effective treatments.
• Take any medicinal drugs or herbs including aspirin, laxatives, cold and cough remedies, antacids, vitamins, minerals, amino acids, supplements, other prescription or non-prescription drugs.

Pregnancy:
• Problems in pregnant women taking small or usual doses have not been proved. But the chance of problems does exist. Don't use unless prescribed by your doctor.

Breast-feeding:
• Problems in breast-fed infants of lactating mothers taking small or usual doses have not been proved. But the chance of problems does exist. Don't use unless prescribed by your doctor.

Infants and children:
• Treating infants and children under 2 with any herbal preparation is hazardous.

Others:
• Don't use. This product will not help you and may cause toxic symptoms.

Storage:
• Keep cool and dry, but don't freeze. Store safely away from children.

Safe dosage:
• At present no "safe" dosage has been established.

Toxicity

Rated relatively safe when taken in appropriate quantities for short periods of time.

For symptoms of toxicity: See *Adverse Reactions, Side Effects or Overdose Symptoms* section below.

Adverse Reactions, Side Effects or Overdose Symptoms

Signs and symptoms:	What to do:
Burning sensation in throat and mouth	Discontinue. Call doctor when convenient.
Cramping abdominal pain	Discontinue. Call doctor when convenient.
Nausea	Discontinue. Call doctor immediately.
Vomiting	Discontinue. Call doctor immediately.
Watery diarrhea	Discontinue. Call doctor immediately.

Basic Information

Biological name (genus and species):
    Carica papaya
Parts used for medicinal purposes:
    Berries/fruits
    Inner bark
    Stems
Chemicals this herb contains:
    Amlylolytic enzyme
    Caricin
    Myrosin
    Peptidase
    Vitamins C and E

Known Effects

- Stimulates stomach to increase secretions.
- Releases histamine from body tissues.
- Depresses central nervous system.
- Kills some intestinal parasites.

Miscellaneous information:
- When eaten as a common food, no problems are expected.
- Is used as a meat tenderizer

Unproved Speculated Benefits

- Aids digestion.
- Liquifies excessive mucus in mouth and stomach.
- Inner bark treats sore teeth.

Warnings and Precautions

Don't take if you:
- Are pregnant, think you may be pregnant or plan pregnancy in the near future.
- Have any chronic disease of the gastrointestinal tract, such as stomach or duodenal ulcers, esophageal reflux (reflux esophagitis), ulcerative colitis, spastic colitis, diverticulosis, diverticulitis.

Consult your doctor if you:
- Take this herb for any medical problem that doesn't improve in 2 weeks. There may be safer, more-effective treatments.
- Take any medicinal drugs or herbs including aspirin, laxatives, cold and cough remedies, antacids, vitamins, minerals, amino acids, supplements, other prescription or non-prescription drugs.

Pregnancy:
- Pregnant women should experience no problems taking usual amounts as part of a balanced diet. Other products extracted from this herb have not been proved to cause problems.

Breast-feeding:
- Breast-fed infants of lactating mothers should experience no problems when mother takes usual amounts as part of a balanced diet. Other products extracted from this herb have not been proved to cause problems.

Infants and children:
- Treating infants and children under 2 with any herbal preparation is hazardous.

Others:
- None expected if you are beyond childhood and under 45, basically healthy and take for only a short time.

Storage:
- Keep cool and dry, but don't freeze. Store safely away from children.

Safe dosage:
- At present no "safe" dosage has been established.

Toxicity

Generally regarded as safe when taken in appropriate quantities for short periods of time.

For symptoms of toxicity: See *Adverse Reactions, Side Effects or Overdose Symptoms* section below.

Adverse Reactions, Side Effects or Overdose Symptoms

Signs and symptoms:	What to do:
Heartburn caused by irritation of lower part of esophagus	Discontinue. Call doctor when convenient.

Parsley

Basic Information
Biological name (genus and species):
  Petroselinum sativum
Parts used for medicinal purposes:
  Berries/fruits
  Leaves
  Roots
  Stems
Chemicals this herb contains:
  Apiin (Also called parsley camphor)
  Apiol
  Pinene
  Volatile oils (See Glossary)

Known Effects

- Decreases blood pressure.
- Decreases pulse rate.
- Aids digestion.
- Helps body dispose of excess fluid by increasing amounts of urine produced.

Miscellaneous information:
- When fresh sprigs are eaten, no problems are expected.

Unproved Speculated Benefits

- Treats painful menstruation.
- Causes abortions.
- Treats dyspepsia.

Warnings and Precautions

Don't take if you:
- Are pregnant, think you may be pregnant or plan pregnancy in the near future.
- Have any chronic disease of the gastrointestinal tract, such as stomach or duodenal ulcers, esophageal reflux (reflux esophagitis), ulcerative colitis, spastic colitis, diverticulosis, diverticulitis.

Consult your doctor if you:
- Take this herb for any medical problem that doesn't improve in 2 weeks. There may be safer, more-effective treatments.
- Take any medicinal drugs or herbs including aspirin, laxatives, cold and cough remedies, antacids, vitamins, minerals, amino acids, supplements, other prescription or non-prescription drugs.

Pregnancy:
- Dangers outweigh any possible benefits. Avoid taking *any* herbal medication made from parsley. It is all right to eat fresh parsley as a condiment.

Breast-feeding:
- Dangers outweigh any possible benefits. Avoid taking *any* herbal medication made from parsley. It is all right to eat fresh parsley as a condiment.

Infants and children:
- Treating infants and children under 2 with any herbal preparation is hazardous.

Others:
- None expected if you are beyond childhood and under 45, not pregnant, basically healthy and take for only a short time.

Storage:
- Keep cool and dry, but don't freeze. Store safely away from children.

Safe dosage:
- At present no "safe" dosage has been established.

Toxicity

Rated relatively safe when taken in appropriate quantities for short periods of time.

For symptoms of toxicity: See *Adverse Reactions, Side Effects or Overdose Symptoms* section below.

Adverse Reactions, Side Effects or Overdose Symptoms

Signs and symptoms:	What to do:
Dizziness	Discontinue. Call doctor immediately.
Jaundice (yellow skin and eyes)	Discontinue. Call doctor immediately.
Nausea	Discontinue. Call doctor immediately.
Vomiting	Discontinue. Call doctor immediately.

Partridgeberry (Squawvine)

Basic Information

Biological name (genus and species):
  Mitchella repens
Parts used for medicinal purposes:
  Stems
Chemicals this herb contains:
  Dextrin
  Mucilage (See Glossary)
  Saponins (See Glossary)
  Wax (See Glossary)

Known Effects

- Helps body dispose of excess fluid by increasing amount of urine produced.
- Shrinks tissues.
- Prevents secretion of fluids.

Unproved Speculated Benefits

- Makes labor less difficult.
- Helps flow of milk in lactating women.
- Treats insomnia.
- Decreases diarrhea.
- Treats congestive heart failure, kidney failure, liver failure.

Warnings and Precautions

Don't take if you:
- Are pregnant, think you may be pregnant or plan pregnancy in the near future.

Consult your doctor if you:
- Take this herb for any medical problem that doesn't improve in 2 weeks. There may be safer, more-effective treatments.
- Take any medicinal drugs or herbs including aspirin, laxatives, cold and cough remedies, antacids, vitamins, minerals, amino acids, supplements, other prescription or non-prescription drugs.

Pregnancy:
- Problems in pregnant women taking small or usual amounts have not been proved. But the chance of problems does exist. Don't use unless prescribed by your doctor.

Breast-feeding:
- Problems in breast-fed infants of lactating mothers taking small or usual amounts have not been proved. But the chance of problems does exist. Don't use unless prescribed by your doctor.

Infants and children:
- Treating infants and children under 2 with any herbal preparation is hazardous.

Others:
- None expected if you are beyond childhood and under 45, basically healthy and take for only a short time.

Storage:
- Keep cool and dry, but don't freeze. Store safely away from children.

Safe dosage:
- At present no "safe" dosage has been established.

Toxicity

Rated relatively safe when taken in appropriate quantities for short periods of time.

Adverse Reactions, Side Effects or Overdose Symptoms

None expected

MEDICINAL HERB

Pasque Flower (May Flower, Pulsatilla)

Basic Information

Biological name (genus and species):
  Anemone pulsatilla
Parts used for medicinal purposes:
  Petals/flower
  Roots
Chemicals this herb contains:
  Anemone camphor
  Ranunculin
  Tannins (See Glossary)
  Volatile oils (See Glossary)

Known Effects

- Irritates mucous membranes.
- Shrinks tissues.
- Prevents secretion of fluids.
- Decreases thickness and increases fluidity of mucus from lungs and bronchial tubes.
- Interferes with absorption of iron and other minerals when taken internally.

Unproved Speculated Benefits

- Treats menstrual disorders.
- Depresses sexual excitement.
- Increases sexual strength.

Warnings and Precautions

Don't take if you:
- Are pregnant, think you may be pregnant or plan pregnancy in the near future.
- Have any chronic disease of the gastrointestinal tract, such as stomach or duodenal ulcers, esophageal reflux (reflux esophagitis), ulcerative colitis, spastic colitis, diverticulosis, diverticulitis.

Consult your doctor if you:
- Take this herb for any medical problem that doesn't improve in 2 weeks. There may be safer, more-effective treatments.
- Take any medicinal drugs or herbs including aspirin, laxatives, cold and cough remedies, antacids, vitamins, minerals, amino acids, supplements, other prescription or non-prescription drugs.

Pregnancy:
- Dangers outweigh any possible benefits. Don't use.

Breast-feeding:
- Dangers outweigh any possible benefits. Don't use.

Infants and children:
- Treating infants and children under 2 with any herbal preparation is hazardous.

Others:
- None expected if you are beyond childhood and under 45, basically healthy and take for only a short time.

Storage:
- Keep cool and dry, but don't freeze. Store safely away from children.

Safe dosage:
- At present no "safe" dosage has been established.

Toxicity

Rated slightly dangerous, particularly in children, persons over 55 and those who take larger than appropriate quantities for extended periods of time.

For symptoms of toxicity: See *Adverse Reactions, Side Effects or Overdose Symptoms* section below.

Adverse Reactions, Side Effects or Overdose Symptoms

Signs and symptoms:	What to do:
Abdominal pain	Discontinue. Call doctor when convenient.
Diarrhea	Discontinue. Call doctor immediately.
Kidney damage characterized by blood in urine, decreased urine flow, swelling of hands and feet	Seek emergency treatment.
Nausea	Discontinue. Call doctor immediately.
Vomiting	Discontinue. Call doctor immediately.

Passion Flower (Maypop)

Basic Information

Biological name (genus and species):
   Passiflora incarnata
Parts used for medicinal purposes:
   Flowers
   Fruit
Chemicals this herb contains:
   Cyanogenic glycosides (See Glossary)
   Harmaline
   Harman
   Harmine
   Harmol ·

Known Effects

- Depresses nerve transfer in spinal cord and brain.
- Increases respiratory rate.
- Slightly depresses central nervous system.
- Causes hallucinations.

Miscellaneous information:
- Smoking passion flower reportedly causes mental changes similar to marijuana.

Unproved Speculated Benefits

- Reduces headaches.
- Treats epilepsy.
- Treats convulsions.
- Treats insomnia.
- Is used as "nerve tonic."
- Acts as a tranquilizer.

Warnings and Precautions

Don't take if you:
- Are pregnant, think you may be pregnant or plan pregnancy in the near future.

Consult your doctor if you:
- Take this herb for any medical problem that doesn't improve in 2 weeks. There may be safer, more-effective treatments.
- Take any medicinal drugs or herbs including aspirin, laxatives, cold and cough remedies, antacids, vitamins, minerals, amino acids, supplements, other prescription or non-prescription drugs.

Pregnancy:
- Dangers outweigh any possible benefits. Don't use.

Breast-feeding:
- Dangers outweigh any possible benefits. Don't use.

Infants and children:
- Treating infants and children under 2 with any herbal preparation is hazardous.

Others:
- This product will not help you. It may cause toxic symptoms.

Storage:
- Keep cool and dry, but don't freeze. Store safely away from children.

Safe dosage:
- At present no "safe" dosage has been established.

Toxicity

Rated relatively safe when taken in appropriate quantities for short periods of time.

For symptoms of toxicity: See *Adverse Reactions, Side Effects or Overdose Symptoms* section below.

Adverse Reactions, Side Effects or Overdose Symptoms

Signs and symptoms:	What to do:
Convulsions	Seek emergency treatment.
Decreased body temperature	Discontinue. Call doctor immediately.
Hallucinations	Seek emergency treatment.
Muscle paralysis, including muscles used in breathing	Seek emergency treatment.

Peach

Basic Information

Biological name (genus and species):
  Prunus persica or other Prunus species
Parts used for medicinal purposes:
  Bark
  Leaves
  Roots
  Seeds
Chemicals this herb contains:
  Cyanide, especially in kernels
  Phloretin
  Volatile oils (See Glossary)

Known Effects

• Irritates and stimulates gastrointestinal tract.

Miscellaneous information:
• North-American Indians made tea from bark.
• Fruit, except for peach pit, is safe.

Unproved Speculated Benefits

• Treats constipation (leaves).
• Treats systemic infections (bark and roots).

Warnings and Precautions

Don't take if you:
• Are pregnant, think you may be pregnant or plan pregnancy in the near future.
• Have any chronic disease of the gastrointestinal tract, such as stomach or duodenal ulcers, esophageal reflux (reflux esophagitis), ulcerative colitis, spastic colitis, diverticulosis, diverticulitis.

Consult your doctor if you:
• Take this herb for any medical problem that doesn't improve in 2 weeks. There may be safer, more-effective treatments.
• Take any medicinal drugs or herbs including aspirin, laxatives, cold and cough remedies, antacids, vitamins, minerals, amino acids, supplements, other prescription or non-prescription drugs.

Pregnancy:
• Dangers of taking this as a medicinal herb outweigh any possible benefits. Avoid pits! There should be no problems with fruit.

Breast-feeding:
• Dangers of taking this as a medicinal herb outweigh any possible benefits. Avoid pits! There should be no problems with fruit.

Infants and children:
• Treating infants and children under 2 with any herbal preparation is hazardous.

Others:
• Pits will not help you. They may cause toxic symptoms.

Storage:
• Keep cool and dry, but don't freeze. Store safely away from children.

Safe dosage:
• At present no "safe" dosage has been established.

Toxicity

Comparative-toxicity rating not available from standard references.

For symptoms of toxicity: See *Adverse Reactions, Side Effects or Overdose Symptoms* section below.

Adverse Reactions, Side Effects or Overdose Symptoms

Signs and symptoms:	What to do:
Diarrhea	Discontinue. Call doctor immediately.
Nausea	Discontinue. Call doctor immediately.
Vomiting	Discontinue. Call doctor immediately.

Basic Information

Biological name (genus and species):
   Anacyclus pyrethrum
Parts used for medicinal purposes:
   Various parts of the entire plant, frequently differing by country and/or culture
Chemical this herb contains:
   Pellitorine

Known Effects

- Kills insects.

Miscellaneous information:
- Tastes bitter.

Unproved Speculated Benefits

- Relieves pain from toothache or gum infections.
- Relieves facial pain.
- Increases saliva flow.

Warnings and Precautions

Don't take if you:
- Are pregnant, think you may be pregnant or plan pregnancy in the near future.
- Have any chronic disease of the gastrointestinal tract, such as stomach or duodenal ulcers, esophageal reflux (reflux esophagitis), ulcerative colitis, spastic colitis, diverticulosis, diverticulitis.

Consult your doctor if you:
- Take this herb for any medical problem that doesn't improve in 2 weeks. There may be safer, more-effective treatments.
- Take any medicinal drugs or herbs including aspirin, laxatives, cold and cough remedies, antacids, vitamins, minerals, amino acids, supplements, other prescription or non-prescription drugs.

Pregnancy:
- Problems in pregnant women taking small or usual amounts have not been proved. But the chance of problems does exist. Don't use unless prescribed by your doctor.

Breast-feeding:
- Problems in breast-fed infants of lactating mothers taking small or usual amounts have not been proved. But the chance of problems does exist. Don't use unless prescribed by your doctor.

Infants and children:
- Treating infants and children under 2 with any herbal preparation is hazardous.

Others:
- None expected if you are beyond childhood and under 45, basically healthy and take for only a short time.

Storage:
- Keep cool and dry, but don't freeze. Store safely away from children.

Safe dosage:
- At present no "safe" dosage has been established.

Toxicity

Comparative-toxicity rating not available from standard references.

For symptoms of toxicity: See *Adverse Reactions, Side Effects or Overdose Symptoms* section below.

Adverse Reactions, Side Effects or Overdose Symptoms

Signs and symptoms:	What to do:
Diarrhea	Discontinue. Call doctor immediately.
Nausea	Discontinue. Call doctor immediately.
Vomiting	Discontinue. Call doctor immediately.

Pennyroyal

Basic Information

Biological name (genus and species):
  Mentha pulegium, Hedeoma pulegioides
Parts used for medicinal purposes:
  Entire plant
Chemical this herb contains:
  Puligone (yellow or green-yellow oil)

 ## Known Effects

- Stimulates uterine contractions.
- Depresses central nervous system.
- Irritates mucous membranes.
- Reddens skin by increasing blood supply to it.
- Increases salivation.
- Can cause severe liver and kidney damage.

Miscellaneous information:
- Pennyroyal is used as a flavoring agent.
- As little as 2 ounces of the essential oil can cause severe liver and kidney damage.

 ## Unproved Speculated Benefits

- Causes abortions.
- Decreases intestinal cramps and flatulence.
- "Purifies" blood.
- Treats colds.
- Regulates menstruation.
- Increases perspiration.

 ## Warnings and Precautions

Don't take if you:
- Are pregnant, think you may be pregnant or plan pregnancy in the near future.

Consult your doctor if you:
- Take this herb for any medical problem that doesn't improve in 2 weeks. There may be safer, more-effective treatments.
- Take any medicinal drugs or herbs including aspirin, laxatives, cold and cough remedies, antacids, vitamins, minerals, amino acids, supplements, other prescription or non-prescription drugs.

Pregnancy:
- Dangers outweigh any possible benefits. Don't use.

Breast-feeding:
- Dangers outweigh any possible benefits. Don't use.

Infants and children:
- Treating infants and children under 2 with any herbal preparation is hazardous.

Others:
- Don't use in an attempt to induce abortion. Pennyroyal can be deadly.

Storage:
- Keep cool and dry, but don't freeze. Store safely away from children.

Safe dosage:
- At present no "safe" dosage has been established.

 ## Toxicity

Rated relatively safe when taken in appropriate quantities for short periods of time.

For symptoms of toxicity: See *Adverse Reactions, Side Effects or Overdose Symptoms* section below.

 ## Adverse Reactions, Side Effects or Overdose Symptoms

Signs and symptoms:	What to do:
Bleeding from gastrointestinal tract	Seek emergency treatment.
Blood in urine	Seek emergency treatment.
Jaundice (yellow skin and eyes)	Discontinue. Call doctor immediately.
Seizures	Seek emergency treatment.
Unusual vaginal bleeding	Seek emergency treatment.

Basic Information

Biological name (genus and species):
Mentha piperita
Parts used for medicinal purposes:
Flowering tops
Leaves
Chemicals this herb contains:
Menthol
Menthone
Methyl acetate
Tannic acid
Terpenes (See Glossary)
Volatile oils (See Glossary)

Known Effects

- Increases stomach acidity.
- Irritates and stimulates gastrointestinal tract.
- Irritates mucous membranes.
- Interferes with absorption of iron and other minerals when taken internally.

Miscellaneous information:
- Peppermint is used to add flavor to medical and non-medical preparations.
- No effects are expected on the body, either good or bad, when herb is used in very small amounts to enhance the flavor of food.

Unproved Speculated Benefits

- Treats colic.
- Treats abdominal cramps.
- Aids in expelling gas from intestinal tract.

Warnings and Precautions

Don't take if you:
- Are pregnant, think you may be pregnant or plan pregnancy in the near future.
- Have any chronic disease of the gastrointestinal tract, such as stomach or duodenal ulcers, esophageal reflux (reflux esophagitis), ulcerative colitis, spastic colitis, diverticulosis, diverticulitis.

Consult your doctor if you:
- Take this herb for any medical problem that doesn't improve in 2 weeks. There may be safer, more-effective treatments.
- Take any medicinal drugs or herbs including aspirin, laxatives, cold and cough remedies, antacids, vitamins, minerals, amino acids, supplements, other prescription or non-prescription drugs.

Pregnancy:
- Problems in pregnant women taking small or usual amounts have not been proved. But the chance of problems does exist. Don't use unless prescribed by your doctor.

Breast-feeding:
- Problems in breast-fed infants of lactating mothers taking small or usual amounts have not been proved. But the chance of problems does exist. Don't use unless prescribed by your doctor.

Infants and children:
- Treating infants and children under 2 with any herbal preparation is hazardous.

Others:
- None expected if you are under 45, not pregnant, basically healthy, take it for only a short time and do not exceed manufacturer's recommended dosage.

Storage:
- Keep cool and dry, but don't freeze. Store safely away from children.

Safe dosage:
- At present no "safe" dosage has been established.

Toxicity

Comparative-toxicity rating not available from standard references.

For symptoms of toxicity: See *Adverse Reactions, Side Effects or Overdose Symptoms* section below.

Adverse Reactions, Side Effects or Overdose Symptoms

Signs and symptoms:	What to do:
Drowsiness	Discontinue. Call doctor when convenient.
Vomiting	Discontinue. Call doctor immediately.

Periwinkle (Madagascar or Cape Periwinkle, Old Maid)

Basic Information

Biological name (genus and species):
   Catharanthus roseus, Vinca rosea
Parts used for medicinal purposes:
   Leaves
Chemicals this herb contains:
   Vinblastine
   Vincristine
   Vinleurosine
   Vinrosidine

Known Effects

- Inhibits growth and development of germs.
- Depresses bone-marrow production, damaging body's blood-cell-manufacturing processes.
- Effective in treatment of several different types of malignant tumors.
- Reduces granulocytes—white blood cells—in body.

Miscellaneous information:
- When purified, derivatives of Vinca (vincristine sulfate, vinblastine sulfate) are used to treat cancer under rigidly controlled supervision.

Unproved Speculated Benefits

- Ointment decreases inflammation.
- Treats sore throats and inflamed tonsils.
- Treats diabetes mellitus.
- Causes hallucinations when smoked.

Warnings and Precautions

Don't take if you:
- Are pregnant, think you may be pregnant or plan pregnancy in the near future.
- Have any chronic disease of the gastrointestinal tract, such as stomach or duodenal ulcers, esophageal reflux (reflux esophagitis), ulcerative colitis, spastic colitis, diverticulosis, diverticulitis.

Consult your doctor if you:
- Take this herb for any medical problem that doesn't improve in 2 weeks. There may be safer, more-effective treatments.
- Take any medicinal drugs or herbs including aspirin, laxatives, cold and cough remedies, antacids, vitamins, minerals, amino acids, supplements, other prescription or non-prescription drugs.

Pregnancy:
- Dangers outweigh any possible benefits. Don't use.

Breast-feeding:
- Dangers outweigh any possible benefits. Don't use.

Infants and children:
- Treating infants and children under 2 with any herbal preparation is hazardous.

Others:
- This product will not help you. It may cause toxic symptoms.

Storage:
- Keep cool and dry, but don't freeze. Store safely away from children.

Safe dosage:
- At present no "safe" dosage has been established.

Toxicity

Rated slightly dangerous, particularly in children, persons over 55 and those who take larger than appropriate quantities for extended periods of time.

For symptoms of toxicity: See *Adverse Reactions, Side Effects or Overdose Symptoms* section below.

Adverse Reactions, Side Effects or Overdose Symptoms

Signs and symptoms:	What to do:
Drowsiness	Discontinue. Call doctor when convenient.
Hair loss	Discontinue. Call doctor when convenient.
Nausea	Discontinue. Call doctor immediately.
Seizures	Seek emergency treatment.
Yellow eyes, dark urine and yellow skin resulting from destruction of some liver cells	Seek emergency treatment.

Pipsissewa

Basic Information

Biological name (genus and species):
   Chimaphila
Parts used for medicinal purposes:
   Leaves
Chemicals this herb contains:
   Arbutin
   Chimaphilin
   Chlorophyll
   Ericolin
   Minerals
   Pectic acid
   Tannins (See Glossary)
   Urson

Known Effects

- Helps body dispose of excess fluid by increasing amount of urine produced.
- Interferes with absorption of iron and other minerals when taken internally.

Unproved Speculated Benefits

- Treats indigestion or mild stomach upsets.
- Treats irritations of urinary tract (kidney, bladder, urethra).

Warnings and Precautions

Don't take if you:
- Are pregnant, think you may be pregnant or plan pregnancy in the near future.
- Have any chronic disease of the gastrointestinal tract, such as stomach or duodenal ulcers, esophageal reflux (reflux esophagitis), ulcerative colitis, spastic colitis, diverticulosis, diverticulitis.

Consult your doctor if you:
- Take this herb for any medical problem that doesn't improve in 2 weeks. There may be safer, more-effective treatments.
- Take any medicinal drugs or herbs including aspirin, laxatives, cold and cough remedies, antacids, vitamins, minerals, amino acids, supplements, other prescription or non-prescription drugs.

Pregnancy:
- Dangers outweigh any possible benefits. Don't use.

Breast-feeding:
- Dangers outweigh any possible benefits. Don't use.

Infants and children:
- Treating infants and children under 2 with any herbal preparation is hazardous.

Others:
- None expected if you are under 45, not pregnant, basically healthy, take it for only a short time and do not exceed manufacturer's recommended dosage.

Storage:
- Keep cool and dry, but don't freeze. Store safely away from children.

Safe dosage:
- At present no "safe" dosage has been established.

Toxicity

Comparative-toxicity rating not available from standard references.

For symptoms of toxicity: See *Adverse Reactions, Side Effects or Overdose Symptoms* section below.

Adverse Reactions, Side Effects or Overdose Symptoms

Signs and symptoms:	What to do:
Diarrhea	Discontinue. Call doctor immediately.
Nausea	Discontinue. Call doctor immediately.
Skin eruptions	Discontinue. Call doctor when convenient.
Vomiting	Discontinue. Call doctor immediately.

Pitcher Plant

Basic Information
Biological name (genus and species):
  Sarracenia
Parts used for medicinal purposes:
  Roots
Chemicals this herb contains:
  Resin (See Glossary)
  Yellow dye

Known Effects

• Irritates gastrointestinal tract.
• Has diuretic properties.

Unproved Speculated Benefits

• Treats constipation.
• Treats indigestion.
• Increases amount of urine kidneys produce.

Warnings and Precautions

Don't take if you:
• Are pregnant, think you may be pregnant or plan pregnancy in the near future.

Consult your doctor if you:
• Take this herb for any medical problem that doesn't improve in 2 weeks. There may be safer, more-effective treatments.
• Take any medicinal drugs or herbs including aspirin, laxatives, cold and cough remedies, antacids, vitamins, minerals, amino acids, supplements, other prescription or non-prescription drugs.

Pregnancy:
• Problems in pregnant women taking small or usual amounts have not been proved. But the chance of problems does exist. Don't use unless prescribed by your doctor.

Breast-feeding:
• Problems in breast-fed infants of lactating mothers taking small or usual amounts have not been proved. But the chance of problems does exist. Don't use unless prescribed by your doctor.

Infants and children:
• Treating infants and children under 2 with any herbal preparation is hazardous.

Others:
• None expected if you are under 45, not pregnant, basically healthy, take it for only a short time and do not exceed manufacturer's recommended dosage.

Storage:
• Keep cool and dry, but don't freeze. Store safely away from children.

Safe dosage:
• At present no "safe" dosage has been established.

Toxicity

Comparative-toxicity rating not available from standard references.

Adverse Reactions, Side Effects or Overdose Symptoms

None expected

Pleurisy Root (Butterfly Weed)

Basic Information

Biological name (genus and species):
Asclepias tuberosa

Parts used for medicinal purposes:
Roots

Chemicals this herb contains:
Asclepiadin
Asclepion
Galitoxin
Volatile oils (See Glossary)

 ## Known Effects

- Decreases thickness and increases fluidity of mucus from lungs and bronchial tubes.
- Irritates mucous membranes.
- Stimulates and irritates gastrointestinal tract.

 ## Unproved Speculated Benefits

- Acts as a strong laxative to cause watery, explosive bowel movements.
- Increases perspiration.

 ## Warnings and Precautions

Don't take if you:
- Are pregnant, think you may be pregnant or plan pregnancy in the near future.
- Have any chronic disease of the gastrointestinal tract, such as stomach or duodenal ulcers, esophageal reflux (reflux esophagitis), ulcerative colitis, spastic colitis, diverticulosis, diverticulitis.

Consult your doctor if you:
- Take this herb for any medical problem that doesn't improve in 2 weeks. There may be safer, more-effective treatments.
- Take any medicinal drugs or herbs including aspirin, laxatives, cold and cough remedies, antacids, vitamins, minerals, amino acids, supplements, other prescription or non-prescription drugs.

Pregnancy:
- Dangers outweigh any possible benefits. Don't use.

Breast-feeding:
- Dangers outweigh any possible benefits. Don't use.

Infants and children:
- Treating infants and children under 2 with any herbal preparation is hazardous.

Others:
- Dangers outweigh any possible benefits. Don't use.

Storage:
- Keep cool and dry, but don't freeze. Store safely away from children.

Safe dosage:
- At present no "safe" dosage has been established.

 ## Toxicity

Comparative-toxicity rating not available from standard references.

For symptoms of toxicity: See *Adverse Reactions, Side Effects or Overdose Symptoms* section below.

 ## Adverse Reactions, Side Effects or Overdose Symptoms

Signs and symptoms:	What to do:
Appetite loss	Discontinue. Call doctor when convenient.
Coma	Seek emergency treatment.
Diarrhea	Discontinue. Call doctor immediately.
Lethargy	Discontinue. Call doctor when convenient.
Muscle weakness	Discontinue. Call doctor immediately.
Nausea	Discontinue. Call doctor immediately.
Vomiting	Discontinue. Call doctor immediately.

Poke (Pokeweed, Scoke)

Basic Information

Biological name (genus and species):
Phytolacca americana
Parts used for medicinal purposes:
Leaves
Roots
Seeds
Chemicals this herb contains:
Asparagine
Mitogen
Phytolaccigenin
Resin (See Glossary)
Saponins (See Glossary)

Known Effects

• Stimulates and irritates gastrointestinal tract.

Miscellaneous information:
• All parts of native plants are poisonous. Don't take it. Children are especially vulnerable to toxic effects.
• Leaves are boiled and eaten as flavoring in some areas, particularly the southern United States. Used this way, pokeberry may be toxic. Don't use!

Unproved Speculated Benefits

• Treats chronic arthritis.
• Treats constipation.

Warnings and Precautions

Don't take if you:
• Are pregnant, think you may be pregnant or plan pregnancy in the near future.
• Have any chronic disease of the gastrointestinal tract, such as stomach or duodenal ulcers, esophageal reflux (reflux esophagitis), ulcerative colitis, spastic colitis, diverticulosis, diverticulitis.

Consult your doctor if you:
• Take this herb for any medical problem that doesn't improve in 2 weeks. There may be safer, more-effective treatments.
• Take any medicinal drugs or herbs including aspirin, laxatives, cold and cough remedies, antacids, vitamins, minerals, amino acids, supplements, other prescription or non-prescription drugs.

Pregnancy:
• Dangers outweigh any possible benefits. Don't use.

Breast-feeding:
• Dangers outweigh any possible benefits. Don't use.

Infants and children:
• Treating infants and children under 2 with any herbal preparation is hazardous.

Others:
• Handling roots may cause skin abrasions.

Storage:
• Keep cool and dry, but don't freeze. Store safely away from children.

Safe dosage:
• At present no "safe" dosage has been established.

Toxicity

Comparative-toxicity rating not available from standard references.

For symptoms of toxicity: See *Adverse Reactions, Side Effects or Overdose Symptoms* section below.

Adverse Reactions, Side Effects or Overdose Symptoms

Signs and symptoms:	What to do:
Decreased heart rate	Seek emergency treatment.
Diarrhea	Discontinue. Call doctor immediately.
Nausea	Discontinue. Call doctor immediately.
Skin eruptions	Discontinue. Call doctor when convenient.
Vomiting	Discontinue. Call doctor immediately.

Pomegranate

Basic Information

Biological name (genus and species):
  Punica granatum
Parts used for medicinal purposes:
  Bark
  Berries/fruits, including rind
Chemicals this herb contains:
  Isopelletierine
  Methylisopelletierine
  Pelletierine
  Pseudopelletierine
  Tannins (See Glossary)

 ## Known Effects

Rind and bark:
• Shrinks tissues.
• Prevents secretion of fluids.
• Destroys intestinal worms.
• Interferes with absorption of iron and other minerals when taken internally.

Miscellaneous information:
• Fruits are edible and non-toxic. Bark and rind contain herbal-medicinal properties.

 ## Unproved Speculated Benefits

• Treats stasis ulcers and "bed sores."

 ## Warnings and Precautions

Don't take if you:
• Are pregnant, think you may be pregnant or plan pregnancy in the near future.
• Have any chronic disease of the gastrointestinal tract, such as stomach or duodenal ulcers, esophageal reflux (reflux esophagitis), ulcerative colitis, spastic colitis, diverticulosis, diverticulitis.

Consult your doctor if you:
• Take this herb for any medical problem that doesn't improve in 2 weeks. There may be safer, more-effective treatments.
• Take any medicinal drugs or herbs including aspirin, laxatives, cold and cough remedies, antacids, vitamins, minerals, amino acids, supplements, other prescription or non-prescription drugs.

Pregnancy:
• Taken internally as a medicinal herb, dangers outweigh any possible benefits. Don't use. Eating fruit as part of the diet will not cause problems.

Breast-feeding:
• Taken internally as a medicinal herb, dangers outweigh any possible benefits. Don't use. Eating fruit as part of the diet will not cause problems.

Infants and children:
• Treating infants and children under 2 with any herbal preparation is hazardous.

Others:
• Taken internally, dangers outweigh any possible benefits. Don't use.

Storage:
• Keep cool and dry, but don't freeze. Store safely away from children.

Safe dosage:
• At present no "safe" dosage has been established.

 ## Toxicity

Comparative-toxicity rating not available from standard references.

For symptoms of toxicity: See *Adverse Reactions, Side Effects or Overdose Symptoms* section below.

 ## Adverse Reactions, Side Effects or Overdose Symptoms

Signs and symptoms:	What to do:
Diarrhea	Discontinue. Call doctor immediately.
Dilated pupils	Seek emergency treatment.
Dizziness	Discontinue. Call doctor immediately.
Double vision	Seek emergency treatment.
Nausea	Discontinue. Call doctor immediately.
Vomiting	Discontinue. Call doctor immediately.
Weakness	Discontinue. Call doctor immediately.

Poplar Bud

Basic Information

Biological name (genus and species):
  Populus candicans
Parts used for medicinal purposes:
  Leaf bud
Chemicals this herb contains:
  Chrysin
  Gallic acid
  Humulene
  Malic acid
  Mannite
  Populin
  Resin (See Glossary)
  Salicin
  Tectochrysin

Known Effects

- Blocks pain impulses to brain.
- Changes fever-conrol "thermostat" in brain.

Miscellaneous information:

- Anti-oxidant effect helps prevent rancidity in ointments.
- Poplar bud is used as an additive in several pharmaceutical preparations.

Unproved Speculated Benefits

- Reduces pain of sprains and bruises when applied to skin.
- Treats coughs and colds when taken internally.
- Reduces fever.

Warnings and Precautions

Don't take if you:

- Are pregnant, think you may be pregnant or plan pregnancy in the near future.
- Have any chronic disease of the gastrointestinal tract, such as stomach or duodenal ulcers, esophageal reflux (reflux esophagitis), ulcerative colitis, spastic colitis, diverticulosis, diverticulitis.

Consult your doctor if you:

- Take this herb for any medical problem that doesn't improve in 2 weeks. There may be safer, more-effective treatments.
- Take any medicinal drugs or herbs including aspirin, laxatives, cold and cough remedies, antacids, vitamins, minerals, amino acids, supplements, other prescription or non-prescription drugs.

Pregnancy:

- Problems in pregnant women taking small or usual amounts have not been proved. But the chance of problems does exist. Don't use unless prescribed by your doctor.

Breast-feeding:

- Problems in breast-fed infants of lactating mothers taking small or usual amounts have not been proved. But the chance of problems does exist. Don't use unless prescribed by your doctor.

Infants and children:

- Treating infants and children under 2 with any herbal preparation is hazardous.

Others:

- None expected if you are beyond childhood and under 45, basically healthy and take for only a short time.

Storage:

- Keep cool and dry, but don't freeze. Store safely away from children.

Safe dosage:

- At present no "safe" dosage has been established.

Toxicity

Comparative-toxicity rating not available from standard references.

For symptoms of toxicity: See *Adverse Reactions, Side Effects or Overdose Symptoms* section below.

Adverse Reactions, Side Effects or Overdose Symptoms

Signs and symptoms:	What to do:
Itching and redness of skin	Apply hydrocortisone ointment, available without prescription.
Skin rash	Apply hydrocortisone ointment, available without prescription.

Basic Information

Biological name (genus and species):
Xanthoxylum americanum *(northern)*
Xanthoxylum clava-herculus *(southern)*
Parts used for medicinal purposes:
Bark
Berries/fruits
Chemicals this herb contains:
Acid amide
Asarinin
Berberine
Herculin
Xanthoxyletin
Xanthyletin

Known Effects

- Stimulates and irritates gastrointestinal tract.
- Increases perspiration.

Unproved Speculated Benefits

- Stimulates appetite.
- Treats arthritis.
- Decreases flatulence.

Warnings and Precautions

Don't take if you:
- Are pregnant, think you may be pregnant or plan pregnancy in the near future.
- Have any chronic disease of the gastrointestinal tract, such as stomach or duodenal ulcers, esophageal reflux (reflux esophagitis), ulcerative colitis, spastic colitis, diverticulosis, diverticulitis.

Consult your doctor if you:
- Take this herb for any medical problem that doesn't improve in 2 weeks. There may be safer, more-effective treatments.
- Take any medicinal drugs or herbs including aspirin, laxatives, cold and cough remedies, antacids, vitamins, minerals, amino acids, supplements, other prescription or non-prescription drugs.

Pregnancy:
- Problems in pregnant women taking small or usual amounts have not been proved. But the chance of problems does exist. Don't use unless prescribed by your doctor.

Breast-feeding:
- Problems in breast-fed infants of lactating mothers taking small or usual amounts have not been proved. But the chance of problems does exist. Don't use unless prescribed by your doctor.

Infants and children:
- Treating infants and children under 2 with any herbal preparation is hazardous.

Others:
- None expected if you are beyond childhood and under 45, basically healthy and take for only a short time.

Storage:
- Keep cool and dry, but don't freeze. Store safely away from children.

Safe dosage:
- At present no "safe" dosage has been established.

Toxicity

Comparative-toxicity rating not available from standard references.

For symptoms of toxicity: See *Adverse Reactions, Side Effects or Overdose Symptoms* section below.

Adverse Reactions, Side Effects or Overdose Symptoms

Signs and symptoms:	What to do:
Diarrhea	Discontinue. Call doctor immediately.
Nausea	Discontinue. Call doctor immediately.
Vomiting	Discontinue. Call doctor immediately.

MEDICINAL HERB

Prickly Poppy (Thistle Poppy, Mexican Poppy)

Basic Information

Biological name (genus and species):
  Argemone mexicana
Parts used for medicinal purposes:
  Seeds
Chemicals this herb contains:
  Berberine
  Dihydrosanquinarine
  Protopine
  Sanquinarine

Known Effects

• Depresses central nervous system very mildly.

Miscellaneous information:
• This poppy is not the origin of morphine, codeine or other narcotics.

Unproved Speculated Benefits

• Smoking prickly poppy produces euphoria.
• Reduces pain.

Warnings and Precautions

Don't take if you:
• Are pregnant, think you may be pregnant or plan pregnancy in the near future.
• Have any chronic disease of the gastrointestinal tract, such as stomach or duodenal ulcers, esophageal reflux (reflux esophagitis), ulcerative colitis, spastic colitis, diverticulosis, diverticulitis.

Consult your doctor if you:
• Take this herb for any medical problem that doesn't improve in 2 weeks. There may be safer, more-effective treatments.
• Take any medicinal drugs or herbs including aspirin, laxatives, cold and cough remedies, antacids, vitamins, minerals, amino acids, supplements, other prescription or non-prescription drugs.

Pregnancy:
• Dangers outweigh any possible benefits. Don't use.

Breast-feeding:
• Dangers outweigh any possible benefits. Don't use.

Infants and children:
• Treating infants and children under 2 with any herbal preparation is hazardous.

Others:
• Dangers outweigh any possible benefits. Don't use.

Storage:
• Keep cool and dry, but don't freeze. Store safely away from children.

Safe dosage:
• At present no "safe" dosage has been established.

Toxicity

Rated slightly dangerous, particularly in children, persons over 55 and those who take larger than appropriate quantities for extended periods of time.

For symptoms of toxicity: See *Adverse Reactions, Side Effects or Overdose Symptoms* section below.

Adverse Reactions, Side Effects or Overdose Symptoms

Signs and symptoms:	What to do:
Diarrhea	Discontinue. Call doctor immediately.
Dizziness	Discontinue. Call doctor immediately.
Fluid retention	Discontinue. Call doctor when convenient.
Loss of consciousness	Seek emergency treatment.
Nausea	Discontinue. Call doctor immediately.
Swollen abdomen	Discontinue. Call doctor when convenient.
Vision disturbances	Discontinue. Call doctor immediately.
Vomiting	Discontinue. Call doctor immediately.

Prostrate Knotweed (Pigweed)

Basic Information

Biological name (genus and species):
    Polygonum aviculare
Parts used for medicinal purposes:
    Various parts of the entire plant, frequently
    differing by country and/or culture
Chemicals this herb contains:
    Avicularin
    Emodin
    Quercetin 3-arabinoside

Known Effects

- Reduces capillary fragility.
- Reduces capillary permeability.
- Retards destruction of epinephrine.

Unproved Speculated Benefits

- Causes watery, explosive bowel movements.
- Treats kidney and bladder stones.

Warnings and Precautions

Don't take if you:
- Are pregnant, think you may be pregnant or plan pregnancy in the near future.
- Have any chronic disease of the gastrointestinal tract, such as stomach or duodenal ulcers, esophageal reflux (reflux esophagitis), ulcerative colitis, spastic colitis, diverticulosis, diverticulitis.

Consult your doctor if you:
- Take this herb for any medical problem that doesn't improve in 2 weeks. There may be safer, more-effective treatments.
- Take any medicinal drugs or herbs including aspirin, laxatives, cold and cough remedies, antacids, vitamins, minerals, amino acids, supplements, other prescription or non-prescription drugs.

Pregnancy:
- Problems in pregnant women taking small or usual amounts have not been proved. But the chance of problems does exist. Don't use unless prescribed by your doctor.

Breast-feeding:
- Problems in breast-fed infants of lactating mothers taking small or usual amounts have not been proved. But the chance of problems does exist. Don't use unless prescribed by your doctor.

Infants and children:
- Treating infants and children under 2 with any herbal preparation is hazardous.

Others:
- None expected if you are under 45, not pregnant, basically healthy, take it for only a short time and do not exceed manufacturer's recommended dosage.

Storage:
- Keep cool and dry, but don't freeze. Store safely away from children.

Safe dosage:
- At present no "safe" dosage has been established.

Toxicity

Rated relatively safe when taken in appropriate quantities for short periods of time.

For symptoms of toxicity: See *Adverse Reactions, Side Effects or Overdose Symptoms* section below.

Adverse Reactions, Side Effects or Overdose Symptoms

Signs and symptoms:	What to do:
Abdominal pain	Discontinue. Call doctor when convenient.
Diarrhea	Discontinue. Call doctor immediately.
Nausea	Discontinue. Call doctor immediately.
Skin eruptions	Discontinue. Call doctor when convenient.
Vomiting	Discontinue. Call doctor immediately.

MEDICINAL HERB

Psyllium

Basic Information
Biological name (genus and species):
  Plantago psyllium
Parts used for medicinal purposes:
  Seeds
Chemicals this herb contains:
  Glycosides (See Glossary)
  Mucilage (See Glossary)

Known Effects

- Produces bulky bowel movements (1 gram swells 8-14 times its size when placed in water).
- Softens stools.

Miscellaneous information:
- Psyllium is a popular product and available over-the-counter without prescription.

Unproved Speculated Benefits

- Treats constipation.
- Protects scraped tissues.

Warnings and Precautions

Don't take if you:
- Are pregnant, think you may be pregnant or plan pregnancy in the near future.

Consult your doctor if you:
- Take this herb for any medical problem that doesn't improve in 2 weeks. There may be safer, more-effective treatments.
- Take any medicinal drugs or herbs including aspirin, laxatives, cold and cough remedies, antacids, vitamins, minerals, amino acids, supplements, other prescription or non-prescription drugs.

Pregnancy:
- Problems in pregnant women taking small or usual amounts have not been proved. But the chance of problems does exist. Don't use unless prescribed by your doctor.

Breast-feeding:
- Problems in breast-fed infants of lactating mothers taking small or usual amounts have not been proved. But the chance of problems does exist. Don't use unless prescribed by your doctor.

Infants and children:
- Treating infants and children under 2 with any herbal preparation is hazardous.

Others:
- None expected if you are beyond childhood and under 45, basically healthy and take for only a short time.

Storage:
- Keep cool and dry, but don't freeze. Store safely away from children.

Safe dosage:
- At present no "safe" dosage has been established.

Toxicity

Comparative-toxicity rating not available from standard references.

Adverse Reactions, Side Effects or Overdose Symptoms

None expected

Basic Information

Biological name (genus and species):
Echinacea angustifolia, E. pallida
Parts used for medicinal purposes:
Various parts of the entire plant, frequently
differing by country and/or culture
Chemicals this herb contains:
Betaine
Echinacin
Echinoside
Fatty acids
Inulin
Resin (See Glossary)
Sucrose

 ## Known Effects

• Kills insects, especially houseflies.
• Possible anti-tumor activity.

Miscellaneous information:
• Another herb, *Rudbeckia laciniata,* is also
called *coneflower* and has been reported to be
toxic. If you take *any* coneflower, be sure it is
Echinacea angustifolia.

 ## Unproved Speculated Benefits

• Acts as natural anti-toxin for internal and
external infections.
•"Blood purifier."
• Helps heal wounds.

 ## Warnings and Precautions

Don't take if you:
• Are pregnant, think you may be pregnant or
plan pregnancy in the near future.

Consult your doctor if you:
• Take this herb for any medical problem that
doesn't improve in 2 weeks. There may be
safer, more-effective treatments.
• Take any medicinal drugs or herbs including
aspirin, laxatives, cold and cough remedies,
antacids, vitamins, minerals, amino acids,
supplements, other prescription or non-
prescription drugs.

Pregnancy:
• Problems in pregnant women taking small or
usual amounts have not been proved. But the
chance of problems does exist. Don't use
unless prescribed by your doctor.

Breast-feeding:
• Problems in breast-fed infants of lactating
mothers taking small or usual amounts have
not been proved. But the chance of problems
does exist. Don't use unless prescribed by your
doctor.

Infants and children:
• Treating infants and children under 2 with any
herbal preparation is hazardous.

Others:
• None expected if you are beyond childhood
and under 45, basically healthy and take for
only a short time.

Storage:
• Keep cool and dry, but don't freeze. Store
safely away from children.

Safe dosage:
• At present no "safe" dosage has been
established.

 ## Toxicity

Comparative-toxicity rating not available from
standard references.

 ## Adverse Reactions, Side Effects or Overdose Symptoms

None reported

Rauwolfia (Snakeroot, Chandra, Sarpaganda)

Basic Information

Biological name (genus and species):
Rauwolfia serpentina
Parts used for medicinal purposes:
Roots
Chemicals this herb contains:

Amajaline	Serpentine
Deserpidine	Serpentinine
Reserpine	Yohimbine
Rescinnamine	

Known Effects

- Reduces blood pressure.
- Depresses activity of central nervous system.
- Acts as a hypnotic.

Miscellaneous information:

- Snakeroot depletes catecholamines and serotonin from nerves in central nervous system.
- Refined snakeroot has been used extensively in recent years to treat hypertension.
- Animal studies suggest snakeroot may *produce* cancers.

Unproved Speculated Benefits

- Decreases anxiety.
- Decreases fever.
- Kills intestinal parasites.
- In India, it is used as antidote for snakebites.

Warnings and Precautions

Don't take if you:

- Are pregnant, think you may be pregnant or plan pregnancy in the near future.
- Have any chronic disease of the gastrointestinal tract, such as stomach or duodenal ulcers, esophageal reflux (reflux esophagitis), ulcerative colitis, spastic colitis, diverticulosis, diverticulitis.

Consult your doctor if you have:

- Take this herb for any medical problem that doesn't improve in 2 weeks. There may be safer, more-effective treatments.
- Take any medicinal drugs or herbs including aspirin, laxatives, cold and cough remedies, antacids, vitamins, minerals, amino acids, supplements, other prescription or non-prescription drugs.

Pregnancy and breast-feeding:

- Dangers outweigh any benefits. Don't use.

Infants and children:

- Treating infants and children under 2 with any herbal preparation is hazardous.

Others:

- Dangers outweigh any benefits. Don't use.

Storage:

- Keep cool and dry, but don't freeze. Store safely away from children.

Safe dosage:

- At present no "safe" dosage established.

Toxicity

Rated slightly dangerous, particularly in children, persons over 55 and those who take larger than appropriate quantities for extended periods of time.

For symptoms of toxicity: See *Adverse Reactions, Side Effects or Overdose Symptoms* section below.

Adverse Reactions, Side Effects or Overdose Symptoms

Signs and symptoms:	What to do:
Bizarre dreams	Discontinue. Call doctor when convenient.
Decreased libido and sexual performance	Discontinue. Call doctor when convenient.
Diarrhea	Discontinue. Call doctor immediately.
Drowsiness	Discontinue. Call doctor when convenient.
Nasal congestion	Discontinue. Call doctor when convenient.
Precipitous blood-pressure drop—symptoms include faintness, cold sweat, paleness, rapid pulse	Seek emergency treatment.
Slow heartbeat	Seek emergency treatment.
Stupor	Seek emergency treatment.
Upper-abdominal pain	Discontinue. Call doctor when convenient.

Red Clover (Pavine Clover, Cowgrass)

Basic Information
Biological name (genus and species):
Trifolium pratense
Parts used for medicinal purposes:
Flowers
Chemical this herb contains:
Glycosides (See Glossary)

 ## Known Effects

- Decreases irritation and muscular movement (peristalsis) of gastrointestinal tract.
- Decreases activity of central nervous system.

 ## Unproved Speculated Benefits

- Reduces upper-abdominal cramps.
- Treats indigestion.
- Loosens secretions in bronchial tubes due to infections or chronic lung disease.
- Suppresses appetite.
- Treats cancers. (Controlled studies show no evidence of benefit. Using red clover for this purpose delays obtaining proper medical care.)

 ## Warnings and Precautions

Don't take if you:
- Are pregnant, think you may be pregnant or plan pregnancy in the near future.

Consult your doctor if you:
- Take this herb for any medical problem that doesn't improve in 2 weeks. There may be safer, more-effective treatments.
- Take any medicinal drugs or herbs including aspirin, laxatives, cold and cough remedies, antacids, vitamins, minerals, amino acids, supplements, other prescription or non-prescription drugs.

Pregnancy:
- Problems in pregnant women taking small or usual amounts have not been proved. But the chance of problems does exist. Don't use unless prescribed by your doctor.

Breast-feeding:
- Problems in breast-fed infants of lactating mothers taking small or usual amounts have not been proved. But the chance of problems does exist. Don't use unless prescribed by your doctor.

Infants and children:
- Treating infants and children under 2 with any herbal preparation is hazardous.

Others:
- None expected if you are beyond childhood and under 45, basically healthy and take for only a short time.

Storage:
- Keep cool and dry, but don't freeze. Store safely away from children.

Safe dosage:
- At present no "safe" dosage has been established.

 ## Toxicity

Generally regarded as safe when taken in appropriate quantities for short periods of time.

 ## Adverse Reactions, Side Effects or Overdose Symptoms

None expected

Red Raspberry

Basic Information

Biological name (genus and species):
  Rubus strigosus, R. idaeus
Parts used for medicinal purposes:
  Bark
  Leaves
  Roots
Chemicals this herb contains:
  Citric acid
  Tannins (See Glossary)

Known Effects

- Relaxes uterine spasms.
- Relaxes intestinal spasms.
- Interferes with absorption of iron and other minerals when taken internally.

Miscellaneous information:
- Berries are delicious, nutritious and non-toxic.
- When eaten as a common food, no problems are expected for anyone.

Unproved Speculated Benefits

- Regulates labor pains.
- Decreases excessive menstrual bleeding.
- Is used as gargle for sore throats.

Warnings and Precautions

Don't take if you:
- Are pregnant, think you may be pregnant or plan pregnancy in the near future.

Consult your doctor if you:
- Take this herb for any medical problem that doesn't improve in 2 weeks. There may be safer, more-effective treatments.
- Take any medicinal drugs or herbs including aspirin, laxatives, cold and cough remedies, antacids, vitamins, minerals, amino acids, supplements, other prescription or non-prescription drugs.

Pregnancy:
- Problems in pregnant women taking small or usual amounts have not been proved. But the chance of problems does exist. Don't use unless prescribed by your doctor.

Breast-feeding:
- Problems in breast-fed infants of lactating mothers taking small or usual amounts have not been proved. But the chance of problems does exist. Don't use unless prescribed by your doctor.

Infants and children:
- Treating infants and children under 2 with any herbal preparation is hazardous.

Others:
- None expected if you are beyond childhood and under 45, basically healthy and take for only a short time.

Storage:
- Keep cool and dry, but don't freeze. Store safely away from children.

Safe dosage:
- At present no "safe" dosage has been established.

Toxicity

Comparative-toxicity rating not available from standard references.

Adverse Reactions, Side Effects or Overdose Symptoms

None expected

Basic Information

Biological name (genus and species):
Krameria triandra
Parts used for medicinal purposes:
Various parts of the entire plant, frequently differing by country and/or culture
Chemicals this herb contains:
Calcium oxalate
Gum (See Glossary)
Lignin
N-Methyltyrosine
Saccharine
Starch
Tannins (See Glossary)

Known Effects

- Shrinks tissues.
- Prevents secretion of fluids.
- Causes protein molecules to clump together.
- Interferes with absorption of iron and other minerals when taken internally.

Unproved Speculated Benefits

- Treats sore throat.
- Treats hemorrhoids.
- Treats chronic bowel inflammations.
- Treats diarrhea.

Warnings and Precautions

Don't take if you:
- Are pregnant, think you may be pregnant or plan pregnancy in the near future.
- Have any chronic disease of the gastrointestinal tract, such as stomach or duodenal ulcers, esophageal reflux (reflux esophagitis), ulcerative colitis, spastic colitis, diverticulosis, diverticulitis.

Consult your doctor if you:
- Take this herb for any medical problem that doesn't improve in 2 weeks. There may be safer, more-effective treatments.
- Take any medicinal drugs or herbs including aspirin, laxatives, cold and cough remedies, antacids, vitamins, minerals, amino acids, supplements, other prescription or non-prescription drugs.

Pregnancy:
- Dangers outweigh any possible benefits. Don't use.

Breast-feeding:
- Dangers outweigh any possible benefits. Don't use.

Infants and children:
- Treating infants and children under 2 with any herbal preparation is hazardous.

Others:
- None expected if you are beyond childhood and under 45, basically healthy and take for only a short time.

Storage:
- Keep cool and dry, but don't freeze. Store safely away from children.

Safe dosage:
- At present no "safe" dosage has been established.

Toxicity

Comparative-toxicity rating not available from standard references.

For symptoms of toxicity: See *Adverse Reactions, Side Effects or Overdose Symptoms* section below.

Adverse Reactions, Side Effects or Overdose Symptoms

Signs and symptoms:	What to do:
Diarrhea	Discontinue. Call doctor immediately.
Kidney damage characterized by blood in urine, decreased urine flow, swelling of hands and feet	Seek emergency treatment.
Nausea	Discontinue. Call doctor immediately.
Vomiting	Discontinue. Call doctor immediately.

MEDICINAL HERB

Rheumatism Root (Wild-Yam Root)

Basic Information

Biological name (genus and species):
Dioscarea villosa
Parts used for medicinal purposes:
Roots
Chemicals this herb contains:
Dioscin
Diosgenin
Resin (See Glossary)

Known Effects

- Breaks membranous covering, destroying red blood cells (toxic to fish and amoeba).
- Decreases thickness and increases fluidity of mucus from lungs and bronchial tubes.
- Helps body dispose of excess fluid by increasing amount of urine produced.

Miscellaneous information:
- Diosgenin is a steroid base used to synthesize cortisone and progesterone (hormones).

Unproved Speculated Benefits

- Treats arthritis by allegedly removing accumulated waste in joints.

Warnings and Precautions

Don't take if you:
- Are pregnant, think you may be pregnant or plan pregnancy in the near future.
- Have any chronic disease of the gastrointestinal tract, such as stomach or duodenal ulcers, esophageal reflux (reflux esophagitis), ulcerative colitis, spastic colitis, diverticulosis, diverticulitis.

Consult your doctor if you:
- Take this herb for any medical problem that doesn't improve in 2 weeks. There may be safer, more-effective treatments.
- Take any medicinal drugs or herbs including aspirin, laxatives, cold and cough remedies, antacids, vitamins, minerals, amino acids, supplements, other prescription or non-prescription drugs.

Pregnancy:
- Dangers outweigh any possible benefits. Don't use.

Breast-feeding:
- Dangers outweigh any possible benefits. Don't use.

Infants and children:
- Treating infants and children under 2 with any herbal preparation is hazardous.

Others:
- None expected if you are beyond childhood and under 45, basically healthy and take for only a short time.

Storage:
- Keep cool and dry, but don't freeze. Store safely away from children.

Safe dosage:
- At present no "safe" dosage has been established.

Toxicity

Generally regarded as safe when taken in appropriate quantities for short periods of time.

For symptoms of toxicity: See *Adverse Reactions or Side Effects* section below.

Adverse Reactions or Side Effects

Signs and symptoms:	What to do:
Diarrhea	Discontinue. Call doctor immediately.
Nausea	Discontinue. Call doctor immediately.
Vomiting	Discontinue. Call doctor immediately.

Basic Information
Biological name (genus and species):
Rosa
Parts used for medicinal purposes:
Berries/fruits
Petals/flower
Chemicals this herb contains:
Ascorbic acid
Cyanogenic glycoside (See Glossary)
Quercitrin
Tannins (See Glossary)
Vitamins A and C
Volatile oils (See Glossary)

 Known Effects

- Shrinks tissues.
- Prevents secretion of fluids.
- Interferes with absorption of iron and other minerals when taken internally.

Miscellaneous information:
- North-American Indians formerly used fruit as a food source. Leaves are used to make tea or salad and smoked like tobacco.
- Rose hips are used in vitamin-C supplements.
- Adds flavor to foods during cooking.

 Unproved Speculated Benefits

- None

 Warnings and Precautions

Don't take if you:
- Are pregnant, think you may be pregnant or plan pregnancy in the near future.

Consult your doctor if you:
- No contraindications if you are not pregnant and do not take amounts larger than manufacturer's recommended dosage.

Pregnancy:
- Problems in pregnant women taking small or usual amounts have not been proved. But the chance of problems does exist. Don't use unless prescribed by your doctor.

Breast-feeding:
- Problems in breast-fed infants of lactating mothers taking small or usual amounts have not been proved. But the chance of problems does exist. Don't use unless prescribed by your doctor.

Infants and children:
- Treating infants and children under 2 with any herbal preparation is hazardous.

Others:
- None expected if you are beyond childhood and under 45, basically healthy and take for only a short time.

Storage:
- Keep cool and dry, but don't freeze. Store safely away from children.

Safe dosage:
- At present no "safe" dosage has been established.

 Toxicity

Comparative-toxicity rating not available from standard references.

 Adverse Reactions, Side Effects or Overdose Symptoms

None expected

Rosemary

Basic Information

Biological name (genus and species):
   Rosmarinus officinalis
Parts used for medicinal purposes:
   Berries/fruits
   Leaves
Chemicals this herb contains:
   Bitters (See Glossary)
   Borneol   Pinene
   Camphene   Resin (See Glossary)
   Camphor   Tannins (See Glossary)
   Cineole   Volatile oils (See Glossary)

Known Effects

- Volatile oils irritate tissue and kill bacteria.
- Acts as an astringent.
- Increases perspiration.
- Increases stomach acidity.

Miscellaneous information:
- Non-medical uses of rosemary include as an ingredient in perfumes, hair lotions and soaps.
- No effects are expected on the body, either good or bad, when herb is used in very small amounts to enhance the flavor of food.

Unproved Speculated Benefits

- Aids in expelling gas from intestinal tract.
- Triggers onset of menstrual period.
- Reddens skin by increasing blood supply to it.
- Stimulates appetite.

Warnings and Precautions

Don't take if you:
- Are pregnant, think you may be pregnant or plan pregnancy in the near future.
- Have any chronic disease of the gastrointestinal tract, such as stomach or duodenal ulcers, esophageal reflux (reflux esophagitis), ulcerative colitis, spastic colitis, diverticulosis, diverticulitis.

Consult your doctor if you:
- Take this herb for any medical problem that doesn't improve in 2 weeks. There may be safer, more-effective treatments.
- Take any medicinal drugs or herbs including aspirin, laxatives, cold and cough remedies, antacids, vitamins, minerals, amino acids, supplements, other prescription or non-prescription drugs.

Pregnancy:
- Problems in pregnant women taking small or usual amounts have not been proved. But the chance of problems does exist. Don't use unless prescribed by your doctor.

Breast-feeding:
- Problems in breast-fed infants of lactating mothers taking small or usual amounts have not been proved. But the chance of problems does exist. Don't use unless prescribed by your doctor.

Infants and children:
- Treating infants and children under 2 with any herbal preparation is hazardous.

Others:
- None expected if you are beyond childhood and under 45, basically healthy and take for only a short time.

Storage:
- Keep cool and dry, but don't freeze. Store safely away from children.

Safe dosage:
- At present no "safe" dosage has been established.

Toxicity

Rated relatively safe when taken in appropriate quantities for short periods of time.

For symptoms of toxicity: See *Adverse Reactions, Side Effects or Overdose Symptoms* section below.

Adverse Reactions, Side Effects or Overdose Symptoms

Signs and symptoms:	What to do:
Diarrhea	Discontinue. Call doctor immediately.
Nausea	Discontinue. Call doctor immediately.
Skin eruptions	Discontinue. Call doctor when convenient.
Vomiting	Discontinue. Call doctor immediately.

Basic Information

Biological name (genus and species):
   Ruta graveolens
Parts used for medicinal purposes:
   Entire plant
Chemicals this herb contains:
   Esters
   Methyl-nonylketone
   Phenols
   Rutin
   Tannins (See Glossary)
   Volatile oils (See Glossary)

Known Effects

- Stimulates uterine contractions.
- Prolongs action of epinephrine.
- Relieves spasm in skeletal or smooth muscle.
- Decreases capillary fragility.
- Interferes with absorption of iron and other minerals when taken internally.

Unproved Speculated Benefits

- Causes onset of menstruation.
- May cause abortion.
- Treats hysteria.
- Treats intestinal parasites (worms).
- Treats colic.
- Controls bleeding after delivering a baby.

Warnings and Precautions

Don't take if you:
- Are pregnant, think you may be pregnant or plan pregnancy in the near future.
- Have any chronic disease of the gastrointestinal tract, such as stomach or duodenal ulcers, esophageal reflux (reflux esophagitis), ulcerative colitis, spastic colitis, diverticulosis, diverticulitis.

Consult your doctor if you:
- Take this herb for any medical problem that doesn't improve in 2 weeks. There may be safer, more-effective treatments.
- Take any medicinal drugs or herbs including aspirin, laxatives, cold and cough remedies, antacids, vitamins, minerals, amino acids, supplements, other prescription or non-prescription drugs.

Pregnancy:
- Dangers outweigh any possible benefits. Don't use.

Breast-feeding:
- Dangers outweigh any possible benefits. Don't use.

Infants and children:
- Treating infants and children under 2 with any herbal preparation is hazardous.

Others:
- None expected if you are beyond childhood and under 45, not pregnant, basically healthy and take for only a short time.

Storage:
- Keep cool and dry, but don't freeze. Store safely away from children.

Safe dosage:
- At present no "safe" dosage has been established.

Toxicity

Rated relatively safe when taken in appropriate quantities for short periods of time.

For symptoms of toxicity: See *Adverse Reactions, Side Effects or Overdose Symptoms* section below.

Adverse Reactions, Side Effects or Overdose Symptoms

Signs and symptoms:	What to do:
Abdominal pain	Discontinue. Call doctor when convenient.
Abortion	Seek emergency treatment.
Confusion	Discontinue. Call doctor immediately.
Diarrhea	Discontinue. Call doctor immediately.
Jaundice (yellow skin and eyes)	Discontinue. Call doctor immediately.
Nausea	Discontinue. Call doctor immediately.
Skin rashes	Discontinue. Call doctor when convenient.
Vomiting	Discontinue. Call doctor immediately.

MEDICINAL HERB

Saffron (Saffron Crocus)

Basic Information

Biological name (genus and species):
Crocus sativus
Parts used for medicinal purposes:
Berries/fruits
Chemicals this herb contains:
Glycosides (See Glossary)
Volatile oils (See Glossary)

Known Effects

- Reduces irritation of gastrointestinal tract.
- Increases perspiration.
- Increases fluidity of bronchial secretions.

Miscellaneous information:
- No effects are expected on the body, either good or bad, when herb is used in very small amounts to enhance the flavor of food.

Unproved Speculated Benefits

- Stimulates respiration in asthma, whooping cough.
- Causes abortions.
- Arouses or enhances instinctive sexual desire.

Warnings and Precautions

Don't take if you:
- Are pregnant, think you may be pregnant or plan pregnancy in the near future.
- Have any chronic disease of the gastrointestinal tract, such as stomach or duodenal ulcers, esophageal reflux (reflux esophagitis), ulcerative colitis, spastic colitis, diverticulosis, diverticulitis.

Consult your doctor if you:
- Take this herb for any medical problem that doesn't improve in 2 weeks. There may be safer, more-effective treatments.
- Take any medicinal drugs or herbs including aspirin, laxatives, cold and cough remedies, antacids, vitamins, minerals, amino acids, supplements, other prescription or non-prescription drugs.

Pregnancy:
- Dangers outweigh any possible benefits. Don't use.

Breast-feeding:
- Dangers outweigh any possible benefits. Don't use.

Infants and children:
- Treating infants and children under 2 with any herbal preparation is hazardous.

Others:
- None expected if you are under 45, not pregnant, basically healthy, take it for only a short time and do not exceed manufacturer's recommended dosage.

Storage:
- Keep cool and dry, but don't freeze. Store safely away from children.

Safe dosage:
- At present no "safe" dosage has been established.

Toxicity

Rated relatively safe when taken in appropriate quantities for short periods of time.

For symptoms of toxicity: See *Adverse Reactions, Side Effects or Overdose Symptoms* section below.

Adverse Reactions, Side Effects or Overdose Symptoms

Signs and symptoms:	What to do:
Diarrhea	Discontinue. Call doctor immediately.
Dizziness	Discontinue. Call doctor immediately.
Nosebleeds	Discontinue. Call doctor when convenient.
Slow heart rate	Seek emergency treatment.
Stupor	Seek emergency treatment.
Vomiting	Discontinue. Call doctor immediately.

Basic Information

Biological name (genus and species):
Salvia officinalis
Parts used for medicinal purposes:
Leaves
Chemicals this herb contains:
Terpene
Thujone
Camphor
Resin (See Glossary)
Salvene
Tannins (See Glossary)
Volatile oils (See Glossary)

 Known Effects

- Depresses fever-control center in brain.
- Relieves spasm in skeletal or smooth muscle.
- Stimulates gastrointestinal tract.
- Stimulates central nervous system.
- Interferes with absorption of iron and other minerals when taken internally.

Miscellaneous information:

- Sage is used as a flavoring agent and in perfume.
- Salvia is *not* the brush sage of the desert or red sage.
- No effects are expected on the body, either good or bad, when herb is used in very small amounts to enhance the flavor of food. However, prolonged use of large amounts can cause seizures and unconsciousness.

 Unproved Speculated Benefits

- Aids in expelling gas from intestinal tract.
- Repels insects.
- Decreases salivation.
- Treats coughs and colds.

 Warnings and Precautions

Don't take if you:

- Are pregnant, think you may be pregnant or plan pregnancy in the near future.

Consult your doctor if you:

- Take this herb for any medical problem that doesn't improve in 2 weeks. There may be safer, more-effective treatments.
- Take any medicinal drugs or herbs including aspirin, laxatives, cold and cough remedies, antacids, vitamins, minerals, amino acids, supplements, other prescription or non-prescription drugs.

Pregnancy:

- Problems in pregnant women taking small or usual amounts have not been proved. But the chance of problems does exist. Don't use unless prescribed by your doctor.

Breast-feeding:

- May reduce milk flow. Don't use.

Infants and children:

- Treating infants and children under 2 with any herbal preparation is hazardous.

Others:

- None expected if you are beyond childhood and under 45, basically healthy and take for only a short time.

Storage:

- Keep cool and dry, but don't freeze. Store safely away from children.

Safe dosage:

- At present no "safe" dosage has been established.

 Toxicity

Generally regarded as safe when taken in appropriate quantities for short periods of time.

For symptoms of toxicity: See *Adverse Reactions, Side Effects or Overdose Symptoms* section below.

 Adverse Reactions, Side Effects or Overdose Symptoms

Signs and symptoms:	What to do:
Dry mouth	Discontinue. Call doctor when convenient.

St. John's Wort (Klamath Weed)

Basic Information

Biological name (genus and species):
Hypericum perforatum
Parts used for medicinal purposes:
Petals/flower
Chemicals this herb contains:
Hypericin
Resin (See Glossary)
Tannins (See Glossary)
Volatile oils (See Glossary)

Known Effects

- Causes photosensitization.
- Interferes with absorption of iron and other minerals when taken internally.
- Slightly depresses central nervous system.

Miscellaneous information:
- Appears on the FDA list of unsafe herbs.

Unproved Speculated Benefits

- Acts as an anti-depressant.
- Repels or destroys "demons."
- Relieves anxiety.

Warnings and Precautions

Don't take if you:
- Are pregnant, think you may be pregnant or plan pregnancy in the near future.

Consult your doctor if you:
- Take this herb for any medical problem that doesn't improve in 2 weeks. There may be safer, more-effective treatments.
- Take any medicinal drugs or herbs including aspirin, laxatives, cold and cough remedies, antacids, vitamins, minerals, amino acids, supplements, other prescription or non-prescription drugs.

Pregnancy:
- Problems in pregnant women taking small or usual amounts have not been proved. But the chance of problems does exist. Don't use unless prescribed by your doctor.

Breast-feeding:
- Problems in breast-fed infants of lactating mothers taking small or usual amounts have not been proved. But the chance of problems does exist. Don't use unless prescribed by your doctor.

Infants and children:
- Treating infants and children under 2 with any herbal preparation is hazardous.

Others:
- None expected if you are beyond childhood and under 45, basically healthy and take for only a short time.

Storage:
- Keep cool and dry, but don't freeze. Store safely away from children.

Safe dosage:
- At present no "safe" dosage has been established.

Toxicity

Rated slightly dangerous, particularly in children, persons over 55 and those who take larger than appropriate quantities for extended periods of time.

For symptoms of toxicity: See *Adverse Reactions, Side Effects or Overdose Symptoms* section below.

Adverse Reactions, Side Effects or Overdose Symptoms

Signs and symptoms:	What to do:
Abnormal skin coloring	Discontinue. Call doctor when convenient.

Sassafras

Basic Information

Biological name (genus and species):
  Sassafrass albidum
Parts used for medicinal purposes
  Bark
  Roots
Chemicals this herb contains:
  Cadinene
  Camphor
  Eugenol
  Phennandrene
  Pinene
  Safrol

Known Effects

- Depresses central nervous system.
- Irritates mucous membranes.

Miscellaneous information:
- Banned in United States as a flavoring agent because of proved carcinogenic potential.

Unproved Speculated Benefits

- Is used as spring "tonic" and "stimulant."
- Is used as "blood thinner."
- Treats common cold.

Warnings and Precautions

Don't take if you:
- Are pregnant, think you may be pregnant or plan pregnancy in the near future.
- Have any chronic disease of the gastrointestinal tract, such as stomach or duodenal ulcers, esophageal reflux (reflux esophagitis), ulcerative colitis, spastic colitis, diverticulosis, diverticulitis.

Consult your doctor if you:
- Take this herb for any medical problem that doesn't improve in 2 weeks. There may be safer, more-effective treatments.
- Take any medicinal drugs or herbs including aspirin, laxatives, cold and cough remedies, antacids, vitamins, minerals, amino acids, supplements, other prescription or non-prescription drugs.

Pregnancy:
- Risk outweighs potential benefits. Don't take.

Breast-feeding:
- Risk outweighs potential benefits. Don't take.

Infants and children:
- Treating infants and children under 2 with any herbal preparation is hazardous.

Others:
- None expected if you are beyond childhood and under 45, basically healthy and take for only a short time.

Storage:
- Keep cool and dry, but don't freeze. Store safely away from children.

Safe dosage:
- At present no "safe" dosage has been established.

Toxicity

Has carcinogenic potential. Don't use.

For symptoms of toxicity: See *Adverse Reactions, Side Effects or Overdose Symptoms* section below.

Adverse Reactions, Side Effects or Overdose Symptoms

Signs and symptoms:	What to do:
Breathing difficulties	Seek emergency treatment.
Coma	Seek emergency treatment.
Dilated pupils	Discontinue. Call doctor immediately.
Fainting	Discontinue. Call doctor immediately.
Frequent nosebleeds	Discontinue. Call doctor when convenient.
Heart, liver, kidney damage characterized by swelling of extremities, shortness of breath, jaundice (yellow skin and eyes), blood in urine	Seek emergency treatment.
Nausea	Discontinue. Call doctor immediately.
Vomiting	Discontinue. Call doctor immediately.

Saw Palmetto (Sabal)

Basic Information

Biological name (genus and species):
Serenoa repens
Parts used for medicinal purposes:
Seeds
Chemicals this herb contains:
Capric
Caproic
Caprylic
Lauric
Oleic
Palmitic
Resin (See Glossary)

Known Effects

- Irritates mucous membranes.
- Helps body dispose of excess fluid by increasing amount of urine produced.

Miscellaneous information:
- Berries are edible but don't taste good.
- Reported to enlarge female breasts and to treat benign prostatic hypertrophy. Studies have conclusively shown this herb will *not* accomplish either.

Unproved Speculated Benefits

- Treats chronic cystitis.
- Treats urethritis and other inflammations of male genito-urinary tract, including prostatitis.
- Reduces accumulated fluid in the body resulting from heart, kidney or liver diseases.

Warnings and Precautions

Don't take if you:
- Are pregnant, think you may be pregnant or plan pregnancy in the near future.
- Have any chronic disease of the gastrointestinal tract, such as stomach or duodenal ulcers, esophageal reflux (reflux esophagitis), ulcerative colitis, spastic colitis, diverticulosis, diverticulitis.

Consult your doctor if you:
- Take this herb for any medical problem that doesn't improve in 2 weeks. There may be safer, more-effective treatments.
- Take any medicinal drugs or herbs including aspirin, laxatives, cold and cough remedies, antacids, vitamins, minerals, amino acids, supplements, other prescription or non-prescription drugs.

Pregnancy:
- Problems in pregnant women taking small or usual amounts have not been proved. But the chance of problems does exist. Don't use unless prescribed by your doctor.

Breast-feeding:
- Problems in breast-fed infants of lactating mothers taking small or usual amounts have not been proved. But the chance of problems does exist. Don't use unless prescribed by your doctor.

Infants and children:
- Treating infants and children under 2 with any herbal preparation is hazardous.

Others:
- None expected if you are under 45, not pregnant, basically healthy, take it for only a short time and do not exceed manufacturer's recommended dosage.

Storage:
- Keep cool and dry, but don't freeze. Store safely away from children.

Safe dosage:
- At present no "safe" dosage has been established.

Toxicity

Rated relatively safe when taken in appropriate quantities for short periods of time.

For symptoms of toxicity: See *Adverse Reactions, Side Effects or Overdose Symptoms* section below.

Adverse Reactions, Side Effects or Overdose Symptoms

Signs and symptoms:	What to do:
Diarrhea	Discontinue. Call doctor immediately.
Nausea	Discontinue. Call doctor immediately.
Vomiting	Discontinue. Call doctor immediately.

Basic Information

Biological name (genus and species):
 Cytisus scoparius
Parts used for medicinal purposes:
 Leaves
Chemicals this herb contains:
 Cytisine
 Genisteine
 Hydroxytyramine
 Sarothamnine
 Scaparin
 Sparteine

Known Effects

- Stimulates uterine contractions.
- Helps body dispose of excess fluid by increasing amount of urine produced.
- Sometimes causes sharp rise in blood pressure.

Unproved Speculated Benefits

- Treats congestive heart failure.
- Produces sedative-hypnotic effect when smoked.

Warnings and Precautions

Don't take if you:
- Are pregnant, think you may be pregnant or plan pregnancy in the near future.
- Have any chronic disease of the gastrointestinal tract, such as stomach or duodenal ulcers, esophageal reflux (reflux esophagitis), ulcerative colitis, spastic colitis, diverticulosis, diverticulitis.

Consult your doctor if you:
- Take this herb for any medical problem that doesn't improve in 2 weeks. There may be safer, more-effective treatments.
- Take any medicinal drugs or herbs including aspirin, laxatives, cold and cough remedies, antacids, vitamins, minerals, amino acids, supplements, other prescription or non-prescription drugs.

Pregnancy:
- Problems in pregnant women taking small or usual amounts have not been proved. But the chance of problems does exist. Don't use unless prescribed by your doctor.

Breast-feeding:
- Problems in breast-fed infants of lactating mothers taking small or usual amounts have not been proved. But the chance of problems does exist. Don't use unless prescribed by your doctor.

Infants and children:
- Treating infants and children under 2 with any herbal preparation is hazardous.

Others:
- None expected if you are beyond childhood and under 45, basically healthy and take for only a short time.

Storage:
- Keep cool and dry, but don't freeze. Store safely away from children.

Safe dosage:
- At present no "safe" dosage has been established.

Toxicity

Rated slightly dangerous, particularly in children, persons over 55 and those who take larger than appropriate quantities for extended periods of time.

For symptoms of toxicity: See *Adverse Reactions, Side Effects or Overdose Symptoms* section below.

Adverse Reactions, Side Effects or Overdose Symptoms

Signs and symptoms:	What to do:
Diarrhea	Discontinue. Call doctor immediately.
Nausea	Discontinue. Call doctor immediately.
Vomiting	Discontinue. Call doctor immediately.

MEDICINAL HERB

Silverwood (Goose-tansy)

Basic Information

Biological name (genus and species):
Potentilla answerina
Parts used for medicinal purposes:
Entire plant
Chemicals this herb contains:
Ellagic acid
Kinovic acid
Tannins (See Glossary)

Known Effects

- Shrinks tissues.
- Prevents secretion of fluids.
- Causes protein molecules to clump together.
- Stimulates uterine contractions.
- Interferes with absorption of iron and other minerals when taken internally.

Unproved Speculated Benefits

- Treats dysmenorrhea (painful menstruation).
- When used with lobelia, treats tetanus in absence of medical help.

Warnings and Precautions

Don't take if you:
- Are pregnant, think you may be pregnant or plan pregnancy in the near future.
- Have any chronic disease of the gastrointestinal tract, such as stomach or duodenal ulcers, esophageal reflux (reflux esophagitis), ulcerative colitis, spastic colitis, diverticulosis, diverticulitis.

Consult your doctor if you:
- Take this herb for any medical problem that doesn't improve in 2 weeks. There may be safer, more-effective treatments.
- Take any medicinal drugs or herbs including aspirin, laxatives, cold and cough remedies, antacids, vitamins, minerals, amino acids, supplements, other prescription or non-prescription drugs.

Pregnancy:
- Dangers outweigh any possible benefits. Don't use.

Breast-feeding:
- Dangers outweigh any possible benefits. Don't use.

Infants and children:
- Treating infants and children under 2 with any herbal preparation is hazardous.

Others:
- None expected if you are under 45, not pregnant, basically healthy, take it for only a short time and do not exceed manufacturer's recommended dosage.

Storage:
- Keep cool and dry, but don't freeze. Store safely away from children.

Safe dosage:
- At present no "safe" dosage has been established.

Toxicity

Comparative-toxicity rating not available from standard references.

For symptoms of toxicity: See *Adverse Reactions, Side Effects or Overdose Symptoms* section below.

Adverse Reactions, Side Effects or Overdose Symptoms

Signs and symptoms:	What to do:
Diarrhea	Discontinue. Call doctor immediately.
Nausea	Discontinue. Call doctor immediately.
Painful urination	Discontinue. Call doctor when convenient.
Vomiting	Discontinue. Call doctor immediately.

Basic Information

Biological name (genus and species):
Ulmus fulva
Parts used for medicinal purposes:
Inner bark
Chemicals this herb contains:
Calcium
Calcium oxalate
Mucilage (See Glossary)
Polysaccharide
Starch
Tannins (See Glossary)

 ## Known Effects

- Decreases thickness and increases fluidity of mucus from lungs and bronchial tubes.
- Protects scraped tissues.
- Interferes with absorption of iron and other minerals when taken internally.

 ## Unproved Speculated Benefits

- Decreases discomfort of cough.

 ## Warnings and Precautions

Don't take if you:
- Are pregnant, think you may be pregnant or plan pregnancy in the near future.

Consult your doctor if you:
- Take this herb for any medical problem that doesn't improve in 2 weeks. There may be safer, more-effective treatments.
- Take any medicinal drugs or herbs including aspirin, laxatives, cold and cough remedies, antacids, vitamins, minerals, amino acids, supplements, other prescription or non-prescription drugs.

Pregnancy:
- Dangers outweigh any possible benefits. Don't use.

Breast-feeding:
- Dangers outweigh any possible benefits. Don't use.

Infants and children:
- Treating infants and children under 2 with any herbal preparation is hazardous.

Others:
- None expected if you are beyond childhood and under 45, basically healthy and take for only a short time.

Storage:
- Keep cool and dry, but don't freeze. Store safely away from children.

Safe dosage:
- At present no "safe" dosage has been established.

 ## Toxicity

Rated relatively safe when taken in appropriate quantities for short periods of time.

For symptoms of toxicity: See *Adverse Reactions, Side Effects or Overdose Symptoms* section below.

 ## Adverse Reactions, Side Effects or Overdose Symptoms

Signs and symptoms:	What to do:
Skin rash	Discontinue. Call doctor when convenient.

MEDICINAL HERB

Snakeplant

Basic Information

Biological name (genus and species):
Rivea corymbosa
Parts used for medicinal purposes:
Seeds
Chemicals this herb contains:
Five related LSD-like alkaloids:
Chanoclavine
D-isolysergic acid amide
D-lysergic acid amide
Elymoclavine
Lysergol

Known Effects

• Depresses central nervous system.

Miscellaneous information:
• Snakeplant is used primarily by Mexican Indians in religious ceremonies. They call it *badah.*

Unproved Speculated Benefits

• Changes mood.
• Causes hallucinations.

Warnings and Precautions

Don't take if you:
• Are pregnant, think you may be pregnant or plan pregnancy in the near future.
• Have any chronic disease of the gastrointestinal tract, such as stomach or duodenal ulcers, esophageal reflux (reflux esophagitis), ulcerative colitis, spastic colitis, diverticulosis, diverticulitis.

Consult your doctor if you:
• Take this herb for any medical problem that doesn't improve in 2 weeks. There may be safer, more-effective treatments.
• Take any medicinal drugs or herbs including aspirin, laxatives, cold and cough remedies, antacids, vitamins, minerals, amino acids, supplements, other prescription or non-prescription drugs.

Pregnancy:
• Dangers outweigh any possible benefits. Don't use.

Breast-feeding:
• Dangers outweigh any possible benefits. Don't use.

Infants and children:
• Treating infants and children under 2 with any herbal preparation is hazardous.

Others:
• Dangers outweigh any possible benefits. Don't use.

Storage:
• Keep cool and dry, but don't freeze. Store safely away from children.

Safe dosage:
• At present no "safe" dosage has been established.

Toxicity

Rated slightly dangerous, particularly in children, persons over 55 and those who take larger than appropriate quantities for extended periods of time.

For symptoms of toxicity: See *Adverse Reactions, Side Effects or Overdose Symptoms* section below.

Adverse Reactions, Side Effects or Overdose Symptoms

Signs and symptoms:	What to do:
Blurred vision	Discontinue. Call doctor immediately.
Coma	Seek emergency treatment.
Confusion	Discontinue. Call doctor immediately.
Hallucinations	Seek emergency treatment.
Nausea	Discontinue. Call doctor immediately.
Stupor	Discontinue. Call doctor immediately.
Vomiting	Discontinue. Call doctor immediately.

Snakeroot (Virginia Snakeroot, Serpentaria)

Basic Information

Biological name (genus and species):
  Aristolochia serpentaria
Parts used for medicinal purposes:
  Roots
Chemicals this herb contains:
  Borneol
  Serpentaria
  Terpene
  Volatile oils (See Glossary)

Known Effects

- Stimulates stomach secretions.
- Stimulates smooth-muscle contractions of gastrointestinal tract and heart.

Unproved Speculated Benefits

- Increases circulation.
- Stimulates heart action.
- Treats dyspepsia.
- Reduces fever.
- Treats sores on skin.

Warnings and Precautions

Don't take if you:
- Are pregnant, think you may be pregnant or plan pregnancy in the near future.
- Have any chronic disease of the gastrointestinal tract, such as stomach or duodenal ulcers, esophageal reflux (reflux esophagitis), ulcerative colitis, spastic colitis, diverticulosis, diverticulitis.

Consult your doctor if you:
- Take this herb for any medical problem that doesn't improve in 2 weeks. There may be safer, more-effective treatments.
- Take any medicinal drugs or herbs including aspirin, laxatives, cold and cough remedies, antacids, vitamins, minerals, amino acids, supplements, other prescription or non-prescription drugs.

Pregnancy:
- Dangers outweigh any possible benefits. Don't use.

Breast-feeding:
- Dangers outweigh any possible benefits. Don't use.

Infants and children:
- Treating infants and children under 2 with any herbal preparation is hazardous.

Others:
- None expected if you are under 45, not pregnant, basically healthy, take it for only a short time and do not exceed manufacturer's recommended dosage.

Storage:
- Keep cool and dry, but don't freeze. Store safely away from children.

Safe dosage:
- At present no "safe" dosage has been established.

Toxicity

Rated relatively safe when taken in appropriate quantities for short periods of time.

For symptoms of toxicity: See *Adverse Reactions, Side Effects or Overdose Symptoms* section below.

Adverse Reactions, Side Effects or Overdose Symptoms

Signs and symptoms:	What to do:
Diarrhea	Discontinue. Call doctor immediately.
Nausea	Discontinue. Call doctor immediately.
Tenesmus (spasm of rectal sphincter)	Discontinue. Call doctor when convenient.
Vomiting	Discontinue. Call doctor immediately.

Spanish Broom

Basic Information
Biological name (genus and species):
  Spartium junceum
Parts used for medicinal purposes:
  Petals/flower
Chemicals this herb contains:
  Anagyrine
  Cytisine
  Methylcystinine

Known Effects

- Stimulates uterine contractions.
- Helps body dispose of excess fluid by increasing amount of urine produced.
- Stimulates gastrointestinal tract.
- Causes vomiting.

Unproved Speculated Benefits

- Induces labor.
- Causes watery, explosive bowel movements.

Warnings and Precautions

Don't take if you:
- Are pregnant, think you may be pregnant or plan pregnancy in the near future.
- Have any chronic disease of the gastrointestinal tract, such as stomach or duodenal ulcers, esophageal reflux (reflux esophagitis), ulcerative colitis, spastic colitis, diverticulosis, diverticulitis.

Consult your doctor if you:
- Take this herb for any medical problem that doesn't improve in 2 weeks. There may be safer, more-effective treatments.
- Take any medicinal drugs or herbs including aspirin, laxatives, cold and cough remedies, antacids, vitamins, minerals, amino acids, supplements, other prescription or non-prescription drugs.

Pregnancy:
- Dangers outweigh any possible benefits. Don't use.

Breast-feeding:
- Dangers outweigh any possible benefits. Don't use.

Infants and children:
- Treating infants and children under 2 with any herbal preparation is hazardous.

Others:
- None expected if you are under 45, not pregnant, basically healthy, take it for only a short time and do not exceed manufacturer's recommended dosage.

Storage:
- Keep cool and dry, but don't freeze. Store safely away from children.

Safe dosage:
- At present no "safe" dosage has been established.

Toxicity

Comparative-toxicity rating not available from standard references.

For symptoms of toxicity: See *Adverse Reactions, Side Effects or Overdose Symptoms* section below.

Adverse Reactions, Side Effects or Overdose Symptoms

Signs and symptoms:	What to do:
Diarrhea	Discontinue. Call doctor immediately.
Kidney damage characterized by blood in urine, decreased urine flow, swelling of hands and feet	Seek emergency treatment.
Muscle weakness	Discontinue. Call doctor immediately.
Nausea	Discontinue. Call doctor immediately.
Vomiting	Discontinue. Call doctor immediately.

Basic Information

Biological name (genus and species):
  Mentha spicata
Parts used for medicinal purposes:
  Leaves
  Petals/flower
Chemicals this herb contains:
  Carvone
  Resin (See Glossary)
  Volatile oils (See Glossary)

 ## Known Effects

• Stimulates muscular action of gastrointestinal tract.

Miscellaneous information:
• Spearmint is used as a flavoring agent in many foods.

 ## Unproved Speculated Benefits

• Aids in expelling gas from intestinal tract.

 ## Warnings and Precautions

Don't take if you:
• Are pregnant, think you may be pregnant or plan pregnancy in the near future.
• Have any chronic disease of the gastrointestinal tract, such as stomach or duodenal ulcers, esophageal reflux (reflux esophagitis), ulcerative colitis, spastic colitis, diverticulosis, diverticulitis.

Consult your doctor if you:
• Take this herb for any medical problem that doesn't improve in 2 weeks. There may be safer, more-effective treatments.
• Take any medicinal drugs or herbs including aspirin, laxatives, cold and cough remedies, antacids, vitamins, minerals, amino acids, supplements, other prescription or non-prescription drugs.

Pregnancy:
• Problems in pregnant women taking small or usual amounts have not been proved. But the chance of problems does exist. Don't use unless prescribed by your doctor.

Breast-feeding:
• Problems in breast-fed infants of lactating mothers taking small or usual amounts have not been proved. But the chance of problems does exist. Don't use unless prescribed by your doctor.

Infants and children:
• Treating infants and children under 2 with any herbal preparation is hazardous.

Others:
• None expected if you are under 45, not pregnant, basically healthy, take it for only a short time and do not exceed manufacturer's recommended dosage.

Storage:
• Keep cool and dry, but don't freeze. Store safely away from children.

Safe dosage:
• At present no "safe" dosage has been established.

 ## Toxicity

Comparative-toxicity rating not available from standard references.

For symptoms of toxicity: See *Adverse Reactions, Side Effects or Overdose Symptoms* section below.

 ## Adverse Reactions, Side Effects or Overdose Symptoms

Signs and symptoms:	What to do:
Convulsions and coma in children	Seek emergency treatment.
Diarrhea	Discontinue. Call doctor immediately.
Nausea	Discontinue. Call doctor immediately.
Vomiting	Discontinue. Call doctor immediately.

MEDICINAL HERB

Strawberry (Earth Mulberry)

Basic Information

Biological name (genus and species):
Fragaria vesa, F. americana
Parts used for medicinal purposes:
Berries
Leaves
Roots
Chemicals this herb contains:
Catechins
Leucoanthocyanin
Minerals
Vitamin C

Known Effects

- Shrinks tissues.
- Prevents secretion of fluids.
- Prevents scurvy.
- Inhibits production of histamines.
- Precipitates proteins.

Miscellaneous information:
- Wild strawberry is a member of the rose family.

Unproved Speculated Benefits

- Increases effectiveness of antihistamines.
- Berries treat kidney stones.
- Roots and leaves treat eczema, diarrhea, toothache, skin ulcers.

Warnings and Precautions

Don't take if you:
- Are allergic to strawberries.

Consult your doctor if you:
- Take this herb for any medical problem that doesn't improve in 2 weeks. There may be safer, more-effective treatments.
- Take any medicinal drugs or herbs including aspirin, laxatives, cold and cough remedies, antacids, vitamins, minerals, amino acids, supplements, other prescription or non-prescription drugs.

Pregnancy:
- Pregnant women should experience no problems taking usual amounts as part of a balanced diet. Other products extracted from this herb have not been proved to cause problems.

Breast-feeding:
- Breast-fed infants of lactating mothers should experience no problems when mother takes usual amounts as part of a balanced diet. Other products extracted from this herb have not been proved to cause problems.

Infants and children:
- Treating infants and children under 2 with any herbal preparation is hazardous.

Others:
- None expected if you are beyond childhood and under 45, basically healthy and take for only a short time.

Storage:
- Keep cool and dry, but don't freeze. Store safely away from children.

Safe dosage:
- At present no "safe" dosage has been established.

Toxicity

Generally regarded as safe when taken in appropriate quantities for short periods of time.

Adverse Reactions, Side Effects or Overdose Symptoms

None expected

Basic Information

Biological name (genus and species):
Rhus glabra, R. blabrum
Parts used for medicinal purposes:
Bark
Berries
Leaves
Chemicals this herb contains:
Albumin
Malic acid
Resin (See Glossary)
Tannins (See Glossary)
Volatile oils (See Glossary)

 ## Known Effects

Bark:
- Shrinks tissues.
- Prevents secretion of fluids.
- Inhibits growth and development of germs.

Berries:
- Help body dispose of excess fluid by increasing amount of urine produced.
- Interfere with absorption of iron and other minerals when taken internally.

Miscellaneous information:
- In same plant family as poison ivy and poison oak.

 ## Unproved Speculated Benefits

- Treats diarrhea.
- Treats rectal bleeding.
- Treats asthma when leaves are smoked.

 ## Warnings and Precautions

Don't take if you:
- Are pregnant, think you may be pregnant or plan pregnancy in the near future.

Consult your doctor if you:
- Take this herb for any medical problem that doesn't improve in 2 weeks. There may be safer, more-effective treatments.
- Take any medicinal drugs or herbs including aspirin, laxatives, cold and cough remedies, antacids, vitamins, minerals, amino acids, supplements, other prescription or non-prescription drugs.

Pregnancy:
- Problems in pregnant women taking small or usual amounts have not been proved. But the chance of problems does exist. Don't use unless prescribed by your doctor.

Breast-feeding:
- Problems in breast-fed infants of lactating mothers taking small or usual amounts have not been proved. But the chance of problems does exist. Don't use unless prescribed by your doctor.

Infants and children:
- Treating infants and children under 2 with any herbal preparation is hazardous.

Others:
- None expected if you are beyond childhood and under 45, basically healthy and take for only a short time.

Storage:
- Keep cool and dry, but don't freeze. Store safely away from children.

Safe dosage:
- At present no "safe" dosage has been established.

 ## Toxicity

Comparative-toxicity rating not available from standard references.

 ## Adverse Reactions, Side Effects or Overdose Symptoms

None expected

Sundew

Basic Information
Biological name (genus and species):
  Drosea rotundifolia
Parts used for medicinal purposes:
  Various parts of the entire plant, frequently differing by country and/or culture
Chemicals this herb contains:
  Citric acid
  Droserone
  Malic acid
  Resin (See Glossary)
  Tannins (See Glossary)

Known Effects

• Interferes with absorption of iron and other minerals when taken internally.
• Loosens bronchial secretions.

Unproved Speculated Benefits

• Treats whooping cough.
• Treats laryngitis.
• Treats smoker's cough.

Warnings and Precautions

Don't take if you:
• Are pregnant, think you may be pregnant or plan pregnancy in the near future.

Consult your doctor if you:
• Take this herb for any medical problem that doesn't improve in 2 weeks. There may be safer, more-effective treatments.
• Take any medicinal drugs or herbs including aspirin, laxatives, cold and cough remedies, antacids, vitamins, minerals, amino acids, supplements, other prescription or non-prescription drugs.

Pregnancy:
• Problems in pregnant women taking small or usual amounts have not been proved. But the chance of problems does exist. Don't use unless prescribed by your doctor.

Breast-feeding:
• Problems in breast-fed infants of lactating mothers taking small or usual amounts have not been proved. But the chance of problems does exist. Don't use unless prescribed by your doctor.

Infants and children:
• Treating infants and children under 2 with any herbal preparation is hazardous.

Others:
• None expected if you are beyond childhood and under 45, basically healthy and take for only a short time.

Storage:
• Keep cool and dry, but don't freeze. Store safely away from children.

Safe dosage:
• At present no "safe" dosage has been established.

Toxicity

Comparative-toxicity rating not available from standard references.

Adverse Reactions, Side Effects or Overdose Symptoms

None expected

Sunflower

Basic Information

Biological name (genus and species):
Helianthus annuus

Parts used for medicinal purposes:
Leaves
Petals/flower
Seeds

Chemicals this herb contains:
Arachidic acid
Behenic acid
Linoleic acid
Linolenic acid
Oleic acid
Palmitic acid
Stearic acid
Vitamin E

Known Effects

• Reduces blood sugar.
• Decreases inflammation in bronchi.

Miscellaneous information:
• Sunflower is a food source.

Unproved Speculated Benefits

• Treats vitamin-E deficiency.
• Treats bronchial irritation and common colds.
• Reduces fevers.
• Relieves pain of arthritis.
• Increases urine flow.
• Increases perspiration.

Warnings and Precautions

Don't take if you:
• No contraindications if you are not pregnant and do not take amounts larger than manufacturer's recommended dosage.

Consult your doctor if you:
• Take this herb for any medical problem that doesn't improve in 2 weeks. There may be safer, more-effective treatments.
• Take any medicinal drugs or herbs including aspirin, laxatives, cold and cough remedies, antacids, vitamins, minerals, amino acids, supplements, other prescription or non-prescription drugs.

Pregnancy:
• Problems in pregnant women taking small or usual amounts have not been proved. But the chance of problems does exist. Don't use unless prescribed by your doctor.

Breast-feeding:
• Problems in breast-fed infants of lactating mothers taking small or usual amounts have not been proved. But the chance of problems does exist. Don't use unless prescribed by your doctor.

Infants and children:
• Treating infants and children under 2 with any herbal preparation is hazardous.

Others:
• None expected if you are beyond childhood and under 45, basically healthy and take for only a short time.

Storage:
• Keep cool and dry, but don't freeze. Store safely away from children.

Safe dosage:
• At present no "safe" dosage has been established.

Toxicity

Comparative-toxicity rating not available from standard references.

Adverse Reactions, Side Effects or Overdose Symptoms

None expected

Sweet Violet

Basic Information
Biological name (genus and species):
  Viola odorata, V. pedapa
Parts used for medicinal purposes:
  Leaves
  Seeds
Chemicals this herb contains:
  Glycosides (See Glossary)
  Myrosin

Known Effects

- Irritates mucous membranes.
- Stimulates gastrointestinal tract.

Miscellaneous information:
- Sweet violet was used to treat cancer as early as 500 B.C., but evidence of real benefit is lacking.

Unproved Speculated Benefits

- Is used as poultice to treat cancer. (See Glossary.)
- Treats skin disease.
- Is used as a mild laxative.
- Causes vomiting.
- Decreases thickness and increases fluidity of mucus from lungs and bronchial tubes.
- Treats coughs.

Warnings and Precautions

Don't take if you:
- Are pregnant, think you may be pregnant or plan pregnancy in the near future.
- Have any chronic disease of the gastrointestinal tract, such as stomach or duodenal ulcers, esophageal reflux (reflux esophagitis), ulcerative colitis, spastic colitis, diverticulosis, diverticulitis.

Consult your doctor if you:
- Take this herb for any medical problem that doesn't improve in 2 weeks. There may be safer, more-effective treatments.
- Take any medicinal drugs or herbs including aspirin, laxatives, cold and cough remedies, antacids, vitamins, minerals, amino acids, supplements, other prescription or non-prescription drugs.

Pregnancy:
- Problems in pregnant women taking small or usual amounts have not been proved. But the chance of problems does exist. Don't use unless prescribed by your doctor.

Breast-feeding:
- Problems in breast-fed infants of lactating mothers taking small or usual amounts have not been proved. But the chance of problems does exist. Don't use unless prescribed by your doctor.

Infants and children:
- Treating infants and children under 2 with any herbal preparation is hazardous.

Others:
- None expected if you are beyond childhood and under 45, basically healthy and take for only a short time.

Storage:
- Keep cool and dry, but don't freeze. Store safely away from children.

Safe dosage:
- At present no "safe" dosage has been established.

Toxicity

Comparative-toxicity rating not available from standard references.

For symptoms of toxicity: See *Adverse Reactions, Side Effects or Overdose Symptoms* section below.

Adverse Reactions, Side Effects or Overdose Symptoms

Signs and symptoms:	What to do:
Seeds:	
Diarrhea	Discontinue. Call doctor immediately.
Nausea	Discontinue. Call doctor immediately.
Vomiting	Discontinue. Call doctor immediately.

Tansy

Basic Information

Biological name (genus and species):
   Tanacetum vulgare
Parts used for medicinal purposes:
   Entire plant
Chemicals this herb contains:
   Bitters (See Glossary)
   Borneol
   Camphor
   Resin (See Glossary)
   Tanacetone
   Thujone

Known Effects

- Stimulates uterine contractions.
- Stimulates appetite.
- Kills intestinal parasites.

Miscellaneous information:
- Tansy is a powerful herb that should be avoided or used *only* under strict medical supervision.

Unproved Speculated Benefits

- Treats pain.
- Causes euphoria.
- Treats roundworms and pinworms.
- Treats menstrual difficulties.

Warnings and Precautions

Don't take if you:
- Are pregnant, think you may be pregnant or plan pregnancy in the near future.
- Have any chronic disease of the gastrointestinal tract, such as stomach or duodenal ulcers, esophageal reflux (reflux esophagitis), ulcerative colitis, spastic colitis, diverticulosis, diverticulitis.

Consult your doctor if you:
- Take this herb for any medical problem that doesn't improve in 2 weeks. There may be safer, more-effective treatments.
- Take any medicinal drugs or herbs including aspirin, laxatives, cold and cough remedies, antacids, vitamins, minerals, amino acids, supplements, other prescription or non-prescription drugs.

Pregnancy:
- Dangers outweigh any possible benefits. Don't use.

Breast-feeding:
- Dangers outweigh any possible benefits. Don't use.

Infants and children:
- Treating infants and children under 2 with any herbal preparation is hazardous.

Others:
- Dangers outweigh any possible benefits. Don't use.

Storage:
- Keep cool and dry, but don't freeze. Store safely away from children.

Safe dosage:
- At present no "safe" dosage has been established.

Toxicity

Rated dangerous, particularly in children, persons over 55 and those who take larger than appropriate quantities for extended periods of time.

For symptoms of toxicity: See *Adverse Reactions, Side Effects or Overdose Symptoms* section below.

Adverse Reactions, Side Effects or Overdose Symptoms

Signs and symptoms:	What to do:
Coma	Seek emergency treatment.
Convulsions	Seek emergency treatment.
Diarrhea	Discontinue. Call doctor immediately.
Dilated pupils	Seek emergency treatment.
Nausea	Discontinue. Call doctor immediately.
Vomiting	Discontinue. Call doctor immediately.
Weak, rapid pulse	Seek emergency treatment.

Thyme, Common

Basic Information

Biological name (genus and species):
Thymus vulgaris
Parts used for medicinal purposes:
Berries/fruits
Leaves
Chemicals this herb contains:
Gum (See Glossary)
Tannins (See Glossary)
Thyme oil

Known Effects

- Inhibits growth and development of germs.
- Stimulates gastrointestinal tract.
- Decreases thickness of bronchial secretions.
- Interferes with absorption of iron and other minerals when taken internally.

Unproved Speculated Benefits

- Reduces flatulence.
- Treats coughs.
- Treats bronchitis.
- Treats hookworm.
- Treats bacterial infections.

Warnings and Precautions

Don't take if you:
- Are pregnant, think you may be pregnant or plan pregnancy in the near future.

Consult your doctor if you:
- Take this herb for any medical problem that doesn't improve in 2 weeks. There may be safer, more-effective treatments.
- Take any medicinal drugs or herbs including aspirin, laxatives, cold and cough remedies, antacids, vitamins, minerals, amino acids, supplements, other prescription or non-prescription drugs.

Pregnancy:
- Problems in pregnant women taking small or usual amounts have not been proved. But the chance of problems does exist. Don't use unless prescribed by your doctor.

Breast-feeding:
- Problems in breast-fed infants of lactating mothers taking small or usual amounts have not been proved. But the chance of problems does exist. Don't use unless prescribed by your doctor.

Infants and children:
- Treating infants and children under 2 with any herbal preparation is hazardous.

Others:
- None expected if you are beyond childhood and under 45, basically healthy and take for only a short time.

Storage:
- Keep cool and dry, but don't freeze. Store safely away from children.

Safe dosage:
- At present no "safe" dosage has been established.

Toxicity

Rated relatively safe when taken in appropriate quantities for short periods of time.

For symptoms of toxicity: See *Adverse Reactions, Side Effects or Overdose Symptoms* section below.

Adverse Reactions, Side Effects or Overdose Symptoms

Signs and symptoms:	What to do:
Diarrhea	Discontinue. Call doctor immediately.
Nausea	Discontinue. Call doctor immediately.
Vomiting	Discontinue. Call doctor immediately.

Tonka Bean (Tonquin Bean)

Basic Information
Biological name (genus and species):
Coumarouna odorata, Dipteryx odorta
Parts used for medicinal purposes:
Seeds
Chemicals this herb contains:
Coumarin
Gum (See Glossary)
Sitosterin
Starch
Stigmasterin
Sugar

Known Effects

- Delays or stops blood clotting.
- Acts as anti-coagulant. Coumarin interferes with synthesis of vitamin K in the human intestines. The absence of adequate vitamin K prevents blood clotting.

Miscellaneous information:
- The tonka bean was once a common adulterant of vanilla extracts.
- Is used as flavoring in tobacco.
- FDA has banned its use as a flavoring agent in foods.

Unproved Speculated Benefits

- Prevents clotting in deep veins.
- Prevents blood clots from breaking away from blood vessels and lodging in vital organs, such as lung or brain. Its use must be monitored carefully with frequent laboratory studies (prothrombin time).

Warnings and Precautions

Don't take if you:
- Are pregnant, think you may be pregnant or plan pregnancy in the near future.

Consult your doctor if you:
- Take this herb for any medical problem that doesn't improve in 2 weeks. There may be safer, more-effective treatments.
- Take any medicinal drugs or herbs including aspirin, laxatives, cold and cough remedies, antacids, vitamins, minerals, amino acids, supplements, other prescription or non-prescription drugs.

Pregnancy:
- Dangers outweigh any possible benefits. Don't use.

Breast-feeding:
- Dangers outweigh any possible benefits. Don't use.

Infants and children:
- Treating infants and children under 2 with any herbal preparation is hazardous.

Others:
- Dangers outweigh any possible benefits. Don't use.

Storage:
- Keep cool and dry, but don't freeze. Store safely away from children.

Safe dosage:
- At present no "safe" dosage has been established.

Toxicity

Comparative-toxicity rating not available from standard references.

For symptoms of toxicity: See *Adverse Reactions, Side Effects or Overdose Symptoms* section below.

Adverse Reactions, Side Effects or Overdose Symptoms

Signs and symptoms:	What to do:
Atrophy of testicles	Discontinue. Call doctor when convenient.
Jaundice (yellow skin and eyes)	Discontinue. Call doctor when convenient.
Retards growth	Discontinue. Call doctor when convenient.
Uncontrollable internal bleeding	Seek emergency treatment.

Tormentil

Basic Information

Biological name (genus and species):
  Potentill erecta, P. tormentil
Parts used for medicinal purposes:
  Roots
Chemicals this herb contains:
  Ellagic acid
  Kinovic
  Tannins (See Glossary)

Known Effects

- Shrinks tissues.
- Prevents secretion of fluids.
- Interferes with absorption of iron and other minerals when taken internally.

Unproved Speculated Benefits

- Treats diarrhea.
- Treats sore throat.
- Is used as poultice for wounds. (See Glossary.)

Warnings and Precautions

Don't take if you:
- Are pregnant, think you may be pregnant or plan pregnancy in the near future.
- Have any chronic disease of the gastrointestinal tract, such as stomach or duodenal ulcers, esophageal reflux (reflux esophagitis), ulcerative colitis, spastic colitis, diverticulosis, diverticulitis.

Consult your doctor if you:
- Take this herb for any medical problem that doesn't improve in 2 weeks. There may be safer, more-effective treatments.
- Take any medicinal drugs or herbs including aspirin, laxatives, cold and cough remedies, antacids, vitamins, minerals, amino acids, supplements, other prescription or non-prescription drugs.

Pregnancy:
- Problems in pregnant women taking small or usual amounts have not been proved. But the chance of problems does exist. Don't use unless prescribed by your doctor.

Breast-feeding:
- Problems in breast-fed infants of lactating mothers taking small or usual amounts have not been proved. But the chance of problems does exist. Don't use unless prescribed by your doctor.

Infants and children:
- Treating infants and children under 2 with any herbal preparation is hazardous.

Others:
- None expected if you are beyond childhood and under 45, basically healthy and take for only a short time.

Storage:
- Keep cool and dry, but don't freeze. Store safely away from children.

Safe dosage:
- At present no "safe" dosage has been established.

Toxicity

Comparative-toxicity rating not available from standard references.

For symptoms of toxicity: See *Adverse Reactions, Side Effects or Overdose Symptoms* section below.

Adverse Reactions, Side Effects or Overdose Symptoms

Signs and symptoms:	What to do:
Diarrhea	Discontinue. Call doctor immediately.
Kidney damage characterized by blood in urine, decreased urine flow, swelling of hands and feet	See emergency treatment.
Nausea	Discontinue. Call doctor immediately.
Vomiting	Discontinue. Call doctor immediately.

Unicorn Root (Star Grass, Colic Root)

Basic Information

Biological name (genus and species):
Aletris farinosa
Parts used for medicinal purposes:
Leaves
Roots
Chemicals this herb contains:
Diosgenin
Resin (See Glossary)
Saponins (See Glossary)
Volatile oils (See Glossary)

Known Effects

• Reduces smooth-muscle spasms.

Miscellaneous information:
• Serves as base substance to produce synthetic progesterone (a female hormone).

Unproved Speculated Benefits

• Treats painful menstruation.
• Decreases chances of miscarriage.
• Soothes sore breasts.
• Relieves flatulence.
• Relieves arthritis.

Warnings and Precautions

Don't take if you:
• Are pregnant, think you may be pregnant or plan pregnancy in the near future.
• Have any chronic disease of the gastrointestinal tract, such as stomach or duodenal ulcers, esophageal reflux (reflux esophagitis), ulcerative colitis, spastic colitis, diverticulosis, diverticulitis.

Consult your doctor if you:
• Take this herb for any medical problem that doesn't improve in 2 weeks. There may be safer, more-effective treatments.
• Take any medicinal drugs or herbs including aspirin, laxatives, cold and cough remedies, antacids, vitamins, minerals, amino acids, supplements, other prescription or non-prescription drugs.

Pregnancy:
• Problems in pregnant women taking small or usual doses have not been proved. But the chance of problems does exist. Don't use unless prescribed by your doctor.

Breast-feeding:
• Problems in breast-fed infants of lactating mothers taking small or usual doses have not been proved. But the chance of problems does exist. Don't use unless prescribed by your doctor.

Infants and children:
• Treating infants and children under 2 with any herbal preparation is hazardous.

Others:
• None expected if you are under 45, not pregnant, basically healthy, take it for only a short time and do not exceed manufacturer's recommended dosage.

Storage:
• Keep cool and dry, but don't freeze. Store safely away from children.

Safe dosage:
• At present no "safe" dosage has been established.

Toxicity

Rated slightly dangerous, particularly in children, persons over 55 and those who take larger than appropriate quantities for extended periods of time.

For symptoms of toxicity: See *Adverse Reactions, Side Effects or Overdose Symptoms* section below.

Adverse Reactions, Side Effects or Overdose Symptoms

Signs and symptoms:	What to do:
Diarrhea	Discontinue. Call doctor immediately.
Lethargy	Discontinue. Call doctor when convenient.
Vomiting	Discontinue. Call doctor immediately.

Valerian (Garden Heliotrope, Tobacco Root)

Basic Information

Biological name (genus and species):
  Valeriana edulis, V. officinalis
Parts used for medicinal purposes:
  Rhizomes
  Roots
Chemicals this herb contains:
  Acetic acid
  Butyric acid
  Camphene
  Chatinine
  Formic acid
  Glycosides (See Glossary)
  Pinene
  Resin (See Glossary)
  Valeric acid
  Valerine
  Volatile oils (See Glossary)

Known Effects

• Depresses central nervous system.

Miscellaneous information:
• Cats are attracted to this herb.

Unproved Speculated Benefits

• Treats anxiety.
• Treats insomnia.
• Treats convulsions.
• Causes sedation.

Warnings and Precautions

Don't take if you:
• Are pregnant, think you may be pregnant or plan pregnancy in the near future.

Consult your doctor if you:
• Take this herb for any medical problem that doesn't improve in 2 weeks. There may be safer, more-effective treatments.
• Take any medicinal drugs or herbs including aspirin, laxatives, cold and cough remedies, antacids, vitamins, minerals, amino acids, supplements, other prescription or non-prescription drugs.

Pregnancy:
• Problems in pregnant women taking small or usual amounts have not been proved. But the chance of problems does exist. Don't use unless prescribed by your doctor.

Breast-feeding:
• Problems in breast-fed infants of lactating mothers taking small or usual amounts have not been proved. But the chance of problems does exist. Don't use unless prescribed by your doctor.

Infants and children:
• Treating infants and children under 2 with any herbal preparation is hazardous.

Others:
• None expected if you are beyond childhood and under 45, basically healthy and take for only a short time.

Storage:
• Use fresh material only. Dried valerian loses potency.

Safe dosage:
• At present no "safe" dosage has been established.

Toxicity

Rated relatively safe when taken in appropriate quantities for short periods of time.

For symptoms of toxicity: See *Adverse Reactions, Side Effects or Overdose Symptoms* section below.

Adverse Reactions, Side Effects or Overdose Symptoms

Signs and symptoms:	What to do:
Diarrhea	Discontinue. Call doctor immediately.
Nausea	Discontinue. Call doctor immediately.
Vomiting	Discontinue. Call doctor immediately.

Vervain (European Vervaine, Verbena)

Basic Information
Biological name (genus and species):
Verbena officinalis
Parts used for medicinal purposes:
Roots
Chemical this herb contains:
Verbenaline

 ## Known Effects

- Stimulates gastrointestinal tract.
- Stimulates parasympathetic branch of autonomic nervous system.

 ## Unproved Speculated Benefits

- Treats coughs.
- Treats upper-abdominal pain.
- Induces vomiting.

 ## Warnings and Precautions

Don't take if you:
- Are pregnant, think you may be pregnant or plan pregnancy in the near future.
- Have any chronic disease of the gastrointestinal tract, such as stomach or duodenal ulcers, esophageal reflux (reflux esophagitis), ulcerative colitis, spastic colitis, diverticulosis, diverticulitis.

Consult your doctor if you:
- Take this herb for any medical problem that doesn't improve in 2 weeks. There may be safer, more-effective treatments.
- Take any medicinal drugs or herbs including aspirin, laxatives, cold and cough remedies, antacids, vitamins, minerals, amino acids, supplements, other prescription or non-prescription drugs.

Pregnancy:
- Problems in pregnant women taking small or usual amounts have not been proved. But the chance of problems does exist. Don't use unless prescribed by your doctor.

Breast-feeding:
- Problems in breast-fed infants of lactating mothers taking small or usual amounts have not been proved. But the chance of problems does exist. Don't use unless prescribed by your doctor.

Infants and children:
- Treating infants and children under 2 with any herbal preparation is hazardous.

Others:
- None expected if you are under 45, not pregnant, basically healthy, take it for only a short time and do not exceed manufacturer's recommended dosage.

Storage:
- Keep cool and dry, but don't freeze. Store safely away from children.

Safe dosage:
- At present no "safe" dosage has been established.

 ## Toxicity

Generally regarded as safe when taken in appropriate quantities for short periods of time.

For symptoms of toxicity: See *Adverse Reactions, Side Effects or Overdose Symptoms* section below.

 ## Adverse Reactions, Side Effects or Overdose Symptoms

Signs and symptoms:	What to do:
Diarrhea	Discontinue. Call doctor immediately.
Nausea	Discontinue. Call doctor immediately.
Vomiting	Discontinue. Call doctor immediately.

Virginian Skullcap

Basic Information

Biological name (genus and species):
Scutellaria lateriflora
Parts used for medicinal purposes:
Entire plant
Chemicals this herb contains:
Cellulose
Fat
Scutellarin
Sugar
Tannins (See Glossary)

Known Effects

- Increases stomach acidity.
- Irritates mucous membranes.
- Relieves spasm in skeletal or smooth muscle.
- Interferes with absorption of iron and other minerals when taken internally.

Unproved Speculated Benefits

- Stimulates appetite.
- Relieves intestinal cramps.

Warnings and Precautions

Don't take if you:
- Are pregnant, think you may be pregnant or plan pregnancy in the near future.

Consult your doctor if you:
- Take this herb for any medical problem that doesn't improve in 2 weeks. There may be safer, more-effective treatments.
- Take any medicinal drugs or herbs including aspirin, laxatives, cold and cough remedies, antacids, vitamins, minerals, amino acids, supplements, other prescription or non-prescription drugs.

Pregnancy:
- Problems in pregnant women taking small or usual amounts have not been proved. But the chance of problems does exist. Don't use unless prescribed by your doctor.

Breast-feeding:
- Problems in breast-fed infants of lactating mothers taking small or usual amounts have not been proved. But the chance of problems does exist. Don't use unless prescribed by your doctor.

Infants and children:
- Treating infants and children under 2 with any herbal preparation is hazardous.

Others:
- None expected if you are under 45, not pregnant, basically healthy, take it for only a short time and do not exceed manufacturer's recommended dosage.

Storage:
- Keep cool and dry, but don't freeze. Store safely away from children.

Safe dosage:
- At present no "safe" dosage has been established.

Toxicity

Rated relatively safe when taken in appropriate quantities for short periods of time.

For symptoms of toxicity: See *Adverse Reactions, Side Effects or Overdose Symptoms* section below.

Adverse Reactions, Side Effects or Overdose Symptoms

Signs and symptoms:	What to do:
Confusion	Discontinue. Call doctor immediately.
Giddiness	Discontinue. Call doctor when convenient.
Irregular heartbeat	Seek emergency treatment.
Stupor	Seek emergency treatment.

Basic Information

Biological name (genus and species):
 Nasturtium officinale
Parts used for medicinal purposes:
 Various parts of the entire plant, frequently
 differing by country and/or culture
Chemicals this herb contains:
 Several trace element minerals, such as
 vanadium and cobalt
 Vitamins A, C, B-1 and B-2

Known Effects

• Provides a good source of vitamins and
 minerals to treat or prevent various
 deficiencies.

Miscellaneous information:
• Watercress is a nutritious food source. Toxicity
 is unlikely.

Unproved Speculated Benefits

• Treats kidney infections.
• Treats urinary bladder stones.
• Increases urine flow.
• Treats heart disease.
• Diminishes pain during childbirth.

Warnings and Precautions

Don't take if you:
• No contraindications if you are not pregnant
 and do not take amounts larger than a
 reputable manufacturer recommends on the
 package.

Consult your doctor if you:
• Take this herb for any medical problem that
 doesn't improve in 2 weeks. There may be
 safer, more-effective treatments.
• Take any medicinal drugs or herbs including
 aspirin, laxatives, cold and cough remedies,
 antacids, vitamins, minerals, amino acids,
 supplements, other prescription or non-
 prescription drugs.

Pregnancy:
• Problems in pregnant women taking small or
 usual amounts have not been proved. But the
 chance of problems does exist. Don't use
 unless prescribed by your doctor.

Breast-feeding:
• Problems in breast-fed infants of lactating
 mothers taking small or usual amounts have
 not been proved. But the chance of problems
 does exist. Don't use unless prescribed by your
 doctor.

Infants and children:
• Treating infants and children under 2 with any
 herbal preparation is hazardous.

Others:
• None expected if you are beyond childhood
 and under 45, basically healthy and take for
 only a short time.

Storage:
• Keep cool and dry, but don't freeze. Store
 safely away from children.

Safe dosage:
• At present no "safe" dosage has been
 established.

Toxicity

Comparative-toxicity rating not available from
standard references.

Adverse Reactions, Side Effects or Overdose Symptoms

None expected

MEDICINAL HERB

White Pine

Basic Information
Biological name (genus and species):
  Pinus strobus, P. alba
Parts used for medicinal purposes:
  Inner bark
Chemicals this herb contains:
  Coniferin
  Coniferyl alcohol
  Mucilage (See Glossary)
  Oleoresin
  Tannic acid
  Vanillin
  Volatile oils (See Glossary)

Known Effects

• Decreases thickness and increases fluidity of mucus from lungs and bronchial tubes.

Unproved Speculated Benefits

• Treats coughs when mixed with other expectorants.

Warnings and Precautions

Don't take if you:
• Are pregnant, think you may be pregnant or plan pregnancy in the near future.

Consult your doctor if you:
• Take this herb for any medical problem that doesn't improve in 2 weeks. There may be safer, more-effective treatments.
• Take any medicinal drugs or herbs including aspirin, laxatives, cold and cough remedies, antacids, vitamins, minerals, amino acids, supplements, other prescription or non-prescription drugs.

Pregnancy:
• Problems in pregnant women taking small or usual amounts have not been proved. But the chance of problems does exist. Don't use unless prescribed by your doctor.

Breast-feeding:
• Problems in breast-fed infants of lactating mothers taking small or usual amounts have not been proved. But the chance of problems does exist. Don't use unless prescribed by your doctor.

Infants and children:
• Treating infants and children under 2 with any herbal preparation is hazardous.

Others:
• None expected if you are beyond childhood and under 45, basically healthy and take for only a short time.

Storage:
• Keep cool and dry, but don't freeze. Store safely away from children.

Safe dosage:
• At present no "safe" dosage has been established.

Toxicity

Rated slightly dangerous, particularly in children, persons over 55 and those who take larger than appropriate quantities for extended periods of time.

Adverse Reactions, Side Effects or Overdose Symptoms

None expected

Willow (Black Willow, Pussy Willow, Yellow Willow)

Basic Information

Biological name (genus and species):
Salix nigra
Parts used for medicinal purposes:
Bark
Chemicals this herb contains:
Salicin
Salinigrin
Tannins (See Glossary)

Known Effects

- Causes protein molecules to clump together.
- Produces puckering.
- Interferes with absorption of iron and other minerals when taken internally.

Miscellaneous information:
- Tannins may help heal *open* wounds.

Unproved Speculated Benefits

- Shrinks tissues.
- Prevents secretion of fluids.
- Acts as an antiseptic for ulcerated surfaces on skin.
- Treats arthritis.

Warnings and Precautions

Don't take if you:
- Are pregnant, think you may be pregnant or plan pregnancy in the near future.

Consult your doctor if you:
- Take this herb for any medical problem that doesn't improve in 2 weeks. There may be safer, more-effective treatments.
- Take any medicinal drugs or herbs including aspirin, laxatives, cold and cough remedies, antacids, vitamins, minerals, amino acids, supplements, other prescription or non-prescription drugs.

Pregnancy:
- Problems in pregnant women taking small or usual doses have not been proved. But the chance of problems does exist. Don't use unless prescribed by your doctor.

Breast-feeding:
- Problems in breast-fed infants of lactating mothers taking small or usual doses have not been proved. But the chance of problems does exist. Don't use unless prescribed by your doctor.

Infants and children:
- Treating infants and children under 2 with any herbal preparation is hazardous.

Others:
- None expected if you are under 45, not pregnant, basically healthy, take it for only a short time and do not exceed manufacturer's recommended dosage.
- Salicylate poisoning is possible. Symptoms include dizziness, vomiting, ringing in ears.

Storage:
- Keep cool and dry, but don't freeze. Store safely away from children.

Safe dosage:
- At present no "safe" dosage has been established.

Toxicity

Comparative-toxicity rating not available from standard references.

For symptoms of toxicity: See *Adverse Reactions, Side Effects or Overdose Symptoms* section below.

Adverse Reactions, Side Effects or Overdose Symptoms

Signs and symptoms:	What to do:
Dizziness	Discontinue. Call doctor immediately.
Ringing in ears	Discontinue. Call doctor when convenient.
Vomiting	Discontinue. Call doctor immediately.

Wintergreen (Boxberry, Teaberry)

Basic Information

Biological name (genus and species):
Gaultheria procumbens
Parts used for medicinal purposes:
Leaves
Roots
Stems
Chemicals this herb contains:
Methyl salicylate
Monotropitoside

Known Effects

- Blocks impulses to pain center in brain.
- Irritates stomach.

Miscellaneous information:
- Toxicity is unlikely unless you consume very large amounts of the entire plant.

Unproved Speculated Benefits

- Relieves headache.
- Treats toothache.
- Treats pain of sprains and bruises.

Warnings and Precautions

Don't take if you:
- Are pregnant, think you may be pregnant or plan pregnancy in the near future.

Consult your doctor if you:
- Take this herb for any medical problem that doesn't improve in 2 weeks. There may be safer, more-effective treatments.
- Take any medicinal drugs or herbs including aspirin, laxatives, cold and cough remedies, antacids, vitamins, minerals, amino acids, supplements, other prescription or non-prescription drugs.

Pregnancy:
- Dangers outweigh any possible benefits. Don't use.

Breast-feeding:
- Dangers outweigh any possible benefits. Don't use.

Infants and children:
- Treating infants and children under 2 with any herbal preparation is hazardous.

Others:
- None expected if you are beyond childhood and under 45, not pregnant, basically healthy and take for only a short time.

Storage:
- Keep cool and dry, but don't freeze. Store safely away from children.

Safe dosage:
- At present no "safe" dosage has been established.

Toxicity

Rated slightly dangerous, particularly in children, persons over 55 and those who take larger than appropriate quantities for extended periods of time.

Adverse Reactions, Side Effects or Overdose Symptoms

None expected

Basic Information

Biological name (genus and species):
Hamamelis virginiana
Parts used for medicinal purposes:
Bark
Leaves
Twigs
Chemicals this herb contains:
Bitters (See Glossary)
Calcium oxalate
Gallic acid
Hamamelitannin
Hexose sugar
Tannins (See Glossary)
Volatile oils (See Glossary)

Known Effects

- Shrinks tissues (ointments, solutions, suppositories).
- Interferes with absorption of iron and other minerals when taken internally.

Unproved Speculated Benefits

- Acts as a sedative.
- Treats diarrhea.
- Soothes irritated skin or hemorrhoids.
- Prevents secretion of fluids.

Warnings and Precautions

Don't take if you:
- Are pregnant, think you may be pregnant or plan pregnancy in the near future.
- Have any chronic disease of the gastrointestinal tract, such as stomach or duodenal ulcers, esophageal reflux (reflux esophagitis), ulcerative colitis, spastic colitis, diverticulosis, diverticulitis.

Consult your doctor if you:
- Take this herb for any medical problem that doesn't improve in 2 weeks. There may be safer, more-effective treatments.
- Take any medicinal drugs or herbs including aspirin, laxatives, cold and cough remedies, antacids, vitamins, minerals, amino acids, supplements, other prescription or non-prescription drugs.

Pregnancy:
- Dangers outweigh any possible benefits. Don't use.

Breast-feeding:
- Dangers outweigh any possible benefits. Don't use.

Infants and children:
- Treating infants and children under 2 with any herbal preparation is hazardous.

Others:
- When used externally, no toxicity is expected.
- When used internally, no toxicity expected if you are beyond childhood and under 45, not pregnant, basically healthy and take for only a short time.

Storage:
- Keep cool and dry, but don't freeze. Store safely away from children.

Safe dosage:
- At present no "safe" dosage has been established.

Toxicity

Rated relatively safe when taken in appropriate quantities for short periods of time.

For symptoms of toxicity: See *Adverse Reactions, Side Effects or Overdose Symptoms* section below.

Adverse Reactions, Side Effects or Overdose Symptoms

Signs and symptoms:	What to do:
Constipation	Discontinue. Call doctor when convenient.
Jaundice (yellow skin and eyes)	Discontinue. Call doctor immediately.
Nausea	Discontinue. Call doctor immediately.
Vomiting	Discontinue. Call doctor immediately.

MEDICINAL HERB

Woodruff (Woodward Herb)

Basic Information
Biological name (genus and species):
   Asperula odorata, Galium odoratum
Parts used for medicinal purposes:
   Entire plant
Chemicals this herb contains:
   Asperuloside
   Bitters (See Glossary)
   Coumarin
   Oil
   Tannins (See Glossary)

Known Effects

- Stimulates gastrointestinal tract.
- Decreases thickness and increases fluidity of mucus from lungs and bronchial tubes.
- Interferes with absorption of iron and other minerals when taken internally.

Miscellaneous information:
- Woodruff is used as a flavoring agent in May wine.
- It is used in sachets for its pleasant odor.

Unproved Speculated Benefits

- Treats coughs.
- Aids in expelling gas from intestinal tract.

Warnings and Precautions

Don't take if you:
- Are pregnant, think you may be pregnant or plan pregnancy in the near future.

Consult your doctor if you:
- Take this herb for any medical problem that doesn't improve in 2 weeks. There may be safer, more-effective treatments.
- Take any medicinal drugs or herbs including aspirin, laxatives, cold and cough remedies, antacids, vitamins, minerals, amino acids, supplements, other prescription or non-prescription drugs.

Pregnancy:
- Problems in pregnant women taking small or usual amounts have not been proved. But the chance of problems does exist. Don't use unless prescribed by your doctor.

Breast-feeding:
- Problems in breast-fed infants of lactating mothers taking small or usual amounts have not been proved. But the chance of problems does exist. Don't use unless prescribed by your doctor.

Infants and children:
- Treating infants and children under 2 with any herbal preparation is hazardous.

Others:
- None expected if you are beyond childhood and under 45, basically healthy and take for only a short time.

Storage:
- Keep cool and dry, but don't freeze. Store safely away from children.

Safe dosage:
- At present no "safe" dosage has been established.

Toxicity

Comparative-toxicity rating not available from standard references.

Adverse Reactions, Side Effects or Overdose Symptoms

None expected

Wormseed (Pigweed)

Basic Information

Biological name (genus and species):
  Chenopodium ambrosioides
Parts used for medicinal purposes:
  Roots
  Berries/fruits
Chemicals this herb contains:

Ascaridol	Saponin (See Glossary)
Calcium	Terpene (See Glossary)
Cymene	Vitamins A and C
d-camphor	Volatile oils (See Glossary)
l-limonene	

 ## Known Effects

- Inhibits growth and development of germs.
- Decreases blood pressure.
- Decreases heart rate.
- Depresses central nervous system.
- Decreases stomach contractions.

 ## Unproved Speculated Benefits

- Is used externally as a poultice. (See Glossary)
- Treats arthritis.
- Kills intestinal parasites.

 ## Warnings and Precautions

Don't take if you:
- Are pregnant, think you may be pregnant or plan pregnancy in the near future.
- Have any chronic disease of the gastrointestinal tract, such as stomach or duodenal ulcers, esophageal reflux (reflux esophagitis), ulcerative colitis, spastic colitis, diverticulosis, diverticulitis.

Consult your doctor if you:
- Take this herb for any medical problem that doesn't improve in 2 weeks. There may be safer, more-effective treatments.
- Take any medicinal drugs or herbs including aspirin, laxatives, cold and cough remedies, antacids, vitamins, minerals, amino acids, supplements, other prescription or non-prescription drugs.

Pregnancy:
- Dangers outweigh any possible benefits. Don't use.

Breast-feeding:
- Dangers outweigh any possible benefits. Don't use.

Infants and children:
- Treating infants and children under 2 with any herbal preparation is hazardous.

Others:
- None expected if you are under 45, not pregnant, basically healthy, take it for only a short time and do not exceed manufacturer's recommended dosage.

Storage:
- Keep cool and dry, but don't freeze. Store safely away from children.

Safe dosage:
- At present no "safe" dosage has been established.

 ## Toxicity

Rated slightly dangerous, particularly in children, persons over 55 and those who take larger than appropriate quantities for extended periods of time.

For symptoms of toxicity: See *Adverse Reactions, Side Effects or Overdose Symptoms* section below.

 ## Adverse Reactions, Side Effects or Overdose Symptoms

Signs and symptoms:	What to do:
Breathing difficulties	Seek emergency treatment.
Drowsiness	Discontinue. Call doctor when convenient.
Headache	Discontinue. Call doctor when convenient.
Hearing problems	Discontinue. Call doctor immediately.
Nausea	Discontinue, Call doctor immediately.
Ringing in ears	Discontinue. Call doctor when convenient.
Slow heartbeat	Seek emergency treatment.
Stomach ulcers	Discontinue. Call doctor immediately.
Vision problems	Discontinue. Call doctor immediately.
Vomiting	Discontinue. Call doctor immediately.

Wormwood (Absinthium, Ajerjo)

Basic Information

Biological name (genus and species):
   Artemisia absinthium
Parts used for medicinal purposes:
   Berries/fruits
   Leaves
Chemicals this herb contains:
   Absinthol
   Thujone
   Volatile oils (See Glossary)

Known Effects

- Depresses central nervous system.
- Thujone causes mind-altering changes and may lead to psychosis.
- Increases stomach acidity.

Miscellaneous information:
- Wormwood can be habit-forming, like ethyl alcohol.

Unproved Speculated Benefits

- Treats anxiety.
- Acts as a mild sedative.
- Stimulates appetite.

Warnings and Precautions

Don't take if you:
- Are pregnant, think you may be pregnant or plan pregnancy in the near future.

Consult your doctor if you:
- Take this herb for any medical problem that doesn't improve in 2 weeks. There may be safer, more-effective treatments.
- Take any medicinal drugs or herbs including aspirin, laxatives, cold and cough remedies, antacids, vitamins, minerals, amino acids, supplements, other prescription or non-prescription drugs.

Pregnancy:
- Dangers outweigh any possible benefits. Don't use.

Breast-feeding:
- Dangers outweigh any possible benefits. Don't use.

Infants and children:
- Treating infants and children under 2 with any herbal preparation is hazardous.

Others:
- This product will *not* help you and may cause toxic symptoms.

Storage:
- Keep cool and dry, but don't freeze. Store safely away from children.

Safe dosage:
- At present no "safe" dosage has been established.

Toxicity

Rated slightly dangerous, particularly in children, persons over 55 and those who take larger than appropriate quantities for extended periods of time.

For symptoms of toxicity: See *Adverse Reactions, Side Effects or Overdose Symptoms* section below.

Adverse Reactions, Side Effects or Overdose Symptoms

Signs and symptoms:	What to do:
Convulsions	Seek emergency treatment.
Stupor	Seek emergency treatment.
Trembling	Discontinue. Call doctor when convenient.

Basic Information

Biological name (genus and species):
  Achillea millefolium
Parts used for medicinal purposes:
  Berries/fruits
  Leaves
Chemicals this herb contains:
  Achilleic acid
  Achilleine
  Bitters (See Glossary)
  Caledivain
  Tannins (See Glossary)
  Volatile oils (See Glossary)

Known Effects

- Reduces blood-clotting time.
- Interferes with absorption of iron and other minerals when taken internally.

Unproved Speculated Benefits

- Acts as a mild sedative to cause drowsiness.
- Treats amenorrhea. (See Glossary.)

Warnings and Precautions

Don't take if you:
- Are pregnant, think you may be pregnant or plan pregnancy in the near future.

Consult your doctor if you:
- Take this herb for any medical problem that doesn't improve in 2 weeks. There may be safer, more-effective treatments.
- Take any medicinal drugs or herbs including aspirin, laxatives, cold and cough remedies, antacids, vitamins, minerals, amino acids, supplements, other prescription or non-prescription drugs.

Pregnancy:
- Problems in pregnant women taking small or usual amounts have not been proved. But the chance of problems does exist. Don't use unless prescribed by your doctor.

Breast-feeding:
- Problems in breast-fed infants of lactating mothers taking small or usual amounts have not been proved. But the chance of problems does exist. Don't use unless prescribed by your doctor.

Infants and children:
- Treating infants and children under 2 with any herbal preparation is hazardous.

Others:
- None expected if you are under 45, not pregnant, basically healthy, take it for only a short time and do not exceed manufacturer's recommended dosage.

Storage:
- Keep cool and dry, but don't freeze. Store safety away from children.

Safe dosage:
- At present no "safe" dosage has been established.

Toxicity

Generally regarded as safe when taken in appropriate quantities for short periods of time.

Adverse Reactions, Side Effects or Overdose Symptoms

None expected

MEDICINAL HERB

Yellow Cedar (Arbor Vitae)

Basic Information

Biological name (genus and species):
  Thuja occidentalis
Parts used for medicinal purposes:
  Leaves
Chemicals this herb contains:
  Fenchone
  Pinipirin
  Tannins (See Glossary)
  Thujetic acid
  Thujone
  Volatile oils (See Glossary)

Known Effects

- Stimulates central nervous system.
- Stimulates heart muscle to contract more efficiently.
- Destroys intestinal worms.
- Causes uterine contractions.
- Interferes with absorption of iron and other minerals when taken internally.

Miscellaneous information:

- Yellow cedar has caused deaths when it was misused to cause abortions.

Unproved Speculated Benefits

- Relieves muscular aches and pains.
- Treats warts.
- Causes abortions (miscarriages).

Warnings and Precautions

Don't take if you:

- Are pregnant, think you may be pregnant or plan pregnancy in the near future.

Consult your doctor if you:

- Take this herb for any medical problem that doesn't improve in 2 weeks. There may be safer, more-effective treatments.
- Take any medicinal drugs or herbs including aspirin, laxatives, cold and cough remedies, antacids, vitamins, minerals, amino acids, supplements, other prescription or non-prescription drugs.

Pregnancy:

- Dangers outweigh any possible benefits. Don't use.

Breast-feeding:

- Dangers outweigh any possible benefits. Don't use.

Infants and children:

- Treating infants and children under 2 with any herbal preparation is hazardous.

Others:

- Dangers outweigh any possible benefits. Don't use.

Storage:

- Keep cool and dry, but don't freeze. Store safely away from children.

Safe dosage:

- At present no "safe" dosage has been established.

Toxicity

Comparative-toxicity rating not available from standard references.

For symptoms of toxicity: See *Adverse Reactions, Side Effects or Overdose Symptoms* section below.

Adverse Reactions, Side Effects or Overdose Symptoms

Signs and symptoms:	What to do:
Abortion	Seek emergency treatment.
Coma	Seek emergency treatment.
Convulsions	Seek emergency treatment.
Precipitous blood-pressure drop—symptoms include, faintness, cold sweat, paleness, rapid pulse	Seek emergency treatment.

Basic Information

Biological name (genus and species):
   Rumex crispus
Parts used for medicinal purposes:
   Leaves
   Roots
Chemical this herb contains:
   Potassium oxalate

Known Effects

- Irritates skin when handled.
- Stimulates gastrointestinal tract.

Miscellaneous information:
- Is used as food in salads.

Unproved Speculated Benefits

- Temporarily relieves constipation.

Warnings and Precautions

Don't take if you:
- Are pregnant, think you may be pregnant or plan pregnancy in the near future.
- Have any chronic disease of the gastrointestinal tract, such as stomach or duodenal ulcers, esophageal reflux (reflux esophagitis), ulcerative colitis, spastic colitis, diverticulosis, diverticulitis.

Consult your doctor if you:
- Take this herb for any medical problem that doesn't improve in 2 weeks. There may be safer, more-effective treatments.
- Take any medicinal drugs or herbs including aspirin, laxatives, cold and cough remedies, antacids, vitamins, minerals, amino acids, supplements, other prescription or non-prescription drugs.

Pregnancy:
- Dangers outweigh any possible benefits. Don't use.

Breast-feeding:
- Dangers outweigh any possible benefits. Don't use.

Infants and children:
- Treating infants and children under 2 with any herbal preparation is hazardous.

Others:
- Dangers outweigh any possible benefits. Don't use.

Storage:
- Keep cool and dry, but don't freeze. Store safely away from children.

Safe dosage:
- At present no "safe" dosage has been established.

Toxicity

Rated slightly dangerous, particularly in children, persons over 55 and those who take larger than appropriate quantities for extended periods of time.

For symptoms of toxicity: See *Adverse Reactions, Side Effects or Overdose Symptoms* section below.

Adverse Reactions, Side Effects or Overdose Symptoms

Signs and symptoms:	What to do:
Diarrhea	Discontinue. Call doctor immediately.
Kidney damage characterized by blood in urine, decreased urine flow, swelling of hands and feet	Seek emergency treatment.
Nausea	Discontinue. Call doctor immediately.
Skin eruptions	Discontinue. Call doctor when convenient.
Vomiting	Discontinue. Call doctor immediately.

MEDICINAL HERB

Yellow Lady's Slipper

Basic Information

Biological name (genus and species):
  Cypripedium pubescens
Parts used for medicinal purposes:
  Roots
Chemicals this herb contains:
  Resin (See Glossary)
  Tannins (See Glossary)
  Volatile acid
  Volatile oils (See Glossary)

Known Effects

- Irritates mucous membranes.
- Stimulates gastrointestinal tract.
- Increases perspiration.

Miscellaneous information:
- Hairs on stems and leaves irritate body when touched. May produce skin eruption similar to poison ivy.

Unproved Speculated Benefits

- Acts as a sedative to treat anxiety or restlessness.
- Increases perspiration.
- Aids in expelling gas from intestinal tract.
- Relieves spasm in skeletal smooth or smooth muscle.

Warnings and Precautions

Don't take if you:
- Are pregnant, think you may be pregnant or plan pregnancy in the near future.
- Have any chronic disease of the gastrointestinal tract, such as stomach or duodenal ulcers, esophageal reflux (reflux esophagitis), ulcerative colitis, spastic colitis, diverticulosis, diverticulitis.

Consult your doctor if you:
- Take this herb for any medical problem that doesn't improve in 2 weeks. There may be safer, more-effective treatments.
- Take any medicinal drugs or herbs including aspirin, laxatives, cold and cough remedies, antacids, vitamins, minerals, amino acids, supplements, other prescription or non-prescription drugs.

Pregnancy:
- Problems in pregnant women taking small or usual amounts have not been proved. But the chance of problems does exist. Don't use unless prescribed by your doctor.

Breast-feeding:
- Problems in breast-fed infants of lactating mothers taking small or usual amounts have not been proved. But the chance of problems does exist. Don't use unless prescribed by your doctor.

Infants and children:
- Treating infants and children under 2 with any herbal preparation is hazardous.

Others:
- None expected if you are beyond childhood and under 45, basically healthy and take for only a short time.

Storage:
- Keep cool and dry, but don't freeze. Store safely away from children.

Safe dosage:
- At present no "safe" dosage has been established.

Toxicity

Rated relatively safe when taken in appropriate quantities for short periods of time.

For symptoms of toxicity: See *Adverse Reactions, Side Effects or Overdose Symptoms* section below.

Adverse Reactions, Side Effects or Overdose Symptoms

Signs and symptoms:	What to do:
Drowsiness	Discontinue. Call doctor when convenient.
Nausea	Discontinue. Call doctor immediately.
Vomiting	Discontinue. Call doctor immediately.

Yerba Mate (Paraguay Tea, South American Holly)

Basic Information
Biological name (genus and species):
  Ilex paraguariensis St. Hill
Parts used for medicinal purposes:
  Leaves
Chemical this herb contains:
  Caffeine

Known Effects

- Stimulates central nervous system.
- Helps body dispose of excess fluid by increasing amount of urine produced.
- Causes hallucinations.

Unproved Speculated Benefits

- Is used as a laxative.
- Increases perspiration.

Warnings and Precautions

Don't take if you:
- Are pregnant, think you may be pregnant or plan pregnancy in the near future.
- Have any chronic disease of the gastrointestinal tract, such as stomach or duodenal ulcers, esophageal reflux (reflux esophagitis), ulcerative colitis, spastic colitis, diverticulosis, diverticulitis.

Consult your doctor if you:
- Take this herb for any medical problem that doesn't improve in 2 weeks. There may be safer, more-effective treatments.
- Take any medicinal drugs or herbs including aspirin, laxatives, cold and cough remedies, antacids, vitamins, minerals, amino acids, supplements, other prescription or non-prescription drugs.

Pregnancy:
- Dangers outweigh any possible benefits. Don't use.

Breast-feeding:
- Dangers outweigh any possible benefits. Don't use.

Infants and children:
- Treating infants and children under 2 with any herbal preparation is hazardous.

Others:
- None expected if you are beyond childhood and under 45, not pregnant, basically healthy and take for only a short time.

Storage:
- Keep cool and dry, but don't freeze. Store safely away from children.

Safe dosage:
- At present no "safe" dosage has been established.

Toxicity

Rated relatively safe when taken in appropriate quantities for short periods of time.

For symptoms of toxicity: See *Adverse Reactions, Side Effects or Overdose Symptoms* section below.

Adverse Reactions, Side Effects or Overdose Symptoms

Signs and symptoms:	What to do:
Confusion	Seek emergency treatment.
Excessive urination	Discontinue. Call doctor when convenient.
Hallucinations	Seek emergency treatment.
Heartburn	Discontinue. Call doctor when convenient.
Insomnia	Discontinue. Call doctor when convenient.
Irritability	Discontinue. Call doctor when convenient.
Nausea	Discontinue. Call doctor immediately.
Nervousness	Discontinue. Call doctor when convenient.
Rapid heartbeat	Seek emergency treatment.

MEDICINAL HERB

Yerba Santa (Bear's Weed)

Basic Information
Biological name (genus and species):
  Eriodictyon californicum
Parts used for medicinal purposes:
  Leaves
Chemicals this herb contains:
  Formic acid
  Pentatriacontane eriodicytyol
  Resin (See Glossary)
  Tannic acid
  Tannins (See Glossary)

Known Effects

- Masks taste of bitter medicines.
- Decreases thickness and increases fluidity of mucus from lungs and bronchial tubes.
- Interferes with absorption of iron and other minerals when taken internally.

Unproved Speculated Benefits

- Treats hay fever and other nasal allergies.
- Treats hemorrhoids.

Warnings and Precautions

Don't take if you:
- Are pregnant, think you may be pregnant or plan pregnancy in the near future.
- Have any chronic disease of the gastrointestinal tract, such as stomach or duodenal ulcers, esophageal reflux (reflux esophagitis), ulcerative colitis, spastic colitis, diverticulosis, diverticulitis.

Consult your doctor if you:
- Take this herb for any medical problem that doesn't improve in 2 weeks. There may be safer, more-effective treatments.
- Take any medicinal drugs or herbs including aspirin, laxatives, cold and cough remedies, antacids, vitamins, minerals, amino acids, supplements, other prescription or non-prescription drugs.

Pregnancy:
- Dangers outweigh any possible benefits. Don't use.

Breast-feeding:
- Dangers outweigh any possible benefits. Don't use.

Infants and children:
- Treating infants and children under 2 with any herbal preparation is hazardous.

Others:
- None expected if you are beyond childhood and under 45, not pregnant, basically healthy and take for only a short time.

Storage:
- Keep cool and dry, but don't freeze. Store safely away from children.

Safe dosage:
- At present no "safe" dosage has been established.

Toxicity

Comparative-toxicity rating not available from standard references.

For symptoms of toxicity: See *Adverse Reactions, Side Effects or Overdose Symptoms* section below.

Adverse Reactions, Side Effects or Overdose Symptoms

Signs and symptoms:	What to do:
Diarrhea	Discontinue. Call doctor immediately.
Nausea	Discontinue. Call doctor immediately.
Vomiting	Discontinue. Call doctor immediately.

Basic Information

Biological name (genus and species):
Corynanthe yohimbe
Parts used for medicinal purposes:
Bark
Chemical this herb contains:
Yohimbine, also called quebrachine, aphrodine or corynine

 Known Effects

- Blocks responses of parts of autonomic nervous system.
- Increases blood pressure.
- Acts as a local anesthetic.
- Inhibits monamine oxidase and may cause alarming blood-pressure rise—even strokes—when taken with cheese, red wine or other foods or supplements containing tyromines.
- Causes hallucinations.

Miscellaneous information:
- Yohimbe can produce severe anxiety when given intravenously.

 Unproved Speculated Benefits

- Arouses or enhances sexual desire.
- Treats impotence.
- Treats painful menstrual cramps.
- Treats chest pain due to coronary artery disease (angina).
- Treats arteriosclerosis.

 Warnings and Precautions

Don't take if you:
- Are pregnant, think you may be pregnant or plan pregnancy in the near future.
- Have kidney or liver disease.

Consult your doctor if you:
- Take this herb for any medical problem that doesn't improve in 2 weeks. There may be safer, more-effective treatments.
- Take any medicinal drugs or herbs including aspirin, laxatives, cold and cough remedies, antacids, vitamins, minerals, amino acids, supplements, other prescription or non-prescription drugs.

Pregnancy:
- Dangers outweigh any possible benefits. Don't use.

Breast-feeding:
- Dangers outweigh any possible benefits. Don't use.

Infants and children:
- Treating infants and children under 2 with any herbal preparation is hazardous.

Others:
- This product will *not* help you and may cause toxic symptoms.

Storage:
- Keep cool and dry, but don't freeze. Store safely away from children.

Safe dosage:
- At present no "safe" dosage has been established.

 Toxicity

Rated dangerous, particularly in children, persons over 55 and those who take larger than appropriate quantities for extended periods of time.

For symptoms of toxicity: See *Adverse Reactions, Side Effects or Overdose Symptoms* section below.

 Adverse Reactions, Side Effects or Overdose Symptoms

Signs and symptoms:	What to do:
Abdominal pain	Discontinue. Call doctor when convenient.
Fatigue	Discontinue. Call doctor when convenient.
Hallucinations	Seek emergency treatment.
High blood pressure	Discontinue. Call doctor immediately.
Muscle paralysis	Seek emergency treatment.
Weakness	Discontinue. Call doctor immediately.

Toxicity Ratings for Herbs

The following list has been compiled from several sources. The toxicity rating is an average of ratings previously published by other experts. The list is included here for your consideration before you take a medicinal herb or suggest it for someone else.

Achuma *Trichocereus pachanoi*
Rated slightly dangerous, particularly for children, people over 55 and those who take larger-than-appropriate quantities for extended periods of time.

Agrimony *Agrimonia eupatoria*
Rated relatively safe when taken in appropriate quantities for short periods of time.

Akee *Blighia sapida*
Rated dangerous, particularly for children, people over 55 and those who take larger than appropriate quantities for extended periods.

Alder *Alnus glutinosa*
Rated relatively safe when taken in appropriate quantities for short periods of time.

Alexandrian senna *Cassia senna*
Rated slightly dangerous, particularly for children, people over 55 and those who take larger-than-appropriate quantities for extended periods of time.

Alfalfa *Medicago sativa*
Generally regarded as safe when taken in appropriate quantities for short periods of time.

Allspice *Pimenta dioica*
Rated relatively safe when taken in appropriate quantities for short periods of time.

Aloe *Aloe barbadensis*
Generally regarded as safe when taken in appropriate quantities for short periods of time.

American ginseng
Panax quinquefolius
Generally regarded as safe when taken in appropriate quantities for short periods of time.

American hellebore
Veratrum viride
Rated dangerous, particularly for children, people over 55 and those who take larger-than-appropriate quantities for extended periods of time.

American mistletoe
Phoradendron serotinum
Rated slightly dangerous, particularly for children, people over 55 and those who take larger-than-appropriate quantities for extended periods of time.

Angelica   *Angelica archangelica*
Rated relatively safe when taken in appropriate quantities for short periods of time.

Anise   *Pimpinella anisum*
Rated relatively safe when taken in appropriate quantities for short periods of time.

Annual mercury
Mercurialis annua
Rated dangerous, particularly for children, people over 55 and those who take larger-than-appropriate quantities for extended periods of time.

Apple   *Malus sylvestris*
Generally regarded as safe when taken in appropriate quantities for short periods of time.

Apricot   *Prunus armeniaca*
Rated slightly dangerous, particularly for children, people over 55 and those who take larger-than-appropriate quantities for extended periods of time.

Arrowpoison tree
Acokanthera schimperi
Rated dangerous, particularly for children, people over 55 and those who take larger-than-appropriate quantities for extended periods of time.

Arrowroot   *Maranta arundinacea*
Generally regarded as safe when taken in appropriate quantities for short periods of time.

Asafetida   *Ferula assa-foetida*
Rated relatively safe when taken in appropriate quantities for short periods of time.

Ashwagandha
Withania somniferum
Rated dangerous, particularly for children, people over 55 and those who take larger-than-appropriate quantities for extended periods of time.

Autumn crocus
Colchicum autumnale
Rated dangerous, particularly for children, people over 55 and those who take larger-than-appropriate quantities for extended periods of time.

Aveloz   *Euphorbia tirucalli*
Rated dangerous, particularly for children, people over 55 and those who take larger-than-appropriate quantities for extended periods of time.

Aztec tobacco   *Nicotiana rustica*
Rated dangerous, particularly for children, people over 55 and those who take larger-than-appropriate quantities for extended periods of time.

Balsam of Peru
Myroxylon balsamum var.
Rated slightly dangerous, particularly for children, people over 55 and those who take larger-than-appropriate quantities for extended periods of time.

Balsam pear   *Momordica charantia*
Rated slightly dangerous, particularly for children, people over 55 and those who take larger-than-appropriate quantities for extended periods of time.

Baneberry   *Actaea pachypoda*
Rated slightly dangerous, particularly for children, people over 55 and those who take larger-than-appropriate quantities for extended periods of time.

Barbasco   *Dioscorea composita*
Rated slightly dangerous, particularly for children, people over 55 and those who take larger-than-appropriate quantities for extended periods of time.

Barberry   *Berberis vulgaris*
Rated slightly dangerous, particularly for children, people over 55 and those who take larger-than-appropriate quantities for extended periods of time.

Basil   *Ocimum basilicum*
Rated relatively safe when taken in appropriate quantities for short periods of time.

Bay   *Laurus nobilis*
Rated relatively safe when taken in appropriate quantities for short periods of time.

Bayberry *Myrica cerifera*
Rated relatively safe when taken in appropriate quantities for short periods of time.

Bayrum tree *Pimenta racemosa*
Rated relatively safe when taken in appropriate quantities for short periods of time.

Bean *Phaseolus vulgaris*
Generally regarded as safe when taken in appropriate quantities for short periods of time.

Bearberry *Arctostaphylos uva-ursi*
Rated relatively safe when taken in appropriate quantities for short periods of time.

Belladonna *Atropa bella-donna*
Rated dangerous, particularly for children, people over 55 and those who take larger-than-appropriate quantities for extended periods of time.

Benzoin *Styrax benzoin*
Rated slightly dangerous, particularly for children, people over 55 and those who take larger-than-appropriate quantities for extended periods of time.

Betel pepper *Piper betel*
Rated slightly dangerous, particularly for children, people over 55 and those who take larger-than-appropriate quantities for extended periods of time.

Betel-nut *Areca catechu*
Rated relatively safe when taken in appropriate quantities for short periods of time.

Betony *Stachys officinalis*
Rated relatively safe when taken in appropriate quantities for short periods of time.

Bitter root
Apocynum androsaemifolium
Rated slightly dangerous, particularly for children, people over 55 and those who take larger-than-appropriate quantities for extended periods of time.

Bittersweet *Solanum dulcamara*
Rated slightly dangerous, particularly for children, people over 55 and those who take larger-than-appropriate quantities for extended periods of time.

Black cohosh *Cimicifuga racemosa*
Rated slightly dangerous, particularly for children, people over 55 and those who take larger-than-appropriate quantities for extended periods of time.

Black locust *Robinia pseudoacacia*
Rated slightly dangerous, particularly for children, people over 55 and those who take larger-than-appropriate quantities for extended periods of time.

Black pepper *Piper nigrum*
Rated slightly dangerous, particularly for children, people over 55 and those who take larger-than-appropriate quantities for extended periods of time.

Bloodroot *Sanguinaria canadensis*
Rated slightly dangerous, particularly for children, people over 55 and those who take larger-than-appropriate quantities for extended periods of time.

Blue cohosh
Caulophyllum thalictroides
Rated slightly dangerous, particularly for children, people over 55 and those who take larger-than-appropriate quantities for extended periods of time.

Blue flag *Iris versicolor*
Rated relatively safe when taken in appropriate quantities for short periods of time.

Boldo *Peumus boldus*
Rated dangerous, particularly for children, people over 55 and those who take larger-than-appropriate quantities for extended periods of time.

Bolek hena *Justicia pectoralis*
Rated slightly dangerous, particularly for children, people over 55 and those who take larger-than-appropriate quantities for extended periods of time.

Boneset *Eupatorium perfoliatum*
Rated relatively safe when taken in appropriate quantities for short periods of time.

Borage   *Borago officinalis*
Generally regarded as safe when taken in appropriate quantities for short periods of time.

Borrachero   *Datura candida*
Rated dangerous, particularly for children, people over 55 and those who take larger-than-appropriate quantities for extended periods of time.

Boxwood   *Buxus sempervirens*
Rated slightly dangerous, particularly for children, people over 55 and those who take larger-than-appropriate quantities for extended periods of time.

Brazilian peppertree
Schinus terebinthifolius
Rated slightly dangerous, particularly for children, people over 55 and those who take larger-than-appropriate quantities for extended periods of time.

Buchu   *Barosma betulina*
Rated relatively safe when taken in appropriate quantities for short periods of time.

Buckthorn   *Frangula alnus*
Rated slightly dangerous, particularly for children, people over 55 and those who take larger-than-appropriate quantities for extended periods of time.

Bugleweed   *Ajuga reptans*
Rated relatively safe when taken in appropriate quantities for short periods of time.

Bulbous buttercup
Ranunculus bulbosus
Rated relatively safe when taken in appropriate quantities for short periods of time.

Burdock   *Arctium lappa*
Rated relatively safe when taken in appropriate quantities for short periods of time.

Caapi   *Banisteriopsis caapi*
Rated slightly dangerous, particularly for children, people over 55 and those who take larger-than-appropriate quantities for extended periods of time.

Cabbagebark   *Andira inermis*
Rated slightly dangerous, particularly for children, people over 55 and those who take larger-than-appropriate quantities for extended periods of time.

Cabeza de angel
Calliandra anomala
Rated slightly dangerous, particularly for children, people over 55 and those who take larger-than-appropriate quantities for extended periods of time.

Cajeput   *Melaleuca leucadendron*
Rated relatively safe when taken in appropriate quantities for short periods of time.

California bay
Umbellularia californica
Rated slightly dangerous, particularly for children, people over 55 and those who take larger-than-appropriate quantities for extended periods of time.

California poppy
Eschscholzia californica
Rated slightly dangerous, particularly for children, people over 55 and those who take larger-than-appropriate quantities for extended periods of time.

Camphor   *Cinnamomum camphora*
Rated relatively safe when taken in appropriate quantities for short periods of time.

Canaigre   *Rumex hymenosepalus*
Rated slightly dangerous, particularly for children, people over 55 and those who take larger-than-appropriate quantities for extended periods of time.

Cananga   *Cananga odorata*
Rated slightly dangerous, particularly for children, people over 55 and those who take larger-than-appropriate quantities for extended periods of time.

Candlenut   *Aleurites moluccana*
Rated slightly dangerous, particularly for children, people over 55 and those who take larger-than-appropriate quantities for extended periods of time.

Carrot   *Daucus carota*
Rated relatively safe when taken in appropriate quantities for short periods of time.

Cascara sagrada
Rhamnus purshianus
Rated slightly dangerous, particularly for children, people over 55 and those who take larger-than-appropriate quantities for extended periods of time.

Cascarilla   *Croton eleuteria*
Rated slightly dangerous, particularly for children, people over 55 and those who take larger-than-appropriate quantities for extended periods of time.

Cassava   *Manihot esculenta*
Rated slightly dangerous, particularly for children, people over 55 and those who take larger-than-appropriate quantities for extended periods of time.

Cassie   *Acacia farnesiana*
Generally regarded as safe when taken in appropriate quantities for short periods of time.

Castor   *Ricinus communis*
Rated dangerous, particularly for children, people over 55 and those who take larger-than-appropriate quantities for extended periods of time.

Cat powder   *Actinidia polygama*
Rated slightly dangerous, particularly for children, people over 55 and those who take larger-than-appropriate quantities for extended periods of time.

Catnip   *Nepeta cataria*
Generally regarded as safe when taken in appropriate quantities for short periods of time.

Celandine   *Chelidonium majus*
Rated slightly dangerous, particularly for children, people over 55 and those who take larger-than-appropriate quantities for extended periods of time.

Celery   *Apium graveolens*
Rated relatively safe when taken in appropriate quantities for short periods of time.

Cherry-laurel   *Prunus laurocerasus*
Rated slightly dangerous, particularly for children, people over 55 and those who take larger-than-appropriate quantities for extended periods of time.

Chickweed   *Stellaria media*
Rated relatively safe when taken in appropriate quantities for short periods of time.

Chili   *Capsicum annuum*
Rated relatively safe when taken in appropriate quantities for short periods of time.

Chinaberry   *Melia azedarach*
Rated dangerous, particularly for children, people over 55 and those who take larger-than-appropriate quantities for extended periods of time.

Chinese rhubarb   *Rheum officinale*
Rated relatively safe when taken in appropriate quantities for short periods of time.

Christmas rose   *Helleborus niger*
Rated dangerous, particularly for children, people over 55 and those who take larger-than-appropriate quantities for extended periods of time.

Christthorn   *Ziziphus spina-christi*
Rated relatively safe when taken in appropriate quantities for short periods of time.

Cinnamon   *Cinnamomum verum*
Rated relatively safe when taken in appropriate quantities for short periods of time.

Clary sage   *Salvia sclarea*
Rated relatively safe when taken in appropriate quantities for short periods of time.

Climbing onion   *Bowiea volubilis*
Rated dangerous, particularly for children, people over 55 and those who take larger-than-appropriate quantities for extended periods of time.

Clove   *Syzygium aromaticum*
Rated relatively safe when taken in appropriate quantities for short periods of time.

Coca   *Erythroxylum coca*
Rated slightly dangerous, particularly for children, people over 55 and those who take larger-than-appropriate quantities for extended periods of time.

Cocoa   *Theobroma cacao*
Rated relatively safe when taken in appropriate quantities for short periods of time.

Coffee   *Coffea arabica*
Rated relatively safe when taken in appropriate quantities for short periods of time.

Colocynth   *Citrullus colocynthis*
Rated dangerous, particularly for children, people over 55 and those who take larger-than-appropriate quantities for extended periods of time.

Coltsfoot   *Tussilago farfara*
Rated relatively safe when taken in appropriate quantities for short periods of time.

Columbine   *Aquilegia vulgaris*
Rated slightly dangerous, particularly for children, people over 55 and those who take larger-than-appropriate quantities for extended periods of time.

Comfrey   *Symphytum peregrinum*
Rated relatively safe when taken in appropriate quantities for short periods of time.

Common milkweed
Asclepias syriaca
Rated slightly dangerous, particularly for children, people over 55 and those who take larger-than-appropriate quantities for extended periods of time.

Condurango
Marsdenia reichenbachii
Rated slightly dangerous, particularly for children, people over 55 and those who take larger-than-appropriate quantities for extended periods of time.

Coral bean   *Erythrina fusca*
Rated slightly dangerous, particularly for children, people over 55 and those who take larger-than-appropriate quantities for extended periods of time.

Corkwood   *Duboisia myoporoides*
Rated dangerous, particularly for children, people over 55 and those who take larger-than-appropriate quantities for extended periods of time.

Corncockle   *Agrostemma githago*
Rated slightly dangerous, particularly for children, people over 55 and those who take larger-than-appropriate quantities for extended periods of time.

Cranesbill   *Geranium maculatum*
Rated relatively safe when taken in appropriate quantities for short periods of time.

Creosotebush   *Larrea tridentata*
Rated relatively safe when taken in appropriate quantities for short periods of time.

Culebra   *Methystichodendron amesia*
Rated dangerous, particularly for children, people over 55 and those who take larger-than-appropriate quantities for extended periods of time.

Daffodil   *Narcissus tazetta*
Rated dangerous, particularly for children, people over 55 and those who take larger-than-appropriate quantities for extended periods of time.

Dagga   *Leonotis leonurus*
Rated slightly dangerous, particularly for children, people over 55 and those who take larger-than-appropriate quantities for extended periods of time.

Damiana   *Turnera diffusa*
Rated relatively safe when taken in appropriate quantities for short periods of time.

Dandelion   *Taraxacum officinale*
Generally regarded as safe when taken in appropriate quantities for short periods of time.

Darnel   *Lolium temulentum*
Rated dangerous, particularly for children, people over 55 and those who take larger-than-appropriate quantities for extended periods of time.

Deer's tongue   *Trilisa odoratissima*
Rated relatively safe when taken in appropriate quantities for short periods of time.

Devil's claw
Harpagophytum procumbens
Rated relatively safe when taken in appropriate quantities for short periods of time.

Devil's shoestring
Tephrosia virginiana
Rated slightly dangerous, particularly for children, people over 55 and those who take larger-than-appropriate quantities for extended periods of time.

Digitalis   *Digitalis purpurea*
Rated dangerous, particularly for children, people over 55 and those who take larger-than-appropriate quantities for extended periods of time.

Dill   *Anethum graveolens*
Generally regarded as safe when taken in appropriate quantities for short periods of time.

Dinque pinque
Rauvolfia tetraphylla
Rated slightly dangerous, particularly for children, people over 55 and those who take larger-than-appropriate quantities for extended periods of time.

Dodo   *Elaeophorbia drupifera*
Rated slightly dangerous, particularly for children, people over 55 and those who take larger-than-appropriate quantities for extended periods of time.

Dogwood   *Cornus florida*
Rated relatively safe when taken in appropriate quantities for short periods of time.

Dong quai   *Angelica polymorpha*
Rated slightly dangerous, particularly for children, people over 55 and those who take larger-than-appropriate quantities for extended periods of time.

Dove's dung
Ornithogalum umbellatum
Rated slightly dangerous, particularly for children, people over 55 and those who take larger-than-appropriate quantities for extended periods of time.

Dragon's blood
Daemonorops draco
Rated slightly dangerous, particularly for children, people over 55 and those who take larger-than-appropriate quantities for extended periods of time.

Dumbcane   *Dieffenbachia seguine*
Rated slightly dangerous, particularly for children, people over 55 and those who take larger-than-appropriate quantities for extended periods of time.

Dwarf mallow   *Malva rotundifolia*
Generally regarded as safe when taken in appropriate quantities for short periods of time.

Dyer's broom   *Genista tinctoria*
Rated dangerous, particularly for children, people over 55 and those who take larger-than-appropriate quantities for extended periods of time.

Elderberry   *Sambucus canadensis*
Rated slightly dangerous, particularly for children, people over 55 and those who take larger-than-appropriate quantities for extended periods of time.

Epena   *Virola calophylla*
Rated slightly dangerous, particularly for children, people over 55 and those who take larger-than-appropriate quantities for extended periods of time.

Eucalyptus   *Eucalyptus spp.*
Rated relatively safe when taken in appropriate quantities for short periods of time.

European goldenrod
Solidago virgaurea
Generally regarded as safe when taken in appropriate quantities for short periods of time.

European mistletoe   *Viscum album*
Rated slightly dangerous, particularly for children, people over 55 and those who take larger-than-appropriate quantities for extended periods of time.

Eyebright   *Euphrasis officinalis*
Rated relatively safe when taken in

appropriate quantities for short periods of time.

Fennel   *Foeniculum vulgare*
Generally regarded as safe when taken in appropriate quantities for short periods of time.

Fenugreek
Trigonella foenum-graecum
Rated relatively safe when taken in appropriate quantities for short periods of time.

Feverfew
Chrysanthemum parthenium
Generally regarded as safe when taken in appropriate quantities for short periods of time.

Field horsetail   *Equisetum arvense*
Rated slightly dangerous, particularly for children, people over 55 and those who take larger-than-appropriate quantities for extended periods of time.

Fo-ti   *Polygonum multiflorum*
Rated relatively safe when taken in appropriate quantities for short periods of time.

Fool's parsley   *Aesthusa cynapium*
Rated dangerous, particularly for children, people over 55 and those who take larger-than-appropriate quantities for extended periods of time.

Fringe tree   *Chionanthus virginica*
Rated slightly dangerous, particularly for children, people over 55 and those who take larger-than-appropriate quantities for extended periods of time.

Gambir   *Uncaria gambir*
Rated relatively safe when taken in appropriate quantities for short periods of time.

Gentian   *Gentiana lutea*
Rated relatively safe when taken in appropriate quantities for short periods of time.

German chamomile
Matricaria chamomilla
Generally regarded as safe when taken in appropriate quantities for short periods of time.

Giant milkweed   *Calotropis procera*
Rated slightly dangerous, particularly for children, people over 55 and those who take larger-than-appropriate quantities for extended periods of time.

Glory lily   *Gloriosa superba*
Rated dangerous, particularly for children, people over 55 and those who take larger-than-appropriate quantities for extended periods of time.

Goa   *Andira araroba*
Rated slightly dangerous, particularly for children, people over 55 and those who take larger-than-appropriate quantities for extended periods of time.

Golden chain
Laburnum anagyroides
Rated dangerous, particularly for children, people over 55 and those who take larger-than-appropriate quantities for extended periods of time.

Golden dewdrop   *Duranta repens*
Rated slightly dangerous, particularly for children, people over 55 and those who take larger-than-appropriate quantities for extended periods of time.

Goldenseal   *Hydrastis canadensis*
Rated slightly dangerous, particularly for children, people over 55 and those who take larger-than-appropriate quantities for extended periods of time.

Gotu kola   *Centella asiatica*
Rated slightly dangerous, particularly for children, people over 55 and those who take larger-than-appropriate quantities for extended periods of time.

Granadilla
Passiflora quadrangularis
Rated relatively safe when taken in appropriate quantities for short periods of time.

Ground ivy   *Glechoma hederacea*
Rated relatively safe when taken in appropriate quantities for short periods of time.

Guarana　*Paullina cupana*
Rated relatively safe when taken in appropriate quantities for short periods of time.

Gum arabic　*Acacia senegal*
Generally regarded as safe when taken in appropriate quantities for short periods of time.

Harmel　*Peganum harmala*
Rated slightly dangerous, particularly for children, people over 55 and those who take larger-than-appropriate quantities for extended periods of time.

Hawthorn　*Crataegus oxyacantha*
Rated slightly dangerous, particularly for children, people over 55 and those who take larger-than-appropriate quantities for extended periods of time.

Heart-of-Jesus　*Caladium bicolor*
Rated slightly dangerous, particularly for children, people over 55 and those who take larger-than-appropriate quantities for extended periods of time.

Heliotrope
Heliotropium europaeum
Rated slightly dangerous, particularly for children, people over 55 and those who take larger-than-appropriate quantities for extended periods of time.

Henbane　*Hyoscyamus niger*
Rated dangerous, particularly for children, people over 55 and those who take larger-than-appropriate quantities for extended periods of time.

Henna　*Lawsonia inermis*
Rated relatively safe when taken in appropriate quantities for short periods of time.

Herb paris　*Paris quadrifolia*
Rated slightly dangerous, particularly for children, people over 55 and those who take larger-than-appropriate quantities for extended periods of time.

Holly　*Ilex opaca*
Rated relatively safe when taken in appropriate quantities for short periods of time.

Holy sage　*Salvia divinorum*
Rated relatively safe when taken in appropriate quantities for short periods of time.

Homalomena　*Homalomena sp.*
Rated slightly dangerous, particularly for children, people over 55 and those who take larger-than-appropriate quantities for extended periods of time.

Honey locust　*Gleditsia triacanthos*
Rated relatively safe when taken in appropriate quantities for short periods of time.

Hops　*Humulus lupulus*
Rated relatively safe when taken in appropriate quantities for short periods of time.

Horse chestnut
Aesculus hippocastanum
Rated slightly dangerous, particularly for children, people over 55 and those who take larger-than-appropriate quantities for extended periods of time.

Iboga　*Tabernanthe iboga*
Rated dangerous, particularly for children, people over 55 and those who take larger-than-appropriate quantities for extended periods of time.

Indian acalypha　*Acalypha indica*
Rated relatively safe when taken in appropriate quantities for short periods of time.

Indian hemp
Apocynum cannabinum
Rated slightly dangerous, particularly for children, people over 55 and those who take larger-than-appropriate quantities for extended periods of time.

Indian senna　*Cassia angustifolia*
Rated slightly dangerous, particularly for children, people over 55 and those who take larger-than-appropriate quantities for extended periods of time.

Indian tobacco　*Lobelia inflata*
Rated slightly dangerous, particularly for children, people over 55 and those who take larger-than-appropriate quantities for extended periods of time.

Indigo   *Indigofera tinctoria*
Rated relatively safe when taken in appropriate quantities for short periods of time.

Intoxicating mint
Lagochilus inebrians
Rated slightly dangerous, particularly for children, people over 55 and those who take larger-than-appropriate quantities for extended periods of time.

Ivy   *Hedera helix*
Rated slightly dangerous, particularly for children, people over 55 and those who take larger-than-appropriate quantities for extended periods of time.

Jaborandi   *Pilocarpus spp.*
Rated slightly dangerous, particularly for children, people over 55 and those who take larger-than-appropriate quantities for extended periods of time.

Jack-in-the-pulpit
Arisaema triphyllum
Rated slightly dangerous, particularly for children, people over 55 and those who take larger-than-appropriate quantities for extended periods of time.

Jalap root   *Ipomoea purga*
Rated slightly dangerous, particularly for children, people over 55 and those who take larger-than-appropriate quantities for extended periods of time.

Jamaica dogwood   *Piscidia piscipula*
Rated slightly dangerous, particularly for children, people over 55 and those who take larger-than-appropriate quantities for extended periods of time.

Jamaican quassia   *Picrasma excelsa*
Rated dangerous, particularly for children, people over 55 and those who take larger-than-appropriate quantities for extended periods of time.

Jequerity   *Abrus precatorius*
Rated dangerous, particularly for children, people over 55 and those who take larger-than-appropriate quantities for extended periods of time.

Jimsonweed   *Datura stramonium*
Rated dangerous, particularly for children, people over 55 and those who take larger-than-appropriate quantities for extended periods of time.

Jojoba   *Simmondsia chinensis*
Rated relatively safe when taken in appropriate quantities for short periods of time.

Juniper   *Juniperus communis*
Rated slightly dangerous, particularly for children, people over 55 and those who take larger-than-appropriate quantities for extended periods of time.

Jurema   *Mimosa hostilis*
Rated slightly dangerous, particularly for children, people over 55 and those who take larger-than-appropriate quantities for extended periods of time.

Kamyuye   *Hoslundia opposita*
Rated dangerous, particularly for children, people over 55 and those who take larger-than-appropriate quantities for extended periods of time.

Katum   *Mitragyna speciosa*
Rated slightly dangerous, particularly for children, people over 55 and those who take larger-than-appropriate quantities for extended periods of time.

Kava-kava   *Piper methysticum*
Rated slightly dangerous, particularly for children, people over 55 and those who take larger-than-appropriate quantities for extended periods of time.

Khat   *Catha edulis*
Rated slightly dangerous, particularly for children, people over 55 and those who take larger-than-appropriate quantities for extended periods of time

Knotweed   *Polygonum aviculare*
Rated relatively safe when taken in appropriate quantities for short periods of time.

Kola nuts   *Cola acuminata*
Rated relatively safe when taken in appropriate quantities for short periods of time.

Kola *Cola nitida*
Rated relatively safe when taken in appropriate quantities for short periods of time.

Kwashi *Pancratium trianthum*
Rated dangerous, particularly for children, people over 55 and those who take larger-than-appropriate quantities for extended periods of time.

Lance-leaf periwinkle
Catharanthus lanceus
Rated slightly dangerous, particularly for children, people over 55 and those who take larger-than-appropriate quantities for extended periods of time.

Lantana *Lantana camara*
Rated slightly dangerous, particularly for children, people over 55 and those who take larger-than-appropriate quantities for extended periods of time.

Larkspur *Consolida ambigua*
Rated dangerous, particularly for children, people over 55 and those who take larger-than-appropriate quantities for extended periods of time.

Latua *Latua pubiflora*
Rated dangerous, particularly for children, people over 55 and those who take larger-than-appropriate quantities for extended periods of time.

Lavender *Lavandula angustifolia*
Rated relatively safe when taken in appropriate quantities for short periods of time.

Lavender-cotton
Santolina chamaecyparissus
Rated slightly dangerous, particularly for children, people over 55 and those who take larger-than-appropriate quantities for extended periods of time.

Lemon verbena *Aloysia triphylla*
Rated relatively safe when taken in appropriate quantities for short periods of time.

Lettuce *Lactuca virosa*
Rated relatively safe when taken in appropriate quantities for short periods of time.

Licorice *Glycyrrhiza glabra*
Rated slightly dangerous, particularly for children, people over 55 and those who take larger-than-appropriate quantities for extended periods of time.

Life root *Senecio aureus*
Rated slightly dangerous, particularly for children, people over 55 and those who take larger-than-appropriate quantities for extended periods of time.

Lily-of-the-valley
Convallaria majalis
Rated slightly dangerous, particularly for children, people over 55 and those who take larger-than-appropriate quantities for extended periods of time.

Lima bean *Phaseolus lunatus*
Rated relatively safe when taken in appropriate quantities for short periods of time.

Linden *Tilia europaea*
Rated relatively safe when taken in appropriate quantities for short periods of time.

Luckynut *Thevetia peruviana*
Rated dangerous, particularly for children, people over 55 and those who take larger-than-appropriate quantities for extended periods of time.

Madagascar periwinkle
Catharanthus roseus
Rated slightly dangerous, particularly for children, people over 55 and those who take larger-than-appropriate quantities for extended periods of time.

Malabar nut *Adhatoda vasica*
Rated slightly dangerous, particularly for children, people over 55 and those who take larger-than-appropriate quantities for extended periods of time.

Male fern *Dryopteris filix-mas*
Rated relatively safe when taken in appropriate quantities for short periods of time.

Manaca *Brunfelsia uniflorus*
Rated dangerous, particularly for children, people over 55 and those who

take larger-than-appropriate quantities for extended periods of time.

Manchineel   *Hippomane mancinella*
Rated dangerous, particularly for children, people over 55 and those who take larger-than-appropriate quantities for extended periods of time.

Mandrake   *Mandragora officinarum*
Rated dangerous, particularly for children, people over 55 and those who take larger-than-appropriate quantities for extended periods of time.

Mangosteen   *Garcinia hanburyi*
Rated slightly dangerous, particularly for children, people over 55 and those who take larger-than-appropriate quantities for extended periods of time.

Maraba   *Kaempferia galanga*
Rated relatively safe when taken in appropriate quantities for short periods of time.

Marigold   *Calendula officinalis*
Generally regarded as safe when taken in appropriate quantities for short periods of time.

Marijuana   *Cannabis sativa*
Rated relatively safe when taken in appropriate quantities for short periods of time.

Marsh tea   *Ledum palustre*
Rated slightly dangerous, particularly for children, people over 55 and those who take larger-than-appropriate quantities for extended periods of time.

Marula   *Sclerocarya caffra*
Rated slightly dangerous, particularly for children, people over 55 and those who take larger-than-appropriate quantities for extended periods of time.

Mastic   *Pistacia lentiscus*
Rated relatively safe when taken in appropriate quantities for short periods of time.

Mate   *Ilex paraguariensis*
Rated relatively safe when taken in appropriate quantities for short periods of time.

Mayapple   *Podophyllum peltatum*
Rated slightly dangerous, particularly for children, people over 55 and those who take larger-than-appropriate quantities for extended periods of time.

Meadowsweet   *Filipendula ulmaria*
Generally regarded as safe when taken in appropriate quantities for short periods of time.

Mescal   *Agave sisalana*
Generally regarded as safe when taken in appropriate quantities for short periods of time.

Mescal bean   *Sophora secundiflora*
Rated slightly dangerous, particularly for children, people over 55 and those who take larger-than-appropriate quantities for extended periods of time.

Mesquite   *Prosopis juliflora*
Rated relatively safe when taken in appropriate quantities for short periods of time.

Mexican calea   *Calea zacathechichi*
Rated slightly dangerous, particularly for children, people over 55 and those who take larger-than-appropriate quantities for extended periods of time.

Mezereon   *Daphne mezereum*
Rated dangerous, particularly for children, people over 55 and those who take larger-than-appropriate quantities for extended periods of time.

Mohodu   *Cineraria aspera*
Rated slightly dangerous, particularly for children, people over 55 and those who take larger-than-appropriate quantities for extended periods of time.

Mole plant   *Euphorbia lathyris*
Rated dangerous, particularly for children, people over 55 and those who take larger-than-appropriate quantities for extended periods of time.

Monkshood   *Aconitum napellus*
Rated dangerous, particularly for children, people over 55 and those who take larger-than-appropriate quantities for extended periods of time.

Moonseed *Menispermum canadense*
Rated slightly dangerous, particularly for children, people over 55 and those who take larger-than-appropriate quantities for extended periods of time.

Mormon tea *Ephedra nevadensis*
Rated slightly dangerous, particularly for children, people over 55 and those who take larger-than-appropriate quantities for extended periods of time.

Morning Glory *Ipomoea violacea*
Rated slightly dangerous, particularly for children, people over 55 and those who take larger-than-appropriate quantities for extended periods of time.

Motherwort *Leonurus cardiaca*
Rated relatively safe when taken in appropriate quantities for short periods of time.

Mountain laurel *Kalmia latifolia*
Rated dangerous, particularly for children, people over 55 and those who take larger-than-appropriate quantities for extended periods of time.

Mountain tobacco *Arnica montana*
Rated slightly dangerous, particularly for children, people over 55 and those who take larger-than-appropriate quantities for extended periods of time.

Mugwort *Artemisia vulgaris*
Rated slightly dangerous, particularly for children, people over 55 and those who take larger-than-appropriate quantities for extended periods of time.

Muira puama
Ptychopetalum olacoides
Rated slightly dangerous, particularly for children, people over 55 and those who take larger-than-appropriate quantities for extended periods of time.

Musk okra *Abelmoschus moschatus*
Generally regarded as safe when taken in appropriate quantities for short periods of time.

Nene *Coleus blumei*
Rated slightly dangerous, particularly for children, people over 55 and those who take larger-than-appropriate quantities for extended periods of time.

Niando *Alchornea floribunda*
Rated dangerous, particularly for children, people over 55 and those who take larger-than-appropriate quantities for extended periods of time.

Night-blooming cereus
Selenicereus grandiflorus
Rated relatively safe when taken in appropriate quantities for short periods of time.

Nightshade *Solanum nigrum*
Rated slightly dangerous, particularly for children, people over 55 and those who take larger-than-appropriate quantities for extended periods of time.

Niopo *Anadenathera peregrina*
Rated slightly dangerous, particularly for children, people over 55 and those who take larger-than-appropriate quantities for extended periods of time.

Nutmeg *Myristica fragrans*
Rated slightly dangerous, particularly for children, people over 55 and those who take larger-than-appropriate quantities for extended periods of time.

Nux-vomica *Strychnos nux-vomica*
Rated dangerous, particularly for children, people over 55 and those who take larger-than-appropriate quantities for extended periods of time.

Oleander *Nerium oleander*
Rated dangerous, particularly for children, people over 55 and those who take larger-than-appropriate quantities for extended periods of time.

Opium poppy *Papaver somniferum*
Rated slightly dangerous, particularly for children, people over 55 and those who take larger-than-appropriate quantities for extended periods of time.

Ordeal bean
Physostigma venenosum
Rated dangerous, particularly for children, people over 55 and those who take larger-than-appropriate quantities for extended periods of time.

Oregon grape *Mahonia aquifolia*
Rated slightly dangerous, particularly
for children, people over 55 and those
who take larger-than-appropriate
quantities for extended periods of time.

Oriental ginseng *Panax ginseng*
Generally regarded as safe when taken
in appropriate quantities for short
periods of time.

Pakistani ephedra
Ephedra geraridiana
Rated slightly dangerous, particularly
for children, people over 55 and those
who take larger-than-appropriate
quantities for extended periods of time.

Pao d'arco *Tabebuia sp.*
Rated slightly dangerous, particularly
for children, people over 55 and those
who take larger-than-appropriate
quantities for extended periods of time.

Papaya *Carica papaya*
Generally regarded as safe when taken
in appropriate quantities for short
periods of time.

Parsley *Petroselinum crispum*
Rated relatively safe when taken in
appropriate quantities for short periods
of time.

Partridgeberry *Mitchella repens*
Rated relatively safe when taken in
appropriate quantities for short periods
of time.

Pasque flower *Anemone pulsatilla*
Rated slightly dangerous, particularly
for children, people over 55 and those
who take larger-than-appropriate
quantities for extended periods of time.

Passionflower *Passiflora incarnata*
Rated relatively safe when taken in
appropriate quantities for short periods
of time.

Pearly everlasting
Anaphalis margaritacea
Generally regarded as safe when taken
in appropriate quantities for short
periods of time.

Peegee *Hydrangea paniculata*
Rated relatively safe when taken in
appropriate quantities for short periods
of time.

Pennyroyal *Hedeoma pulegioides*
Rated relatively safe when taken in
appropriate quantities for short periods
of time.

Peony *Paeonia officinalis*
Rated dangerous, particularly for
children, people over 55 and those who
take larger-than-appropriate quantities
for extended periods of time.

Perilla *Perilla frutescens*
Rated relatively safe when taken in
appropriate quantities for short periods
of time.

Periwinkle *Vinca minor*
Rated dangerous, particularly for
children, people over 55 and those who
take larger-than-appropriate quantities
for extended periods of time.

Peruvian peppertree *Schinus molle*
Rated slightly dangerous, particularly
for children, people over 55 and those
who take larger-than-appropriate
quantities for extended periods of time.

Peyote *Lophophora williamsii*
Rated relatively safe when taken in
appropriate quantities for short periods
of time.

Physic nut *Jatropha curcas*
Rated dangerous, particularly for
children, people over 55 and those who
take larger-than-appropriate quantities
for extended periods of time.

Pineapple *Ananas comosus*
Rated relatively safe when taken in
appropriate quantities for short periods
of time.

Pink clover *Trifolium pratense*
Generally regarded as safe when taken
in appropriate quantities for short
periods of time.

HERB TOXICITY

Pinkroot   *Spigelia marilandica*
Rated dangerous, particularly for children, people over 55 and those who take larger-than-appropriate quantities for extended periods of time.

Piule   *Rhynchosia pyramidalis*
Rated slightly dangerous, particularly for children, people over 55 and those who take larger-than-appropriate quantities for extended periods of time.

Plantain   *Plantago major*
Generally regarded as safe when taken in appropriate quantities for short periods of time.

Poinsettia   *Euphorbia pulcherrima*
Rated dangerous, particularly for children, people over 55 and those who take larger-than-appropriate quantities for extended periods of time.

Poison hemlock
Conium maculatum
Rated dangerous, particularly for children, people over 55 and those who take larger-than-appropriate quantities for extended periods of time.

Poison ivy   *Rhus toxicodendron*
Rated slightly dangerous, particularly for children, people over 55 and those who take larger-than-appropriate quantities for extended periods of time.

Pokeweed   *Phytolacca americana*
Rated dangerous, particularly for children, people over 55 and those who take larger-than-appropriate quantities for extended periods of time.

Potato   *Solanum tuberosum*
Rated relatively safe when taken in appropriate quantities for short periods of time.

Prickly poppy   *Argemone mexicana*
Rated slightly dangerous, particularly for children, people over 55 and those who take larger-than-appropriate quantities for extended periods of time.

Privet   *Ligustrum vulgare*
Rated slightly dangerous, particularly for children, people over 55 and those who take larger-than-appropriate quantities for extended periods of time.

Pyrethrum
Chrysanthemum cinerariifolium
Rated relatively safe when taken in appropriate quantities for short periods of time.

Queen's delight   *Stillingia sylvatica*
Rated dangerous, particularly for children, people over 55 and those who take larger-than-appropriate quantities for extended periods of time.

Quinine   *Cinchona sp.*
Rated relatively safe when taken in appropriate quantities for short periods of time.

Redroot   *Lachnanthes tinctoria*
Rated slightly dangerous, particularly for children, people over 55 and those who take larger-than-appropriate quantities for extended periods of time.

Roman chamomile
Chamaemelum nobile
Generally regarded as safe when taken in appropriate quantities for short periods of time.

Roselle   *Hibiscus sabdarriffa*
Generally regarded as safe when taken in appropriate quantities for short periods of time.

Rosemary   *Rosmarinus officinalis*
Rated relatively safe when taken in appropriate quantities for short periods of time.

Rosinweed   *Grindelia spp.*
Rated slightly dangerous, particularly for children, people over 55 and those who take larger-than-appropriate quantities for extended periods of time.

Rubber vine
Cryptostegia grandifolia
Rated dangerous, particularly for children, people over 55 and those who take larger-than-appropriate quantities for extended periods of time.

Rue   *Ruta graveolens*
Rated relatively safe when taken in

appropriate quantities for short periods of time.

Sabadilla *Schoenocaulon officinale*
Rated dangerous, particularly for children, people over 55 and those who take larger-than-appropriate quantities for extended periods of time.

Sabine *Juniperus sabina*
Rated slightly dangerous, particularly for children, people over 55 and those who take larger-than-appropriate quantities for extended periods of time.

Saffron *Crocus sativus*
Rated relatively safe when taken in appropriate quantities for short periods of time.

Sage *Salvia officinalis*
Generally regarded as safe when taken in appropriate quantities for short periods of time.

Sago cycas *Cycas revoluta*
Rated dangerous, particularly for children, people over 55 and those who take larger-than-appropriate quantities for extended periods of time.

Sanchi ginseng *Panax notoginseng*
Generally regarded as safe when taken in appropriate quantities for short periods of time.

Sandalwood *Santalum album*
Rated relatively safe when taken in appropriate quantities for short periods of time.

Sandbox tree *Hura crepitans*
Rated dangerous, particularly for children, people over 55 and those who take larger-than-appropriate quantities for extended periods of time.

Sarpaganda *Rauwolfia serpentina*
Rated slightly dangerous, particularly for children, people over 55 and those who take larger-than-appropriate quantities for extended periods of time.

Sarsaparilla *Smilax aristolochiifolia*
Rated relatively safe when taken in appropriate quantities for short periods of time.

Sassafras *Sassafras albidum*
Rated relatively safe when taken in appropriate quantities for short periods of time.

Sassybark
Erythrophleum suaveolens
Rated dangerous, particularly for children, people over 55 and those who take larger-than-appropriate quantities for extended periods of time.

Saw palmetto *Serenoa repens*
Rated relatively safe when taken in appropriate quantities for short periods of time.

Scarlet poppy *Papaver bracteatum*
Rated slightly dangerous, particularly for children, people over 55 and those who take larger-than-appropriate quantities for extended periods of time.

Scopolia *Scopolia carniolica*
Rated dangerous, particularly for children, people over 55 and those who take larger-than-appropriate quantities for extended periods of time.

Scotch broom *Cytisus scoparius*
Rated slightly dangerous, particularly for children, people over 55 and those who take larger-than-appropriate quantities for extended periods of time.

Sea Island cotton
Gossypium barbadense
Rated slightly dangerous, particularly for children, people over 55 and those who take larger-than-appropriate quantities for extended periods of time.

Sea onion *Urginea maritima*
Rated dangerous, particularly for children, people over 55 and those who take larger-than-appropriate quantities for extended periods of time.

Seven barks *Hydrangea arborescens*
Rated relatively safe when taken in appropriate quantities for short periods of time.

Shanshi   *Coriaria thymifolia*
Rated dangerous, particularly for
children, people over 55 and those who
take larger-than-appropriate quantities
for extended periods of time.

Shavegrass   *Equisetum hyemale*
Rated slightly dangerous, particularly
for children, people over 55 and those
who take larger-than-appropriate
quantities for extended periods of time.

Sinicuichi   *Heimia salicifolia*
Rated slightly dangerous, particularly
for children, people over 55 and those
who take larger-than-appropriate
quantities for extended periods of time.

Skunk cabbage
Symplocarpus foetidus
Rated slightly dangerous, particularly
for children, people over 55 and those
who take larger-than-appropriate
quantities for extended periods of time.

Slash pine   *Pinus elliottii*
Rated slightly dangerous, particularly
for children, people over 55 and those
who take larger-than-appropriate
quantities for extended periods of time.

Slippery elm   *Ulmus rubra*
Rated relatively safe when taken in
appropriate quantities for short periods
of time.

Snakeplant   *Rivea corymbosa*
Rated slightly dangerous, particularly
for children, people over 55 and those
who take larger-than-appropriate
quantities for extended periods of time.

Snakeroot   *Aristolochia serpentaria*
Rated relatively safe when taken in
appropriate quantities for short periods
of time.

Soaptree   *Quillaja saponaria*
Rated relatively safe when taken in
appropriate quantities for short periods
of time.

Soksi   *Mirabilis multiflora*
Rated slightly dangerous, particularly
for children, people over 55 and those
who take larger-than-appropriate
quantities for extended periods of time.

Soma   *Sarcostemma acidum*
Rated slightly dangerous, particularly
for children, people over 55 and those
who take larger-than-appropriate
quantities for extended periods of time.

Southernwood
Artemisia abrotanum
Rated slightly dangerous, particularly
for children, people over 55 and those
who take larger-than-appropriate
quantities for extended periods of time.

Spiny ginseng
Eleutherococcus senticosus
Generally regarded as safe when taken
in appropriate quantities for short
periods of time.

Spurge   *Chamaesyee hypericifolia*
Rated dangerous, particularly for
children, people over 55 and those who
take larger-than-appropriate quantities
for extended periods of time.

St. John's wort
Hypericum perforatum
Rated slightly dangerous, particularly
for children, people over 55 and those
who take larger-than-appropriate
quantities for extended periods of time.

Star anise   *Illicium verum*
Rated slightly dangerous, particularly
for children, people over 55 and those
who take larger-than-appropriate
quantities for extended periods of time.

Stinging nettle   *Urtica dioica*
Generally regarded as safe when taken
in appropriate quantities for short
periods of time.

Strawberry tree   *Arbutus unedo*
Rated slightly dangerous, particularly
for children, people over 55 and those
who take larger-than-appropriate
quantities for extended periods of time.

Sumbul   *Ferula sumbul*
Rated relatively safe when taken in
appropriate quantities for short periods
of time.

Surinam quassia   *Quassia amara*
Rated dangerous, particularly for

children, people over 55 and those who take larger-than-appropriate quantities for extended periods of time.

Sweet flag *Acorus calamus*
Rated slightly dangerous, particularly for children, people over 55 and those who take larger-than-appropriate quantities for extended periods of time.

Sweetclover *Melilotus officinalis*
Rated relatively safe when taken in appropriate quantities for short periods of time.

Tansy *Tanacetum vulgare*
Rated relatively safe when taken in appropriate quantities for short periods of time.

Tarragon *Artemisia dracunculus*
Rated relatively safe when taken in appropriate quantities for short periods of time.

Tea *Camellia sinensis*
Rated relatively safe when taken in appropriate quantities for short periods of time.

Thornapple *Datura innoxia*
Rated dangerous, particularly for children, people over 55 and those who take larger-than-appropriate quantities for extended periods of time.

Thyme *Thymus vulgaris*
Rated relatively safe when taken in appropriate quantities for short periods of time.

Tobacco *Nicotiana tabacum*
Rated dangerous, particularly for children, people over 55 and those who take larger-than-appropriate quantities for extended periods of time.

Tomato *Lycopersicon esculentum*
Rated relatively safe when taken in appropriate quantities for short periods of time.

Tonka bean *Dipteryx odorata*
Rated relatively safe when taken in appropriate quantities for short periods of time.

Traveler's joy *Clematis vitalba*
Rated slightly dangerous, particularly for children, people over 55 and those who take larger-than-appropriate quantities for extended periods of time.

Tree tobacco *Nicotiana glauca*
Rated dangerous, particularly for children, people over 55 and those who take larger-than-appropriate quantities for extended periods of time.

Tua-tua *Jatropha gossypiifolia*
Rated dangerous, particularly for children, people over 55 and those who take larger-than-appropriate quantities for extended periods of time.

Tupa *Lobelia tupa*
Rated slightly dangerous, particularly for children, people over 55 and those who take larger-than-appropriate quantities for extended periods of time.

Unicorn root *Aletris farinosa*
Rated slightly dangerous, particularly for children, people over 55 and those who take larger-than-appropriate quantities for extended periods of time.

Unmatal *Datura metel*
Rated dangerous, particularly for children, people over 55 and those who take larger-than-appropriate quantities for extended periods of time.

Upland cotton *Gossypium hirsutum*
Rated slightly dangerous, particularly for children, people over 55 and those who take larger-than-appropriate quantities for extended periods of time.

Valerian *Valeriana officinalis*
Rated relatively safe when taken in appropriate quantities for short periods of time.

Vanilla *Vanilla planifolia*
Rated relatively safe when taken in appropriate quantities for short periods of time.

Verbena *Verbena officinalis*
Generally regarded as safe when taken in appropriate quantities for short periods of time.

Vilca *Anadenathera colubrina*
Rated slightly dangerous, particularly
for children, people over 55 and those
who take larger-than-appropriate
quantities for extended periods of time.

Virginian scullcap
Scutellaria lateriflora
Rated relatively safe when taken in
appropriate quantities for short periods
of time.

Wahoo *Euonymus atropurpureas*
Rated slightly dangerous, particularly
for children, people over 55 and those
who take larger-than-appropriate
quantities for extended periods of time.

Water fennel
Onenanthe phellandrium
Rated dangerous, particularly for
children, people over 55 and those who
take larger-than-appropriate quantities
for extended periods of time.

Water hemlock *Cicuta maculata*
Rated dangerous, particularly for
children, people over 55 and those who
take larger-than-appropriate quantities
for extended periods of time.

Winter savory *Satureja montana*
Rated relatively safe when taken in
appropriate quantities for short periods
of time.

Wintergreen
Gaultheria procumbens
Rated slightly dangerous, particularly
for children, people over 55 and those
who take larger-than-appropriate
quantities for extended periods of time.

Witch-hazel *Hamamelis virginiana*
Rated relatively safe when taken in
appropriate quantities for short periods
of time.

Woodrose *Argyreia nervosa*
Rated slightly dangerous, particularly
for children, people over 55 and those
who take larger-than-appropriate
quantities for extended periods of time.

Woodruff *Galium odoratum*
Generally regarded as safe when taken
in appropriate quantities for short
periods of time.

Wormseed
Chenopodium ambrosioides
Rated slightly dangerous, particularly
for children, people over 55 and those
who take larger-than-appropriate
quantities for extended periods of time.

Wormwood *Artemisia absinthium*
Rated slightly dangerous, particularly
for children, people over 55 and those
who take larger-than-appropriate
quantities for extended periods of time.

Yarrow *Achillea millefolium*
Generally regarded as safe when taken
in appropriate quantities for short
periods of time.

Yellow dock *Rumex crispus*
Rated slightly dangerous, particularly
for children, people over 55 and those
who take larger-than-appropriate
quantities for extended periods of time.

Yellow jessamine
Gelsemium sempervirens
Rated dangerous, particularly for
children, people over 55 and those who
take larger-than-appropriate quantities
for extended periods of time.

Yellow ladyslipper
Cypripedium calceolus
Rated relatively safe when taken in
appropriate quantities for short periods
of time.

Yohimbe *Pausinystalia johimbe*
Rated slightly dangerous, particularly
for children, people over 55 and those
who take larger-than-appropriate
quantities for extended periods of time.

Yoko *Paullinia yoko*
Rated relatively safe when taken in
appropriate quantities for short periods
of time.

Brand Names

This is a list of brand names that various vitamins, minerals and supplements are sold under. Brand names of medicinal herbs are not listed. The list is arranged by *class of substance*. This is followed by the *generic name* and the forms in which the generic substance is manufactured and sold. Following the generic form is a list of the *brand names* on the U.S. market. Other countries may use different brand names.

New brands appear frequently, and brands may be removed from the market. No list can reflect instantaneous changes. Inclusion of a brand name does *not* imply recommendation or endorsement. Exclusion does *not* imply a brand name is less effective or less safe than those listed.

Below is an example of an entry in this list. As you can see, it includes the classification, generic name and form, and brand names.

Class of vitamin mineral or supplement:	**Calcium Glubionate**
Generic substance and form it is sold in:	*Calcium Glubionate Syrup*
Brand name(s) of generic substance:	Neo-Calglucon®

In some cases, a substance is sold only under the generic name—no brand names are available on the market. In this case, the classification and generic substance are listed, but no brand names appear.

Class of vitamin mineral or supplement:	**Calcium Gluconate**
Generic substance and form it is sold in:	*Calcium Gluconate Chewable Tablets*

Vitamin A
Vitamin-A Capsules
  Afaxin®
  Alphalin®
  Aquasol A®
Vitamin-A Oral Solution
  Aquasol A®
Vitamin-A Tablets

Alumina/Magnesia/Calcium Carbonate—
A magnesium supplement
*Alumina/Magnesia/Calcium Carbonate
Chewable Tablets*
  Duracid®
*Alumina/Magnesia/Calcium Carbonate
Oral Suspension*
  Camalox®
*Alumina/Magnesia/Calcium Carbonate/
Simethicone Chewable Tablets*
  Tempo®

Alumina/Magnesium Carbonate—
A magnesium supplement
Alumina/Magnesia Chewable Tablets
  Algenic Alka Improved®
  Aludrox®
  Creamlin®
  Gelusil®
  Gelusil Extra-Strength®
  Maalox®
  Maalox No. 1®
  Maalox No. 2®
  Maalox TC®
  Magmalin®
  Neutralca-S®
  Rulox No. 1®
  Rulox No. 2®
  WinGel®
Alumina/Magnesia Oral Suspension
  Aludrox®
  Alumid®
  Amphojel 500®
  Delcid®
  Diovol Ex®
  Gelamal®
  Gelusil®
  Gelusil Extra Strength®
  Kolantyl®
  Kudrox®
  Maalox®
  Maalox TC®
  Mintox®
  Mylanta-2 Plain®
  Neutralca-S®
  Rolox®
  Rulox®
  Univol®
  WinGel®
Alumina/Magnesia Tablets
  Diovol Ex®
  Univol®

Alumina/Magnesia Wafers
  Kolantyl®
*Alumina/Magnesium Carbonate Chewable
Tablets*
  Algicon®
  Magnagel®
*Alumina/Magnesium Carbonate Oral
Suspension*
  Gaviscon®
  Liquimint®
  Magnagel®

Alumina/Magnesium Trisilicate—
A magnesium supplement
*Alumina/Magnesium Trisilicate Chewable
Tablets*
  Gaviscon®
  Gaviscon-2®

Alumina/Magnesia/Simethicone—
A magnesium supplement
*Alumina/Magnesia/Simethicone Chewable
Tablets*
  Almacone®
  Alma-Mag 4 Improved®
  Diovol®
  Gelusil®
  Gelusil-II®
  Gelusil-M®
  Maalox Plus®
  Mylanta®
  Mylanta-II®
*Alumina/Magnesium/Simethicone Oral
Suspension*
  Almacone®
  Almacone II®
  Alma-Mag Improved®
  Alumid Plus®
  Amphojel Plus®
  AntaGel®
  AntaGel-II®
  Di-Gel®
  Diovol®
  Gelusil®
  Gelusil-II®
  Gelusil-M®
  Maalox Plus®
  Mi-Acid®
  Mygel®
  Mygel II®
  Mylanta®
  Mylanta-II®
  Mylanta-2 Extra Strength®
  Newtrogel II®
  Silain-Gel®
  Simaal Gel®
  Simaal 2 Gel®
  Simeco®

Alumina/Magnesium Trisilicate/Sodium Bicarbonate—
A magnesium supplement
Alumina/Magnesium Trisilicate/Sodium Bicarbonate Chewable Tablets
- Gas-is-gon®
- Triconsil®

Vitamin B-12
Cyanocobalamin Injectable Form—Some doctors prescribe for home use.
- Anacobin®
- Bedoz®
- Berubigen®
- Betalin 12®
- Cyanabin®
- Kaybovite-1000®
- Redisol®
- Rubion®
- Rubramin PC®

Cyanocobalamin Tablets
- Kaybovite®
- Rubramin®

Hydroxocobalamin Injectable Form— Some doctors prescribe for home use.
- Acti-B12®
- alphaREDISOL®
- Codroxomin®
- Droxomin®

Vitamin C (Ascorbic Acid)
Ascorbic-Acid Chewable Tablets
- Apo-C®
- Flavorcee®

Ascorbic-Acid Effervescent Tablets
- Redoxon®

Ascorbic-Acid Extended-Release Capsules
- Ascorbicap®
- Cetane®
- Cevi-Bid®
- Cevita®

Ascorbic-Acid Extended-Release Tablets
- Arco-Cee®
- Cemill®

Ascorbic-Acid Oral Solution
- Cecon®
- Ce-Vi-So®

Ascorbic-Acid Syrup
- Ascorbic-acid Syrup®

Ascorbic-Acid Tablets
- Apo-C®
- Cevalin®
- Cevita®

Calcium Carbonate
Calcium Carbonate Chewable Tablets
- BioCal®
- Cal Sup®
- Os-Cal 500®
- Suplical®

Theracal®
Tums®
Tums E-X®
Calcium Carbonate Tablets
- BioCal®
- Caltrate®
- Os-Cal 500®

Calcium Carbonate/Magnesia
Calcium Carbonate/Magnesia Chewable Tablets
- Bisodol®

Calcium Carbonate/Magnesia/ Simethicone
Calcium Carbonate/Magnesia/Simethicone Tablets
- Advanced Formula Di-Gel®

Calcium Citrate
Calcium Citrate Tablets
- Citracal®

Calcium Glubionate
Calcium Glubionate Syrup
- Neo-Calglucon®

Calcium Gluconate
Calcium Gluconate Chewable Tablets

Calcium Lactate
Calcium Lactate Tablets

Calcium/Magnesium Carbonate—
A magnesium supplement
Calcium/Magnesium Carbonate Chewable Tablets
- Noralac®
- Spastosed®

Calcium/Magnesium Carbonate Oral Suspension
- Marblen®

Calcium/Magnesium Carbonate Tablets
- Marblen®

Calcium/Magnesium Carbonate/ Magnesium Oxide—A magnesium supplement
Calcium/Magnesium Carbonate/ Magnesium Oxide Tablets
- Alkets®

Calcium Phosphate
Tribasic Calcium Phosphate Tablets
- Posture®

Charcoal
Activated-Charcoal Capsules
  Charocaps®
Activated-Charcoal Oral Suspension
  Actidose-Aqua®
  Arma-a-Char®
  Charcoaid®
  Charcodote®
  Insta-Char®
  Liquid-Antidose®
Activated-Charcoal Powder for Slurry
  Charcosalanti Dote®

Children's Multiple Vitamins/Fluorides
Multiple Vitamins/Fluorides
  Adefor®
  Cari-Tab®
  Mulvidren-F®
  Poly-Vi-Flor®
  Tri-Vi-Flor®
  Vi-Daylin/F®
  Vi-Penta F®

Vitamin D
Calcifidiol Capsules
  Calderol®
Calcitrol Capsules
  Rocaltrol®
Dihydrotachysterol Capsules
  Hytakerol®
Dihydrotachysterol Oral Solution
  DHT®
  DHT Intensol®
  Hytakerol®
Dihydrotachysterol Tablets
  DHT®
Ergocalciferol Capsules
  Deltalin®
  Drisdol®
  Ostoforte®
  Radiostol®
Ergocalciferol Oral Solution
  Calciferol®
  Drisdol®
  Radiostol®
  Radiostol Forte®
Ergocalciferol Tablets
  Calciferol®

Dibasic Calcium Phosphate—A calcium supplement
Dibasic Calcium Phosphate Tablets

Dihydroxyaluminum Aminoacetate/Magnesia/Alumina—A magnesium supplement
Dihydroxyaluminum Aminoacetate/Magnesia/Alumina Oral Suspension
  Tralmag®

Vitamin E
Vitamin-E Capsules
  Aquasol E®
  E-Ferol®
  Eprolin®
  Epsilan-M®
  Viterra E®
Vitamin-E Chewable Tablets
  Chew-E®
Vitamin-E Oral Solution
  Aquasol E®
Vitamin-E Tablets
  Pheryl-E®

Ferrous Fumurate
Ferrous-Fumurate Capsules
  Neo-Fer®
Ferrous-Fumurate Chewable Tablets
  Feostat®
Ferrous-Fumurate Extended-Release Tablets
  Feco-T®
Ferrous-Fumurate Oral Suspension
  Feo-Stat®
  Palafer®
Ferrous-Fumurate Tablets
  Femiron®
  Fersamal®
  Fumasorb®
  Fumerin®
  Hemocyte®
  Ircon®
  Novofumar®
  Palmiron®

Ferrous Gluconate
Ferrous-Gluconate Capsules
  Fergon®
  Simron®
Ferrous-Gluconate Elixir
  Fergon®
Ferrous-Gluconate Tablets
  Apo-Ferrous Gluconate®
  Fergon®
  Feralet®
  Fertinic®
  Novoferrogluc®

Ferrous Sulfate
Ferrous-Sulfate Capsules
  Feosol®
  Fer-In-Sol®
Ferrous-Sulfate Enteric-Coated Tablet.
  Apo-Ferrous Sulfate®
  Novoferrosulfa®
Ferrous-Sulfate Extended-Release Capsules
  Ferralyn®
  Fesofor®
Ferrous-Sulfate Extended-Release Tablets
  Fero-Grad®
  Fero-Gradumet®
  Slow Fe®

Ferrous-Sulfate Elixir
  Feosol®
Ferrous-Sulfate Oral Solution
  Fer-In-Sol®
  Fer-Iron®
Ferrous-Sulfate Syrup
  Fer-In-Sol®
Ferrous-Sulfate Tablets
  Feosol®
  Fesofor®
  Hematinic®
  Mol-Iron®
  PMS Ferrous Sulfate®

Fluoride see
  Sodium Fluoride

Folic Acid
  Folic-Acid Tablets
    Apo-Folic®
    Folvite®
    Novofolacid®
    Vitamin B-9®

Gamma-Linolenic Acid (Evening Primrose Oil)
  Primrose-Oil Capsules
    Evening star®
    Fever plant®
    Field primrose®
    King's-cure-all®
    Nature's Way Soft Gel Evening
      Primrose Oil®
    Night Willow Herb®
    Optimax®
    Primrose®
    Scabish®
    Tree Primrose®

Iron-Polysaccharide
  Iron-Polysaccharide Capsules
    Hytinic®
    Niferex®
    Nu-Iron®
  Iron-Polysaccharide Elixir
    Hytinic®
    Niferex®
    Nu-Iron®
  Iron-Polysaccharide Tablets
    Niferex®

Magnesium Carbonate/Sodium Bicarbonate
  Magnesium Carbonate/Sodium
  Bicarbonate for Oral Suspension
    Bisodol®

Magnesium Hydroxide
  Milk of Magnesia
    M.O.M.®
    Phillips' Milk of Magnesia®

Magnesium Hydroxide/Mineral Oil
  Milk of Magnesia and Mineral Oil
  Emulsion
    Haley's M-O®

Magnesium Oxide
  Magnesium-Oxide Capsules
    Par-Mag®
    Uro-Mag®
  Magnesium-Oxide Tablets
    Mag-Ox 400®
    Maox®

Magnesium Sulfate
  Magnesium-Sulfate Tablets
    Bilagog®
  Magnesium-Sulfate Crystals
    Magnesium Sulfate Crystals®

Magnesium Trisilicate/Alumina/Magnesia
  Magnesium Trisilicate/Alumina/Magnesia
  Chewable Tablets
    Magnatril®
  Magnesium Trisilicate/Alumina/Magnesia
  Oral Suspension
    Magnatril®

Magnesium Trisilicate/Alumina/ Magnesium
  Magnesium Trisilicate/Alumina/Magnesium
  Carbonate Capsules
    Escot®

Manganese
  Manganese Tablets
    Mn-Plus®
  Available as a constituent of many
  multivitamin/mineral preparations.

Menadiol
  Menadiol Sodium Diphosphate Tablets
    Synkavite®

Niacin
  Niacin Capsules
  Niacin Extended-Release Capsules
    Diacin®
    Niac®
    Nico-400®
    Nicobid®
    Nico-Span®
    Tega-Span®
  Niacin Extended-Release Tablets
    Span-Niacin®
  Niacin Oral Solution
    Nicotex®
  Niacin Tablets
    Nicolar®

Niacinamide
Niacinamide Capsules
Niacinamide Tablets

Omega-3 Polyunsaturated Fatty Acids
Omega-3 Polyunsaturated Fatty Acids Capsules
  Cardi-Omega 3®
  Marine 500®
  Marine 1000®
  Max-EPA®
  Promega®
  Proto-Chol®
  Sea-Omega 50®

PABA
PABA Capsules
  Hill-Shade®
  Pabagel®
  Pabanol®
  Paraminan®
  Presun 8®
  RV Paba®

Phytonadione
Phytonadione Tablets
  Mephyton®

Potassium Acetate/Potassium Bicarbonate/Potassium Citrate (Trikates)
Trikates Oral Solution
  Tri-K®

Potassium Bicarbonate
Potassium Bicarbonate Effervescent Tablets for Oral Solution
  Klor-Con/EF®
  K-Lyte®

Potassium Bicarbonate/ Potassium Chloride
Potassium Bicarbonate/Potassium Chloride for Effervescent Oral Solution
  Klorvess®
  K-Lyte/Cl®
  Neo-K®
Potassium Bicarbonate/Potassium Chloride Effervescent Tablets for Oral Solution
  Klorvess®
  K-Lyte/Cl®
  Potassium-Sandoz®

Potassium Bicarbonate/Potassium Citrate
Potassium Bicarbonate/Potassium Citrate Effervescent Tablets for Oral Solution
  K-Lyte DS®

Potassium Chloride
Potassium-Chloride Extended-Release Capsules
  Micro-K®
Potassium-Chloride Enteric-Coated Tablets
Potassium-Chloride Extended-Release Tablets
  Apo-K®
  Kalium Durules®
  Kaon-Cl®
  K-Dur®
  K-Long®
  Klor-Con®
  Klotrix®
  K-Tab®
  Novo-Lente-K®
  Slo-Pot®
  Slow-K®
Potassium-Chloride Oral Solution
  Cena-K®
  K-10®
  Kaochlor®
  Kaochlor S-F®
  Kaon-Cl®
  Kay Ciel®
  KCL®
  Klor-10%®
  Klor-Con®
  Klorvess®
  Potachlor®
  Potasalan®
  Potassine®
  Roychlor®
  Rum-K®
  SK-Potassium Chloride®
Potassium-Chloride for Oral Solution
  Kato®
  Kay Ciel®
  K-Lor®
  Klor-Con®
  Potage®

Potassium Chloride/Potassium Bicarbonate/Potassium Citrate
Potassium Chloride/Potassium Bicarbonate/Potassium Citrate Effervescent Tablets for Oral Solution
  Kaochlor-Eff®

Potassium Gluconate
Potassium-Gluconate Elixi
  Bayon®
  Kaon®
  Kaylixir®
  K-G Elixir®
  Potassium-Rougier®
  Royonate®
Potassium-Gluconate Tablets
  Kaon®
  Kao-Nor®

Potassium Gluconate/Potassium Chloride
Potassium Gluconate/Potassium Chloride Oral Solution
Kolyum®

Potassium Gluconate/Potassium Citrate
Potassium Gluconate/Potassium Citrate Oral Solution
Bi-K®
Twin-K®

Potassium Gluconate/Potassium Citrate/ Ammonium Chloride
Potassium Gluconate/Potassium Citrate/ Ammonium Chloride Oral Solution
Twin-K-Cl®

Potassium Phosphate—A phosphate supplement
Monobasic Potassium Phosphate Tablets
K-Phos Original®
Potassium Phosphate Capsules for Oral Solution
Neutra-Phos-K®

Potassium/Sodium Phosphate— A phosphate supplement
Potassium/Sodium Phosphate Capsules for Oral Solution
Neutra-Phos®
Potassium/Sodium Phosphate for Oral Solution
Neutra-Phos®
Potassium/Sodium Phosphate Tablets
Uro-KP-Neutral®
Monobasic Potassium/Sodium Phosphate Tablets
K-Phos M.F.®
K-Phos Neutral®
K-Phos No. 2®

Pyridoxine
Pyridoxine-Hydrochloride Extended-Release Capsules
Rodex®
TexSix T.R.®
Pyridoxine-Hydrochloride Tablets
Hexa-Betalin®
Pyroxin®

Riboflavin (Vitamin B-2)
Vitamin B-2 Tablets
Riboflavin®
Riobin-50®

Selenium
Topical
Excel®
Selsun®
Selsun Blue®
Sul-Blue®
Available as a constituent of many multivitamin/mineral preparations.

Simethicone/Alumina/Magnesium Carbonate/Magnesia
Simethicone/Alumina Magnesium Carbonate/Magnesia Chewable Tablets
Amphojel Plus®
Di-Gel®

Sodium Fluoride
Sodium-Fluoride Chewable Tablets
Denta-FL®
Flo-Tab®
Luride®
Luride-SF®
Nafeen®
Solu-Flur®
Sodium-Fluoride Oral Solution
Denta-FL®
Fluor-A-Day®
Fluoritab®
Flura®
Karidium®
Luride®
Nafeen®
Pediaflor®
Pedi-Dent®
Solu-Flur®
Sodium-Fluoride Tablets
Flo-Tab®
Fluorident®
Fluorodex®
Flura®
Karidium®
Nafeen®
Pedi-Dent®
Stay-Flo®
Studaflor®

Thiamine (Vitamin B-1)
Thiamine-Hydrochloride Elixir
Betalin®
Bewon®
Betalin S®
Thiamine-Hydrochloride Tablets
Betalin S®
Biamine®
Pan-B-1®
Thiamine Hydrochloride®
Vitamin B1®

Tryptophan
Tryptophan Capsules
   Pacitron®
   Trofan®
   Tryptacin®

Zinc
Zinc Tablets
   Medizinc®
   Orazinc®
   Verazinc®
   Zinc-220®
   Zincate®
   ZinKaps-110®
   ZinKaps-220®
   Zinc Sulfate®
*Available as a constituent of many
multivitamin/mineral preparations.*

Glossary

Abortifacient—Induces abortions (miscarriages).

Absorption—Process by which nutrients are absorbed through the lining of the intestinal tract into capillaries and into the bloodstream. Nutrients must be absorbed to affect the body.

Acids—Compounds often found in plant tissues, especially fruits, that shrink tissues and prevent secretion of fluids. They taste sour or tart.

Active principle—Chemical component of a plant or compound that has a therapeutic effect.

Acute—Short, relatively severe. Usually referred to in connection with an illness. Opposite of acute is *chronic*.

Addiction—Psychological or physiological dependence on a drug. With true addictions, severe symptoms appear when the addicted person stops taking the drug on which he is dependent.

Adrenal gland—Gland located immediately adjacent to the kidney that produces epinephrine (adrenaline) and several steroid hormones, including cortisone and hydrocortisone.

Adulterant—Substance that makes another substance impure when the two are mixed together.

Allergen—Capable of producing an allergic response.

Allergy—Excessive sensitivity to a substance.

Alumina—Another term for aluminum oxide or hydrated aluminum oxide.

Amenorrhea—Absence of menstruation.

Amino acid—Chemical building blocks that help produce proteins in the body.

Anabolic—Building up of tissues in the body. It is a destructive metabolism.

Anaphylaxis—Severe allergic response to a substance. Symptoms include wheezing, itching, nasal congestion, hives, immediate intense burning of hands and feet, collapse with severe drop in blood pressure, loss of consciousness and cardiac arrest. Symptoms of anaphylaxis appear within a few seconds or minutes after exposure to substance causing reaction—this can be medication or herbs taken by injection, by mouth, vaginally, rectally, through a breathing apparatus or applied to skin. Anaphylaxis is an uncommon occurrence, but when it occurs, it is a *severe medical emergency*! Without appropriate immediate treatment, it can cause death. Yell for help. Don't leave victim. Begin CPR (cardiopulmonary resuscitation), mouth-to-mouth breathing and external cardiac massage. Have someone dial "0" or 911. Don't stop CPR until help arrives.

Anemia—Too few healthy red blood cells in the bloodstream or too little hemoglobin in the red blood cells. Anemia is usually caused by excessive blood loss, such as excessive bleeding or menstruation, increased blood destruction, such as hemolytic anemia or leukemia, or decreased blood production, such as iron-deficiency anemia.

Anemia, pernicious—Anemia caused by vitamin B-12 deficiency. Symptoms include easy fatigue, weakness, lemon colored skin, numbness and tingling of hands and feet, and symptoms of degeneration of the central nervous system, such as irritability, emotional problems, personality changes and paralysis of extremities.

Anesthetic—Used to abolish pain.

Angina (angina pectoris)—Chest pain, with sensation of impending death. Pain may radiate into jaw, ear lobes, between shoulder blades or down shoulder and arm on either side, most frequently the left side. Pain is caused by a temporary reduction in the amount of oxygen to the heart muscle through narrowed, diseased coronary arteries.

Antacid—Neutralizes acid. In medical terms, the neutralized acid is located in the stomach, esophagus or first part of the duodenum.

Anti-bacterial—Destroys bacteria (germs) or suppresses their growth or reproduction.

Antibiotic—Inhibits growth of germs or kills germs. When it inhibits growth, it is called *bacteriostatic*. When it kills germs, it is called *bacteriocidal*.

Anti-cholinergic—Reduces nerve impulses through the part of the autonomic nervous system called *parasympathetic*.

Anti-coagulant—Delays or stops blood clotting.

Anti-emetic—Prevents or stops nausea and vomiting.

Anti-helmintic—Destroys intestinal worms.

Antihistamine—Prevents histamine, the chemical in body tissues that dilates smallest blood vessels, constricts smooth muscle surrounding bronchial tubes and stimulates stomach secretions, from acting on tissues of the body.

Anti-hypertensive—Reduces blood pressure.

Anti-mitotic—Inhibits or prevents cell division.

Anti-neoplastic—Inhibits or prevents growth of neoplasms (cancers).

Anti-oxidant—Prevents oxidation (combining with oxygen). Anti-oxidant substances include superoxide dismutase, selenium, vitamins C and E, and zinc.

Anti-pyretic—Reduces fevers.

Antiseptic—Prevents or retards growth of germs.

Anti-spasmodic—Relieves spasm in skeletal or smooth muscle.

Apertive—Stimulates the appetite.

Aphrodisiac—Arouses or enhances instinctive sexual desire.

Aromatic—Chemical with a spicy fragrance and stimulant characteristics used to relieve various symptoms.

Artery—Blood vessel that carries blood away from the heart.

Asthma—Disease with recurrent attacks of breathing difficulty characterized by wheezing. It is caused by spasms of the bronchial tubes, which can be caused by many factors including adverse reactions to drugs, vitamins, minerals or medicinal herbs.

Astringent—Shrinks tissues and prevents secretion of fluids.

Bacteria—Microscopic germs. Some bacteria contribute to health; others cause disease.

Bitters—Medicine with a bitter taste. Used as a tonic or appetizer.

Blepharitis—Inflammation of eyelid.

Blood sugar (blood glucose)—Necessary element in blood to sustain life. The blood level of glucose is determined by insulin, a hormone secreted by the pancreas. When the pancreas no longer satisfies this function, the disease *diabetes mellitus* results.

Bronchitis—Inflammation of the breathing tubes.

Bulb—Modified plant bulb with scaly leaves that grows beneath the soil.

Carcinogen—Chemical or substance that can cause cancer.

Cardiac arrhythmias—Abnormal heart rate or rhythm.

Cardiac—Pertaining to the heart.

Carminative—Aids in expelling gas from the intestinal tract.

Cathartic—Very strong laxative that produces explosive, watery bowel movements.

Cell—Unit of protoplasm, the essential living matter of all plants and animals.

Central nervous system—Brain and spinal cord and their nerve endings.

Central-nervous-system depressant—Causes changes in the body, including changes in consciousness, lethargy, loss of judgment or coma.

Chronic—Disease of long standing. Opposite of *acute*.

Co-enzyme—Heat-stable molecule that must be loosely associated with an enzyme for the enzyme to perform its function.

Colic—Abdominal pain that recurs in a pattern every few seconds or minutes.

Collagen—Gelatinous protein used to make body tissues.

Congestive—Excess accumulation of blood. In congestive heart failure, blood congregates in lungs, liver, kidney and other parts to cause shortness of breath, swelling of ankles, sleep disturbances, rapid heartbeat and easy fatigue.

Conjunctivitis—Inflammation of the outer membrane of the eye.

Constriction—Tightness or pressure.

Contraceptive—Prevents pregnancy.

Contraindication—Inadvisability of using a substance that may cause harm under specific circumstances. For example, high-caloric intake in someone who is overweight is contraindicated.

Convulsion—Violent, uncontrollable contraction of the voluntary muscles.

Corticosteroid (adrenocorticosteroid)—Hormones produced by the body or manufactured synthetically.

Counterirritant—Process of applying an irritating substance to the skin to produce increased blood circulation to the area. Classic example (now considered an outdated treatment) is mustard plaster applied to the chest to relieve bronchial congestion or cough.

Cyanogenic glycoside(s)—Sugars that have the capacity to be used in the production of cyanide.

Cystitis—Inflammation of the urinary bladder.

DNA (desoxyribonucleic acid)—Complex protein chemical in genes that determines the type of life form into which a cell will develop.

Decoction—Extract of a crude drug obtained by boiling the substance in water.

Dehiscent—Fruit that splits open when ripe.

Delirium—Temporary mental disturbance accompanied by hallucinations, agitation, incoherence.

Demonic—Destroys or repels demons.

Demulcent—Mucilagenous or oily substance capable of protecting scraped tissues.

Dermatitis—Skin inflammation or irritation.

Diaphoretic—Increases perspiration.

Diuretic—Increases urine flow. Most diuretics force kidneys to excrete more than the usual amount of sodium. Sodium forces more water and urine to be excreted.

Dosage—The amount of medicine to be taken for a specific problem. Dosages may be listed as liquids (ml or milliliters, cc or cubic centimeters, teaspoons, tablespoons), dry weight (kg or kilograms, mg or milligrams, g or grams) or by biological assay (Retinol Units, International Units).

Drupe—Fleshy fruit with a hard stone, such as an apricot or peach.

Duodenum—First 12 inches of small intestine.

Dysentery—Disorder with inflammation of the intestines, especially the colon, accompanied by pain, a feeling of urgent need to have bowel movements and frequent stools containing blood or mucus.

Dysmenorrhea—Painful or difficult menstruation.

Dyspepsia—Digestion impairment causing uncomfortable feeling of indigestion.

Eczema—Non-contagious disease of skin characterized by redness, itching, scaling and lesions with discharge. Frequently becomes encrusted. Eczema primarily affects young children. The underlying cause is usually an allergy to many things, including foods, wool, skin lotions. The disorder may begin in month-old babies. It usually subsides by age 3 but may flare again at age 10 to 12 and last through puberty.

Electrolyte—Chemical substance with an available electron in its atomic structure that can transmit electrical impulses when dissolved in fluids.

Emetic—Causes vomiting.

Emmenagogue—Triggers onset of menstrual period.

Emollient—Softens or soothes.

Emphysema—Lung disease characterized by loss of elasticity of muscles surrounding air sacs. Lungs cannot supply adequate oxygen to body cells for normal function.

Endometriosis—Medical condition in which uterine tissue is found outside the uterus. Symptoms include pain, abnormal menstruation, infertility.

Enzyme—Protein chemical that accelerates a chemical reaction in the body without being consumed in the process.

Epilepsy—Symptom or disease characterized by episodes of brain disturbance that cause convulsions and loss of consciousness.

Essential oils—Same as *volatile oils*. Oils evaporate at room temperature.

Estrogens—Female sex hormones that must be present for secondary sexual characteristics of the female to develop. Estrogens serve many functions in the body, including preparation of the uterus to receive a fertilized egg.

Eupeptic—Promotes optimum digestion.

Expectorant—Decreases thickness and increases fluidity of mucus from the lungs and bronchial tubes.

Extract—Solution prepared by soaking plant in solvent, then allowing solution to evaporate.

Extremity—Arm, hand, leg, foot.

Fat-soluble—Dissolves in fat.

Fatty acids—Nutritional substances found in nature that are fats or lipids. These include triglycerides, cholesterol, fatty acids and prostaglandins. *Fatty acids* include stearic, palmitic, linoleic, linolenic, eicosapentaenoic (EPA), decosahexanoic acid. Other lipids of nutritional importance include lecithin, choline, gamma-linoleic acid and inositol.

Fixed oil(s)—Lipids, fats or waxes often made from seeds of plants.

Flatulence—Distention of the stomach or other parts of the intestinal tract with air or other gases.

Fluid extract—Alcoholic solution of a chemical or drug of plant origin. Fluid extracts usually contain 1g of dry drug in each milliliter.

Free radicals—Highly reactive molecules with an unpaired free electron that combines with any other molecule that accepts it. Free radicals are usually toxic oxygen molecules that damage cell membranes and fat molecules. To protect against possible damage from free radicals, the body has several defenses. The most important appears at present to be anti-oxidant substances, such as superoxide dismutase, selenium, vitamin C, vitamin E, zinc and others.

G6PD—Deficiency of glucose 6-phosphate, a chemical necessary for glucose metabolism. Some people have inherited deficiencies of this substance and have added risks when taking some drugs.

Gastritis—Inflammation of the lining of the stomach.

Gastroenteritis—Inflammation of stomach and intestines characterized by pain, nausea and diarrhea.

Gastrointestinal—Pertaining to stomach, small intestine, large intestine, colon, rectum and sometimes the liver, pancreas and gallbladder.

Generic—Relating to or descriptive of an entire group or class.

Gingivitis—Inflammation of the gums surrounding teeth.

Gland—Cells that manufacture and excrete materials not required for their own metabolic needs.

Glossitis—Inflammation of the tongue.

Gluten—Mixture of plant proteins occurring in grains, chiefly corn and wheat. People who are sensitive to gluten develop gastrointestinal symptoms that can be controlled only by eating a gluten-free diet.

Glycoside(s)—Plant substance that produces a sugar and other substances when combined with oxygen and hydrogen.

Griping—Intestinal cramps.

Gums—Translucent substances without form. Usually a decomposition product of cellulose. Gums dissolve in water.

Hallucinogen—Produces hallucinations—apparent sights, sounds or other sensual experiences that do not actually exist or do not exist for other people.

Heart block—An electrical disturbance in the controlling system of the heartbeat. Heart block can cause unconsciousness and in its worst form can lead to cardiac arrest.

Hematuria—Blood in the urine.

Hemoglobin—Pigment necessary for red cells to transport oxygen. Iron is a necessary component of hemoglobin.

Hemolysis—Breaking a membranous covering or destroying red blood cells.

Hemorrhage—Extensive bleeding.

Hemostatic—Prevents bleeding and promotes clotting of blood.

Hepatitis—Inflammation of liver cells, usually accompanied by *jaundice*.

Herb—Plant or plant part valued for its medicinal qualities, pleasant aroma or pleasing taste.

Histamine—Chemical in the body tissues that constricts the smooth muscle surrounding bronchial tubes, dilates small blood vessels, allows leakage of fluid to form itching skin and hives and increases secretion of acid in stomach.

Hives—Elevated patches on skin usually caused by an allergic reaction accompanied by a release of histamine into the body tissues. Patches are redder or paler than the surrounding skin and itch intensely.

Homeopathy—Practice of using extremely small doses of medicines and herbs to cause the same symptoms the disease causes. Homeopaths (practitioners of homeopathy) acknowledge no diseases, only symptoms.

Hormone—Chemical substance produced by endocrine glands—thymus, pituitary, thyroid, parathyroid, adrenal, ovaries, testicles, pancreas—that regulates many body functions to maintain homeostasis (a steady state).

Humectant—Moistens or dilutes.

Hypercalcemia—Abnormally high level of calcium in the blood.

Hypertension—High blood pressure.

Hypocalcemia—Abnormally low level of calcium in the blood.

Hypoglycemia—Abnormally low blood sugar.

Impotence—Inability of a male to achieve and maintain an erection of the penis to allow satisfying sexual intercourse.

Indehiscent—Fruit that remains closed upon reaching maturity.

Inflorescence—Flowerhead of a plant.

Infusion—Product that results when a drug or herb is steeped to extract its medicinal properties.

Insomnia—Inability to sleep.

Interaction—Change in body's response to one substance when another is taken. Interactions may increase the response, decrease the response, cause toxicity or completely change the response expected from either substance. Interactions may occur between drugs and drugs, drugs and vitamins, drugs and herbs, drugs and foods, vitamins and vitamins, minerals and minerals, vitamins and foods, minerals and foods, vitamins and herbs, herbs and herbs.

International units—Measurement of biological activity. In the case of vitamin E, 1 International Unit (IU) equals 1 milligram (mg) (1IU = 1mg).

I.U. or IU—International units.

Jaundice—Symptom of liver damage, bile obstruction or excessive red-blood-cell destruction. Jaundice is characterized by yellowing of the whites of the eyes, yellow skin, dark urine and light stool.

Kidney stones—Small, solid stones made from calcium, cysteine, cholesterol and other chemicals in the bloodstream. They are produced in the kidneys.

Lactagogue—Increases the flow of breast milk in a woman.

Lactase—Enzyme that helps body convert lactose to glucose and galactose.

Lactase deficiency—Lack of adequate supply of enzyme *lactase*. People with lactase deficiency have difficulty digesting milk and milk products.

Larvacide—Kills larvae.

Latex—Milky juice produced by plants.

Laxative—Stimulates bowel movements.

LDH—Abbreviation for lactic dehydrogenase, a blood test to measure liver function and to detect damage to the heart muscle.

Libido—Sex drive.

Lipid—Fat or fatty substance.

Lymph glands—Glands located in the lymph vessels of the body that trap foreign material, including infectious material, and protect the bloodstream from becoming infected.

Maceration—Softening of a plant by soaking.

Magnesia—Another term for magnesium hydroxide.

Malabsorption—Poor absorption of nutrients from the intestinal tract into the bloodstream.

Mcg—Abbreviation for microgram, which is 1/1,000,000th (1/1-millionth) of a gram or 1/1,000th of a milligram.

Megadose—Very large dose. In terms of *recommended dietary allowance (RDA),* anything 10 or more times the RDA is considered megadose. Nutritionists urge no one take megadoses of *any* substance because these doses may be toxic, cause an imbalance of other nutrients, cause damage to an unborn child and do not provide benefits beyond rational doses.

Menopause—End of menstruation in the female caused by decreased production of female hormones. Symptoms include hot flushes, irritability, vaginal dryness, changes in the skin and bones.

Metabolism—Chemical and physical processes in the maintenance of life.

Mg—Abbreviation for milligram, which is 1/1,000th of a gram.

Migraine—Periodic headaches caused by constriction of arteries in the skull. Symptoms include visual disturbances, nausea, vomiting, light sensitivity and severe pain.

Milk sickness—Intolerance to milk and milk products due to a deficiency of an enzyme called *lactase.*

Mitogen—Causes nucleus of cell to divide; leads to a new cell.

Mucilage—Gelatinous substance that contains proteins and polysaccharides.

Narcotic—Depresses the central nervous system, reduces pain and causes drowsiness and euphoria. Narcotics are addicting substances.

Naturopathy—Medical practice that uses herbs and various methods to return body to healthy state by stimulating innate defenses—never supplanting them—with drugs. In early years, many naturopathic physicians were ill-prepared to practice a healing profession. Many received mail-order degrees and had little training. However by the 1950s, some degree of academic acceptability returned. Several accredited schools award degrees for training, and many states now require examinations and licensure to ensure competence.

Neuropathy—Group of symptoms caused by abnormalities in sensory or motor nerves. Symptoms include tingling and numbness in hands or feet, followed by gradually progressive muscular weakness.

Oleoresin—*Resins* and *volatile oils* in a homogenous mixture.

Osteoporosis—Softening of bones.

Oxidation—Combining a substance with oxygen.

Parasympathetic—Division of the autonomic (also called *automatic*) nervous system. Parasympathetic nerves control functions of digestion, heart and lung activity, constriction of eye pupils and many other normal functions of the body.

Parkinson's disease—Disease of the central nervous system characterized by a fixed, emotionless expression of the face, slower-than-normal muscle movements, tremor (particularly when attempting to reach or hold objects), weakness, changed gait and a forward-leaning posture.

Paronychia—Infection around a fingernail bed.

Peduncle—Stalk attached to a flower.

Pellagra—Disease caused by a deficiency of thiamine (vitamin B-1). Symptoms include diarrhea, skin inflammation and dementia (brain disturbance).

Peristalsis—Wave of contractions of the intestinal tract.

Pernicious anemia—See *Anemia, pernicious.*

Pharyngitis—Inflammation of the throat.

Phenylketonuria—Inherited disease caused by lack of an enzyme necessary for converting phenylalanine into a form the body can use. Accumulation of too much phenylalanine can cause poor mental and physical development in a newborn. Most states require a test at birth to detect the disease. When detected early and treated, phenylketonuria symptoms can be prevented by dietary control.

Phosphates—Salts of phosphoric acid. Important part of the body system that controls acid-base balance. Other chemicals involved in acid-base balance include sodium, potassium, bicarbonate and proteins.

Photosensitization—Process by which a substance or organism becomes sensitive to light.

Photosensitizing pigment—Pigment that makes a substance sensitive to light.

Potassium—Important element found in body tissue that plays a critical role in electrolyte and fluid balance in the body.

Poultice—Applied to a body surface to provide heat and moisture. Material is held between layers of muslin or other cloth. Poultices contain an active substance and a base. They are placed on any part of the body and changed when cool. Purpose is to relieve pain and reduce congestion or inflammation.

Prostate—Gland in the male that surrounds the neck of the bladder and urethra. In older men, it may become infected (prostatitis) or obstructed (prostatic hypertrophy), cause urinary difficulties or become cancerous.

Psoriasis—Chronic, recurrent skin disease characterized by patches of flaking skin with discoloration.

Psychosis—Mental disorder characterized by deranged personality, loss of contact with reality, delusions and hallucinations.

Purgative—Powerful laxative usually leading to explosive, watery diarrhea.

Purine foods—Foods metabolized into uric acid; these include anchovies, brains, liver, sweetbreads, sardines, meat extracts, oysters, lobster and other shellfish.

RDA—See *Recommended dietary allowance*

Recommended dietary allowance—Recommendations based on data derived from different population groups and ages. The quoted RDA figures represent the *average* amount of a particular nutrient needed per day to maintain good health in the average healthy person. Data for these recommendations have been collected and analyzed by the Food and Nutritional Board of the National Research Council. These figures serve as a reference point for comparison. The latest revised amounts were published in 1980, with a new revision promised soon. It is only within the framework of statistical probability that RDA can be used legitimately and meaningfully.

RNA (ribonucleic acid)—Complex protein chemical in genes that determines the type of life form into which a cell will develop.

Renal—Pertaining to the kidneys.

Resin—Complex chemicals, usually hard, transparent or translucent, that frequently cause adverse effects in the body.

Retina—Inner covering of the eyeball on which images form to be perceived in the brain via the optic nerve.

Rhizome—Root-like, horizontal-growing stem growing just below the surface of the soil.

Rickets—Bone disease caused by vitamin-D deficiency. Bones become ben and distorted during infancy or childhood if there is insufficient vitamin D for normal growth and development.

Rubefacient—Reddens skin by increasing blood supply to it.

Saponin(s)—Chemicals from plants, frequently associated with adverse or toxic reactions. They uniformly produce soapy lathers.

Sedative—Reduces excitement or anxiety.

SGOT—Abbreviation for serum glutamic oxaloacetic transaminase, a blood test to measure liver function or detect damage to the heart muscle.

Spasmolytic—Decreases spasm of smooth muscle or skeletal (striated) muscle.

Steroidal chemicals—Group of chemicals with same properties as steroids. Steroids are fat-soluble compounds with carbon and acid components. They are found in nature in the form of hormones and bile acids, and in plants as naturally occurring drugs, such as digitalis.

Stimulant—Stimulates (temporarily arouses or accelerates) physiological activity of an organ or organ system.

Stomachic—Promotes increased contraction of stomach muscles.

Stomatitis—Inflammation of the mouth.

Stroke—Sudden, severe attack that results in brain damage. Usually sudden paralysis or speech difficulty results from injury to the brain or spinal cord by a blood clot, hemorrhage or occlusion of blood supply to the brain from a narrowed or blocked artery.

Tannins—Complex acidic mixtures of chemicals.

Tenesmus—Urgent feeling of having to have a bowel movement or to urinate.

Terpenes—Complex hydrocarbons ($C_{10}H_{16}$). Most volatile oils are mostly terpenes.

Thrombophlebitis—Inflammation of a vein, usually caused by a blood clot. If the clot becomes detached and travels to the lung, the condition is called *thromboembolism*.

Tincture—Solution of chemicals in a highly alcoholic solvent made by simple solution or by methods described in the *United States Pharmacopeia* or the *National Formulary*.

Tonic—Medicinal preparations used to restore normal tone to tissues or to stimulate the appetite.

Toxicity—Poisonous reaction that impairs body functions or damages cells.

Toxin—Poison in dead or live organism.

Tranquilizer—Calms a person without clouding mental function.

Tremor—Involuntary trembling.

Tyramine—Chemical component of the body. In normal quantities, without interference from other chemicals, tyramine helps sustain normal blood pressure. In the presence of some drugs—monamine-oxidase inhibitors and some rauwolfia compounds—tyramine levels can rise and cause toxic or fatal levels in the blood.

Urethra—Hollow tube through which urine (and semen in men) is transported from the bladder to outside the body.

Uterus—Hollow, muscular organ in the female in which an embryo develops into a fetus. Menstruation occurs when the lining sloughs periodically.

Vein—Blood vessel that returns blood to the heart.

Virus—Infectious organism that reproduces in the cells of an infected host.

Volatile oils—Chemicals that evaporate at room temperature.

Water-soluble—Dissolves in water.

Wax—High-molecular-weight hydrocarbons; they are insoluble in water.

Yeast—Single-cell organism that can cause infection of the skin, mouth, vagina, rectum and other parts of the gastrointestinal system. The terms *yeast, fungus* and *monilia* are used interchangeably.

Metric Chart

The following units of measurement and weight are commonly used in establishing doses of vitamins, minerals, supplements and medicinal herbs.

Unit	Abbreviation	Volume	Approximate U.S. Equivalent
Cubic centimeter	cc	0.000001 cubic meters	0.061 cubic inch
Liter	l	1 liter	1.057 quarts
Deciliter	dl	0.10 liter	0.21 quarts
Centiliter	cl	0.01 liter	0.338 fluid ounce
Milliliter	ml	0.001 liter	0.27 fluid dram
Kilogram	kg	1,000 grams	2.2046 pounds
Gram	g or gr	1 gram	0.035 ounce
Milligram	mg	0.001 gram	0.015 grain
Microgram	mcg	0.000001 gram	0.00015 grain

Index